S0-ADW-875

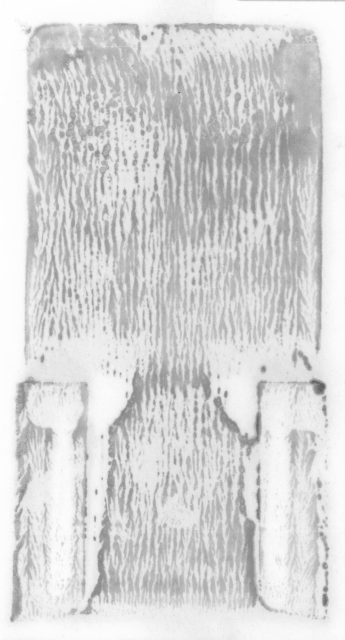

Design of Steel Structures

Boris Bresler

Professor of Civil Engineering
University of California
Berkeley, California

T. Y. Lin

Professor of Civil Engineering
University of California
Berkeley, California

John B. Scalzi

General Manager Marketing Technical Services
United States Steel Corporation
Pittsburgh, Pennsylvania

SECOND EDITION

Design of Steel Structures

JOHN WILEY & SONS, INC.
New York · London · Sydney

Preface

During the last four decades structural engineering has undergone many significant changes. The breadth of knowledge of structural theory has expanded through the elastic range and into the inelastic range of the material behavior. The application of the principles of theoretical mechanics has encouraged greater sophistication in the analysis and design of steel structures. Laboratory investigations and field studies have been conducted to correlate actual behavior with theoretical predictions. Improved constructional steels and fabrication methods have provided further impetus for the development of sound rational design techniques.

The objective of this textbook is to present a rational approach to the design of steel structures and, wherever possible, to correlate this approach with current design practice. We believe that good design practice must, of necessity, be based on a sound knowledge of the fundamental principles of structural mechanics, a thorough understanding of the behavior of actual structures, an appreciation of their relationship to an idealized structure, and an awareness of the practical requirements such as fabrication, feasibility, safety, and economy. Therefore the reader will find these concepts emphasized throughout.

This book is intended for all engineering courses concerned with the design of steel structures. The contents permit its use in elementary and advanced courses in structural design and, as such, also serve the needs of the practicing engineer as a reference source for special topics.

The reader is expected to have a knowledge of the general principles of statics, dynamics, mechanics of material including material behavior, and structural theory. It is possible that the student can study structural theory concurrently with the elementary material presented. Various provisions of pertinent specifications and codes are examined in the light of rational analysis, empirical evidence, and practical requirements. Assumptions and limitations of theory and design procedures are discussed. Each chapter contains a reference list of other sources which provide a more complete treatment of particular topics.

The text contains numerous illustrative examples, each of which is intended to assist the reader in his understanding of a certain principle or particular design method. The examples cited are intentionally focused on a specific topic to enable him to master the point presented and, subsequently, to relate it to other more complicated design situations. Problems included at the end of chapters provide the student an opportunity to develop his facility and confidence in the principles discussed in that chapter.

An elementary course in design of steel structures would include the major portion of Chapters 1, 2, 5, 6, 7, 8, 9, 10, and 11. Some of the material in these together with the remaining chapters 3, 4, 12, 13, 14, and 15 usually is taught in more advanced courses in design.

The first chapter discusses the basic principles of structural design, the behavior and safety of structures, and general considerations of design procedures.

Chapters 2 and 3 deal with the properties of the constructional steels and their behavior with respect to applications in weldments and fatigue conditions.

Chapter 4 discusses the fundamental principles of the behavior of steel as an elastic and plastic material. This chapter is basic to a full understanding of the structural concepts presented in later chapters.

Chapters 5 and 6 discuss bolted, riveted, pinned, and welded connections. The use of high-strength bolts is noted in Chapter 5. Attention is directed to the use of simple connections in order to illustrate the basic principle. Complex connections are discussed in later chapters in conjunction with their use in different types of members and structural frames.

Chapter 7 is concerned with various types of tension members including rods, bars, shapes, and cables. Built-up members are included as well as connections and splices.

Chapter 8 discusses the theoretical aspects of the bending and torsion of beams. The discussion includes bending of simple, unsymmetrical, and open sections. Torsion of free and restrained sections of open, tubular, and cellular cross section is included. Special topics of tapered and curved beams and combined bending for the elastic and plastic range are also discussed.

Chapter 9 discusses the fundamental problem of buckling of prismatic members, frames, and plates. Basic theory is presented for the usual members used in structural design.

Chapter 10 deals with the design of compression members using the principles evolved in Chapter 9.

Chapter 11 discusses the design of beams and girders with respect to bending, lateral instability, deflections and economy. Elastic and plastic designs of continuous beams and composite beam design are included in this chapter and the development of the criteria for designing a thin-web plate girder is presented in detail.

The subject of the design of buildings and bridges is so broad that only an introductory treatment is included in Chapters 12 and 13, respectively. Loads, types of structures, and special topics as well as practical examples are included in each chapter.

Chapter 14 deals with the principles of design when using sheet steel as the principal structure material. This is a relatively new development in steel design. Basic principles and practical design expressions are discussed in relation to compression members, beams, and panels.

Chapter 15 introduces several types of structures of relatively recent development which are being designed with steel components. These structures include orthotropic bridges, curved bridges, cable-supported bridges and buildings, space structures, prestressed steel structures, tubular shell and semimonocoque high rise structures. Full discussion of design methods for these complex structures is beyond the scope of this book and the material presented is of an introductory nature.

Most of the material presented is not original and although individual acknowledgments of the many sources of information are evidently not possible, full credit is noted whenever the source can be identified.

Every effort has been made to eliminate errors, but we would appreciate notification from the reader of any errors.

We are indebted to many organizations and individuals for their help and assistance and permission to reproduce photographs, designs, tables, and graphs. Wherever possible, credit is noted in the text.

We wish to acknowledge the help of our associates and to thank Mrs. Andrea G. Vance and Miss R. Meade for typing the revised manuscript.

Boris Bresler
T. Y. Lin
December 1967   *John B. Scalzi*

Contents

ix

Brittle Fracture and Fatigue 57

Elastic and Plastic Concepts of Structural Behavior 79

CHAPTER 5

Bolted, Riveted, and Pinned Connections   103

CHAPTER 6

Welded Connections   176

CHAPTER 7

Tension Members   244

CHAPTER 8

Bending and Torsion of Beams   285

CHAPTER 9

Buckling of Prismatic Members, Frames, and Plates   351

CHAPTER 10

Design of Compression Members   403

CHAPTER 11

Design of Beams and Girders   **460**

CHAPTER 12

Design of Buildings   **562**

CHAPTER 13

Design of Bridges 627

CHAPTER 14

Design of Light Gage Members 693

CHAPTER 15

Special Structures   **732**

J. W. Gillespie, J. F. McDermott, W. Podolny, Jr.*

* J. W. Gillespie is Manager, Marketing Technical Services, United States Steel Corporation, Pittsburgh, Pa.; J. F. McDermott is Senior Research Engineer, Structural Mechanics, United States Steel Corporation, Applied Research Laboratory, Pittsburgh, Pa.; and W. Podolny, Jr., is Design Engineer, Marketing Technical Services, United States Steel Corporation, Pittsburgh, Pa.

Index   **825**

1

General Principles
of Structural Design

1-1 INTRODUCTION

The aim of the structural designer is to produce a safe and economical structure to meet certain functional and esthetic requirements. In order to achieve this goal the designer must have a thorough knowledge of the properties of materials, structural behavior, structural analysis and mechanics, and the correlation between the layout and the function of a structure. He must also have an appreciation of esthetic values in order to work closely with architects and contribute to the development of desired functional and environmental qualities of a structure.

Structural design is to a great extent an art based on creative ability, imagination, and the experience of the designer. To the extent that structural design involves these qualifications, it will always remain an art. It should not remain a pure art, however, because the public should receive the greatest benefit within its economic reach. This requires development of new types of structures and new construction techniques, which often necessitate a more exacting and scientific solution. Thus, engineering mechanics and economic analysis must aid in the art of creating better buildings, bridges, machines, and equipment. In a broad sense the term "design" includes both creative art and scientific analysis.

The building of Egyptian monuments, Greek temples, and Roman bridges was an art based primarily on empirical rules, intuition, and experience. A rational approach to structural design, which began its development in the

1

seventeenth century, represents a compromise between art and science, between experience and theory.

Theory of structures and experimental evidence are valuable tools for structural design, but they are not sufficient for establishment of a completely scientific design procedure. First, to make a theoretical analysis possible, the structural behavior is idealized a great deal by sound engineering assumptions so that the computed internal forces and displacements represent only approximations of the actual ones in the structures. The resistance of actual structures to imposed loads and deformations can be determined only approximately. Furthermore, actual structures are often subjected to forces and service conditions that cannot be accurately predicted. Thus, experience and judgment always play an important role in the practice of structural design. Experience and judgment alone, however, are generally insufficient; they must be guided by scientific analysis, based upon a complete understanding of the theory of structures and structural mechanics.

1-2 CLASSIFICATION OF STRUCTURES

Structures can be divided into two principal groups: (*a*) shell structures, which are made largely of plates or sheets, such as tanks, bins, ship hulls,

Fig. 1-1 Verrazano-Narrows Bridge, New York. (Courtesy of AISC.)

Fig. 1-2 United Air Lines jet hangar, San Francisco, California. Each girder 365 ft long, cantilevers 145 ft. Architects: Skidmore, Owings and Merrill. (Courtesy of Stanray Pacific Corporation, formerly P.I. Steel Corporation.)

railroad cars, airplanes, and shell roof coverings for large buildings, and (*b*) framed structures, which are characterized by assemblies of elongated members, such as truss frames, rigid frames, girders, tetrahedrons, or three-dimensional framed structures.

In shell structures, the plating or sheeting is used as a functional covering as well as a main load-carrying element. For this purpose the plating is stiffened by frames which may or may not carry the principal loads. But in the framed structures, the main members are generally not functional and are used solely for load transmission. Therefore additional material must be installed for functional requirements, such as walls, roofs, floors, and pavements. It may appear that shell structures are more efficient than framed structures because the plating or "skin" is used for a dual purpose: functional and structural. Heretofore, in steel structures, shells have not been used widely, especially in the United States. This is attributed to several factors: (*a*) the savings introduced by this type of design are mainly weight savings, and are effected only for certain layouts and spans, (*b*) weight savings may be accompanied by corresponding increases in construction costs, and (*c*) in order to reduce the construction cost of metal shell structures, a reorganization and re-equipment of structural shops and construction crews is required. These conditions are currently being remedied and a great variety of steel structural systems is becoming possible.

Figures 1-1 through 1-3 illustrate some of the typical steel structures. The

Fig. 1-3 Steel-framed apartment buildings, Chicago. (Courtesy of Hedrich-Blessing and AISC.)

famed Verrazano-Narrows bridge (Fig. 1-1) utilizes the high tensile strength of steel wires in its cables and suspenders. The two steel towers, 690 ft high, carry a vertical load of 105,000 tons per tower and resist horizontal loads at the same time. The trusses along the roadway stiffen the bridge against moving traffic and dynamic forces of wind and earthquake.

Modern construction makes use of heavy erection equipment which enables the handling of large fabricated pieces, such as illustrated in Fig. 1-2. Here steel plate girders cantilever 145 ft, supporting the roof for a jet airplane hanger. Each cantilever weighing 125 tons was fabricated in three pieces and connected in the field.

Two multistory steel framed buildings are shown in Fig. 1-3. Note the architectural expression of the vertical columns transmitting forces to the foundation.

1-3 STRUCTURAL MEMBERS AND CONNECTIONS

A conventional framed structure is composed of members assembled by means of connections. A member can be a standard rolled-steel shape, or it may be built up by riveting, welding, or bolting together several shapes (Fig. 1-4). Members may transmit four fundamental types of loads and are classified accordingly (Fig. 1-5): (*a*) ties, transmitting tensile loads, (*b*) columns, transmitting compressive loads, (*c*) beams or girders, transmitting transverse loads, and (*d*) shafts, transmitting torsional loads.

In practice, a member rarely transmits loads of one type only. Even a

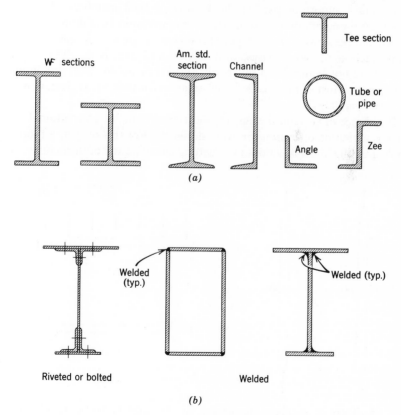

Fig. 1-4 Typical steel shapes: (*a*) rolled shapes and (*b*) built-up members.

pin-connected horizontal or diagonal member in tension is subjected to a small amount of bending produced by its own weight. Hence, most members transmit some combination of bending, torque, and axial tension or compression load. For bridges and buildings, it is very rare that a member is designed primarily for torsion, but quite frequently members designed for other types of loads are also subjected to some torque.

Often, when members are subjected to combined loadings, one of the loads is the most important and governs the design of the member. Therefore, members can be classified and studied according to their predominant load. Members in tension, in compression, in bending and in torsion are separately examined in Chapters 7, 8, 9, 10 and 11.

There are four principal types of connections: riveted, bolted, pinned, and arc- or resistance-welded connections. These are discussed in detail in Chapters 5 and 6. Although riveted connections have been most commonly employed, modern developments in welding and bolting are playing an increasingly important part in the joining of steel members. Figure 1-6 illustrates the common types of structural connections. In addition to the four principal types, various other types of connections are used in special applications, such as screws, clevises, turnbuckles, and rivet bolts. These special devices are not examined in this book, since their use in structural design is infrequent. Most of the necessary information for their use is given in manufacturers' catalogs.

The choice of structural elements and connections for conventional structures is a routine operation for every designer. For this reason, a thorough understanding of the functions of structural elements and of their connections

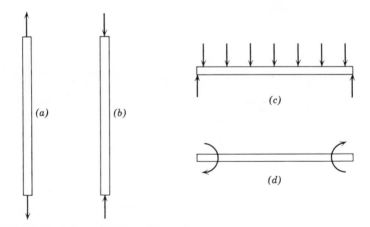

Fig. 1-5 Types of members. (*a*) tension tie. (*b*) column. (*c*) beam. (*d*) torsional shaft.

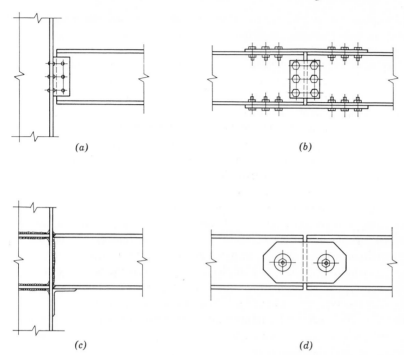

Fig. 1-6 Typical structural connections. (*a*) riveted connection. (*b*) bolted splice. (*c*) welded connection. (*d*) pinned connection.

is of primary importance. Study of the design of complete structures becomes profitable only when proportioning of structural members and their connections is well understood.

The layout and design of complete structures are much more difficult than the choice and design of their components, and several years of experience are needed to master these problems. A general description of some basic considerations in design of selected types of structures encountered in building and bridge design is given in Chapters 13 and 14.

1-4 DESIGN PROCEDURES

The structural design procedure consists of six principal steps: (*a*) selection of type and layout of structure, (*b*) determination of loads on the structure, (*c*) determination of internal forces and moments in the structural components, (*d*) selection of material and proportioning of members and connections for safety and economy, (*e*) checking the performance of the structure under service conditions, and (*f*) final review.

Selection of Type of Structure.   The type of structure is selected on the basis of functional, economic, service, and esthetic requirements. In some instances other considerations, such as customer's wishes, designer's preference, or established precedents, dictate the type of structure to be adopted. Often it is necessary to investigate several layouts and the final choice is made only after several comparative designs are fairly well advanced.

In selecting a type of structure the following questions arise: What are the nature, magnitude, distribution, and frequency of the forces to be transmitted by the structure? What is the effect of temperature variations or foundation settlements on the behavior of the structure? For a selected structure, if analysis discloses that some elements of the structure are overstressed, what can be done about it? Can it best be remedied by changing the proportions of the members, or by changing their arrangements, or must the type of structure be changed completely?[1]* What is the best way of building the given type of structure and what effect might the method of construction have on the type of structure selected and its design? In answering these questions the structural engineer must be aware of his creative role as a designer, which he often shares with his coworkers, such as the architect, the fabricator, and the contractor.

Determination of Service Loads.   When the general type of structure is chosen, or at least several alternatives are defined, a small-scale layout sketch can be made. The arrangement of the members is naturally governed by the magnitudes of the loads in the members, which are not yet known. Hence, experience plays an important role in this step and enables the designer to avoid the necessity of considering too many variations. From the general layout, an estimate of the applied loads can be made. These loads are of various types: traffic loads on railway or highway bridges; loads on floors in buildings, including people, furnishings, machinery, equipment, and stored materials; loads of roof coverings, floors, walls, and partitions; live loads of wind, snow, earthquake, and blast loads. The loads can be static or dynamic, temporary or permanent, occasional or repetitive. To these loads must be added the weight of the structure, which is unknown at this stage of the design but can be estimated fairly accurately for conventional structures. Various formulas and tables have been established for that purpose and can be found in technical journals and structural handbooks.[2] Further discussion of the nature of loads encountered in design are included in Chapters 12 and 13.

Dynamic loads, such as earthquakes, mechanically induced vibrations,

---

* Numbers indicate references listed at the end of each chapter.

impact, blast, and wind gusts, are difficult to define. The conventional procedure has been to replace the nonstatic loads by "equivalent" static loads. The combinations of loads that act simultaneously are then defined as "loading conditions" and are used for computation of stress in the members. In many instances response of structures to dynamic loads must be investigated on the basis of dynamic behavior of materials and structural systems.

Internal Forces and Moments. For statically determinate structures, subjected to static loads, the forces and moments in all the members can be computed simply by using conditions of equilibrium. For statically indeterminate structures, it is necessary to make some estimates of the member sizes in order to determine the stresses in the structure. Sometimes only the relative stiffnesses of structural members are required in order to proceed with the analysis, and these can be more easily approximated than the absolute sizes.

In order to make a preliminary analysis of a statically indeterminate frame, the locations of points of inflection are often estimated by experience, and a preliminary analysis is made on this basis. For statically indeterminate trusses, the distribution of loads between the members can be assumed and preliminary proportioning of the members determined before a more accurate analysis is made.

Proportioning of Members and Connections. When internal forces in the members and materials are known, the size of each member can be selected, with the following criteria in mind: (*a*) adequate strength and rigidity, (*b*) ease of connection, and (*c*) economy.

In choosing the over-all shape and size of a member, the designer must consider its connection to the adjacent members. The connections should be arranged so as to minimize eccentricities which may introduce secondary bending or torsional effects. In addition, the rigidity of the connection must correspond to the condition assumed in the analysis. For example, if in the analysis the beam is assumed to be fixed at the ends, then rigid supporting elements and connections must be provided.

The economy of the member is determined by the costs of material and labor, so that ease of fabrication, handling, connection, and maintenance must be considered. The lightest section may not be the most economical since it may require special connections or fabrication which increases the over-all cost of the structure. Economic design requires a minimum of random arrangement of holes and of handling. Sometimes the designer's choice is restricted by the availability of shapes, fabricating facilities, or skilled labor. For example, in some localities high-tensile-strength bolts are used in field erection because skilled field riveting or welding crews are not readily available.

Frequently the proportions of each member can be determined independently of other members. Sometimes this procedure is not theoretically correct as, for instance, when some members are used to restrain other members against buckling. A typical example of this condition is the analysis of a light rigid frame. Theoretically, the girders restrain the columns from buckling, and the effectiveness of this restraint is a function of the relative stiffness of the members and the type of loading on the frame. Such refinement, however, is often disregarded, and designers use average restraining values which are selected to be on the conservative side.

Performance under Service Conditions.   After the size of a member has been determined from known loads, it should be checked for service requirements, such as limiting deflections, undue twisting, vibration, fatigue, corrosion, stresses due to temperature variations or support settlements, and other conditions that might affect the functioning of the structure.

Final Review.   When the section properties are finally known, it is necessary to verify whether or not the assumed weights of the structure correspond to the final weights. For structures having small spans, the weight of the structure proper is a small portion of the total load, so that, if the first weight estimate is incorrect, even by an appreciable amount, the change in the total load is insignificant and it is unnecessary to recalculate and redesign. However, for large spans, the structural weight is a major part of the total load and small errors in the estimated weight may have an appreciable influence on the total loads.

For statically indeterminate structures, it is necessary to verify whether the relative stiffnesses of the selected sections correspond to the assumed values. If the differences are small, it is not necessary to repeat the analysis. Experience will help to determine what magnitudes of variations can be neglected and no general rules can be given. It is possible, when the differences are small but not negligible, to modify the design without repeating all the calculations. After verification of loads and internal forces and moments, the members should be rechecked for stresses, limiting deformations, and other service requirements, such as possible effects of support settlement, vibration, fatigue, effects of temperature variations, corrosion, and fire resistance.

It follows from these considerations that the design of any structure of importance is essentially a trial-and-error procedure. For statically determinate structures, the trial-and-error process consists of two stages: determination of the structural weights, and proportioning of members. For statically indeterminate structures, there is an additional stage involved in the assumption and determination of the relative stiffnesses of all components of the structure.

1-5 STANDARD SPECIFICATIONS

The building of steel structures involves the owners, the designers (engineers and architects), the fabricators, and the constructors. First, the owners and the designers must reach an understanding concerning the over-all requirements of the project. On the basis of these requirements, the designers prepare plans and specifications describing the project in detail; then the fabricators and the constructors build the structure from these plans and specifications. Specifications play an important role in this process by defining acceptable standards of quality of materials and of workmanship in fabrication and erection.

Three types of specifications are utilized: project specifications, material standards, and design codes or specifications.

Project specifications together with the drawings furnish the contractors (fabricators, erectors, and other service subcontractors) with complete information as to the owner's and engineer's precise requirements for the completed structure. They also form the basis of a legal contract between the owner and the contractor and therefore their accuracy, completeness, and clarity of meaning are of great importance.

Material standards are established primarily by the American Society for Testing Materials. In special cases American Standards Association and various federal, state, and local governmental agencies prepare standards for specific types of products.

Design codes and *specifications* are prepared by various professional associations as well as governmental agencies and prescribe minimum acceptable criteria for design. These include recommended loads, limits of stress, and deformation, as well as special requirements governing proportioning of members and connections. Often such specifications apply to a limited class of structures. For steel buildings the most widely accepted general specifications are those of the American Institute of Steel Construction (AISC). For special aspects of design those of American Welding Society (AWS), American Iron and Steel Institute (AISI), Steel Joist Institute (SJI), and others are used. Regional and local specifications which have legal validity, such as the Uniform Building Code and city building codes, usually are based on the foregoing specifications. For bridges, the American Association of State Highway Officials (AASHO), U.S. Bureau of Public Roads (USBPR), American Railway Engineering Association (AREA), and AWS specifications are widely used, although governmental agencies often have codes which may in some respect vary from the foregoing specifications.

A list of selected Design Specifications and Material Standards is given in Appendix A.

During the years 1850–1910, many different specifications existed for the same type of structure. For example, many railroad companies had their own specifications for the design of steel railroad bridges; some consulting engineers prepared and published their specifications for general use—such as Bouscaren's (1887), Cooper's (1884), or Schneider's (1887) specifications. No legally enforceable standards existed. As a result of some cases of inferior design, material, or workmanship, a few structures failed, resulting in loss of life and property. Public bodies concerned with public safety began to legislate ordinances regulating the design, erection, and construction of various structures. Under these ordinances construction permits were granted only after designs were reviewed and found adequately safe, in accordance with provisions in the ordinances. Partly as a result of such legislation and partly to protect their own interest, professional and trade associations also became concerned with the establishment of standard codes and specifications.

The American Railway Engineering Association, founded in 1900, adopted the first AREA Specifications for Steel Bridges in 1906. Office of Public Roads, USDA, issued Typical Specifications for Steel Highway Bridges in 1913. In 1921 the American Association of Highway Officials began developing its own standard specifications for highway bridges and the first set was completed and adopted in 1926. The London County Council Building Act was passed in 1909 and New York passed a building ordinance in 1916. In 1923 the American Institute for Steel Construction adopted a Standard Specification for Design, Fabrication, and Erection of Structural Steel Buildings.

In April 1946, the American Iron and Steel Institute adopted the Specification for the Design of Light Gage Cold-Formed Steel structural members. As new technology and techniques developed, all the specifications were revised and are constantly being reviewed and updated.

The building codes adopted in different cities vary in some requirements, because of special local problems. For example, in California, provisions for earthquake resistance are much more important than in New York and, even then, the requirement of different localities in California vary, depending on the severity of earthquake hazards and on the judgment of the local engineering departments.

Although specifications may differ in certain items, all of them are based on the following general requirements for a satisfactory structure: (*a*) the material must be suitable and its quality adequate, (*b*) proper loads and service conditions must be considered in the design, (*c*) computations and design must be made so that the structure and its details possess the required strength and rigidity, and (*d*) the workmanship must be good. The loads prescribed in the various specifications are only approximations of the actual conditions for design purposes. In some instances actual maximum loads may exceed the specified loadings, as, for example, wind loads on isolated buildings, or unusual concentrations of heavy commercial or military vehicles on short-span highway bridges. Also, methods of stress analysis used in general practice are not based on exact theories and may be either too conservative or unconservative. Moreover, fabrication and erection practices vary and cannot always be controlled to assure a high quality of workmanship. All these contingencies are generally offset by specifying an allowable stress for design considerably smaller than the yield strength of the material.

The diversity of allowable stresses in the different specifications is caused partly by the individual traditions and policies of the different organizations and partly by the specific nature of the structure to which the specifications are applicable. AISC Specifications, intended for buildings, allow relatively higher stresses, mainly because of the relatively small amount of live load and impact and the better weather protection afforded to buildings; the AREA Specifications, intended for railroad bridges, which are subject to more severe exposure and heavier live loads, tend to allow lower values.

It follows that specifications for structural design represent a compromise between theoretical considerations and practical requirements. Therefore, the specifications are not entirely rational and, for some structures or loading conditions, they may lead to more conservative results than for others. Also, the values of loads and allowable stresses are generally based on past experience and test data and have to be revised periodically to agree with the latest findings.

In the final analysis the engineer has the responsibility for the adequacy of the safety of the structure whether the specification applies or not.

1-6 FABRICATION

Ease of fabrication and erection has an important influence on the economy of the design. Although it is advisable that the structural engineer have the knowledge of all the details of fabrication and erection, he should have at least a fair understanding of the processes involved in these operations.

A factor in the engineer's consideration of economy of design is the unavoidable fact that fabrication costs money. This cost evolves from the use of manpower, tools, and machinery. Therefore, to reduce the costs of fabrication, the engineer must minimize the amount of fabricating work to be done and must balance the costs of reduced material weight with the costs of increased fabrication.

In some designs it may be advantageous to use higher strength steels for reduced weight, smaller architectural sections, or because of strength requirements. Although the weight of the steel is less, the cost of fabrication and erection is not necessarily reduced because most fabricating operations are relatively independent of the weight or thickness of the part worked on.

Fabricating costs may not be reduced by substituting higher strength steels for structural carbon steels because slower speeds of punches and drills are required and increased precautions are needed for welding operations.

Other items for consideration in fabrication are the following.

(*a*) Dimensional accuracy and tolerances of the various parts. If these are excessively rigid, the costs will, of necessity, be increased.

(*b*) Stiffness of large members. Because of the large size of the members, it is not possible to keep them precisely straight. The deviations from straightness and camber can be kept within reasonable limits that will not affect structural usefulness, but the stiff members cannot be connected too easily to other parts.

(*c*) Methods of straightening material and fabricated members. The common method is the use of a press which works on the material at atmospheric temperature, referred to as "cold pressing." Less commonly used is the oxy-gas torch method of applying heat locally to the part. Both of these methods leave residual stresses in the bent member.

The engineer should be aware of the various methods used in fabricating steel and be ever conscious of the effect of his design on fabrication costs.

Based on the designer's drawings and specifications, several selected fabricators prepare bids for the fabrication of the structure. In determining their bid price the following costs must be estimated: mill cost of raw materials, shipping cost from mill to fabricator, cost of shop drawings and templates, shop fabrication cost, cost of shipping fabricated steel from the shop to the site, erection cost (if included in the contract), overhead, and profit. The bid states the time of delivery and the price, either as a lump sum or a price per unit weight of material. The contract is usually awarded to the lowest responsible bidder, although in some instances a higher price may be paid to secure quick delivery.

When the contract is awarded, the engineering staff of the fabricator receives the design drawings and specifications. The engineer in charge may suggest changes in detail in order to simplify fabrication or erection and, when all information is complete, shop drawings are prepared detailing each fabricated piece in the structure. The shop detail drawings show identifying mark or piece number, number of pieces required, length of piece, location and size of holes in the piece, details of cuts and of shop-attached connections.

The shop drawings must conform to the design and require careful checking by an experienced engineer to see that all parts fit and all dimensions and details are correctly indicated. Full-scale cardboard or wood templates showing locations of all the holes and cuts for each piece are made from shop drawings. A bill of materials is prepared and sent to the mill. The usual practice is to order material for main members cut to length at the mill, and materials for other members and detail pieces are ordered to stock lengths.

Material arriving at the fabricator's receiving yard is checked against the mill order and stored until needed for fabrication. The fabricating shop may be divided into several operating bays; for example, in a shop organized primarily for steel buildings, there may be a column bay, a beam bay, and a miscellaneous bay for plate girders, trusses, and details. In each bay, appropriate machinery is located to carry out the special operations in the bay in a proper sequence.

The first operation in the shop is "laying out." Each piece is marked with its job name and piece mark, number of pieces required, and special instructions regarding fabrication procedures. If necessary, the pieces are cut to lengths, or portions of flange or web are cut away. All the duplicate pieces are laid out together. Then the pieces are drilled, punched, or milled as required. When the main and detail pieces of an assembly have been fabricated, they are transferred to the fitting skids. Here the parts are fitted together, holes are matched and reamed if necessary, and the assemblies are riveted, bolted, or welded together. For example, base plates and beam seats are attached to the columns, and plate girders, roof trusses, and large built-up members are assembled. Fitting up the pieces for assembly is an important job because correction of assembly errors is very expensive. Therefore, inspection for correct fit and sound workmanship in riveting and welding are extremely important at this stage. After assemblies are completed, they are transferred to the shipping yard where they are cleaned, painted, and stored ready for shipment to the erection site.

A major part of fabrication in a shop consists of transporting material from one operation to another. An overhead traveling crane is used to transport the material, and a hoist or a jib crane is used to service the area adjacent to a particular operation. Economy of fabrication often depends on the amount of handling in the shop. For example, one advantage of welded design is the elimination of punching and reaming, or drilling, and the corresponding reduction in handling operations.

Readers interested in a more detailed description of shop fabrication procedures are referred to the *Structural Steel Detailing Textbook*, published by the American Institute of Steel Construction.

1-7   ERECTION

After fabrication, the structural parts are transported to the erection site by flatcars, trucks, or barges. There they are unloaded and either stored at the site or directly positioned in place. The pieces or assemblies are rolled, hoisted, or jacked into place, fitted to supports or adjacent parts of the structure, and then permanently fastened in place. In all these operations the safety of men and materials, economy and rapidity of construction are the important considerations.

Safe construction of large structures often requires a detailed analysis of stresses and deflections during the various stages of erection. Frequently, special handling equipment must be built and special temporary stiffening and bracing frames have to be provided during construction.

Methods used in erection of steel structures vary, depending on the type and size of structure, the site conditions, the availability of equipment, and the erector's preference.[3] Erection procedures cannot be fully standardized. Each problem has special features which must be taken into account in developing the most advantageous erection plan. Some typical erection procedures will be briefly described.

Erection of Multistory Buildings.   Multistory buildings are usually erected in two-story tiers. After foundations are completed, the columns are raised, set on the base plates, and bolted in place. During erection, it is customary to brace the columns laterally until the structure is completed. After the columns are set, beams and girders are hoisted, fitted to the columns, and temporarily bolted to them (Fig. 1-7). As soon as the beams in one story across the entire building are in place, the columns are plumbed, the beams are leveled, and the parts are fastened together permanently by means of rivets, high-strength bolts, or welds. When one tier is completed, erection of the next one begins, repeating the sequence of the first tier. Buildings from 100 to 200 ft high can be erected by using truck cranes. Taller buildings require the use of special derricks, which are raised to the top of the completed frame as the building construction progresses upwards.

Erection of Industrial Buildings.   Industrial buildings of one or two stories are usually erected by cranes, as illustrated in Fig. 1-2. Each bay is assembled and connected while the cranes move along the length of the building. The bracing members are also placed in position in pieces convenient for handling and connecting.

Erection of Truss Bridges.   A common procedure for truss bridge erection is to assemble the truss in place, using falsework underneath, and to build up

Fig. 1-7 Erection of an office building. (Courtesy of AISC.)

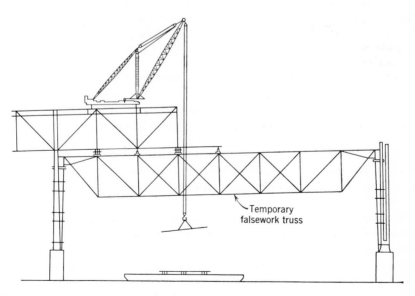

Temporary
falsework truss

Fig. 1-8 Truss bridge erection on false span, Richmond-San Rafeal Bridge. (Courtesy of San Francisco Bay Toll Crossings.)

the truss one member at a time. The bottom chords are placed first, then the floor system is fitted to the chords, followed by web members, top-chord members, and bracing. Instead of building falsework for each span, it is sometimes economical to use a light erection truss which can be floated into position for each span. Figure 1-8 shows an erection truss which was used 11 times for supporting the steel truss erection. Another alternative is to assemble each truss span on shore, float it to the site, and erect it into position.

Cantilever Erection for Bridges.   For long spans of the continuous cantilever or arch types, it is often economical to erect the bridge by cantilevering from the shores or approach spans. The bridge is built from the supports outward, member by member. The omission of falsework over deep gorges or waters effects great savings, although the bridge members often have to be reinforced for erection stresses.

1-8   FIREPROOFING

Although a building structure of steel is usually classified as incombustible and provides reasonable safety for light occupancy structures, it must be protected against fire hazards. The objectives of fire protection are to provide for the safe exit of occupants during the fire, the safety of the firemen, the safety of the adjoining property, to prevent the spread of the fire, and to minimize the economic loss to the property damaged by the fire.

In general, the various types of construction can be classified as follows.

(*a*) Fireproof construction—masonry, reinforced concrete, steel with special exterior fireproofing.

(*b*) Noncombustible construction—steel without exterior protection.

(*c*) Ordinary construction—wood frame and other combustible materials.

The fire safety of the given types of construction is measured in terms of hours of fire resistance based on standardized test procedures. Steel construction is classified as incombustible and provides reasonable safety for light occupancy structures.

The fire resistance of steel can be increased by the application of fire-protective covers such as concrete, gypsum, vermiculite, asbestos sprays, and special paints.

Specifications state the number of hours of fire protection (fire rating) required for various parts of the structure such as floors, beams, girders, columns, partitions, etc. Engineers should become familiar with these requirements and ratings and pay proper attention to them in their designs.

1-9 CORROSION PROTECTION—ARCHITECTURALLY EXPOSED STEEL

The most important factors in determining corrosion resistance are the physical and chemical environment of the material, the composition of the material, and the shielding or protection of the material against contact with the deleterious elements of its environment. It has been suggested that thin steel elements are more susceptible to corrosion than the conventional relatively thick shapes. This does not appear to be reasonable, for thickness is no deterrent to corrosion once the latter attacks the steel. Although increased thickness may, in the presence of corrosion, increase the durability of steel structures somewhat, the only effective means of preventing corrosion is the use of alloying elements, such as chromium or copper, and/or painting with lead, chromate, or aluminum paint, or use of special coatings such as zinc or asphalt.

Where steel is exposed to severe corrosion conditions, it must be protected with a special coating, and this protective coating or paint must be reapplied periodically. Where steel members are not exposed to the effects of alternate wetting and drying, and extreme changes in temperatures, a light coat of properly applied paint is adequate to assure excellent durability.

The rebuilding following World War II brought a new era of architectural and structural design. The rapid growth of technology and the public acceptance of new concepts of design brought forth new shapes and forms in buildings and bridges. Architects and engineers from all parts of the world were quick to apply the advanced technology in creating new building forms in which the structure often becomes the dominant feature of the building form. In some instances the structure is exposed to accent the lines of the building and painted to protect it from corrosion. Figure 1-9 illustrates the concepts of structural form and exposed steel.

In 1933, United States Steel Corporation developed a chemical composition for a high-strength low-alloy steel to meet the corrosion resistance requirements of the coal handling box cars of the railroad. This particular grade of steel oxidized and formed a tightly adherent film which prevented further oxidation. The process produced a deep reddish-brown-black patina. Following the acceptance of exposed painted steel structures, architects began to use this special unpainted exposed steel in different types of buildings. Figure 1-10 illustrates the use of unpainted steel in a large office building in Chicago.

In buildings with exposed bare or painted structural frames the engineer must consider the added design consideration of temperature differentials

Fig. 1-9   Bank building, exposed steel.   (Courtesy of United States Steel Corp.)

between the inside and outside of the building and its effect on the stresses in the structure.

With the acceptance of the new corrosion resistant steel as an architectural material, it was not long before it found its way into the bridge field. The cost of bridge construction and maintenance was increasing and in some locations maintenance was impossible due to hazards caused by the volume of traffic constantly using the roads.   The corrosion resistant steel has possibilities of providing an economical answer to the maintenance problem.   Once the oxidized film has formed, no further maintenance is required.

1-10   ECONOMIC DESIGN

The economy of a structure depends on many factors. Sometimes only the initial cost of a structure is to be considered, but more often the cost of maintenance and operation plays an important part.   In some instances the cost of the structural frame represents only a small part of the total, for example, in a building where heating, plumbing, lighting, and other equipment, as well as interior and exterior finish, constitute a major share of the project cost. Fabrication and erection methods also have some effect on the total cost of the structure. To obtain real economy for a project it may be necessary to modify the structural layout to achieve savings in fabrication and erection or maintenance and operation cost.

For a given class of structures in a given locality, a unit cost index is often used for rough estimating. For example, the cost of a steel frame in a building may be estimated on the basis of cost per pound of steel in place or on the basis of cost per square foot of building area. This method of estimating is very approximate because the cost per unit weight of material in place may vary within wide limits, depending on details of fabrication and erection.

Fig. 1-10 Chicago Civic Center. (Courtesy of United States Steel Corp.)

The fabrication and erection cost may depend on the number of pieces involved, the methods used in fabrication and erection, and the site conditions. Significant reductions in material and labor cost cannot be made once the type of construction and the layout have been determined, although some economies are possible through proper attention to detail design and field operations.

1-11 SAFETY OF STRUCTURES

The safety of structures depends upon a great many factors such as the type of building, fire rating of the fireproofing material, construction details, durability, probability of failure in the structural members and connections, inspection methods for quality control, and the amount of supervision. Some of these factors have already been discussed and this section will examine the safety of steel structures from the point of view of the uncertainties in design considerations.

Structural failure does not always mean collapse. Often excessive distortion of the structural frame prevents the structure from proper functioning, and constitutes a failure as serious as collapse. Collapse or rupture in the structure occurs when some of the principal members or connections fail by slip (shear), cleavage (tension), buckling, or crushing. Complete collapse may occur under severe blast or impact, or as a result of fatigue after a large number of alternating stresses, or after buckling of the main members. Excessive distortions occur under sustained conditions of extreme overload or under moderate impact. These distortions result from extensive yielding of the material in tension or compression, or buckling in compression. In some instances local distortions may be sufficiently serious to be classified as structural failure, although in many cases the ultimate strength of the structure may not be seriously impaired by these local distortions.

A rational definition of the *safety factor n*[4] can be expressed in terms of ratio of computed strength S of the structure or any of its elements to the corresponding computed internal load R, that is, $n = S/R$. Uncertainties in the mechanism of failure, properties of material, and workmanship influence the magnitude of lowest probable strength $\bar{S}$; uncertainties in loading conditions and structural behavior influence the magnitude of highest probable internal load $\bar{R}$. These uncertainties cause deviation of probable values from the computed values, so that the lowest probable value of strength is $\bar{S} = S - \Delta S$, and the highest probable value of the load is $\bar{R} = R + \Delta R$, where ΔS and ΔR are the probable maximum deviations of actual values from computed values.

In order to just prevent failure under most adverse conditions, $\bar{S}$ should be at least equal to or slightly greater than $\bar{R}$. Then

$$S - \Delta S \geq R + \Delta R \quad \text{or} \quad S\left(1 - \frac{\Delta S}{S}\right) \geq R\left(1 + \frac{\Delta R}{R}\right) \quad (1\text{-}1)$$

The minimum safety factor n is then

$$n = \frac{S}{R} = \frac{1 + \Delta R/R}{1 - \Delta S/S} \quad (1\text{-}2)$$

If the maximum deviations ΔS and ΔR are assumed to be 25% of the computed values S and R respectively, then the minimum required safety factor is $n = (1 + 0.25)/(1 - 0.25) = 1.67$.

A quantitative evaluation of the deviations ΔS and ΔR is extremely difficult. The complexity of this task can be appreciated from a brief summary of the various factors influencing it. Let us first consider the mechanism of failure. In this respect, results of classical theory of elasticity are often erroneous since structural members behave inelastically before failure. In fact, at least three principal mechanisms of failure should be considered: gradual yielding leading to collapse, general or local buckling (which may be sudden or gradual), and brittle fracture. Furthermore, the ultimate strength depends a great deal on the rigidity of end connections, and the effects of stiffening elements and collateral material such as walls, partitions, floors, and fire protection or other encasement. The effect of these stiffening elements is particularly difficult to estimate. Determination of strength S is further complicated by the variations in the mechanical properties of material. Although the quality control is high, there still exists a range of variations in the yield strength of structural steel. The state of stress and ambient temperature greatly affect the type of failure and the ultimate strength. In welded structures, the presence of stress concentrations and residual stresses combined with low temperatures may cause a brittle failure to occur where it would not be expected under normal conditions. Manufacturing tolerances, faulty workmanship, and corrosion also contribute to the uncertainties of the computed value of strength S.

Determination of internal load R in a structural element is also subject to numerous uncertainties. Although the least uncertain of the loads on the structure is its own dead load, its estimation and distribution are subject to a variation of at least 10%. Under adverse conditions greater variations have been observed. For example, the dead loads computed after the collapse of the original Quebec Bridge exceeded the design loads[5] by 20 to 30%. In another case[6] stresses measured in the columns of a 48-story building were 20 to 30% in excess of the computed stresses.

Live-load variations depend on the type of structure. On highway bridges, variation in live load depends on the composition of traffic and on the possibility of future changes in types of vehicles and the nature of traffic. On railway bridges and in industrial buildings, live loads can be determined more closely, while buildings for storage and human occupancy are subject to a greater variation of live loads. Dynamic effect of live loads and possible earthquake or blast load effects further complicate determination of the loads on the structure. The conventional practice of using an equivalent static load cannot account for the type of dynamic impulse and the response characteristics of the structure which determine the dynamic effect. Also, the frequency and the number of cycles of load repetitions that may occur during the service life of the structure should be considered in evaluating fatigue effects of live loads.

Conventional wind loads do not always represent realistic maximum conditions and sometimes err on the unconservative side. For example, a 20-psf wind load corresponds to a wind velocity of about 60 mph, whereas exposed structures may be subjected to winds of 90 or 100 mph velocity. Hurricane velocities of 120 to 150 mph have been observed. Also, the occasional practice of neglecting suction forces due to wind introduces some inaccuracy into the computed value of load in a member.

In addition to loading uncertainties, the values of R are affected by the service conditions, such as settlement of foundations and extreme changes in temperature, and secondary effects due to deflections, torsion, or partial rigidity of connections. Also, if safety of the structure is to be determined by comparison of internal load R with ultimate strength S, then inelastic action preceding failure must be taken into account in computing R. Thus computations based on elastic theory are inadequate.

The primary loads, resulting in the so-called primary stresses, are usually computed for a given set of loading conditions, assuming an idealized action of the structure. For example, primary forces in the members of a truss are computed on the assumption that the members are weightless and the joints are pin-connected (Fig. 1-11). The rigidity of the riveted or welded joints, which prevents free relative rotation of the members, the relative displacement of the joints, and the dead weight of the members induces bending stresses in the members, often termed secondary stresses.

The importance of these secondary stresses in evaluating the safety of structures in terms of their ultimate strength depends on the type of failure. For example, consider a tension and a compression member of the truss (Fig. 1-12), each subjected to some secondary bending moments. As the loads increase, both the primary tension and the secondary moment increase, but the tension tends to straighten the member and thus reduce the effect of secondary bending. The deflection of the compression member, however,

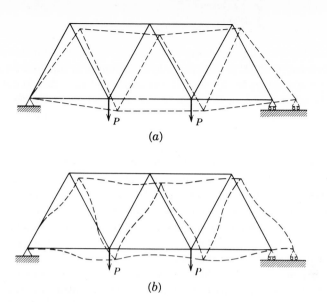

(a)

(b)

Fig. 1-11 Secondary stresses in a truss: (a) deflected shape of truss—joints assumed pinned. (b) deflected shape of truss—joints assumed rigid.

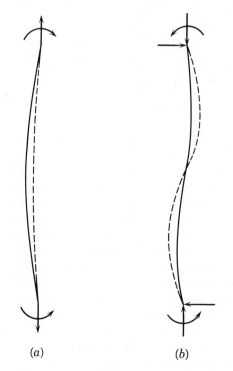

(a)   (b)

Fig. 1-12 Effect of secondary stresses on different types of members: (a) tension member and (b) compression member.

is such that the secondary bending tends to increase with the increase in load and, if the members are slender, the increase of moment due to deflection can be much greater than that due to the increase in the load. Thus the secondary moment has little appreciable effect on the strength of the tension member, but may have a very significant effect on the strength of a slender compression member.

The combined effects of various uncertainties on the computed value of internal load R and on the ultimate strength S are not necessarily directly cumulative. Some of the uncertainties are subject to laws of probability, and statistical determination of ΔS and ΔR has been proposed.[7] Theoretically, a particular value should be determined for each structure which would lead to a particular value of the factor of safety n to be used in its design. Lack of statistical data and analysis for the various influences on ΔS and ΔR precludes the use of this method at present. Therefore, range of fluctuations in S and R must be determined largely on the basis of judgment, guided by some analysis, test results, and past experience. In many instances, values of ΔS and ΔR will be determined subjectively rather than objectively. In spite of this limitation, the value of the minimum safety factor n established by considering all the uncertainties in a qualitative manner and assigning reasonable, although subjective, values to ΔS and ΔR may be preferred to the use of a constant allowable stress which cannot be applicable in all cases with uniform economy. This is especially true if the allowable stress was specified empirically or empirical methods are used for stress computations.

1-12 STRUCTURAL FAILURES

Experience and judgment, which play such an important role in structural design, receive little attention in technical literature. If experience and judgment are to be of real benefit, the designer must learn from the lessons of past failures. Unfortunately, technical literature on the failures of the past is extremely scarce. Understandably, people do not wish to discuss their mistakes. Yet, full discussion of these failures in technical literature could be just as useful to the profession as discussions of great achievements.[8,9]

Structural failures can be caused by unsatisfactory material, fabrication or erection errors, or faulty design. The latter two causes are somewhat related since the designer should attempt to take into account fabrication and erection procedures. The most frequent causes of structural failures can be classified as follows: (a) foundation movements, (b) dynamic resonance and dynamic instability, (c) inadequate connections, (d) incorrect appraisal of buckling strength, (e) lack of adequate bracing against lateral sway or buckling, (f) overloading, and (g) fatigue.

Failures that are due to *movement or settlement of foundations* are probably the most numerous. The detailed appraisal of proper foundation design is entirely beyond the scope of this treatise. The structural designer should be particularly careful to appraise the effect of settlements on the behavior of the structure, especially continuous structures.

Earthquake response of major structures as observed in several recent earthquakes has been on the whole highly satisfactory. Usually the frames withstood the violent ground motion without failure, although in some instances structures not adequately designed for earthquake forces were severely damaged.[10,11,12] In a number of structures, failures of walls, partitions, windows, and local damage to main members or connections were observed. To prevent such damage a rational analysis of dynamic response of a structure should be carried out to determine the forces and displacements which can occur as a result of an earthquake. The members and connections of the frame acting integrally with the other elements of the structure should then be designed to dissipate the energy and to withstand the forces without irreparable damage. The definition of "irreparable" damage is an economic decision: if a high risk of damage is acceptable, a building can be obtained at a relatively low cost. However, certain social costs of damage, for example, loss of life, cannot be evaluated in the same terms as property damage.

The extent of possible damage depends on the design as well as on the estimate of maximum intensity of probable earthquake during the life of a structure. This intensity depends on the seismicity of the region in which the structure is located and on the number of years of useful life of the structure. These cannot be ascertained with precision, and therefore any rational analysis is at best an approximation. However, it seems that empirical designs which bypass the rational analyses place too heavy a burden on the engineer's intuition.

Dynamic resonance and dynamic instability are distinct phenomena which may occur under dynamic loading. Resonance is an increase in vibration which occurs when a periodic motion imposed on a structure has approximately the same frequency as one of the natural modes of the structure. This phenomenon may occur, for example, in electric-power transmission lines when the frequency of the wind gusts is nearly the same as the natural frequency of the cable. The gradual or sudden build-up of catastrophic oscillatory motion in a structure subjected to wind is known as the phenomenon of aerodynamic instability. Before the failure of the Tacoma Narrows Suspension Bridge in 1940, this type of behavior was not considered important in the design of stationary structures. Since then, however, aerodynamic instability has become a factor in the design of slender flexible structures exposed to wind.

Failure of the Tacoma Narrows Suspension Bridge[13] on November 7, 1940, was probably the greatest bridge failure in history. The bridge was in service for about four months before its collapse and, from the first day of operation, excessive vibration of the bridge was observed. Some efforts to prevent these oscillations were made, but were not successful; other possible remedies were under study. The collapse occurred when a moderate 40-mph wind caused vibrations which were gradually increasing in violence. The bridge was oscillating so severely that it was closed to traffic about $1\frac{1}{2}$ hours before collapse, and finally the predominantly twisting motion caused breaking of the suspenders and buckling of the main girders in the center span. Then the floor system fell from the bridge, followed shortly by the girders of the center span.

The most important error in *design of connections* is the disregard of some of the forces acting on them. For example, if the members of the truss joint (Fig. 1-13) are designed for axial forces only, then the rivets at A are subjected to an eccentric load. Failure to provide for this eccentricity would result in an inadequate connection. In a tension rod attached by an angle (Fig. 1-14), failure to check the angle for bending at the heel and failure to provide adequate connection for resisting both shear and moment due to pull P would also result in an inadequate connection.

Another common error is the provision of inadequate anchorage and bearing support of steel beams framing into masonry walls. This was illustrated during a serious failure of a New York apartment house[14] in 1936. The sixth-story wall and the steel frame supporting a 25 ft masonry tower collapsed during construction. Eighteen workmen were killed. The owners, the contractor, and the city building inspector were tried and sentenced to prison

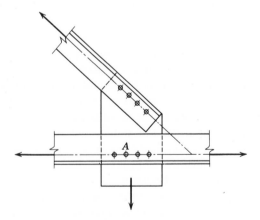

Fig. 1-13 Eccentric load on a truss connection.

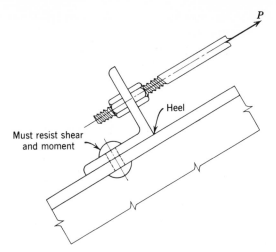

Fig. 1-14   Bending moment in an angle connection.

terms ranging from six months to five years. The architect and the city plans examiner were also found guilty of manslaughter but received suspended sentences.

Incorrect appraisal of *buckling strength* of compression members or beam flanges is a much too frequent cause of serious failures. Determination of strength of an isolated compression member with known conditions of loading and end support is a relatively simple problem. But when the member is considered as a part of the structure as a whole, it is very difficult to establish the amount of load eccentricity, the rigidity of end supports, and the effectiveness of lateral bracing. In addition to problems related to the strength of columns, buckling of beam or girder compression flanges, buckling of plates under various conditions of loading and support, and torsional buckling of light columns sometimes have to be considered.

Buckling of compression members have contributed to many structural failures. One of the most widely known examples is the failure of the Quebec Bridge under construction in 1907. As recently as 1954 failure of columns resulted in a serious accident.[16] Nine 6-in.-diameter standard pipe columns 16 ft long collapsed during the lifting of a 250-ton $8\frac{1}{2}$-in.-thick concrete slab. Before the slab reached its position, the columns leaned 3 in. out of plumb. When cables were attached to the columns and pulled to plumb the structure, the columns collapsed, the slab falling 15 ft west of its original position. Eight persons were injured. The following reasons are said to be largely responsible for the collapse: (*a*) the columns were not designed to resist lateral loads without bracing during lifting, and (*b*) the contractors did not brace the columns in both directions during lifting. Need for adequate bracing against lateral sway or buckling cannot be overemphasized. In finished buildings, the floor and the wall diaphragms, if sufficiently strong and properly attached to the frame, provide such bracing. During construction, however, temporary bracing should be provided to stabilize the columns and frames.

Overloading often occurs as a result of a change in the use of the structure. For example, in the 1930s a number of old bridges failed largely because the traffic became heavier and denser than that assumed in the original design. Buildings converted from one occupancy to another without adequate investigation of structural safety often are overloaded, and a number of such failures have been recorded. In bridges, overloading is sometimes combined with an increased number of loading cycles, resulting in *fatigue failures*. One such failure occurred in 1953 in the lower chord member of the stiffening truss of the Manhattan Bridge in New York. The failure was discovered during a routine inspection, and was not sufficiently serious to close the bridge to traffic. This failure was attributed to fatigue overloading on the northern side of the bridge which had been subjected to heavy subway-train traffic. The chord was repaired by splicing the cracked portion. Discovery of the crack prevented a catastrophic failure in which many lives might have been lost.

Minor failures in riveted bridges carrying railroad traffic are not infrequent, and in general these failures have been attributed to fatigue under repeated cycles of loading.

To prevent all failures is not humanly possible. But if major disasters are to be prevented in the future, the lesson of each failure must be learned by all.

REFERENCES

1. Cross, Hardy, "Relation of Analysis to Structural Design," *Trans. ASCE* **101** (1936).
2. Ketchum, M. S., *Structural Engineering Handbook*, 3rd ed., McGraw-Hill Book Co., New York, 1941.
3. Oppenheimer, S. P., *Erecting Structural Steel,* McGraw-Hill Book Co., New York, 1960.
4. Freudenthal, A. M., "The Safety of Structures," *Trans. ASCE* **112** (1947).
5. Royal Commission on Quebec Bridge Inquiry Report, Canada, Parliam. Session Paper 154 (1907–1908).
6. "Dead Load Stresses in the Columns of Tall Buildings," *Ohio State Univ. Eng. Expt. Sta. Bull.* **40**.
7. Freudenthal, A. M., J. M. Garrelts, and M. Shinozuka, "The Analysis of Structural Safety," *Proc. ASCE, J. Str. Div.*, **92**, ST1, Feb. 1966.
8. Feld, J., "Structural Success or Failure?" *Proc. ASCE*, separate **632** (1955).
9. Godfrey, E., "Engineering Failures and Their Lessons," Akron, Ohio, Superior Printing Co., 1924.

10. Steinbrugge, K. V., and D. F. Moran, "An Engineering Study of the Southern California Earthquake of July 21, 1952 and Its Aftershocks," *Bulletin of Seismological Society of America* **44,** No. 2B, April 1954.
11. Rosenbleuth, E., "The Earthquake of July 1957 in Mexico City," Second World Conference on Earthquake Engineering, Tokyo, 1960.
12. Clough, R. W., "Report on Agadir Earthquake," Second World Conference on Earthquake Engineering, Tokyo, 1960.
13. Amman, O. H., T. von Karman, and G. B. Woodruff, "The Failure of the Tacoma Narrows Bridge," Report to Federal Works Agency, 1941.
14. Seelye, E. E., "Lessons from a Building Collapse," *Eng. News-Record* **119,** 93 (1937).
15. Strehan, G. E., "Designing Lateral Supports for Compression Members," *Eng. News-Record* **120,** 467 (1938).
16. "Lift Slab Collapsed When Pipe Column Failed," *Eng. News-Record* **153,** 25 (1954).

2

Materials

2-1 INTRODUCTION

The advancements in metallurgy and the making of higher strength steels which were developed during the period from 1940 to 1950 soon found their way into steels for bridge and building designs. Engineers had been searching for stronger steels which could carry more load with reduced weights. At first these higher-strength steels were used in special applications of buildings and bridges because they had not been incorporated in the design specifications at the time. Special permission was granted by building officials and highway agencies to use the stronger steels. As experience was gained and more designers sought to use these stronger steels, they were incorporated in building and bridge specifications.

The American Institute of Steel Construction (AISC) included several of the high-strength low-alloy steels in the Specification adopted in November 1961 and revised in April 1963. The adoption of this Specification by municipal, state, and federal agencies permitted the use of these higher-strength steels as a normal part of the design procedure. The American Association of State Highway Officials (AASHO) also incorporated these steels into their bridge Specification.

The steels included in these Specifications have been codified by the American Society for Testing and Materials (ASTM) and have been given a standard designation. Although there are many steels available for structural design, some may not be included in the standard specification because they are regarded as proprietary steels produced by only one steel company. However, such steels may be used in design with the permission of the

32

approving agency or bureau. Therefore the mechanical properties of the grades of steels which may be used in design will be discussed in this text.

2-2 MECHANICAL PROPERTIES

Mechanical properties depend primarily upon the chemical composition, rolling processes, and heat treatment of the steels. Other factors which may influence the mechanical properties are the techniques of testing, such as the rate of loading the specimen, the conditions and geometry of the specimen, cold work, and the temperature at the time of testing. These influences may produce varying results for the same grade of steel.

The usual test coupon is a tension specimen and for all practical purposes the behavior in compression is assumed to be similar to that in tension. Because the tension test is easier to perform, most mechanical properties are taken from the tension stress-strain diagram.

A typical stress-strain diagram for structural carbon steel (Figs. 2-1 and 2-2) is characterized by an initially linear stress-strain relationship, followed by the so-called plastic region of considerable deformation with practically no increase in stress, and then terminating in a region of strain hardening where an increase in deformation is again accompanied by some increase in stress. The plastic strain ϵ_p preceding strain hardening is about ten to twenty times greater than initial yield strain ϵ_y, and therefore a member developing this plastic strain undergoes large deformations.

Initial yielding is not necessarily an indication of failure. On the contrary, ability to yield locally is a valuable characteristic of steel structural elements.

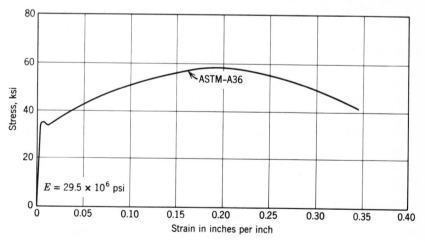

Fig. 2-1 Stress-strain curve for specified minimum values of A36 steel.

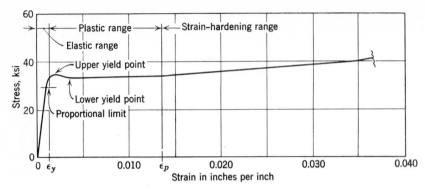

Fig. 2-2   Partial stress-strain curve for A36 steel.

When yielding spreads over a large portion of the member, that is, when it is no longer local, then the deformation increases rapidly and failure occurs in the form of excessive deflection or total collapse of the structure.

A portion of the stress-strain diagram up to the strain-hardening point is indicated in Fig. 2-2 and is applicable to all the constructional steels having a definite yield point. The quenched and tempered carbon and alloy steels do not exhibit a pronounced yield point as indicated in Fig. 2-3.

Yield Point.   The yield point is defined as the stress in the material at which the strain exhibits a large increase without an increase in stress. This is indicated by the flat portion of the stress-strain diagram indicated as the plastic or inelastic range. Some steels have an initial upper yield value which reduces to a plateau referred to as the lower yield stress.

For all steels the upper yield point stress is the one reported in design specifications.

Yield Strength.   The stress-strain diagram of the heat-treated higher-strength steels indicates that these steels do not have the extended yield plateau but instead exhibit a continuing curve up the tensile strength. Therefore, the yield strength of the steel is defined as a specified point on the curve established by constructing a 0.2% offset of strain parallel to the initial elastic portion of the curve. The point at which this line crosses the stress-strain curve is taken as the yield strength.

In some instances, especially in production tests, the yield strength is taken to be the value at 0.005 strain extension under load.

Tensile Strength.   The tensile strength is defined as the maximum axial load on the specimen divided by the original cross-sectional area. In some respects this is an arbitrary value for reference purposes because the true tensile strength should be based on the true stress-strain curve.

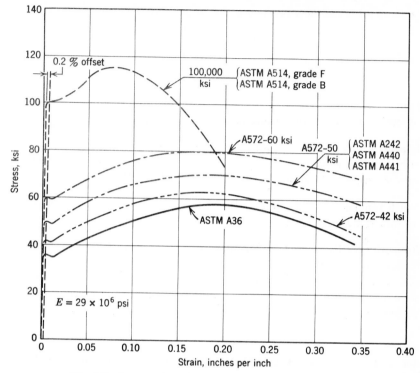

Fig. 2-3 Sress-strain curves for specified minimum values.

Proportional Limit. The proportional limit is the maximum stress for which the stress is directly proportioned to the strain.

Modulus of Elasticity. The ratio of stress to strain in the elastic region is defined as the modulus of elasticity. This value is determined by the slope of the elastic portion of the stress-strain diagram.

Tangent Modulus of Elasticity. The slope of the tangent at a point on the stress-strain curve above the proportional limit is defined as the tangent modulus of elasticity.

Strain-Hardening Modulus. The slope of the stress-strain curve in the strain-hardening range is referred to ast he strain-hardening modulus E_{st}. It has its highest value at the onset of strain hardening.

Poisson's Ratio. The ratio of transverse strain to longitudinal strain under an axial load is defined as Poisson's ratio μ. For steel, the common value ranges from 0.25 to 0.33 in the elastic range.

Shear Modulus of Elasticity. The ratio of the shearing stress to the shearing strain within the elastic range of the steel is defined as the shearing modulus of elasticity G and may be determined by the expression:

$$G = \frac{E}{2(1 + \mu)} \tag{2-1}$$

For the structural steels G is approximately 12,000 ksi.

Fatigue Strength. The stress at which a steel fails under repeated applications of load is defined as the fatigue strength or endurance limit.

Impact Strength. A measure of the ability of the steel to absorb energy at high rates of loading. A comparative measure of the impact strengths of various steel grades is the notch toughness of the steel.

Notch Toughness. The ability of the material to absorb energy as determined by ASTM standard tests.

Properties for Fabrication or Manufacturing. Other properties of interest to the designer are the manufacturing or fabrication characteristics of weldability, machinability, formability, corrosion resistance, etc. The engineer should acquaint himself with these properties for all steels and consider them when making a specific steel selection for a particular application.

2-3 EFFECT OF TEMPERATURE ON MECHANICAL PROPERTIES

The mechanical properties of steels at high temperatures have been investigated mostly on the basis of a continuous high-temperature service, such as that experienced in chemical process equipment and turbines. Therefore there is a large amount of information available on this type of steel application. Investigations of mechanical properties of structural steels on the basis of short-time elevated temperature have been carried out and more research is underway to determine the behavior of the higher strength steels under conditions of short-time elevated temperatures.

The data on tensile strength and yield strength for A36 and A441 steels are presented as envelope curves in Fig. 2-4, which encompass the test results for a number of specimens of each steel. Specimens were taken from plates varying in thickness from $\frac{1}{2}$ to 2 in. It should be noted that strength properties normally decrease with increasing section thickness. The curve (Fig. 2-4) indicates that the yield and tensile strengths of the higher strength steels follow the same behavior pattern as A36 steel when subjected to elevated temperatures.

The modulus of elasticity of structural steel decreases as the temperature increases. Up to a temperature of 900°F the decrease in the modulus is

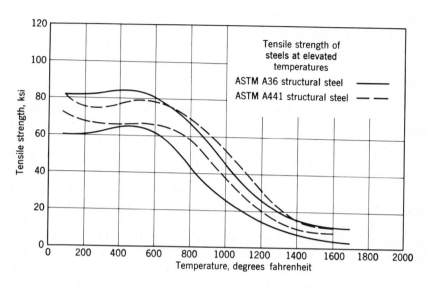

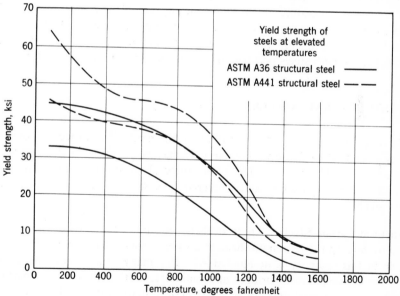

Fig. 2-4 Tensile and yield strengths at elevated temperatures.
(Courtesy of Applied Research Laboratory, United States Steel Corp.)

approximately linear and is equal to about 650,000 psi per 100°F. Above 900°F the modulus of elasticity decreases at a greater rate. The Poisson ratio of steel up to a temperature of 1100°F remains essentially constant.

Although creep (a change in dimension under constant load) of steel at normal temperatures is generally negligible, at elevated temperatures and under sustained loading creep deformations become significant and can no longer be neglected.

It is recognized that information on short-time elevated temperature properties of steel cannot be applied directly in determining the classification for resistance to fire of an assembly which uses columns, beams, and girders. These data are more properly used when appraising the relative performances of different grades of steel.

The performance of building members under fire exposure conditions is usually evaluated on the basis of their performance in standard fire tests of building construction and materials specified in ASTM Standard E-119. The conditions of acceptance for satisfactory performance are such that "the transmission of heat through the protection during the period of fire exposure, for which classification is desired, does not raise the average (arithmetical) temperatures of the steel at any one of the four levels above 1000°F or does not raise the temperature above 1200°F at any one of the measured points."

Higher-strength steels are permitted in assemblies classified by the Standard Fire Test, ASTM E-119. From the data obtained to date, assemblies using the higher-strength steels perform as well as the assemblies using A36 steels.

2-4 STRUCTURAL STEELS

In order to understand the variations in the mechanical properties of the many structural steels now available they may be grouped by strength and grade for ease of discussion. These are structural carbon steels, high-strength low-alloy steels, quenched and tempered carbon steel, and constructional alloy steels.

Structural Carbon Steels. These steels depend upon the amount of carbon used to develop their strength through a wide range of thicknesses. The first grade in this category is designated A7. For many years it was the main steel for construction of bridges and buildings. Although it was primarily developed for riveted and bolted construction, it was also used for welded buildings where the loads could be considered as static or nondynamic loads. Bridge designers preferred a more closely controlled steel as far as the carbon content was concerned and the steel industry developed a grade of steel with improved welding characteristics designated A373.

In 1960 the steel industry announced an improved carbon steel, ASTM A36, with a higher yield point and a carbon content suitable for welding purposes. Since the advent of the A36 grade of steel, the A7 and A373 grades have been annulled by ASTM and therefore are no longer specified in design of structures.

The minimum yield points and tensile strengths are indicated in Table 2-1.

Table 2-1   Mechanical Properties of Structural Carbon Steels

ASTM Grade	Thickness, in.	Min. Yield Point, psi	Tensile Strength, psi
A7		33,000	60,000–75,000
A373	up to 4 in.	32,000	58,000–75,000
A36	up to 8 in.	36,000	58,000–80,000

These minimum strength values are obtained from mill tests in accordance with ASTM established test procedures. The test coupons for rolled shapes are taken from the web of the section.

High-Strength Low-Alloy Steels. This group of steels includes several strength levels and also steels whose chemistry is varied to suit different construction requirements. Alloying elements are added to obtain the strength desired. For example, there may be a specific need for a steel for riveted and bolted construction, for welding, or for enhanced corrosion resistance with welding characteristics.

For these steels the yield points vary for different thicknesses of material and the engineer must take this variable into consideration in his designs. This group of steels derives its increased strength levels from the application of different amounts of alloying elements.

The minimum yield points and tensile strengths are indicated in Table 2-2.

Table 2-2   Mechanical Properties of High-Strength Low-Alloy Steels

ASTM Grade	Thickness, in.	Min. Yield Point, psi	Tensile Strength, psi
A242, A440,	$\frac{3}{4}$ and under	50,000	70,000
and A441	$\frac{3}{4}$ to $1\frac{1}{2}$	46,000	67,000
	$1\frac{1}{2}$ to 4	42,000	63,000
A572-42	up to 4	42,000	60,000
45	up to $1\frac{1}{2}$	45,000	60,000
50	up to $1\frac{1}{2}$	50,000	65,000
55	up to $1\frac{1}{2}$	55,000	70,000
60	up to 1	60,000	75,000
65	up to $\frac{1}{2}$	65,000	80,000

This group also includes many proprietary grades of steels produced by different companies and sold under brand names.

Of the ASTM Grades, A440 is recommended as the economical steel for riveted and bolted construction. The A441 grade is recommended for welded construction, although it may also be used for riveted and bolted applications.

The ASTM A242 Grade is currently considered to be a grade of steel whose corrosion resistance to atmospheric conditions is equal to or greater than twice that of structural carbon steel. Some companies claim four to six times the atmospheric corrosion resistance of carbon steel for their product. The designer should consult with the steel producer before embarking on a design using this grade of steel for its special enhanced corrosion resistance.

Quenched and Tempered Carbon Steels. A new grade of structural steels developed to fill the strength gap from 50,000 to 100,000 psi has been introduced. Some of these steels are proprietary and no ASTM designation is assigned to them at this writing. They are available in a normalized or quenched and tempered condition and depend upon the amount of carbon to develop their strength through a quenching and tempering process. The minimum yield strength, measured by extension under load, is 80,000 psi and the minimum tensile strength is 100,000 psi for plates in thicknesses to $\frac{3}{4}$ in. inclusive.

Quenched and Tempered Alloy Steels. These steels depend upon several alloying elements and heat treatment, in addition to carbon, to obtain their higher yield and tensile strengths. Similar to the high-strength low-alloy steels, these steels have different strength levels for different thicknesses.

These steels and their minimum strength levels follow in Table 2-3:

Table 2-3   Mechanical Properties of Quenched and Tempered Alloy Steels

ASTM Grade	Thickness, in.	Min. Yield Strength, psi	Tensile Strength, psi
A514	to $\frac{3}{4}$ incl.	100,000	115,000–135,000
A514	Over $\frac{3}{4}$ to $2\frac{1}{2}$ incl.	100,000	115,000–135,000
A514	Over $2\frac{1}{2}$ to 4 incl.	90,000	105,000–135,000

The minimum yield strength is determined by 0.5% extension under load or by the 0.2% offset method.

The quenched and tempered alloy steels are weldable, and have an atmospheric corrosion resistance twice that of structural carbon steel. These steels, with a slight modification of chemistry, are also used for conditions requiring resistance to impact abrasion.

The foregoing steels are available in plates, rolled shapes, and bars and the engineer must ascertain which of these he plans to use in his designs and also determine the availability of each in the strength levels he requires. Improvements and new shapes are constantly being made, and it is his responsibility to be well informed.

Other Steels. There are literally thousands of different steels being produced every day to service the many varied and special needs of our manufacturing and construction industries. Many of these steels are not suitable for construction purposes because of the high cost of materials and fabrication, because they lack sufficient ductility or plastic flow, or because they lack sufficient notch toughness. Some steels have been developed for specific applications, such as HY-80 for submarine hulls, missiles and space equipment, or railroad applications, etc. These are specific purpose steels and the designer should be aware of their existence.

One group of these special steels, referred to as "maraging steels," derives its high yield and tensile strengths from a high nickel alloy content which is then heat treated to age the iron-nickel martensite. These steels may have yield strengths from 200,000 to 300,000 psi.

2-5 LIGHT GAGE STEELS

The adoption of the AISI Specification for the Design of Light Gage Cold-Formed Structural Members introduced another group of steels to be used in the design of structures. These steels are of sheet and strip of structural quality and are defined in general by the standard specifications of ASTM.

The specific steels are: Flat-Rolled Carbon Steel Sheets of Structural Quality, ASTM A245; Hot-Rolled Carbon Steel Strip of Structural Quality, ASTM A303; High-Strength Low-Alloy Cold-Rolled Steel Sheets and Strip, ASTM A374; High-Strength Low-Alloy Hot-Rolled Steel Sheets and Strip, ASTM A375, and Zinc-Coated (Galvanized) Steel Sheets of Structural Quality, Coils and Cut Lengths, ASTM A446.

These steels of light gage have minimum yield points ranging from 25,000 to 50,000 psi.

The AISI Design Specification also makes allowance for the use of other material ordered or produced to some other material specification provided the material conforms to the chemical and mechanical requirements of one of the listed specifications or other published specification which establishes its properties and suitability.

The steels specified by the AISI Design Specification are considered to have adequate mechanical properties for structural applications. The mechanical properties of the light gage steels specified in the AISI Specification for structural applications are listed in Table 2-4. The A446 Grade E steel has a yield

Table 2-4   Mechanical Properties of Structural Light Gage Steels

Trade Designation	ASTM Designation	Grade	Thickness, in.	Min. Yield Point or Yield Strength, kips/in.²	Ultimate Strength, kips/in.²	Minimum Elongation in 2 in., percent
Flat-rolled carbon steel sheets of structural quality	A245	A	0.0449	25	45	23–27
		B	to	30	49	21–25
		C	0.2299	33	52	18–23
		D		40	55	15–20
Hot-rolled carbon steel strip of structural quality	A303	A	0.0255	25	45	19–27
		B	to	30	49	18–25
		C	0.2299	33	52	17–23.5
		D		40	55	15–21
High-strength low-alloy cold-rolled steel sheets and strip	A374		0.2499 and under	45	65	20–22
High-strength low-alloy hot-rolled steel sheets and strip	A375		0.0710 to 0.2299	50	70	22
Zinc-coated steel sheets of structural quality	A446	A	0.1756 and under	33	48	20
		B		37	52	18
		C		40	55	16
		D		50	65	12
		E		80	82	1.5

strength of 80 ksi, and is a full hard product used for roofing and is not intended for primary structural applications.

The typical stress-strain curve for the light gage steels is indicated in Fig. 2-5. These steels differ from the AISC structural steels because the stress-strain diagram does not exhibit the usual sharp yield point and elongation at yield stress but instead exhibits a gradual rounded out curve with no definite yield point. For these steels the yield strength is defined by the usual 0.2% offset method or by the stress at a total elongation under load of 0.5%. Either method is acceptable and the results are approximately equal.

The strength of light gage members which fail by buckling depend upon Young's modulus of elasticity E in the range below the proportional limit and the tangent modulus E_t in the range above the proportional limit. The various buckling provisions of the AISI Specification have been written assuming the proportional limit is not lower than 75% of the specified minimum yield strength.

The use of the light gage steels in cold formed members is dependent upon the ductility of the steel as expressed by the minimum elongation. As a result of the cold forming operation, the mechanical properties of the finished product may be substantially different from those of the original sheet. The cold forming process increases the yield and tensile strengths but decreases

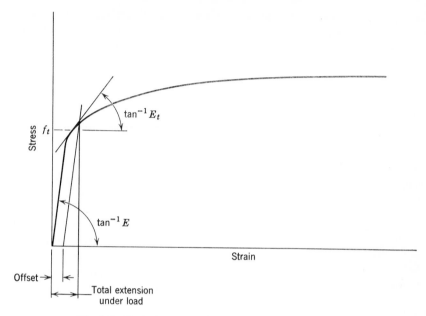

Fig. 2-5 Typical stress-strain curve of sheet and strip steels.

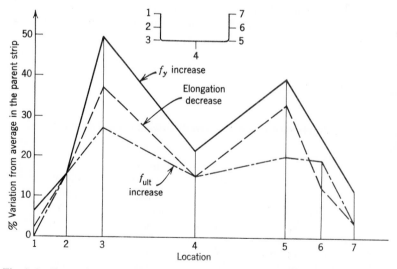

Fig. 2-6 Comparison of mechanical properties of parent material and furnished section at seven specific locations indicated[6].

43

the ductility. Because of the different amount of cold forming in the corners, the mechanical properties are not uniform across the section. A comparison of properties for a channel section is indicated in Fig. 2-6.

2-6 WIRE AND CABLES

Wire and cables are being used for structural applications in buildings for floor hangers and suspended roofs.

Cables are defined as flexible tension members consisting of one or more groups of wires, strands or ropes. A strand is an arrangement of wires helically laid about a center wire to produce a symmetrical section, and a rope is a plurality of strands laid helically around a core composed of a strand, another wire rope, or a fiber core. Wire ropes with fiber cores are used almost entirely for hoisting purposes. Strands and ropes (Fig. 2-7) with strand cores or independent wire rope cores are used in structural applications, and their properties will be discussed here. The strands and ropes are made from wires and as a result their mechanical properties will be considered first.

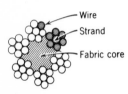

Fig. 2-7 Cross section of a wire rope.

A wire is defined as a single continuous length of metal cold drawn from hot-rolled rods of high carbon steel whose chemical composition is closely controlled. The wires are zinc coated either by the hop dip or the electrolytic process. Although various grades of steel may be used, the most common one for structural applications is galvanized bridge wire, which is also used to make bridge strand and bridge ropes.

The galvanized bridge wire has the tensile strength, yield strength, and elongation indicated in Table 2-5.

Table 2-5   Galvanized Bridge Wire: Tensile Strength, Yield Strength, and Elongation

Coating Class	Diameter, in.	Min. Tensile Strength, ksi	Min. Yield Strength at 0.7% Extension under Load, ksi	Min. Total Elongation in 10 in., percent
A	0.041 and over	220	160	4.0
B	All diameters	210	150	4.0
C	All diameters	200	140	4.0

The minimum yield strength is measured at 0.7% extension under load, and the modulus of elasticity of the wire varies from 28,000,000 to 30,000,000 psi. Wire size is often specified by a gage number rather than a diameter and the most common standard is the U.S. Steel Wire Gage. The bridge wire used in strand and rope is zinc coated to a required minimum coating depending upon the diameter. The minimum weights of coating for three classes A, B, and C are indicated in Table 2-6.

Table 2-6   Galvanized Bridge Wire: Minimum Weights of Coating

Diameter of Coated Wire, in.	Minimum Weight of Coating Ounces per Square Foot of Uncoated Wire Surface		
	Class A	B	C
0.041–0.061	0.40	0.80	1.20
Over 0.061–0.079	0.50	1.00	1.50
Over 0.079–0.092	0.60	1.20	1.80
Over 0.092–0.103	0.70	1.40	2.10
Over 0.103–0.119	0.80	1.60	2.40
Over 0.119–0.142	0.85	1.70	2.55
Over 0.142–0.187	0.90	1.80	2.70
Over 0.187	1.00	2.00	3.00

Strands and ropes for structural purposes are fabricated from helically formed components and therefore their behavior is somewhat different from that of rods, eye bars, or even the individual wires from which they are made. When a tensile load is applied to a strand or wire rope, the resulting elongation will consist of (*a*) a structural stretch caused by the radial and axial adjustment of the wires and strands to the loads and (*b*) the elastic stretch of the wires.

The structural stretch varies with the number of wires per strand, the number of strands per rope, and the length of lay (pitch to helix) of the wires and strands. The stretch also varies with the magnitude of the load imposed and the amount of bending to which the rope may be subjected. Consequently, it can be seen that the structural stretch is gradually removed with an accompanying increase in the modulus of elasticity for the entire rope or strand.

For structural applications in which a limited amount of elongation under load is permissible and where a stable modulus of elasticity is required, removal of the structural stretch is accomplished by prestretching the strand

Table 2-7   Mechanical Properties of Zinc-Coated Bridge Strand
Standards Established by the Wire Rope Technical Board

Nominal Diameter in Inches	Class "A" Coating Throughout	Class "A" Coating Inner Wires Class "B" Coating Outer Wires	Class "A" Coating Inner Wires Class "C" Coating Outer Wires	Approx. Metallic Area Square Inches	Approx. Weight per Foot Pounds
	Minimum Breaking Strength in Tons of 2,000 lb				
$\frac{1}{2}$	15.0	14.5	14.2	0.150	0.52
$\frac{9}{16}$	19.0	18.4	18.0	0.190	0.66
$\frac{5}{8}$	24.0	23.3	22.8	0.234	0.82
$\frac{11}{16}$	29.0	28.1	27.5	0.284	0.99
$\frac{3}{4}$	34.0	33.0	32.3	0.338	1.18
$\frac{13}{16}$	40.0	38.8	38.0	0.396	1.39
$\frac{7}{8}$	46.0	44.6	43.7	0.459	1.61
$\frac{15}{16}$	54.0	52.4	51.3	0.527	1.85
1	61.0	59.2	57.9	0.600	2.10
$1\frac{1}{16}$	69.0	66.9	65.5	0.677	2.37
$1\frac{1}{8}$	78.0	75.7	74.1	0.759	2.66
$1\frac{3}{16}$	86.0	83.4	81.7	0.846	2.96
$1\frac{1}{4}$	96.0	94.1	92.2	0.938	3.28
$1\frac{5}{16}$	106.0	104.0	102.0	1.03	3.62
$1\frac{3}{8}$	116.0	114.0	111.0	1.13	3.97
$1\frac{7}{16}$	126.0	123.0	121.0	1.24	4.34
$1\frac{1}{2}$	138.0	135.0	132.0	1.35	4.73
$1\frac{9}{16}$	150.0	147.0	144.0	1.47	5.13
$1\frac{5}{8}$	162.0	159.0	155.0	1.59	5.55
$1\frac{11}{16}$	176.0	172.0	169.0	1.71	5.98
$1\frac{3}{4}$	188.0	184.0	180.0	1.84	6.43
$1\frac{13}{16}$	202.0	198.0	194.0	1.97	6.90
$1\frac{7}{8}$	216.0	212.0	207.0	2.11	7.39
$1\frac{15}{16}$	230.0	226.0	221.0	2.25	7.89
2	245.0	241.0	238.0	2.40	8.40
$2\frac{1}{16}$	261.0	257.0	253.0	2.55	8.94
$2\frac{1}{8}$	277.0	273.0	269.0	2.71	9.49
$2\frac{3}{16}$	293.0	289.0	284.0	2.87	10.05
$2\frac{1}{4}$	310.0	305.0	301.0	3.04	10.64
$2\frac{5}{16}$	327.0	322.0	317.0	3.21	11.24
$2\frac{3}{8}$	344.0	339.0	334.0	3.38	11.85
$2\frac{7}{16}$	360.0	355.0	349.0	3.57	12.48

Table 2-7 *Continued*

Nominal Diameter in Inches	Class "A" Coating Throughout	Class "A" Coating Inner Wires Class "B" Coating Outer Wires	Class "A" Coating Inner Wires Class "C" Coating Outer Wires	Approx. Metallic Area Square Inches	Approx. Weight per Foot Pounds
		Minimum Breaking Strength in Tons of 2,000 lb			
$2\frac{1}{2}$	376.0	370.0	365.0	3.75	13.13
$2\frac{9}{16}$	392.0	386.0	380.0	3.94	13.80
$2\frac{5}{8}$	417.0	411.0	404.0	4.13	14.47
$2\frac{11}{16}$	432.0	425.0	419.0	4.33	15.16
$2\frac{3}{4}$	452.0	445.0	438.0	4.54	15.88
$2\frac{7}{8}$	494.0	486.0	479.0	4.96	17.36
3	538.0	530.0	522.0	5.40	18.90
$3\frac{1}{8}$	584.0	575.0	566.0	5.86	20.51
$3\frac{1}{4}$	625.0	616.0	606.0	6.34	22.18
$3\frac{3}{8}$	673.0	663.0	653.0	6.83	23.92
$3\frac{1}{2}$	724.0	714.0	702.0	7.35	25.73
$3\frac{5}{8}$	768.0	757.0	745.0	7.88	27.60
$3\frac{3}{4}$	822.0	810.0	797.0	8.44	29.53
$3\frac{7}{8}$	878.0	865.0	852.0	9.01	31.53
4	925.0	911.0	897.0	9.60	33.60

or rope. This is performed by subjecting the strand or rope to a predetermined load for a sufficient length of time to permit adjustment of the individual component parts.

Therefore the modulus of elasticity to determine the elastic elongation of the prestretched bridge strands and wire ropes follows.

Bridge strand of $\frac{1}{2}$ to $2\text{-}\frac{9}{16}$ in. diameter: 24,000,000 psi
Bridge strand of $2\text{-}\frac{5}{8}$ in. and larger diameter: 23,000,000 psi
Bridge rope of $\frac{3}{8}$ to 4 in. diameter: 20,000,000 psi

Because the bridge strands and wire ropes are manufactured from cold drawn wire, they do not have a definite yield point. Therefore, unlike other types of tension members, the working load or allowable design stress is based on the minimum breaking strength or ultimate strength of the rope or strand. The mechanical properties of bridge strand and bridge rope are indicated in Tables 2-7 and 2-8.

Table 2-8   Mechanical Properties of Zinc-Coated Bridge Rope
Standards Established by the Wire Rope Technical Board

Nominal Diameter in Inches	Min. Breaking Strength in Tons of 2,000 lb Class A Coating	Approx. Weight per foot Pounds	Approx. Metallic Area in Square Inches
$\frac{3}{8}$	6.5	0.24	0.065
$\frac{1}{2}$	11.5	0.42	0.119
$\frac{5}{8}$	18.0	0.65	0.182
$\frac{3}{4}$	26.0	0.95	0.268
$\frac{7}{8}$	35.0	1.28	0.361
1	45.7	1.67	0.471
$1\frac{1}{8}$	57.8	2.11	0.596
$1\frac{1}{4}$	72.2	2.64	0.745
$1\frac{3}{8}$	87.8	3.21	0.906
$1\frac{1}{2}$	104.0	3.82	1.076
$1\frac{5}{8}$	123.0	4.51	1.270
$1\frac{3}{4}$	143.0	5.24	1.470
$1\frac{7}{8}$	164.0	6.03	1.690
2	186.0	6.85	1.920
$2\frac{1}{8}$	210.0	7.73	2.170
$2\frac{1}{4}$	235.0	8.66	2.420
$2\frac{3}{8}$	261.0	9.61	2.690
$2\frac{1}{2}$	288.0	10.60	2.970
$2\frac{5}{8}$	317.0	11.62	3.270
$2\frac{3}{4}$	347.0	12.74	3.580
$2\frac{7}{8}$	379.0	13.90	3.910
3	412.0	15.11	4.250
$3\frac{1}{4}$	475.0	18.00	5.040
$3\frac{1}{2}$	555.0	21.00	5.830
$3\frac{3}{4}$	640.0	24.00	6.670
4	730.0	27.00	7.590

2-7   CASTINGS

There are occasions in design for special fittings, special structures such as machinery stands, and parts which cannot be made by other methods and it becomes necessary to design them as a casting or cast steel part. There are two grades of steel for this purpose: (*a*) Mild-to-Medium-Strength Carbon Steel Castings for General Applications, ASTM A27, Grade 65–35 and (*b*) High Strength Steel Castings for Structural Purposes, ASTM A148, Grade 80–50. The minimum yield and tensile strength are as shown in Table 2-9.

Table 2-9   Mechanical Properties of Steel Castings

ASTM Grade	Elongation in 2 in.	Min. Yield Point, psi	Tensile Strength, psi
A27 Grade 65–35	24	35,000	65,000
A148 Grade 80–50	22	50,000	80,000

The A148 Grade is intended for use in those structures which are subjected to higher mechanical stresses than can be carried safely by the A27 Grade. The A27 Grade is considered weldable when using recommended techniques. There are other strength grades of casting steels listed in the respective ASTM Specification, but those noted are the more commonly used grades for structural purposes.

2-8   FORGINGS

There are some structural components such as expansion devices for bridge floors, seats and rockers for bridge beams and girders, and other heavily loaded compact pieces which may be made by a forging process. Two general categories of forging steels are in use for general industrial purposes, which are designated ASTM A235 and A237. The A235 Specification is for Carbon Steel Forgings for General Industrial Use and the specific grade for structural uses comprises Class C1, F, and G. Class C1 includes annealed, normalized, or normalized and tempered materials. Class F is a steel which is normalized and tempered. Class G is a quenched and tempered or normalized, quenched, and tempered material.

Class C1 forgings that are to be welded may have the carbon content limited to 0.35 maximum percent. The yield and tensile strengths and the elongation in 2 in. are indicated in Table 2-10.

Table 2-10   Mechanical Properties of Steel Forgings

ASTM Grade	Class	Thickness	Min. Yield Strength, psi	Tensile Strength, psi	Min. Elongation in 2 in., per cent
A235	Cl	Not over 12 in.	33,000	66,000	23
A235	F	Not over 12 in.	40,000	80,000	21
A235	G	Not over 4 in.	55,000	90,000	20
A237		Not over 12 in.	50,000	80,000	24

The A237 Specification is for Alloy Steel Forgings for General Industrial Use, and covers heat-treated alloy steels. The choice of class of forging is dependent upon the design stress and service imposed. Class A is recommended for general construction purposes. Other classes may be used if the design warrants it.

2-9   RIVET STEELS

There are three ASTM Specifications for rivet steels: A141 Structural Rivet Steel, A195 High-Strength Structural Rivet Steel, and A502 High-Strength Structural Alloy Rivet Steel. The mechanical properties for these steels are indicated in Table 2-11.

Table 2-11   Mechanical Properties of Rivet Steels

ASTM Grade	Min. Yield Point, psi	Tensile Strength, psi	Min. Elongation in 2 in., per cent
A141	28,000	52,000 to 62,000	24
A195	38,000	68,000 to 82,000	20
A502 Grade 1	28,000	52,000 to 62,000	24
A502 Grade 2	38,000	68,000 to 82,000	20

The scope clause of the ASTM Specification for rivet steels describes the recommended application of each type, such that A141 is for structural purposes; A195 is suitable for use with structural silicon (A94) and equivalent steels; and A502 is suitable for A242 and equivalent steels.

2-10   BOLT STEELS

Four types of steels in use for bolts for structural purposes are designated by ASTM as: A325—High-Strength Steel Bolts for Structural Joints; A354 Grade BC—Quenched and Tempered Alloy Steel Bolts and Studs with Suitable Nuts; A307—Specification for Low-Carbon Steel Externally and Internally Threaded Standard Fasteners; and A490—Quenched and Tempered Alloy Steel Bolts for Structural Steel Joints.

The foregoing specifications list tensile requirements for the size of bolts from $\frac{1}{2}$ to $1\frac{1}{2}$ in. and also specify the nuts and washers to be used with each type of bolt.

The strength properties of these steels are equal to or greater than the grade of structural steel for which they are recommended.

2-11 FILLER METAL FOR WELDING

The welding electrodes for shielded metal-arc welding of carbon and low-alloy steels are described in ASTM Specifications for Mild Steel Covered Arc-Welding Electrodes, A233 and Low-Alloy Steel Covered Arc-Welding Electrodes, A316. These specifications prescribe the requirements for (*a*) classification and acceptance, and (*b*) details of tests. Also included in the specification is an appendix describing the classification and use of the many types of electrodes. The engineer should be familiar with all the details of the specification to enable him to understand welding techniques applied to his designs.

For structural applications the E60 or E70 series of electrodes are used for manual shielded metal-arc welding. Bare electrodes and granular flux used in the submerged-arc process conform to the requirements of Grades SAW-1 and SAW-2. The strength properties of the filler metal for design are as follows.

Table 2-12 Mechanical Properties of Weld Metal

		Minimum Values	
Grade	Yield Point, psi	Tensile Strength, psi	Elongation in 2 in., percent
E-60 Series	50,000 and 55,000	62,000 and 67,000	17, 22, 25
E-70 Series	60,000	72,000	17, 22
SAW 1	45,000	62,000 to 80,000	25
SAW 2	50,000	70,000 to 90,000	22

The classification system used for the electrodes is a descriptive one. The letter E designates an electrode. The first two digits, 60 and 70 for example, designate the minimum tensile strength of the deposited metal in the as welded condition in 1000 psi. The third digit is the position for which the electrode is capable of making a satisfactory weld. The code is: digit 1—all positions, flat, vertical, overhead, horizontal; and digit 2—flat and horizontal fillet welds.

The last digit indicates the current to be used and type of covering on the electrode as described in the ASTM Specification. Together, the last two digits designate the position, current, and other pertinent data to be employed when using a particular electrode.

2-12 ECONOMICS OF MATERIAL SELECTION

The engineer is always faced with making a decision on the choice of material and arrangement of structural members to suit a particular application. The choice is usually one of economics, involving costs of material, fabrication, shipping, availability, delivery date, and speed of erection. The time schedule for the job therefore becomes an important consideration.

In selecting steel as the structural material, the designer is faced with the responsibility of making the correct choice for strength, stiffness, and cost.

In general, the cost of a steel design is closely related to the weight of steel used in the design, although such items as fabrication, shipping, and erection are factors not to be ignored. Therefore if the material cost is reduced to a minimum by the selection of the most efficient steel for each member, the total cost of the structure may also be reduced.

Another factor affecting total cost is fabrication, which involves the processing of the material. If labor or shop operations can be reduced or eliminated, the cost of the complete fabrication is less. For example, a rolled shape of any grade of steel is, in most instances, more economical than a built-up shape of the same size and grade of steel. It follows also that a rolled shape of a higher strength steel may be used to replace a lower-strength steel shape requiring flange plates (Fig. 2-8). In this instance, the labor of adding the flange plates is eliminated and cost savings may result.

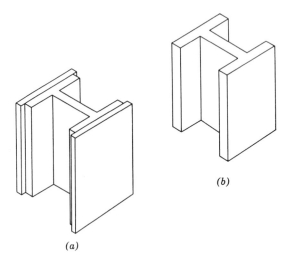

(b)

(a)

Fig. 2-8 Comparison of high-strength section with built-up section. (*a*) A36 section with cover plates riveted or welded. (*b*) higher-strength steel eliminates need for cover plates.

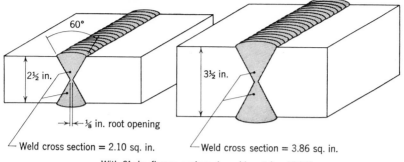

Weld cross section = 2.10 sq. in.   Weld cross section = 3.86 sq. in.

With 2½ in. flange, savings in weld metal = 45.5%

Fig. 2-9 Comparison of butt welds for 2½ and 3½ in.-thick plates.

Still another reduction in total cost, although of lesser amount, may be obtained when thinner plates of high-strength steels replace thicker plates of a lower-strength steel. Figure 2-9 indicates the comparison of a flange butt weld for plate thicknesses of 2½ and 3½ in. Assuming the thinner plate replaces the thicker one, a 45.5% saving in weld metal may be realized. The reduction in thickness will also reflect a saving in welding time and labor.

The three-span Whiskey Creek Bridge in California is a good example of a plate girder design using the various strength grades of steel efficiently for strength requirements and cost reductions. The various steels are placed according to the strength requirements of the loading and are indicated in Fig. 2-10.

In every project, the components must be shipped from the mill to the fabricating shop and thence to the ultimate site or consumer. Thus weight becomes an important factor in the cost of shipment and any reduction in weight is a direct saving in cost.

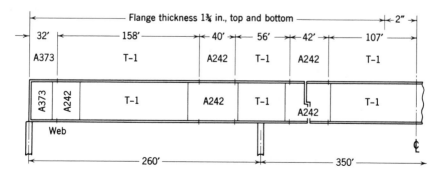

Fig. 2-10 Steel arrangement, Whiskey Creek Bridge.

Most structures requiring field erection may not realize a cost saving if the number of pieces handled remains the same. The equipment needs and time of erection are not necessarily reduced simply by reducing the weight of the pieces handled.

In buildings the effect of using the higher-strength steels is often hidden behind the external appearance. Nevertheless, the esthetics of the structure is in great measure a result of the judicious use of the appropriate steel. For example, an office building with suspended floors using constructional alloy plates as hangers will enable the architect to delineate the thin vertical elements of the hangers in the elevation of the building.

Esthetics is also becoming an important consideration in the design of bridges, and the use of the full range of constructional steels is contributing to this achievement. Slender members combined with a more pleasing outline are now possible and economical. As a result, fabrication and erection costs may be reduced and appearance enhanced.

Maintenance of bridges is a factor to be considered in design, and considerable thought is devoted to reducing the maintenance costs to a minimum. The use of smaller size members reduces paint and painting time required. The uniform depths of trusses and plate girders made possible with the use of the "multiple steel" concept reduce maintenance costs by simplifying the scaffolding for painting.

Another application of the combination of steels concept is in the design of liquid storage tanks. These may be for water, oil, or gasoline storage. Each type has its own specific specification for allowable design stresses and strength grades of steel to be used. The objective in design is to use each grade of steel efficiently, thereby reducing the cost of the tank installation.

When one grade of steel is used to resist the hoop tension caused by the hydrostatic pressures, the wall thickness increases with depth for a constant allowable design stress. However, when applying the combination of steels concept, the thickness of the wall may be kept to a minimum by using higher allowable design stresses for the higher-strength steels (Fig. 2-11).

An example of this type of design is a 10-million gallon steel water storage tank at Monroeville, Pennsylvania.[5] The tank is 33 ft 7 in. high and 226 ft in diameter. The two bottom courses are of A514 Grade B steel, $\frac{15}{32}$ and $\frac{3}{8}$ in. thick, with a design stress of 42,000 psi and a 100% joint efficiency. The third course is of A441 steel, $\frac{13}{32}$ in. thick, with a design stress of 25,000 psi and 100% joint efficiency. The top course is A283 Grade C carbon steel, $\frac{3}{8}$ in. thick, with an allowable stress of 21,000 psi and 100% joint efficiency.

For transportation equipment the principal advantage of the higher-strength steels is in the greater pay-loads realized with the same axle loadings. The

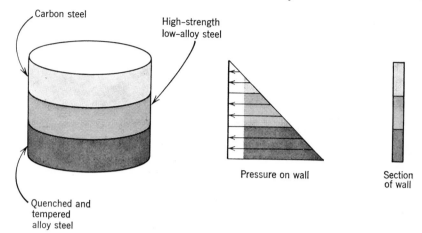

Fig. 2-11 Storage tank design.[3]

higher-strength steels with reduced weight enable the carrier to yield a greater return to the owner. Construction equipment is usually of the type requiring strength and durability of parts and these characteristics may be found in the higher-strength steels.

Although some of the factors enumerated are intangible and are given a credit value when selecting the strength grade of steel to be used, the material prices are usually a major factor in making the choice.

To aid the designer in his choice of steel, a convenient parameter is the strength-to-price ratio for the different strength levels of steel. A typical plot is shown in Fig. 2-12 for plates ranging from $\frac{3}{16}$–$1\frac{1}{2}$ in. The ordinate designates the ratio of strength to price, and the abscissa is the plate thickness. A look at the diagram will indicate that the higher ratio of strength to price is the most economical steel for a specific thickness. The figures are plotted for only a few of the steels available. Because mill prices are subject to change, charts of this type should be reviewed periodically.

The maximum benefit can be obtained only if the material is used to its fullest capacity to satisfy all the design requirements. The strength-to-price ratio plot may be used to evaluate various thicknesses of different possible steels which may be considered for a specific design application. For example, if a plate girder flange requires a $1\frac{1}{2}$-in. plate of a high-strength low-alloy steel of A441, its strength-to-price ratio is 6.0. If the same flange is designed in A514 steel, the required thickness of plate may be 1 in. Its strength-to-price ratio is 7.3. A comparison of the two steels indicates that the A514 steel is the more efficient and economical steel for this specific application.

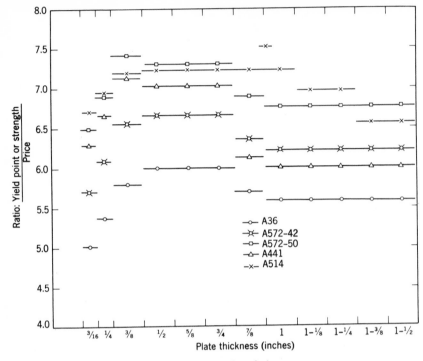

Fig. 2-12 Economics of plates.

REFERENCES

1. American Society for Testing Materials, ASTM Standards 1967, Part 1—Ferrous Metals Specifications, Part 3—Metals Test Methods, Philadelphia, 1967. Part 4—Structural and Boiler Steel; Reinforcing Steel; Ferrous Filler Metal; Ferro-Alloys, 1967.
2. McLean, D., *Mechanical Properties of Metals*, John Wiley and Sons, New York, 1962.
3. "Steels to Match Your Imagination," Design and Engineering Seminar—1964, United States Steel Corporation, Pittsburgh, 1964.
4. "The Making, Shaping and Treating of Steel," 8th ed., United States Steel Corporation, Pittsburgh, 1964.
5. Petsinger, R. E., and H. W. Marsh, "Design Concepts for Petroleum and Chemical Storage Tanks," American Society of Mechanical Engineers, Houston Conference, Sept. 19–22, 1965.
6. Yu, Wei-Wen, "Design of Light Gage Cold-Formed Steel Structures," Engineering Experiment Station, West Virginia University, Morgantown, West Virginia, 1965.

3

Brittle Fracture and Fatigue

🍄🍄

3-1 INTRODUCTION

The general phenomenon of failure of materials is extremely complex and many problems associated with it have not been completely solved yet. Research work in the field of material behavior and the influence of design and fabrication on material failures are constantly adding more knowledge and enlightenment on these subjects.

In the design of bridges and buildings, the definition of "failure" is interpreted to mean "excessive deformations and deflections." Complete failure or general collapse is to be avoided. In case of distress reserve strength is desired in buildings to enable occupants to leave safely. In bridges an excessive deflection becomes visible and is a warning to close the bridge to traffic.

There are other types of structures and special pieces of equipment which may be designed to other failure criteria. These criteria may be established by a specific industry, government, or professional group, or in some instances by the designer's judgment of the service conditions to be satisfied.

Bridges, buildings, ships, and other structures used by people are therefore designed for what may be defined as a ductile or plastic failure. Sudden or abrupt failures are undesirable. However, under certain circumstances sudden failures do occur and these have been attributed to either brittle fracture or fatigue.

Brittle behavior and cleavage-type fractures are caused by high tensile stresses, high carbon content, rapid rate of loading, and the presence of notches. A cleavage failure has a crystalline appearance because each

57

crystal tends to fracture on a single plane. A shear fracture, on the other hand, has a fibrous appearance because of the sliding of the fracture surfaces over each other.

Fatigue is that phenomenon which is associated with metal failing or breaking at a stress considerably smaller than the tensile strength or yield strength when subjected to repeated or cyclic loads.

Some classical examples of failures attributed to brittle fracture have occurred in bridges, e.g., Duplessis Bridge, Quebec, Canada. Others occurred in welded ships during World War II (1941–1945). The possibility of brittle fracture in structures can be minimized, and even eliminated, if the structure is properly designed and fabricated.

3-2 BRITTLE FRACTURE

The phenomenon of brittle fracture is represented fairly accurately by the following simplified theory.[11]

In general, steel behaves in a plastic manner, yielding when slip due to high shear stress occurs along certain crystallographic cleavage planes. The maximum shear stress at which slip occurs will be called shear yield stress, and the maximum principal tension stress at which fracture occurs will be called the brittle strength. The brittle strength of steel does not change appreciably with temperature and is assumed to be constant. The shear yield stress, however, varies with temperature, increasing with lower temperatures. The behavior of the material may be either plastic or brittle, depending on the state of stress and temperature. It is well known, for example, that some steels, unwelded, may behave as ductile steels at room temperature but may become quite brittle at low subzero temperatures.

Consider first the effect of state of stress on failure. Figure 3-1 shows one element of steel subjected to uniaxial tension, and one element to triaxial tension, with corresponding Mohr circles of stress. For uniaxial tension, when normal stresses $f_x = f_y = 0$, yielding will occur when maximum shear stress reaches a value of f_s; that is, when $f_z = 2f_s$. Under normal temperature conditions brittle strength has a value considerably greater than $2f_s$, and the steel element behaves as a ductile element. For triaxial tension, assuming for simplicity that $f_x = f_y$, yielding will occur when $f_z = f_x + 2f_s$, provided this stress f_z is less than brittle strength. If values of f_x and f_y are large, then brittle failure may occur before yielding can take place.

This consideration of yielding or brittle behavior indicates the importance of residual stresses. If high residual stresses exist in a welded structure, then brittle failure may occur under relatively low loads. Consider a case when equal residual stresses exist in x and y directions, both equal to $2f_s$—the tension yield strength in uniaxial tension. If the element is subjected to additional

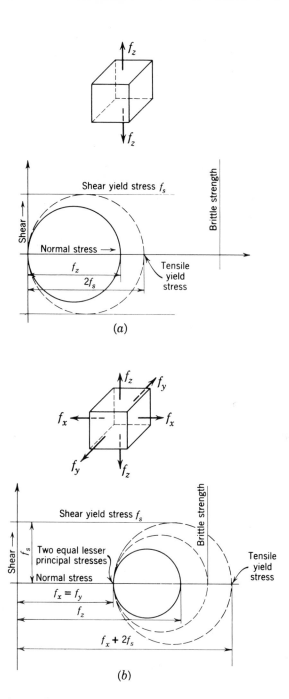

Fig. 3-1 Effect of state of stress on failure: (*a*) uniaxial tension, and (*b*) triaxial tension.

59

tension f in the z direction, yielding will not occur until $f_z = f_x + 2f_s = 2f_s + 2f_s = 4f_s$. This value of $f_z = 4f_s$ may be greater than the brittle strength, and brittle failure may occur when the additional tensile stress f_z is less than $4f_s$.

3-3 EFFECT OF TEMPERATURE ON BRITTLE FRACTURE

The effect of temperature on behavior of the elements shown in Fig. 3-2 can be illustrated by considering the following cases a and b: tension yield occurs when $f_z = 2f_s$ for case a, and when $f_z = 4f_s$ for case b. Shear yield f_s varies with temperature as shown in Fig. 3-2 and therefore tensile yield stresses for cases a and b are obtained as shown in the figure.

It is apparent that, for uniaxial tension, steel will behave in a plastic manner at temperatures greater than T_1, and in a brittle manner at temperatures lower than T_1. For the special case of triaxial tension considered here, steel will behave in a brittle manner at temperatures considerably higher than T_1; the critical temperature for this case is indicated as T_2. If service temperature is higher than this critical temperature T_2, a structure may still behave in a ductile manner.

The foregoing simplified theory indicates the importance of minimizing the residual stresses in a welded structure. It should be pointed out, however, that welded structures are not the only types of structures that are subjected to residual stresses. Residual stresses are set up in plates and shapes by rolling shearing, forming, riveting, etc. If proper design and welding procedures are used, the residual stresses need not be excessive.

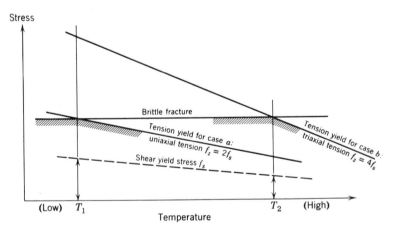

Fig. 3-2 Effect of temperature on failure.

3-4 CHARPY V-NOTCH TEST

A quantitative measure of the ability of a steel strength grade to sustain adverse temperature, and other factors of design and fabrication is the Charpy V-Notch Test. Although there are other rating tests, the Charpy V-Notch Test is the one most commonly used. The test evaluates the notch toughness of the steel which is defined as the resistance to fracture in the presence of a notch. The test results are used qualitatively in the selection of a steel for a specific application.

In this test a small rectangular bar with a specified V-shaped notch at the middle is simply supported on its ends and fractured by a pendulum swung from a fixed height (Fig. 3-3). The amount of energy required to fracture the specimen is calculated from the height to which the pendulum rises after breaking the specimen. The amount of energy required to break the specimen for a range of temperatures is determined and plotted. A typical curve is plotted in Fig. 3-4 indicating variation in energy versus temperature.

The value of 15 ft-lb of energy has been accepted as a reference when comparing steels, and the temperature at which this occurs is called the transition temperature. From the design standpoint the lower the transition temperature, the better will be the rating of the steel for resistance to brittle fracture. The transition temperature will vary with the thickness of the material and care must be exercised in selecting the proper steel and thickness for notch toughness.

A few typical values of the transition temperature corresponding to 15 ft-lb of energy determined by the Charpy V-Notch Test are tabulated below.

Steel	Transition Temperature, F
A36	+30
A441	+15
A440	+30
A514	−50

3-5 BRITTLE FRACTURE DESIGN CONSIDERATIONS

Although a steel is selected for its good notch toughness rating, it is also very important that the design details and fabrication workmanship do not produce notches which could start cracks.

The principal design consideration to avoid or minimize brittle fracture is the elimination of geometrical discontinuities which may act as notches in the bar in the Charpy Test. Sharp re-entrant corners should be avoided by

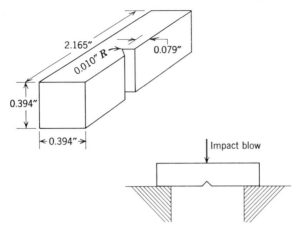

Fig. 3-3 Charpy V-Notch impact test.

providing a generous radius of bend. Changes in cross section along a member should be made gradually to eliminate a stress concentration and a possible notch. Cover plates on beams and girders should be carefully proportioned for length and attachment to the tension flanges.

Fabrication practices should be carefully considered in the design stages in order to avoid undue difficulty in fabrication. Punching and shearing operations may be the start of small cracks and for some types of structures holes should be drilled or reamed to remove the damaged portions.

Cold work operations and the subsequent strain aging, which result in greater sensitivity to brittle fracture, can be reduced by the use of larger bend radii.

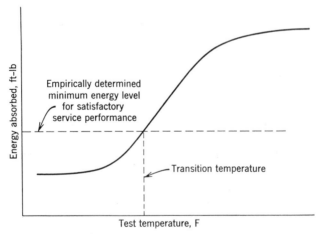

Fig. 3-4 Transition curve obtained from Charpy V-Notch impact tests.

Welding should be designed to minimize residual stresses, should not produce stress raisers by an over-sized weld bead or reinforcement, should not produce cracks or undercuts, and should not leave splatter, arc strikes, or scars from chipping hammers.

Other factors to consider are the low temperatures of the service of the structure, the possibility of shock loading, or a combination of these.

Because it is difficult to evaluate properly all the factors and combinations which may occur to produce a brittle fracture condition, the engineer is still guided by past experience and performance of structures. Even though many structures have performed successfully in adverse conditions when the energy rating of the steel did not justify it, good design and fabrication are always prerequisites for good performance of a structure.

3-6 IMPORTANCE OF FATIGUE

Because fatigue is associated with pulsating load it occurs in structures subjected to moving loads, such as bridges or crane girders, and in parts of machinery and equipment. The consideration of fatigue as a contributing factor for structural failures was established by railway engineers in 1843. These first failures were in the rolling stock and track, but in subsequent years, fatigue cracks were observed in bridges. These cracks were propagated by the accumulated cycles of loading on the bridge extending over a long period of time. Failures in Vierendeel truss bridges in Europe were considered to be the result of stress concentrations subjected to cyclic loading over a period of time.

The parts of machinery which are highly stressed and usually undergo a reversal of stress from tension to compression are most likely to exhibit fatigue failures. These parts should be carefully designed and manufactured.

Stress concentration has a relatively small effect on the static strength of a structure if the material is sufficiently ductile. On the other hand, when a stress concentration is combined with repeated stresses, the strength of the structure at the point of stress concentration is materially reduced. For this reason welds are intrinsically weak spots when repeated stresses are applied, since it is sometimes difficult to avoid stress-raising effects in welded structural connections. And if welds have cooled rapidly and developed a martensite structure, they become more brittle than the base metal, and fatigue effects become more severe. For example, small tack welds which are cooled very rapidly are quite brittle and may develop local cracks. Such spots become potential sources of failure, particularly due to fatigue, unless the tack welds are thoroughly fused during the final welding or chipped out before it.

Fatigue failure generally originates at a localized region which has some stress concentration causing a minute crack. Under repeated loading this minute crack spreads to the surrounding region, until the static strength of the structure is so decreased that it fails suddenly as if under a static load. Because of this mechanism of fatigue failure, the appearance of the part where the crack starts usually can be distinguished from the brittle crystalline fracture of the final break.

The principal factors associated with fatigue failures may be summarized as:

(*a*) A large number of loading cycles.

(*b*) A wide range in stress variation.

(*c*) A high stress in the member with a small range of stress during the cyclic loadings.

(*d*) Local stress concentrations due to design and fabrication details.

The design of structures to overcome all the foregoing factors presents a challenging task. The type and magnitude of cyclic loading for the life of the structure are, at best, uncertain quantities. When these are combined with the indefinite stresses which actually occur, the engineer faced with the task of selecting material to meet the requirements needs considerable experience and judgment.

The fatigue data for steels upon which a judgment of structural performance is made are obtained from test specimens, usually highly polished and tested under laboratory conditions. Although these test data may not be directly applicable for design conditions, they are useful in comparison of different materials.

3-7 FATIGUE LOADINGS

Repeated loadings may be of two types: one in which the direction of stresses is not reversed during the cycle, and the other in which the direction is reversed. When the maximum and minimum stresses of a reversed loading are equal in magnitude but opposite in direction, the condition is called "complete stress reversal." The variables in repeated loadings are: type of structure and loading, maximum stress, minimum stress, frequency of stress cycle, and continuity of loading. In some instances when loading is not continuous, some recovery from fatigue effects may take place between cycles of loading. For a given stress variation it is possible to determine the number of cycles at which failure occurs in a given specimen. Usually there is a value of maximum stress, called fatigue limit, for which no failure occurs even with an extremely large number of cycles.

3-8 FATIGUE TESTS

The data used to evaluate the fatigue resistance of a steel or connection are obtained by one of several test procedures depending upon the type of stress under consideration. The test procedures are described briefly as the following.

(*a*) Rotating beam test—a polished cylindrical specimen with a reduced section is supported as a simple beam and subjected to a bending moment while it is being rotated so that material fibers are alternately at maximum tension and compression of equal values.

(*b*) Flexure test—a polished specimen similar to the one in the rotating beam test or a rectangular specimen cut from the original plate may be used. The specimen is bent in one plane as a simple beam.

(*c*) Axial load test—specimens are similar to those of the flexure test. Tension loads are applied with varying ranges, such as $\frac{1}{2}$ design stress to full design stress, etc.

The tests are carried out to failure, determining the number of cycles of loading for failure to take place under each particular loading condition.

The data thus obtained are plotted with maximum stress S as the ordinate and number of cycles N to failure as the abscissa. Plots of this type may be on log-log or Cartesian scales or combination of scales. The log-log plot approaches straight lines within certain limits.

A plot of the fatigue data for USS "T-1" steel is shown in Fig. 3-5 for axial load and rotating beam tests. The plot has a log scale for the fatigue life, abscissa, and a Cartesian scale for the maximum stress as ordinate.

3-9 *S–N* CURVES

Each *S–N* curve of Fig. 3-5 represents only one type of stress cycle. The top curve is for the condition of axial load applied to the specimen and ranging from a tensile stress of zero to a maximum. Each test is denoted by a small circle and also indicates the maximum stress applied to the specimen. The stress ratio R is the algebraic ratio of the minimum to maximum stress. A ratio of $R = -1$ denotes a range of stress from maximum compression to the same value for maximum tension, as obtained in the rotating beam test. A ratio of $R = 0$ denotes a range of tension stress from zero to a maximum, and $R = -\frac{1}{2}$ denotes a range from a compression stress of one-half the tension to the tension stress.

Each curve must be described for the type of loading used, as indicated in Fig. 3-5.

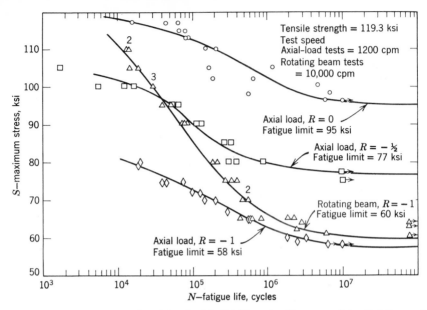

Fig. 3-5 Fatigue results of polished U.S.S. "T-1" steel. (Courtesy of Applied Research Laboratory U.S. Steel Corp.)

The fatigue limit is defined as that stress for which the material or connection can endure an infinite number of cycles. This value is determined from the portion of the curve which is essentially horizontal. Tests on a great number of specimens indicate that the fatigue limit occurs at approximately 2,000,000 cycles.

Fatigue strength is defined as the maximum stress which can be sustained without fracture for a stated number of cycles.

The curves in Fig. 3-5 indicate the number of tests required to plot each curve. It can also be seen that a similar number of tests would be necessary for other stress ratios. For a fuller knowledge of the fatigue behavior of a material or connection more *S–N* curves would be necessary.

3-10 GOODMAN DIAGRAM

An *S–N* curve depicts only one type of stress cycle. Therefore, a complete behavior pattern for a steel or connection can only be obtained from a large number of *S–N* curves. A more convenient method is to summarize all the data from several *S–N* curves in one diagram. This was accomplished by Goodman and the diagram bears his name.

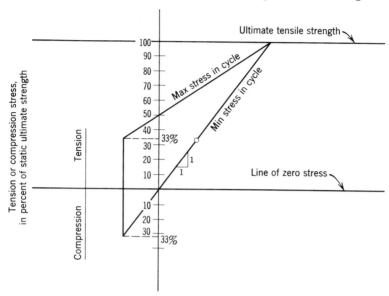

Fig. 3-6   Goodman Diagram fatigue limit for an as-rolled flate plate.

A typical Goodman diagram is illustrated in Fig. 3-6 for an as-rolled flat plate. Each such diagram represents the fatigue behavior for a member or joint for a given fatigue life or number of repetitions of stress for failure.

The diagram has a vertical ordinate of tension and compression stresses in percent of the static ultimate strength, plotted on opposite sides of a zero stress line. At the right of the diagram, the maximum stress line intersects the minimum stress line at the static ultimate tensile strength line of 100%. At the left of the diagram, a complete stress reversal from compression to tension of equal stress magnitude is noted.

The vertical distance between the minimum and maximum stress lines is the permissible range of stress for the fatigue life of the material. At the point where the minimum stress line intersects the line of zero stress, it depicts a stress range from zero tension to a maximum tension as read at the intersection of the maximum stress line with the vertical ordinate. Goodman diagrams are usually plotted for fatigue limits unless a finite life is stated.

3-11   MODIFIED-GOODMAN DIAGRAM

The modified-Goodman diagram was adopted by the American Welding Society-Welding Research Council as a convenient method of interpreting

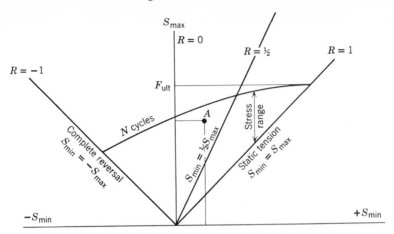

Fig. 3-7 Modified Goodman diagram.

fatigue data. Figure 3-7 is a typical diagram and relates maximum and minimum stresses to the number of cycles for fatigue failure. The curve of N cycles shown could be 2,000,000 cycles, or 6,000,000, etc. There must be a curve for each number of failure cycles investigated.

The use of the diagram may be illustrated by point A, which represents a certain ratio of minimum stress to maximum stress. Because point A falls below the curve of N cycles, fatigue failure will not occur at N cycles. Therefore the curve of N cycles becomes the locus of stress conditions at which failure will occur.

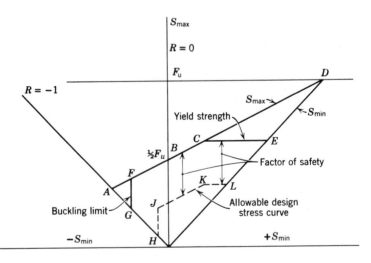

Fig. 3-8 Modified Goodman diagram for allowable design stresses.

The plot is for the condition of fatigue failure, and for design purposes a factor of safety must be applied that will reduce the values of stresses to be used.

For allowable design values the fatigue life curve is assumed to be a straight line which is cut off at the yield strength of the material, such as line *AFBCE* (Fig. 3-8). The portion of the line *CE* is at the yield point or yield strength of the material. In the compression region the maximum *S* line is cut off by the buckling limitation, denoted by *FG*. For design purposes a factor of safety is applied to these limiting values and the line *HJKL* is the envelope of the allowable stresses to be used. Various factors of safety are used by different agencies and specification writers and each has established design equations based on similar diagrams.

3-12 FACTORS AFFECTING FATIGUE STRENGTH

Tests to develop fatigue data have indicated that there are many factors which will influence the fatigue strength of a member or connection. Some of these factors may control during various stages of the project, such as material selection, design, fabrication, erection, and operation of the structure.

A few of the factors are listed in each category.

A. *Material*
   (*a*) Mechanical properties
   (*b*) Surface finish
   (*c*) Residual stresses
   (*d*) Grain size

B. *Design*
   (*a*) Geometrical discontinuities
   (*b*) Type and magnitude of repetitive loads and resulting stresses
   (*c*) Rate of loading
   (*d*) Maximum stress
   (*e*) Stress ratio *R*
   (*f*) Size of member
   (*g*) Stress concentrations

C. *Fabrication*
   (*a*) Welding techniques
   (*b*) Shop practices

D. *Erection*
   (*a*) This phase should not introduce any new items during the erection process. The same care and attention required in fabrication is also required here.

E. *Operation*
    (*a*) The use of the structure in extremes of temperature should be considered in the selection of the material.
    (*b*) The operation of the structure, such as a bridge, equipment, or rolling stock, should be considered in extremes of hot and cold temperatures.

3-13 FATIGUE STRENGTH OF RIVETED CONNECTIONS

The strength of riveted connections under repeated loads is much lower than that under static loading. Nevertheless, this is not usually taken into account in the design of riveted connections for buildings and bridges. Buildings are rarely subjected to repeated loadings with wide range of stress, except some components which may be carrying crane or other moving loads. However, fatigue cracks have been developing in some riveted connections of old bridges as a result of increased traffic and heavier loadings.

Under static loadings, the strength of a riveted connection greatly exceeds the allowable values, primarily because of the reserve strength in the plastic range. Even though the actual stresses at points of stress concentration in a connection may reach the elastic limit at relatively low loads, they are generally increased only slightly with further increase in static load. Under repeated loading, however, high stress concentrations in riveted connections may result in failure at relatively low nominal stresses, provided the stress range and number of repetitions exceed the limiting values.

The factors affecting the fatigue strength of a riveted connection subjected to a particular type of loading are: number of loading cycles, clamping forces of the rivets, degree of rivet-hole filling, length of grip, and rivet pattern in the connection.

Pioneering and classic tests on fatigue strength of riveted joints were reported in 1938[14] and more recent investigations were reported in 1949 and 1954.[15–18] Early test results indicated that:

(*a*) Fatigue *limit* strength F may be arbitrarily defined as the greatest average unit stress on the plate net section (each cycle varying from zero to that stress) to which a joint can be subjected 2,000,000 times before failure.

(*b*) Fatigue *ultimate* strength S of riveted plate in tension depends on the number of load cycles N. An empirical relationship between S and F in terms of N, based on test results, is

$$S = F\left(\frac{2 \times 10^6}{N}\right)^{0.10} \tag{3-1}$$

where both S and F refer to stresses varying from zero to maximum. Note that this relationship applies only when N is a sufficiently large number.

(*c*) Fatigue *ultimate* strength S varies with the range of stress. For complete stress reversal, fatigue limit F' is highly variable, averaging about $\frac{2}{3}F$. For other ranges of stress defined by the ratio $r = $ (minimum stress/maximum stress), the fatigue ultimate strength S' is given by the following approximate relation:

$$S' = S\left(\frac{1}{1 - \frac{1}{2}r}\right) \tag{3-2}$$

where S is defined by Eq. 3-1.

(*d*) Fatigue *limit F* for average tension on net section of plates is about 26 ksi, and seems to be independent of the kind of steel, whether carbon, silicon, or nickel.

Preliminary findings just summarized are now being verified by an extensive investigation sponsored by the Research Council on Riveted and Bolted Joints. Until such time as various codes incorporate results of this research program into specifications, the designers must be guided by the available test data and the present code provisions, such as AISC, 1963, Sec. 1.7 and AASHO, 1965, Sec. 1.7.3.

3-14 FATIGUE STRENGTH OF WELDED CONNECTIONS

Tests of various types of welded joints subjected to repeated loads have been conducted by the Welding Research Council.[19] Some of these results are shown in Fig. 3-9.

Tests indicate that fatigue failures at a welded joint may occur in any one or a combination of the following ways:[20]

(*a*) Failure in the deposited metal. The endurance depends on the welding process and procedure, base metal, and quality of the weld. Porosity, slag inclusions, and poor fusion lower the fatigue (endurance) limit because they act as stress raisers.

(*b*) Failure in the line of fusion, generally caused by poor fusion, lack of penetration, or microscopic cracks.

(*c*) Failure in the heat-affected area. The crystalline change produced by the heat cycle accompanying welding is a function of the welding process and procedure, and of the composition of the base metal; consequently, endurance limit may be either higher or lower than in the original metal.

(*d*) Failure at the edge of the weld. Stress raisers at the edges of the weld, caused by the joint design, weld contour, and undercut, frequently cause fatigue fractures to occur through the base metal. These failures are generally

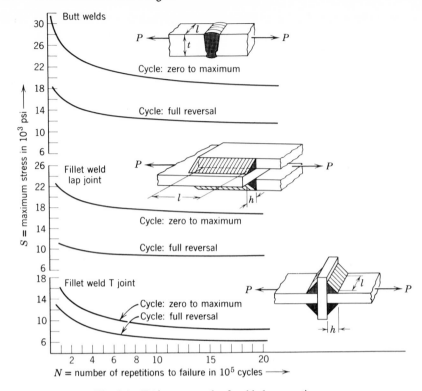

Fig. 3-9 Fatigue strength of welded connections.

the combined result of the effects produced by the stress raisers and the heat-affected area.

Investigations of fatigue strength of test specimens and structures in the field lead to the conclusion that commercial welding made in accordance with American Welding Society Specification on structural grade steel with less than 0.25% carbon and less than 0.7% manganese can be expected to give the fatigue strengths shown in Fig. 3-9.

The most effective factor in improving fatigue strength is elimination of stress raisers. For example, the effect of some treatments on fatigue strength of butt-welded $\frac{7}{8}$-in. plates is shown in Fig. 3-10. It is apparent from this figure that grinding off reinforcement improves the fatigue strength more effectively than stress relieving by heat treatment.

The *impact strength* of structures is often correlated with fatigue strength and both are considered to be related to brittleness, although there are no conclusive data on the validity of this relationship. In general, welded structures in service have a good resistance to impact, but this resistance is

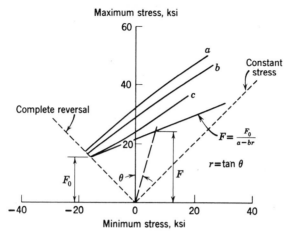

Fig. 3-10 Fatigue strength of butt welds. (*a*) unwelded plate, (*b*) welded reinforcement removed by grinding, and (*c*) welded stress-relieved reinforcement not removed. $N =$ 2,000,000 cycles; $\frac{7}{8}$-in. steel plates.

probably inferior to that of riveted structures because in welded structures a considerable portion of the energy of impact must be absorbed by elastic or plastic deformation of parts close to the weld, whereas the rivet slippage absorbs some energy with a minimum of structural damage. It is suggested that, when a structure is subjected to repeated impact loads, the design stresses be reduced to the values used for fatigue conditions.

3-15 ALLOWABLE STRESSES FOR FATIGUE

Specifications for design of structures include considerations of safety factors applied to the steel mechanical properties of yield point, ultimate bearing stress, ultimate shearing stress, etc., to arrive at allowable stresses for design computations. In some specifications a load factor is applied to the design loadings before the analysis of structural behavior is made and the resulting internal forces are compared with the yield strength and ultimate properties of the steel.

The concept of a single safety factor for design calculations was intended to compensate for (*a*) the necessary design assumptions for analysis, (*b*) the uncertain magnitude of the loads applied to the structure, and (*c*) the uncertain quantities related to the material and its fabrication.

Although the available information on fatigue performance does not always provide the necessary data for all structural applications, the existing data may be used as a guide for specific applications.

Table 3-1 AISC Fatigue Requirements*

AISC Section No.	Application of Design Loads	Calculated Stress Used as Basis for Design	Allowable Stress as Given in Secs. 1.5 and 1.6
1.7.1	Under 10,000 times, with or without stress reversal	Critical static loading (max. static stress produced by any application of specified loads)	Same as for steel and fasteners used
1.7.2	10,000 to 100,000 times, with or without stress reversal	Max. $-\frac{2}{3}$ min., or critical static loading	Same as for steel and fasteners used
1.7.3	100,000 to 2,000,000 times, with or without stress reversal	Max. $-\frac{2}{3}$ min.	Allowable stress for A7 steel,† A141 rivet steel, E 60XX and submerged arc Grade SAW-1 welds
		Critical static loading	Same as for steel and fasteners used
1.7.4	Over 2,000,000 times, with or without stress reversal	Max. $-\frac{3}{4}$ min.	$\frac{2}{3}$ those permitted for A7 steel,† A141 rivet steel, E 60XX and submerged arc Grade SAW-1 welds
		Critical static loading	Same as for steel and fasteners used

* From the Commentary on AISC Specification, 1963.
† Regardless of yield point of steel furnished.

In building design it is generally considered sufficient to design the members and connections on the basis of static loading and they need not be designed for fatigue. However, provision is made for those structures involving fatigue, and allowable stresses are tabulated for various cycles of loading. As indicated in Table 3-1, based on AISC Specification, designers must estimate the number of applications of loading and select the appropriate allowable stress.

Repetitive loadings may occur in buildings under the action of intense earthquake loads, and heavy moving loads, such as cranes or special heavy lifting trucks.

In designing machinery which has moving parts, the engineer must rely on his judgment, experience, and interpretation of available data to establish reasonably safe limits for his design. There are few, if any, design specifications to guide the designer in this field.

In bridge design, fluctuation of loads has to be considered even when no reversal occurs. The AASHO Standard Specifications for Highway Bridges require that, under fluctuating loads, the maximum stress in the weld shall not exceed a reduced value of allowable stress F, which may be defined in terms of variation or reversal of stress in the weld, type of weld, type of stress, and number of cycles of stress fluctuation. The variation of stress may be defined by the ratio $r = P_{min}/P_{max}$. For full reversal $P_{min} = -P_{max}$ and $r = -1$; for static load $P_{min} = P_{max}$ and $r = +1$. Usually it is considered that short railway bridges subjected to critical loadings may undergo 2,000,000 cycles, long railway bridges or short highway bridges subjected to critical loadings may undergo 500,000 cycles, and long highway bridges subjected to critical loadings may undergo only 100,000 cycles of stress fluctuation.

From Fig. 3-8 it can be seen that, for a given number of cycles, the maximum fatigue strength F can be expressed approximately by a linear equation of the following type:

$$F = \frac{F_o}{a - br} \tag{3-3}$$

where F_o is the fatigue strength under complete reversal, and a and b are numerical coefficients depending on type of weld and type of stress. This equation is limited to a range of r values for which F is not greater than the strength F_{st} for static loading condition. A graphical representation of Eq. 3-3 is most convenient, and thus the allowable values of F specified by AASHO are shown in Fig. 3-11. The ratio r is generally known from the consideration of the known loads P_{max} and P_{min} as $r = f_{min}/f_{max} = P_{min}/P_{max}$, and the allowable stress F, for given number of cycles, types of stress, and type of weld, can be easily determined graphically from Fig. 3-11.

Members, fasteners, and welds, subject to repeated variations or reversals of stress, are also designed so that the maximum stress does not exceed the specified allowable stress as given in the standard specifications.

Other specifying bodies, such as the American Welding Society, also include provisions for fatigue stresses in the specification for welded highway and railway bridges and may, in some instances, differ slightly from those given by AASHO.

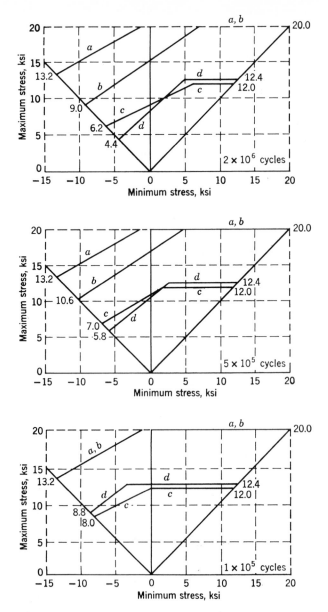

Fig. 3-11 AASAO allowable unit stresses for weld metal, A36 steel. *a*, Base metal in tension. *b*, Weld metal or base metal adjacent to butt weld, tension. *c*, Weld metal in shear. *d*, Base metal, adjacent to or connected by Lillet or plug weld, tension or compression.

3-16 CUMULATIVE FATIGUE DAMAGE

Most of the fatigue data obtained in the laboratory are based on a constant loading cycle. However, in an actual structure this type of loading never occurs. The load range is not constant and it is not applied at regular intervals. Instead, the loadings and resulting stress ratios in the members will vary considerably from day to day, year to year. Recently, laboratory studies have been made on random or variable loadings attempting to simulate actual conditions.

The laboratory tests have been conducted to evaluate the effect of over-stress and under-stress on the fatigue limit of the material. Some studies indicate that damage increases with the number of over-stress cycles applied. Other studies indicate that a large number of cycles of stress just under the fatigue limit, followed by stresses which are repeatedly increased by small increments, produces a substantial increase in fatigue resistance. This process is referred to as "coaxing."

To apply fatigue data properly, the laboratory studies should duplicate the loading spectrum expected or received on the actual structure. Many theories for estimating cumulative damage have been proposed, but Miner's linear cumulative damage rule has been used in engineering design. This rule may be expressed mathematically as

$$N_g = \frac{1}{\sum\limits_{j=1}^{j} \dfrac{\alpha_i}{N_i}} \tag{3-4}$$

where N_g = fatigue life of a component under a spectrum of loading
$\quad\ N_i$ = fatigue life at σ_i
$\quad\ \sigma_i$ = stress in spectrum for N_i
$\quad\ \alpha_i$ = fraction of total cycles applied at each stress level

Thus, with the foregoing expression, it is possible to predict the fatigue life from data on the load spectrum and *S–N* curves for the member or connection under consideration.

REFERENCES

1. Orowan, E., *Fatigue and Fracture of Metals (MIT Symposium)*, John Wiley and Sons, New York, 1952.
2. Tipper, C. F., *The Brittle Fracture Story*, Cambridge University Press, London, 1962.

3. Munse, W. H., and L. M. Grover, "Fatigue of Welded Steel Structures," Welding Research Council, New York, 1964.
4. Martin, C. A., "Fatigue in Constructional Steels," *Machine Design* (August 5, 1965).
5. ASTM, Manual on Fatigue Testing, ASTM Standard No. 91, Philadelphia, 1949.
6. Moore, H. F., and J. B. Kommers, *The Fatigue of Metals*, McGraw-Hill Book Co., New York, 1927.
7. Senes, G., and J. L. Waisman, *Metal Fatigue*, McGraw-Hill Book Co., New York, 1959.
8. Pope, J. A., *Metal, Fatigue*, Chapman & Hall, London, 1959.
9. Reemsynder, H. S., "The Fatigue Behavior of Structural Steel Weldments," A Literature Survey—Fritz Laboratory Report No. 284.1, November 1961.
10. Parker, E. R., *Brittle Behavior of Engineering Structures*, John Wiley and Sons, New York, 1957.
11. Bijlaard, P. P., "Brittle Fractures in Welded Bridges," *Engineering News-Record* **146**, 46 (1951).
12. McLean, D., *Mechanical Properties of Metals*, John Wiley and Sons, New York, 1962.
13. USS Design and Engineering Seminar, "Steels to Match Your Imagination," 1964.
14. Wilson, W. M., and F. P. Thomas, "Fatigue Tests of Riveted Joints," *Univ. Illinois Eng. Expt. Sta. Bull.* **302** (1938).
15. Lenzen, K. H., "The Effect of Various Fasteners on the Fatigue Strength of a Structural Joint," *AREA Bull.* **480** (1949).
16. Baron, F., and E. W. Larson, "Comparative Behavior of Bolted and Riveted Joints," *Proc. ASCE* **80**, 470 (1954).
17. Carter, J. W., K. H. Lenzen, and L. T. Wyly, "Fatigue in Riveted and Bolted Single Lap Joints," *Proc. ASCE* **80**, 469 (1954).
18. Hansen, N. G. "Fatigue Tests of Joints of High Strength Steels," *Proc. ASCE* **85**, ST3, part 1 (March 1959).
19. Committee on Fatigue Testing (Structural), "Fatigue Strength of Butt Welds in Ordinary Bridge Steel," *Welding Research Council Rept.* 3, *Welding J.* (May 1943) and "Fatigue Strength of Fillet, Plug, and Slot Welds in Ordinary Bridge Steel," *Welding Research Council Rept.* **4**, *Welding J.* (July 1943).
20. Spraragen Wm., and G. E. Claussen, "Fatigue Tests of Welded Joints," Welding Research Council, 1937.

4

Elastic and Plastic Concepts of Structural Behavior

ଡ଼

4-1 INTRODUCTION

Design of structures is accomplished by computing the internal forces and moments acting on each component followed by the selection of the appropriate cross section for a given strength grade of steel.

The stresses in a structure are generally computed by the theory of elasticity. However, the load-carrying capacity of a steel structure depends to an appreciable extent on inelastic or plastic action, which is not indicated accurately by the stresses computed on the assumption of elastic behavior.

A good example of divergence between the indications of elastic stresses and the actual load-carrying capacity of a structure is the strength of plate with a small hole (Fig. 4-1). The theory of elasticity indicates that, for a given load P_0, the stress at the hole reaches a value three times that of the stress in the same plate without a hole. If the plate capacity were limited to the load when the maximum stress at any point attained the yield value f_y, and if the plate behaved elastically up to this limit, the capacity of the plate with a hole would be one-third that of a solid plate. However, experiments indicate that the presence of a small hole in a steel plate decreases its static strength only 10 or 15%. This discrepancy between elastic theory and actual behavior is due to inelastic action of the material adjacent to the hole. Initial yielding in a plate with a hole occurs at approximately one-third the yield load in a solid plate. But the yielding is local and it spreads gradually with increase in load until at load P_p practically the entire width of plate is

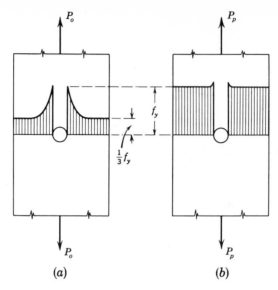

Fig. 4-1   (*a*) Elastic stress distribution and (*b*) plastic stress distribution.

subjected to a constant stress (Fig. 4-1*b*). Provided the load is applied gradually, the effect of the hole is to reduce the load-carrying capacity only slightly; although under impact or repeated loadings, failure by brittle fracture may occur at smaller loads.

The basic principles of the plastic design are not new. Builders of early steel structures intuitively took advantage of the reserve strength inherent in the plastic behavior. The method was formally proposed in Germany in 1920, and considerable theoretical and experimental work was carried out by 1932.[1] In 1936 an intensive investigation of inelastic behavior of structures was started in England under Professor J. F. Baker.[2] By 1945 the method of ultimate design had received wide acceptance in many countries in Europe and South America.

In the United States the question of ultimate design was discussed by Van den Broek[3] in 1940. In 1945 the structural steel committee of the Welding Research Council directed its interest toward a study of elastic and plastic behavior of continuous frames and their components.

Acceptance of plastic design methods in the United States came slowly. Until the adoption in 1958 of the Supplementary Rules for Plastic Design incorporated in the American Institute of Steel Construction Specification for the Design, Fabrication and Erection of Structural Steel for Buildings, practically all steel buildings in the United States were designed on the basis of "allowable stress" concepts, which assumed essentially elastic behavior under normal ("working") loading conditions.

With the adoption of the new rules, analysis and design of steel structures on the basis of inelastic action has become acceptable. In 1961 the AISC adopted a revised Specification which includes "allowable stress design provisions" in Part 1 and "plastic design rules" in Part 2, and provides a basis for design of steel buildings and structures other than bridges. For structures which fall within the scope of AISC Specification the engineer has the option of selecting the most suitable method for the particular structure. Therefore he is expected to know both elastic (allowable stress) and inelastic (plastic) procedures from the standpoint of analysis as well as design of individual members. The AISC rules apply only to continuous beams, and to one- and two-story frames with fully rigid connections. For these structures the rules provide a set of criteria for proportioning members and connections, for design of bracing, and for fabrication; use of these criteria will usually result in safe structures. However, the designer must be familiar with the limitations of the simple plastic-design method,[4,5,6] which result from assumptions of small deflections, and fixed relative proportions of loading. Deflected shape of structure and variations in loading may have important effects on load carrying capacities of frames, particularly where sidesway is present.

Design of structures on the basis of ultimate load has advantages and limitations. In most instances structural behavior in the elastic range will remain an important consideration in design. Depending on the choice of load factors, the economic advantage of structures designed for ultimate load over conventionally designed structures may or may not be significant. But often it will yield a more balanced design. The greatest advantages of the ultimate load design are its rational approach to design, based on the actual strength of structures, and its adaptability to varying requirements of structural safety.

The following sections are intended to introduce some of the basic concepts of plastic behavior. A full treatment of plastic analysis and design methods is beyond the scope of this book.

4-2 SIMPLE PLASTIC THEORY

The "simple plastic theory" takes advantage of the ductility of steel and its ability to yield while continuing to carry stress and undergoing deformation. The plastic deformations are small in most instances, even at ultimate loads, but must be considered in design for the effect on other parts of the structure.

The principal assumptions made in the simple plastic theory are:

(*a*) Steel is a ductile material able to deform plastically without fracture. The stress-strain diagram may be represented by that of an ideal elasto-plastic material. Effects of strain hardening and strain history are neglected.

(*b*) A given cross-sectional shape attains plastification when all fibers develop "plastic" strains. This is an approximation of the actual elasto-plastic behavior, which neglects the small elastic region of the cross section and the effect of shear on local yielding.

(*c*) When full plastification occurs at certain critical sections of a beam or frame, it leads to a development of "plastic hinges" at these sections. In statically indeterminate systems the internal forces and moments which accompany this plastification differ considerably from the corresponding values in an ideally elastic system. The ultimate load is usually defined as the load which produces a sufficient number of plastic hinges to convert the structure into a "mechanism."

(*d*) The loading on the structure is "proportional," that is, all the loads remain in a constant proportion to each other during loading.

(*e*) The deformations of the structure are small and the geometry of the undeformed structure is used in formulating the equations of equilibrium. Members are initially straight and prismatic, and instability does not develop before plastic action begins.

Simple plastic theory may be illustrated by considering the behavior of a steel wide-flange rolled shape subjected to pure flexure.

As the load increases gradually the following four stages of behavior are observed: (*a*) all fibers are stressed below the yield point and the beam behaves elastically, (*b*) extreme fibers at the section of maximum moment begin to yield, and yielding gradually spreads to the web fibers and away from the section of maximum moment, (*c*) practically all the fibers at the section of maximum moment have yielded and the section is said to be fully plastic, and (*d*) the deformation increases rapidly with little or no increase in load and the beam is near collapse.

For a given beam, the moment and deformation limits for any stage of bending can be predicted from the known stress-strain relationship of the material. Assuming that, in all stages of bending, the stress-strain relationship in compression is the same as in tension, the stresses corresponding to the various stages of bending are shown in Fig. 4-2. Initial yielding at the extreme fibers occurs when the maximum stress $f = Mc/I$ has just reached yield point f_y. The corresponding moment $M = M_y = f_y I/c$. As the moment is increased to a value M_2, the strain in the extreme fibers is increased without increase in stress, but a greater portion of the beam is subjected to the yield stress. Increase in moment results in spread of yielding until, at some moment M_p, strain hardening just begins to develop in the extreme fibers. After considerable deformation, the beam may fail either by compressive crushing or buckling or by tensile rupture; the compressive failures are the more common ones.

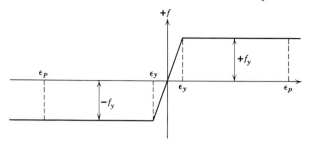

Idealized stress–strain diagram

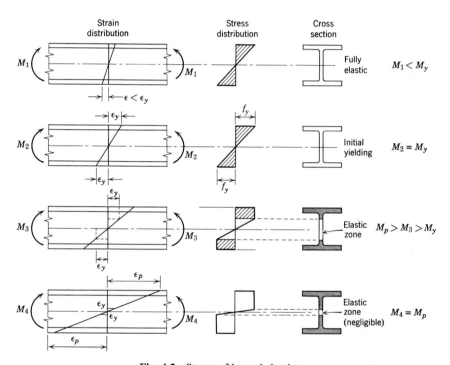

Fig. 4-2 Stages of beam behavior.

The moment at which the deflection begins to increase rapidly corresponds to the plastic moment M_p which can be calculated approximately by idealizing the stress distribution into two rectangular stress blocks. For a symmetrical section, such as an I shape, the plastic moment M_p is (Fig. 4-3)

$$M_p = (f_y A_f)h_f + (f_y \tfrac{1}{2} A_w)\tfrac{1}{2} h_w = f_y(2Q) \tag{4-1}$$

where Q is the static moment of the area above the neutral axis taken about that axis.

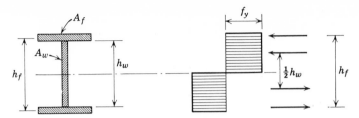

Fig. 4-3   Idealized plastic stress distribution.

The moments M_y and M_p just defined are only approximations. Test results[4] indicate that, because of the effects of residual stresses, stress concentrations, and local distortions of plane sections, local initial yielding occurs at moments considerably lower than the theoretical value M_y—in some instances at $M = 0.25 M_y$—and the theoretical value of plastic moment M_p is realized only when deformation has progressed nearly to the strain-hardening range.

4-3   SHAPE FACTOR

The ratio of the plastic moment M_p to the initial yielding moment M_y, $k = M_p/M_y$, is a measure of the reserve strength of the beam after initial yielding. This ratio varies for different shapes of cross section and is called the shape factor. Generally the shape factor is determined as follows:

$$k = \frac{M_p}{M_y} = \frac{f_y(2Q)}{f_y(I/c)} = \frac{2Q}{S} \tag{4-2}$$

It is clear that the shape factor becomes the ratio of the static moment of area for the full section, defined as plastic modulus, to the section modulus of the cross section. It is therefore independent of the yield point of the steel and is a function of the shape of the cross-sectional area only.

For wide flange shapes the increase in plastic moment over the yield moment ranges from 10 to 20%, depending upon the particular size of the section. There is a relatively small increase because the material is concentrated at some distance from the neutral axis. In general, it can be said, at least for static loads, that the effect of ductility is such that the plastic stress distribution tends to make the best use of the material.

For a rectangular cross section having width b and depth h, k is

$$k = \frac{2Q}{S} = \frac{2b(h/2)(h/4)}{bh^2/6} = \frac{6bh^2}{4bh^2} = 1.5 \tag{4-3}$$

The 50% increase in moment due to inelastic action is valid only for a rectangular cross section. Values of shape factor for other typical sections are shown in Fig. 4-4.

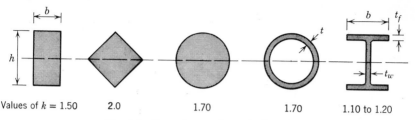

Values of $k = 1.50$ 2.0   1.70   1.70   1.10 to 1.20

Fig. 4.4 Shape factors for typical sections.

4-4 LENGTH OF PLASTIC HINGE

The length of the beam over which the moment is greater than the yield moment may be described as the plastic "hinge length." This hinge length will vary for different cross sections and for different arrangements of loading. For a simply supported beam loaded at midspan (Fig. 4-5), the hinge length L_p is

$$L_p = \left(1 - \frac{1}{k}\right)L \qquad (4\text{-}4)$$

where k is the shape factor $= M_p/M_y$.

For a rectangular section the yield moment M_y is equal to $\frac{2}{3}M_p$ because the shape factor $k = \frac{3}{2}$. Therefore the length of the plastic hinge, that is, the portion of the beam which is yielding, is the center third of the beam (Fig. 4.5a). The amount of the cross section which is yielding will vary from the extreme fiber only at the third-points to all the fibers at midspan.

For a wide flange shape (Fig. 4.5b) with a k of $\frac{9}{8}$ the plastic hinge length is seen to be $L/9$ in the center of the span.

Although the effect of a plastic hinge extends over a finite length of the beam, it will be assumed to be concentrated at a section for analysis purposes (Fig. 4.5c). However, for deflection and design considerations of bracing, the extended length of the hinge must be taken into account.

4-5 ELASTIC AND PLASTIC DEFLECTIONS

The deformed shape of a beam is defined geometrically in terms of angle change $d\phi$ per unit length dz (Fig. 4-6a). If the rate of this angle change,

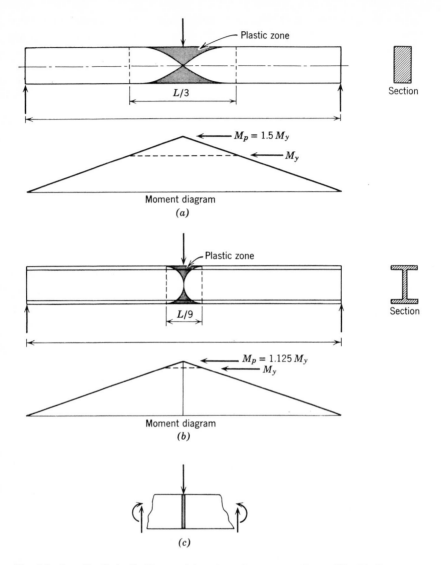

Fig. 4-5 Length of plastic hinge. (*a*) rectangular cross section. (*b*) wide-flange cross section. (*c*) idealized length of plastic hinge.

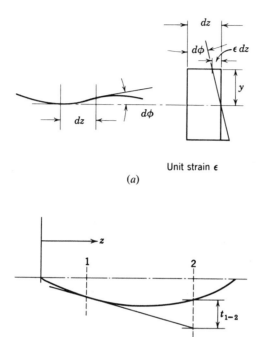

Fig. 4-6 Geometry of beam deflection. (*a*) relation between angle change and unit strain; (*b*) tangential deviation between two points.

$\phi' = d\phi/dz$, along the span of the beam is known, then slope and deflection at any point of the beam can be computed, provided that conditions of restraint at the supports—usually called the boundary conditions—are known. The change in slope $\Delta\phi$ between two points z_1 and z_2 is:

$$\Delta\phi = \phi_1 - \phi_2 = \int_{z_1}^{z_2} \left(\frac{d\phi}{dz}\right) dz \tag{4-5}$$

and the tangential deviation (Fig. 4-6*b*) between these points is

$$t_{1-2} = \int_{z_1}^{z_2} \left(\frac{d\phi}{dz}\right) z\, dz \tag{4-6}$$

To calculate deflections caused by specific loading conditions it is necessary to relate the rate of angle change ϕ' with the applied loads. Considering the deformation of the beam due to bending based on the assumption of plane section (Fig. 4-6), it is apparent that

$$\phi' = \left(\frac{d\phi}{dz}\right) = \left(\frac{\epsilon}{y}\right) \tag{4-7}$$

where ϵ is unit strain at a distance y from the neutral axis. This relationship between the rate of angle change ϕ' and the rotation of plane section is purely a geometric one, and is independent of elastic or plastic behavior of the beam.

In the elastic range, this relation becomes $\epsilon/y = (f/E)/y = (My/IE)/y = M/EI$. Substitution of $\phi' = M/EI$ into Eqs. 4-5 and 4-6 and use of the known boundary conditions permit an evaluation of slope and deflection at any point of the beam.

A convenient method for evaluating the integrals in Eqs. 4-5 and 4-6 is the moment-area method, whereby the following rules are defined: (*a*) the change in slope between two sections is equal to the area under the M/EI diagram between these sections, and (*b*) the tangential deviation between two sections is equal to the static moment of the area under the M/EI diagram taken about the section at which the deviation is measured.

In the plastic range ϕ' can no longer be defined as M/EI because $f \neq My/I$. To obtain an accurate relationship between ϕ' and M, a numerical procedure is usually employed. Consider a stage of bending defined by the condition that the fiber just yielding is located a distance y from the neutral axis (Fig. 4-7*a*). The angle change is

$$\phi' = \epsilon_y/y \tag{4-8}$$

and the corresponding moment M is (Fig. 4-7*a*)

$$M = F_y(2\bar{y}) + F_e\left(\frac{4y}{3}\right) = 2f_y A_y \bar{y} + f_y A_e\left(\frac{2y}{3}\right) \tag{4-9}$$

where F_y is the normal force in the yielded zone and F_e is the normal force in the elastic zone. If we assume different values of y, corresponding values of ϕ' and M can be computed and a M–ϕ' diagram can be easily constructed. A typical diagram is shown in Fig. 4-7*b*.

Once M–ϕ' relationship for the beam is established both in the elastic and in the plastic range, deflections for a given loading condition can be computed if the bending moments are known. A ϕ' diagram, corresponding to the M/EI diagram, is plotted and the moment-area principles are applied to this diagram to compute the slopes and deflections.

For loading conditions producing a locally concentrated maximum bending moment (Fig. 4-8*a*) the load-deflection relationship can be approximated by assuming that a local "plastic hinge" forms at the section of maximum moment when the magnitude of this moment reaches the value M_p. This concept of plastic hinge assumes that, for statically determinate beams,

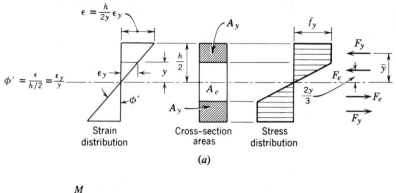

(a)

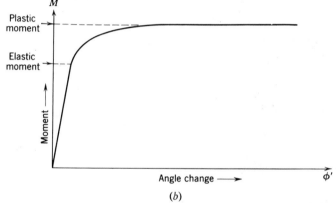

(b)

Fig. 4-7 Moment-curvature for elasto-plastic bending. (*a*) stress and strain distribution for section at partial yielding and (*b*) typical *M-ϕ* diagram for steel beam curvature.

load-deflection relationship is linear all the way up to the load producing maximum moment equal to M_p. At this point the deflection is assumed to increase without any further increase in load (Fig. 4-8). For example, for a beam loaded by a single concentrated load at midspan the approximate plastic-hinge load-deflection relationship is obtained:

$$\Delta_p = \frac{P_p L^3}{48EI} \tag{4-10}$$

where

$$P_p = \frac{4M_p}{L} = \frac{4kM_y}{L} = \frac{4f_y k(I/c)}{L}$$

The plastic-hinge method agrees well with test results where the maximum moment is concentrated locally, as in a cantilever beam or a beam with a

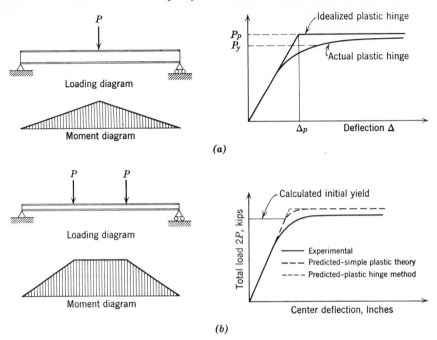

Fig. 4-8 Load-deflection diagrams. (*a*) maximum moment, concentrated and (*b*) maximum moment, distributed.

single concentrated load. Its great advantage lies in the fact that deflections can be estimated without computing M–ϕ' relationship in the plastic range. For beams where maximum moment is distributed over an appreciable portion of the beam span, as in Fig. 4-8*b*, the plastic-hinge method gives only an approximate load-deflection relationship. At loads approaching or slightly greater than initial yield P_y, the computed deflections are smaller than the actual beam deflection. This error becomes significant for sections with relatively thick flanges and for deep sections, particularly when a considerable length of the beam is subjected to constant moment. The problem of determining deflections by more accurate methods and for more complex structures is outside the scope of this text.

It must be noted that the principle of superposition does not apply to structures stressed in the plastic range. Consider the beam loaded as shown in Fig. 4-8*b*; the magnitude of loads P is such that the beam behaves plastically when both loads are applied. When each of the loads P is applied separately, the beam behavior may still be in the elastic range. Superposition of the effects of the two loads assumes that the beam behaves elastically under the total load $2P$, which is contrary to the actual behavior of the beam.

4-6 ELASTIC BEHAVIOR OF STATICALLY INDETERMINATE STRUCTURES

Most theories of structural analysis are based on the assumption of elastic behavior of the system. The internal forces and moments are predicated on this assumption and the designer must be cognizant of this fact when he designs the individual members and their connections. The joints of the framework must be sufficient to insure the behavior of the system to satisfy the analysis assumptions. Any variations of this action must be reflected in the analysis and design details.

For example, if a connection of beams to columns is considered rigid in the analysis, the fabrication must be such that the required rigidity is assured.

A truss frame is often fabricated with essentially rigid connections, although the analysis of internal forces is based on elastic behavior and frictionless pins at the joints. The designer must be sufficiently knowledgeable to know when the assumptions of the analysis approximate the fabricated structure with sufficient accuracy. Merely guessing at the result is insufficient in view of the powerful tools available for analysis of complex elastic structures. Most of these are numerical methods which have become practical because of the availability of large digital computers.

The basic assumptions in the analysis of elastic systems are: (*a*) the relationship between load and deformation is linear and independent of load history and (*b*) the deformations are small. For steel structures these assumptions are valid when stress levels are below some fraction of the ultimate capacity; determination of the ultimate capacity, however, requires an analysis based on inelastic behavior.

4-7 PLASTIC BEHAVIOR OF STATICALLY INDETERMINATE STRUCTURES

Although attempts are still being made to incorporate inelastic behavior within the scope of elastic analysis by increasing allowable stresses, the designer would be well advised to investigate the plastic behavior of structures.

In statically determinate structures inelastic action does not modify the distribution of forces and moments in the members, but in statically indeterminate structures inelastic action modifies the load distribution and often simplifies the method of analysis. For example, consider the behavior of a trussed beam (Fig. 4-9). The load P on the frame is resisted partly by the beam carrying load $P_1 = P - P_2$ and partly by the tension rod carrying load P_2. In the elastic range, loads P_1 and P_2 are proportional to the deflection of

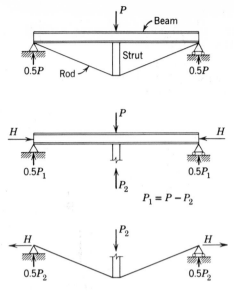

Fig. 4-9   Plastic behavior of a statically indeterminate trussed beam.

the frame, and the proportion of P resisted by each component is statically indeterminate.   Usually a virtual work method of analysis is used in the solution of this problem.   With increase in load P beyond the elastic range, the rod may yield first and, with further increase in total load, the rod will carry a constant load equal to its yield capacity P_2'.   The increase in P is taken up by the beam until it also yields and reaches its load-carrying capacity P_1', provided the plastic deformations of the two components are of the same order of magnitude.   If the load-carrying capacity of the rod P_2' and of the beam P_1' can be determined separately, then the total load-carrying capacity P' of the trussed beam is merely the sum of these capacities, that is, $P' = P_1' + P_2'$.   The problem of distribution of total load P' to the beam and to the rod is no longer statically indeterminate; each carries a load up to its capacity.   Apparent simplification results and statically indeterminate analysis does not appear to be necessary.   In this particular example the moment capacity of the beam is only slightly reduced by the axial compression and this effect may be disregarded.

The redistribution of moments in continuous beams and rigid frames because of local plastic action effectively increases their ultimate strength. For instance, in a uniformly loaded beam with fixed ends (Fig. 4-10), the elastic theory indicates that the maximum moment—at the supports—is equal to $M_o = -wL^2/12$ where w is the intensity of the uniformly distributed load, and L the span of the beam.   When the stresses at the supports reach the yield point, the outer fibers deform without increase in stress and, with further increase in load, the end moments increase less quickly than the moment at midpoint. The ultimate load-carrying capacity is reached when the moments at the supports and at midspan are both equal to the plastic moment capacity of the beam M_p. The beam then behaves as a three-hinge mechanism and the beam deflection increases very rapidly. For prismatic members of constant section the moments at the supports and at the midspan are equal to M_p and must satisfy the condition of equilibrium which requires that $2M_p = w'L^2/8$, and

$$M_p = \frac{\frac{1}{2}w'L^2}{8} = \frac{w'L^2}{16} \tag{4-11}$$

Thus, using simple plastic theory, the maximum moment is $w'L^2/16$, only $\frac{3}{4}$ of $M = w'L^2/12$ obtained by elastic theory. This again illustrates that the load-carrying capacity defined by plastic action can be determined in a simple manner, as in Eq. (4-11), without a statically indeterminate analysis.

The ratio of the maximum M_p obtained by plastic theory to the maximum moment M_o obtained by the elastic theory varies with conditions of loading and support and can be designated as a "redistribution factor" $r = M_p/M_o$. For statically determinate beams the moment is independent of beam yielding

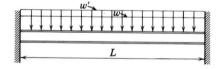

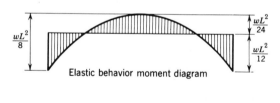

Elastic behavior moment diagram

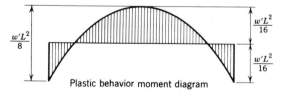

Plastic behavior moment diagram

Fig. 4-10 Redistribution of moments in a fixed end beam.

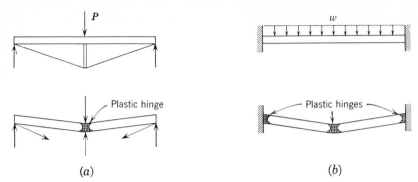

Fig. 4-11   Plastic hinge behavior of simple indeterminate structures.

and r is equal to unity; for a uniformly loaded fixed-end beam, as in Fig. 4-10, r is $\frac{3}{4}$.

The two examples just discussed illustrate special simple cases of determining load-carrying capacity of structures as limited by large deformations. This condition occurs when a sufficient number of elements of the structure behaves plastically so that the structure itself behaves as a linkage mechanism. When the load on the trussed beam (Fig. 4-11a) reaches a maximum value $P_{\max}$, the beam behaves as a simple double link with a plastic hinge at midspan, which can transmit a limited magnitude of moment M_p. The fixed-end beam behaves in a similar fashion, with plastic hinges at the ends and at midspan (Fig. 4-11b) when the load on the beam reaches the maximum value w'. The particular mechanical model corresponding to the plastic behavior of an indeterminate structure is called the "collapse mechanism." Whenever the collapse mechanism of a structure can be established by simple reasoning, as in the foregoing examples, the determination of collapse loads is a simple problem of equilibrium. For complex structures and/or for complex loading conditions, the determination of collapse mechanism requires more elaborate analytical procedures.[6,7,8]

4-8   MECHANISM METHOD

Although the procedures for analyzing complex structures by the plastic theory are outside the scope of this textbook, an application of the method to a simple problem will be illustrated.

The task is to determine the ultimate load which may be applied to a two-span continuous beam fixed at one end, having a uniform cross section and supporting two concentrated loads, as indicated in Fig. 4-12.

For a beam with a uniform cross section the ultimate moment M_p will be the same at all points. Therefore, it is necessary to investigate all possible

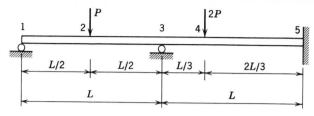

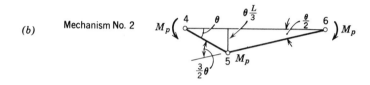

(a)   Mechanism No. 1

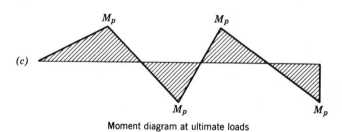

(b)   Mechanism No. 2

(c)

Moment diagram at ultimate loads

Fig. 4-12   Collapse mechanisms.

mechanisms which may form in the structure. The one which requires the least loading will be the collapse mechanism.

A mechanism may be formed by three plastic hinges or plastic hinges acting in conjunction with actual simple supports acting as hinges. For the problem under consideration there are two possible mechanisms and they are indicated in Fig. 4-12a and b.

Applying the principle that internal work is equal to external work the ultimate load may be determined easily.

Mechanism No. 1:

$$P\theta \frac{L}{2} = M_p(\theta + 2\theta) = 3\theta M_p \tag{4-12}$$

and

$$P_{\text{ult}} = \frac{6M_p}{L} \qquad (4\text{-}13)$$

Mechanism No. 2:

$$2P\theta \frac{L}{3} = M_p\left(\theta + \frac{\theta}{2} + \frac{3}{2}\theta\right) = 3\theta M_p \qquad (4\text{-}14)$$

$$P_{\text{ult}} = \frac{9}{2}\frac{M_p}{L} = 4.5\frac{M_p}{L} \qquad (4\text{-}15)$$

The ultimate loading P_{ult} is the lowest value calculated and in this instance is equal to $4.5\,M_p/L$ which occurs when a collapse mechanism forms in span 3, 4, 5. The required beam shape may be selected from tables of plastic moment capacities.

The ultimate moment diagram is illustrated in Fig. 4-12c with plastic moments at critical supports and load points.

4-9 ALLOWABLE LOAD VS. ULTIMATE LOAD DESIGN

There are two usual approaches to structural design. The first, which is the more conventional approach, is based on the concept of "allowable stress" and elastic behavior, and the second, which appears to be more rational and is gradually becoming accepted, is based on the concept of "plastic design" and ultimate load.

Allowable load is a fraction of the ultimate strength of the member determined on the basis of a limiting value of maximum stress, called the allowable stress. Allowable stresses are usually defined in the code applicable to the particular structure. The magnitude of allowable stress is a fraction of the yield stress, and the ratio f_y/f_a is often called the "factor of safety." This concept of safety is based on the assumption that the first yielding is the useful limit of the structure and for adequate safety the allowable load must be equal to or greater than the computed design load. The design load on the member, corresponding to the conditions under service loads, is computed using the elastic theory.

This method of design, based on *service loads, elastic behavior,* and *allowable stresses,* is widely accepted because it developed as an integral part of rational stress analysis and has the authority of experience and tradition behind it. Many empirical rules have been included in specifications to make the use of this method practical.

The principal disadvantage of this method is that it fails to provide uniform overload capacity for all parts and types of structures. Consider a beam that carries a dead load w_d and a design live load w_l. The beam is so proportioned that, when subjected to load $(w_d + w_l)$, it behaves elastically, and because of the maximum bending moment M_a, develops a maximum fiber stress M_a/S just equal to the allowable stress f_a. Thus:

$$M_a = q(w_d + w_l) \times L^2 \quad \text{and} \quad f_a = \frac{M_a}{S} \qquad (4\text{-}16)$$

where q is a numerical coefficient defining the maximum bending moment in the beam based on elastic analysis. The over-load capacity of the beam is defined by the magnitude of the live load that the beam is capable of carrying up to plastic collapse. The maximum plastic moment that the beam can carry is $M_p = k f_y S$ and the live load corresponding to collapse condition is $w_c = m w_l$. The overload capacity is measured in terms of the factor m. Because of plastic redistribution of moments in the beam $M_p = rq(w_d + m w_l)L^2$, so that ratio M_p/M_a can be expressed as follows:

$$\frac{M_p}{M_a} = \frac{k f_y S}{f_a S} \qquad (4\text{-}17)$$

and

$$\frac{M_p}{M_a} = \frac{r(w_d + m w_l)}{w_d + w_l} \qquad (4\text{-}18)$$

The value of m from the foregoing equations is

$$m = \frac{k f_y}{r f_a}\left(1 + \frac{w_d}{w_l}\right) - \frac{w_d}{w_l} \qquad (4\text{-}19)$$

It is seen that, for a constant value of f_a, the overload capacity m varies with the shape factor k, yield-to-allowable stress ratio f_y/f_a, redistribution factor r, and dead-to-live-load ratio w_d/w_l.

The variation in the overload capacity m for two simple shapes (an I shape and a rectangular bar), two support conditions (simple span and fixed ends), and two dead-to-live-load ratios ($\frac{1}{3}$ and 3) is shown in Table 4-1 for two f_y/f_a ratios (1.65 and 1.0).

The wide variation in the overload capacities indicates the limitation of the use of constant value of allowable stress f_a. For example, at the usual level of $(f_y/f_a) = 1.65$ and for a typical I-shaped beam the overload capacity m may vary from a value of 2.21 (simply supported beam with relatively low dead load) to a value of 7.12 (fixed-ended beam with relatively high dead load).

If a constant overload capacity m is desired, then a variable allowable stress f_a must be used, as obtained from Eq. 4-19:

$$f_a = f_y \frac{k(1 + w_d/w_l)}{r(m + w_d/w_l)} \qquad (4\text{-}20)$$

Table 4-1   Values of Overload Capacity m

Support condition	$r=1.0$				$r=0.75$			
Load ratio w_d/w_l	$\frac{1}{3}$		3.0		$\frac{1}{3}$		3.0	
Shape factor k	1.15 I	1.50 ∎	1.15 I	1.50 ∎	1.15 I	1.50 ∎	1.15 I	1.50 ∎
$f_y/f_a = 1.00$	1.21	1.67	1.60	3.00	1.72	2.33	3.14	5.00
$f_y/f_a = 1.65$	2.21	2.97	4.60	6.90	3.05	4.06	7.12	10.2

$$m = \frac{kf_y}{rf_a}\left(1 + \frac{w_d}{w_l}\right) - \frac{w_d}{w_l}$$

k = shape factor
f_y = yield stress
f_a = allowable stress
w_d = dead load
w_l = live load
r = moment redistribution factor

The use of different values of allowable stresses for different loading conditions is proposed in several specifications. For example, AISC permits a 20% increase in the allowable bending stress for negative moments at interior supports of fully continuous beams for compact sections. It also permits a 33.3% increase in allowable stresses for members subject to wind stresses only, or to a combination of wind and other loads. These procedures still do not account for all the factors affecting f_a and do not necessarily give uniform overload capacity m. Therefore, designs based on allowable stress method, although usually safe, will not always be uniformly economical. On the other hand, specifying such a "tried and true" procedure, does to a certain extent, protect the public from failures that might result if no restriction is placed on the judgment of inexperienced designers.

The plastic design procedure differs from the conventional allowable stress method in three important respects: (*a*) ultimate loads are used instead of service loads, (*b*) forces and moments in members subjected to the ultimate loads are determined on a more realistic basis of including inelastic action, and (*c*) the members are so proportioned that their ultimate strength exceeds, or at least equals, the forces and moments produced by ultimate loads.

In arriving at the ultimate loads, the dead and live loads are considered separately and each is increased by a different factor to account for the most severe service loading. The dead loads, estimated from a preliminary design, are not likely to change during the life of the structure. The dead-load factor must account only for minor deviations from the estimated value due to variations in the density of the materials, the structural dimensions, the approximate nature of the distribution assumed in the analysis, and some possible future additions; a 20% variation in the estimated value of dead load would normally allow for these variations. Live load, on the other hand, is subject to considerable variations; future increase, such as change in the nature and density of traffic over a bridge, or change in the type of occupancy or equipment in a building, might increase live loads materially. The possibility of local concentration must also be considered. For example, concentration of steel filing cabinets in one area gives rise to a load of 120 psf on an office floor usually designed for a 50-psf average load. In some instances, dynamic or impact effects may be included in the live-load factor. However, when these effects are of major importance, as in supports for an elevator or heavy vibrating machinery, a special evaluation of dynamic effect should be made for these conditions. Although the live-load factor need not provide for *any possible* condition, it must provide for the *rare but probable* loading which should not be permitted to destroy the usefulness of the structure. Generally, a live-load factor between 1.5 and 2.0 is considered minimum as far as load increase itself is concerned; a higher factor is specified to allow for other uncertainties.

Other loads, such as wind and earthquake, should be estimated and then increased by a suitable load factor for ultimate design. Several combinations of loading conditions might be considered as critical; for example, the AISC Rules for Plastic Design of Buildings specify that the minimum ultimate loads shall be 1.70 (dead + live) loads for simple and continuous beams, 1.85 (dead + live) loads for continuous frames, and 1.40 (dead + live + wind or earthquake) loads.

The concept that load distribution in statically indeterminate structures is based on the load-carrying capacity of the members is basic to ultimate-load-design philosophy. It implies that the members and connections have to be proportioned and their load-carrying capacity determined before the ultimate-load distribution is defined. For the trussed beam (Fig. 4-9), for example, the designer can proportion the members by any one of the following procedures:

(*a*) Select, more or less arbitrarily, a convenient size of beam. Determine its load-carrying capacity P_1'. The rods have to resist P_2', which is the balance of the ultimate load P', that is, $P_2' = P' - P_1'$. Proportion the rods to yield at a load just slightly higher than the load P_2'.

(*b*) Reverse the above procedure, that is, select rods first and then proportion the beam to carry the balance of the ultimate load.

(*c*) Decide what proportion of total ultimate load each element is to carry. Say, proportion the beam and the rods so that the capacity of each is half the ultimate load. Other proportions than half can be used.

This concept of ultimate load design clearly illustrates what Professor Hardy Cross[9] called "the pragmatic concept of structural action . . . that is when one tells the structure what to do, it will try to do it . . . . Questions now arise: Can it do it? Can it do it efficiently? How can the designer determine how to make it do it?" The trussed beam chosen to illustrate this concept was a relatively simple structure. In this instance, the nature of structural action up to failure was easily visualized. In more complex structures, the action up to failure could be very difficult to visualize directly and the determination of ultimate load distribution is no longer simpler than the conventional statically indeterminate analysis. It is, however, no more complicated, although it involves some new concepts and techniques.

After safety of the members against failure under ultimate loads has been verified, they must be checked for performance under service loads. This includes consideration of deflections, fatigue, dynamic response, initial local yielding, and other structural characteristics which may have a bearing on functional performance. For example, with a large dead-to-live-load ratio and a small live-load factor, the design may be governed by the conventional limitation of preventing yield under normal dead plus live load, instead of the ultimate capacity. Temperature changes and support settlements should also be considered insofar as they affect the stresses and deflections.

Although plastic design is a rational method which considers inelastic behavior of the structure, it will not replace all other analysis and design methods. To be sure, the method has many advantages to encourage its use, but it also has some limitations. Among the advantages are (*a*) ability to prescribe overload capacity under simple loading conditions, (*b*) effective use of material, (*c*) simplicity of plastic analysis calculations for simple frame structures, and (*d*) design of more economical details reflecting plastic behavior.

Several limitations of the plastic theory as a criterion of beam strength have already been mentioned. These are that plane sections are assumed to remain plane, and residual and local stress concentrations are neglected. Other important considerations in evaluating plastic strength criteria are material properties, effects of shear and axial loads, local and lateral buckling, effect of encasement, effect of repeated loads, and the possibility of brittle fracture. The special design provisions which account for these considerations are discussed in detail in other books devoted to plastic analysis and design.[4,5,6]

PROBLEMS

1. Determine the shape factor k defining plastic moment M_p for (a) a circular bar having diameter d, (b) 4-in. standard pipe section, (c) 8 I 23, (d) 8 ⊏ 11.5, (e) 8 W 31; consider bending about the xx axis only.

2. Determine the shape factor k for sections c, d, and e in Prob. 1, bending about the yy axis.

3. Using simple plastic theory, determine the maximum deflection when M_{max} reaches the plastic moment M_p for the beams shown in Fig. P-3.

4. Determine the collapse load P for beams b and c in Prob. 3.

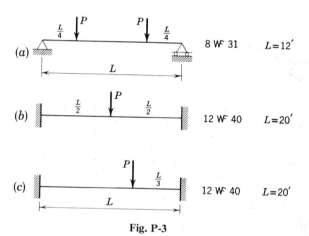

Fig. P-3

5. Using simple plastic theory, determine the maximum deflection for beams b and c in Prob. 3 when P reaches collapse load.

6. Determine the moment redistribution factor r for beams b and c in Prob. 3.

7. Determine the collapse load P in terms of the plastic moment capacity M_p of the beam in Fig. P-7.

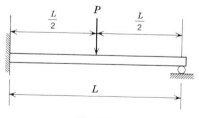

Fig. P-7

8. For Prob. 7, if P is 20 kips and L is 40 ft, what size W beam is required for (a) A36 steel, (b) A441 steel.

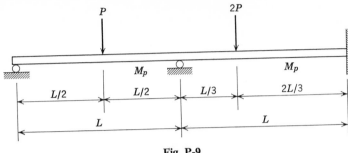

Fig. P-9

9. Determine the collapse load P, in terms of M_p, for the beam in Fig. P-9.
10. For Prob. 9, select a W beam of A36 steel if $P = 15$ kips and L is 30 ft.

REFERENCES

1. Maier-Leibnitz, H., "Contribution to the Problem of Ultimate Carrying Capacity of Simple and Continuous Beams of Structural Steel and Timber," *Bautechnik* (1927).
2. Baker, J. F., "A Review of Recent Investigations into the Behavior of Steel Frames in the Plastic Range," *J. Inst. Civil Engrs.* **31** (1948–49).
3. Van den Broek, J. A., "Theory of Limit Design," *Trans. ASCE* **105** (1940).
4. "Commentary on Plastic Design in Steel," *ASCE Manual of Eng. Pract.* **41** (1961).
5. Beedle, L. S., *Plastic Design of Steel Frames*, John Wiley and Sons, New York, 1958.
6. Massonnet, C. E., and M. A. Save, *Plastic Analysis and Design*, Vol. 1, Beams and Frames, Blaisdell, 1965.
7. Hodge, P. G., *Plastic Analysis of Structures*, McGraw-Hill Book Co., New York, 1959.
8. Neal, B. G., *The Plastic Methods of Structural Analysis*, John Wiley and Sons, New York, 1956.
9. Cross, H., "Relation of Analysis to Structural Design," *Trans. ASCE* **101** (1936).

5

Bolted, Riveted, and Pinned Connections

5-1 INTRODUCTION

The designer has many types of fasteners available to him (Fig. 5-1) and it is his prerogative to select the one that he feels best satisfies the design requirements. Considerations that may influence his choice are: connection strength required, space limitations of the connection, available technicians to fabricate and erect the structure, service conditions, and finally, total cost of installation.

Connections using bolts, rivets, or pins serve essentially the same function in transferring loads from one component to another and therefore will be

Fig. 5-1   (Courtesy of Screw and Bolt Corp., Pittsburgh, Pa.)

103

subject to similar analysis and design considerations. For simplicity, hereafter the terms fasteners or connectors will be used as general terms applying to rivets, bolts, or pins. Although evaluation of force distribution will be similar, the load-carrying ability of each type of fastener will be different and will be discussed separately.

5-2 TYPES OF FASTENERS

Rivets. A piece of round ductile steel forged in place to join several pieces of steel together is called a rivet. The rivet is manufactured with a special head, referred to as the manufactured head, and is installed using a riveting hammer which forms another head during the installation. The complete process is called riveting, and its essential steps are illustrated in Fig. 5-2.

Riveting is essentially a forging process which has developed from a handhammering operation into a machine-driving process. It is believed that the art of riveting is as old as the use of ductile metals, but the development of the process as it is known today has occurred with the growth of the steel construction industry during the XIX century.[1,2]

Rivets are described by the manner and location of installation. For example, most rivets used in structural steel work are hot-driven either in the shop or field and thus become known as hot-driven shop or field rivets.

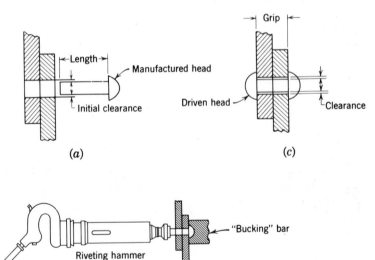

Fig. 5-2 Essential steps in riveting.

The rivets are heated by an oil or gas-fired furnace or an electrical-resistance rivet heater. They are then inserted in the punched or drilled holes of the steel pieces to be connected and the riveting hammer shapes the head while pressure is exerted on the opposite side to hold the rivet in place. As the hot rivet is driven in its plastic state, it usually swells under the hammer load and fills the hole completely. Then, as the rivet cools, it tends to shrink in size both lengthwise and diametrically. The tendency of the rivet to shrink in length is largely prevented by the plates, thus producing tension in the shank of the rivet and compression between the plates. This compression action is referred to as "clamping action" and sets up a frictional resistance to sliding between the plates. The decrease in rivet diameter is due partly to the shrinkage as it cools and partly to the Poisson effect of the rivet in tension. Thus hot-driven rivets may become smaller than the hole, although in many instances the shrinkage is imperceptible.

Cold-driven rivets are driven at room temperature and require large pressures to form the head and complete the process. Cold driving can be applied more conveniently to rivets of small sizes, ranging from $\frac{1}{2}$ to $\frac{7}{8}$ in. in diameter. Although cold driving increases the strength of the rivet and eliminates the need for heating, the process is limited by the equipment required and the inconvenience of using it in the field.

Nominal rivet diameters for structural purposes range from $\frac{1}{2}$ to $1\frac{1}{2}$ in., by $\frac{1}{8}$-in. increments. Sizes most frequently used in structures are $\frac{3}{4}$-in. for buildings and $\frac{7}{8}$-in. for bridges. In especially heavy connections larger sizes are used.

The most common grade of rivet steel is ASTM A141, discussed in Sec. 2.9, which is used to connect components of structural carbon steel or steels of higher strength. High-strength rivets made of ASTM A195 or A502 may also be used for structures made of high-strength steel. The choice of rivet steel is dependent upon the application and installation cost.

Properly driven shop and field rivets must be tight and grip the connected parts securely. Their heads must be full size, neatly formed, and concentric with the shank. Loose or otherwise defective rivets can be detected by tapping them with a light hammer and noting the sound or "ring" of the metal under the blow. Defective rivets, loose, or cock-headed ones must be drilled out and replaced.

Bolts. A bolt is a metal pin with a head formed at one end and the shank threaded at the other in order to receive a nut (Fig. 5-1). Bolts are used for joining together pieces of metal by inserting them through holes in the metal and tightening the nut at the threaded ends. Structural bolts may be classified according to the following: type of shank—unfinished or turned; material and strength—ordinary structural or high-strength steel; shape of head and nut—square or hexagonal, regular or heavy; and pitch and fit of thread—standard, coarse, or fine.

Unfinished bolts are forged from rolled-steel round rods and have large tolerances on shank and thread dimensions. Therefore, punched or drilled holes with diameters of $\frac{1}{16}$ in. greater than the nominal bolt diameter are used. In some structures, when a tight fit is desired, the bolt holes are reamed or drilled, and the bolts are turned or finished to the size necessary for the desired fit.

Ordinary structural bolts are made of mild steel (A307) with an ultimate tensile strength of approximately 65 ksi. However, high-tensile-strength bolts have been gaining favor for use in structural connections. These bolts are made of quenched and tempered steels (ASTM A325 and A490) having an ultimate tensile strength of about 105 to 150 ksi, and a yield strength of about 77 to 125 ksi.[3,4]

Structural bolts usually have either square or hexagonal heads, and are available in "regular" and "heavy" sizes, as shown in the AISC Manual. Square heads cost slightly less and are generally used, but hexagonal heads are easier to turn or hold with a wrench, require less turning space, and may occasionally be desirable. The nuts are also either square or hexagonal and available in "regular" and "heavy" sizes. The usual practice is to use bolts with hexagonal heads and square or hexagonal nuts. Heavy-size nuts may be required for bolts carrying tension loads or when high initial tension is to be developed in the bolts by tightening, as with high-tensile-strength bolts (Fig. 5-3).

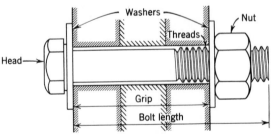

Fig. 5-3 A bolt assembly.

The threaded portion of structural and high-tensile-strength bolts has American Standard screw threads. The gross and the net diameters and areas of the bolts, as well as the pitch of the screw threads, are given in the AISC Manual, pp. 4-91.

Steel washers (Table 5-1) are usually used under the bolt head and the nut in order to distribute the clamping pressure on the bolted member and also to prevent the threaded portion of the bolt from bearing on the connecting

Table 5-1 Washer Dimensions[a]

Bolt Diameter D	Circular Washers				Square or Rectangular Beveled Washers for American Standard Beams and Channels		
	Nominal Outside Diameter[b]	Nominal Diameter of Hole	Thickness		Minimum Side Dimension	Mean Thickness	Slope of Taper in Thickness
			Min.	Max.			
$\frac{1}{2}$	$1\frac{1}{16}$	$\frac{17}{32}$	0.097	0.177	$1\frac{3}{4}$	$\frac{5}{16}$	1:6
$\frac{5}{8}$	$1\frac{5}{16}$	$\frac{21}{32}$	0.122	0.177	$1\frac{3}{4}$	$\frac{5}{16}$	1:6
$\frac{3}{4}$	$1\frac{15}{32}$	$\frac{13}{16}$	0.122	0.177	$1\frac{3}{4}$	$\frac{5}{16}$	1:6
$\frac{7}{8}$	$1\frac{3}{4}$	$\frac{15}{16}$	0.136	0.177	$1\frac{3}{4}$	$\frac{5}{16}$	1:6
1	2	$1\frac{1}{16}$	0.136	0.177	$1\frac{3}{4}$	$\frac{5}{16}$	1:6
$1\frac{1}{8}$	$2\frac{1}{4}$	$1\frac{1}{4}$	0.136	0.177	$2\frac{1}{4}$	$\frac{5}{16}$	1:6
$1\frac{1}{4}$	$2\frac{1}{2}$	$1\frac{3}{8}$	0.136	0.177	$2\frac{1}{4}$	$\frac{5}{16}$	1:6
$1\frac{3}{8}$	$2\frac{3}{4}$	$1\frac{1}{2}$	0.136	0.177	$2\frac{1}{4}$	$\frac{5}{16}$	1:6
$1\frac{1}{2}$	3	$1\frac{5}{8}$	0.136	0.177	$2\frac{1}{4}$	$\frac{5}{16}$	1:6
$1\frac{3}{4}$	$3\frac{3}{8}$	$1\frac{7}{8}$	0.178[c]	0.28[c]			
2	$3\frac{3}{4}$	$2\frac{1}{8}$	0.178	0.28			
Over 2 to 4 incl.	$2D - \frac{1}{2}$	$D + \frac{1}{8}$	0.24[d]	0.34[d]			

[a] Dimensions in inches.
[b] May be exceeded by $\frac{1}{4}$ in.
[c] $\frac{3}{16}$ in. nominal.
[d] $\frac{1}{4}$ in. nominal.

pieces. For high-strength bolts, washers with a hardened surface may be required.

In order to assure proper functioning of bolted connections under load, the parts must be tightly clamped between the bolt head and the nut. When bolted connections with ordinary bolts are subjected to alternating loads or vibrations, the nuts may become loose and thus reduce the strength of the connection. To prevent this the nuts must be locked in position. A cotter pin with a castellated nut and a hole drilled in the bolts has been widely used to prevent the nut from turning on the bolt. Jam nuts, with the heavy nut on the outside, accomplish the same purpose. Various special nuts, generally known as "locknuts" are available commercially and serve to prevent loosening. A typical locknut system, known as rib-bolt has also been used. High initial tension in the bolts, as in high-strength bolts, also prevents nuts from loosening.

High-strength bolts depend upon the clamping action produced when the nut or bolt is tightened to a predetermined tension load, as indicated in Table 5-2. This tension is developed by tightening the nut by properly calibrated torque wrenches or by the turn-of-nut method. Impact wrenches may be used and should be of adequate capacity and sufficiently supplied with air to perform the required tightening of each bolt in approximately 10 sec. The Specification for Structural Joints Using ASTM A325 or A490 Bolts imposes the following requirements for tightening procedures.

Table 5-2   Bolt Tension

Bolt Diameter Inches	Minimum Bolt Tension[a] in Kips	
	A325 Bolts	A490 Bolts
$\frac{1}{2}$	12	15
$\frac{5}{8}$	19	24
$\frac{3}{4}$	28	35
$\frac{7}{8}$	39	49
1	51	64
$1\frac{1}{8}$	56	80
$1\frac{1}{4}$	71	102
$1\frac{3}{8}$	85	121
$1\frac{1}{2}$	103	148

[a] Equal to the proof load (length measurement method) given in ASTM A325 and A490

Calibrated Wrench Tightening. When calibrated wrenches are used to provide the bolt tension specified in Table 5-2 their setting shall be such as to induce a bolt tension 5 to 10% in excess of this value. These wrenches shall be calibrated at least once each working day by tightening, in a device capable of indicating actual bolt tension, not less than three typical bolts of each diameter from the bolts to be installed. Power wrenches shall be adjusted to stall or cut-out at the selected tension. If manual torque wrenches are used the torque indication corresponding to the calibrating tension shall be noted and used in the installation of all bolts of the tested lot. Nuts shall be in tightening motion when torque is measured. When using calibrated wrenches to install several bolts in a single joint, the wrench shall be returned to "touch up" bolts previously tightened, which may have been loosened by the tightening of subsequent bolts, until all are tightened to the prescribed amount.

Turn-of-Nut Tightening. When the turn-of-nut method is used to provide the bolt tension specified in Table 5-2, there shall first be enough bolts brought to a "snug tight" condition to insure that the parts of the joint are brought into full contact with each other. Snug tight shall be defined as the tightness attained by a few impacts of an impact wrench or the full effort of a man using an ordinary spud wrench. Following this initial operation, bolts shall be placed in any remaining holes in the connection and brought to snug tightness. All bolts in the joint shall then be tightened additionally by

Table 5-3   Nut Rotation[a] from Snug Tight Condition

Disposition of Outer Faces of Bolted Parts		
Both Faces Normal to Bolt Axis, or One Face Normal to Axis and Other Face Sloped 1:20 (Bevel Washer Not Used)		Both Faces Sloped 1:20 from Normal to Bolt Axis (Bevel Washers Not Used)
Bolt Length[b] Not Exceeding 8 Diameters or 8 in.	Bolt Length[b] Exceeding 8 Diameters or 8 in.	For All Lengths of Bolts $\frac{3}{4}$ Turn
$\frac{1}{2}$ Turn	$\frac{2}{3}$ Turn	

[a] Nut rotation is rotation relative to bolt regardless of the element (nut or bolt) being turned.   Tolerance on rotation, $\frac{1}{6}$ turn (60°) over and nothing under. For coarse thread heavy hexagon structural bolts of all sizes and length and heavy hexagon semifinished nuts.

[b] Bolt length is measured from underside of head to extreme end of point.

the applicable amount of nut rotation specified in Table 5-3, with tightening progressing systematically from the most rigid part of the joint to its free edges.   During this operation there shall be no rotation of the part not turned by the wrench.

Bolts for structural connections ordinarily vary from $\frac{5}{8}$ to $1\frac{1}{4}$in. in diameter, although larger or smaller sizes are occasionally used.   Special anchor bolts or swedge bolts are employed for connecting steel members to masonary structures, as shown in the AISC Manual.

Anchor bolts are cast or grouted into concrete footings.   For steel-column bases, they generally vary from $\frac{1}{2}$ to 4 in. in diameter.   Larger sizes require large embedded lengths and are often not economical.

Pins.   A single cylindrical steel pin is sometimes used to connect members which must rotate relative to each other.   The pin is assumed to turn freely in the connection;   therefore, clamping action due to initial tension is undesirable.   Since only one pin is used at such a connection where several bolts or rivets may be required, the size of the pin is usually larger than that of bolts or rivets.

Pins for structural purposes are made of structural carbon steel, forged and machined to accurate dimensions.   Cold-rolled pins with suitable surfaces are sometimes employed, especially in alloy steel.   The usual diameter of structural pins varies from $1\frac{1}{2}$ to 12 in. in diameter, although larger sizes up to 24 in. are available.   When pins are specially machined, there is no necessity to limit the diameter to any particular standard set of sizes.

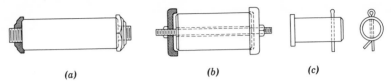

Fig. 5-4 Types of pins; (*a*) pin with recessed nuts, (*b*) pin caps with bolt, and (*c*) pin with cotter.

The most common type of pin has threaded ends and two recessed nuts screwed on the ends to hold the pin in place (Fig. 5-4). For pins over 10 in. in diameter it is preferable to use a long bolt passing through recessed caps and the pin, thus holding them together (Fig. 5-4*b*). This eliminates the use of large locknuts. For smaller pins carrying light loads a head may be forged at one end and a cotter pin inserted at the other (Fig. 5-4*c*) or two cotters may be used, one at each end. Typical details of such pins are shown on p. 4-97 of the AISC Manual, 1963.

5-3 STRUCTURAL USES OF RIVETS, BOLTS, AND PINS

Until 1948, when the use of high-strength bolts was approved for steel construction, practically all permanent fastener connections were made with rivets. These were installed hot for shop and field connections and required a four-man team in the field. With the advent of the high-strength bolt, field riveting has been largely replaced with field bolting, primarily because of reduced cost with fewer men required. Rivets are therefore used only for shop connections for some members and rarely for field connections.

The high-strength bolts are currently being used in practically all types of structures. The advantage of these bolts is obtained by developing high initial tension in them, which clamps the joining plates between the bolt head and the nut. The clamping action enables the load to be transmitted from one plate to the other by friction with little slip and hence produces a very rigid joint. Since the frictional resistance depends on the amount of initial tension, it is most important that adequate clamping force is developed by proper tightening of the nut on the bolt. Because of clamping action, all the load is transmitted by friction and the bolts are not subjected to any shearing or bearing stresses. Since the frictional resistance is effective outside of the hole area, the load transmitted by the plate at the section of the holes is reduced and the possibility of failure at the net section is minimized. The fatigue strength of the joint is also high, since alternating loads produce little change in the bolt stresses. In addition, the nuts are also prevented from loosening.

Thus, the advantages of high-strength bolts may be summarized as follows:

(*a*) Rigid joint—no slip between plates at working loads.
(*b*) High static strength due to high frictional resistance.
(*c*) Smaller load transmitted at net section of plates.
(*d*) No shearing or bearing stress in bolts.
(*e*) High fatigue strength.
(*f*) Nuts prevented from loosening.

Although high-strength bolts are generally designed to develop full frictional resistance, there are occasions when slip is permitted and the bolts are designed to resist the load in direct bearing.

Structural pins can be classified into two types: the pins on which the connecting members turn through large angles, and the pins on which the members might turn only through small angles, primarily because of elastic deformations of the members.

The first type of pin, sometimes called a trunnion, is used on bascule bridges, crane booms, etc. These pins should be constantly lubricated to prevent rusting and to reduce wear. The second type of pin is used for arch hinges; hinge plates, such as in cantilever bridges; expansion joints; rocker supports; etc. Depending on the maintenance conditions and the magnitude of elastic deformations in the members, the rotation of the members on these pins may be somewhat less than predicted by theory. Before about 1920, many bridge trusses were built using pin connections. This was done in order to make the frames statically determinate, and particularly to eliminate the secondary stresses due to rigid connections at the joints. It was later found that these pin connections rust rapidly and actually do not permit free rotation. Furthermore, pin connections may at best provide hinges in one plane only, and do not eliminate joint rigidity in other planes. Thus, due to transverse and longitudinal loads (wind, traffic, etc), secondary stresses in the joints are not eliminated. Experience indicates that pin connections in trusses are subject to considerable vibration, wear, and in general are not desirable except for very long spans.

5-4 TYPES OF CONNECTIONS

Connections can be classified according to the mode of load transmission. Connections loaded as shown in Fig. 5-5*a*, *b*, and *c* tend to shear the fasteners and are called "shear connections"; connections loaded as shown in Fig. 5-5*d* tend to fail the fasteners in tension and are called "tension connections."

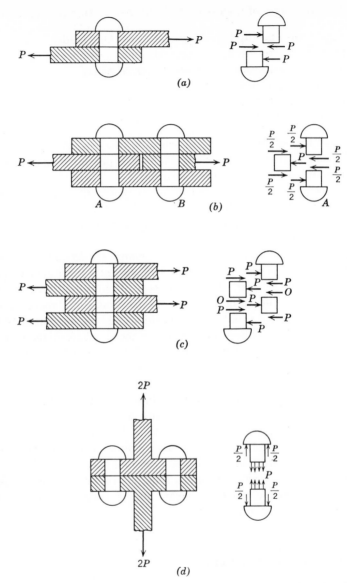

Fig. 5-5 Types of connections. (*a*) rivet in single shear, lap joint, (*b*) rivet in double shear, butt joint, (*c*) rivet in multiple shear, and (*d*) rivet in tension.

If the load in a shear connection is transmitted solely by friction produced by large clamping forces between the plates, the connection is called "friction-type connection" and no slip between the plates can be tolerated. If slip occurs between the plates and the load is transmitted by bearing between the fasteners and the plates with shear stresses induced in the fasteners, the connection is called "bearing-type connection." Ordinarily all riveted connections and those made with common A307 bolts are classified as bearing type connections. The high-strength bolts may be used in either friction-type or bearing-type connections.

A friction-type connection is recommended for shear connections subjected to stress reversal, severe stress fluctuations, or in those applications where slip would be undesirable. A bearing-type connection may be used in applications where the load is considered to be essentially static. In friction-type connections the bearing of the bolt on the plate need not be considered. In bearing-type connections the bolt is assumed to bear on the plate with a nominal bearing pressure computed on the basis of uniform stress distribution over an area equal to the nominal bolt diameter times the thickness of the connected plate.

Tests of bearing-type connections have indicated that failure occurs at 15% less load when threading is present at one of the two shear planes of an enclosed part and at 30% less load when threads are present in both shear planes. This latter failure load also occurs for single shear joints having threads in the shear plane. Ordinary bolts have been observed to behave in a similar manner. These tests merely reflect the ratio of area at the root of thread to the nominal area. This action is also reflected in the allowable stresses assigned to the bolts.

Ordinary bolts behave in a manner similar to the bearing-type high-strength bolts but at reduced allowable stresses.

Bolted and riveted connections can also be classified according to the nature and location of load with respect to the fastener group (Fig. 5-6). When the load passes through the centroid of the fastener cross-sectional areas, the

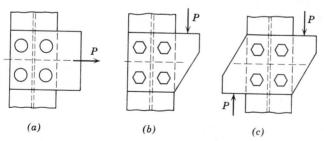

(a)   (b)   (c)

Fig. 5-6 Types of connections. (a) direct load connection. (b) eccentric load connection. and (c) pure moment connection.

connection is said to be carrying direct load. When the load does not pass through the centroid of the fastener group, it is an eccentric load connection. When the load transmitted consists of a pure torque or moment, it is a pure moment connection.

Sometimes a connection transmits loads so that the fasteners are both in shear and in tension. For example, in an ordinary beam-to-column connection, there is usually a shear at the end of the beam and there also exists a certain amount of end moment between the beam and the column. Such a connection is called a moment connection or shear-moment connection.

5-5 NOMINAL STRESSES

The strength of a fastener connection depends on the type of failure, such as:

(*a*) Tension failure in the plates (Fig. 5-7*a*).

(*b*) Shearing failure across one or more planes of the fastener (Fig. 5-7*b*).

(*c*) Bearing failure between the plates and the fastener; this can be a bearing failure of the plates, the fastener, or both (Fig. 5-7*c*).

(*d*) Shear tear-out failure in the plates (Fig. 5-7*d*); this, however, is generally prevented by providing a sufficient edge distance beyond the fastener and the "shear-out" stresses are not calculated.

Historically speaking, the design of riveted and bolted joints is based essentially on consideration of the foregoing failures with little or no regard for the elastic stresses. When specifications are written, a factor of safety is applied to the ultimate values to obtain the allowable stresses. These low values of allowable stresses are somewhat deceiving, since they appear to be actual stresses within the elastic range, whereas in fact they are only nominal stresses to serve as empirical guides in design.

According to the conventional method of design, the nominal stress in the plates or fasteners is defined as the total load divided by the area involved in a particular type of failure (Fig. 5-8).

(*a*) **Tensile unit stress across the net section** (Fig. 5-8*b*):

$$f_t = P/A_n \qquad (5\text{-}1)$$

where P = load acting on the connection, and A_n = net section area of the plate, equal to the gross area A_g less the deduction for the bolt or rivet holes. The deduction in the gross area due to a rivet hole is the product of hole diameter D and plate thickness t. Thus $A_n = A_g - \Sigma\, Dt$.

The majority of bolt or rivet holes in structural work are punched $\frac{1}{16}$ in. greater than the nominal diameter; sometimes because of poor fit of the connecting parts, additional reaming or drilling may be required to match

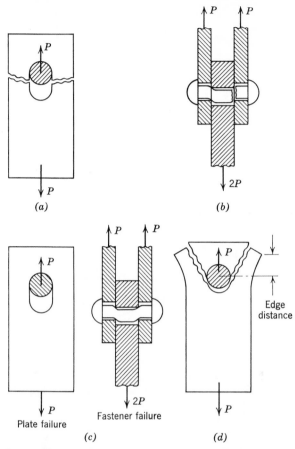

Fig. 5-7 Types of connection failures. (*a*) tension failure in the plates, (*b*) shearing failure in the fastener, (*c*) bearing failure, and (*d*) shear-out failure in the plates.

the holes, thus further increasing the hole diameter. Also, owing to punching of the plate, the material around the hole may be damaged and therefore not fully effective in tension. Thus it is usually specified that, in computing net area, the diameter of a punched hole shall be taken $\frac{1}{8}$ in. greater than the nominal diameter of the bolt or rivet. If the holes are sub-punched and reamed to size after various parts are assembled and bolted together, the holes will be very nearly the size of the fastener, and a deduction equal to the nominal rivet diameter (sometimes plus $\frac{1}{16}$ in.) will be considered sufficient.

(*b*) **Shearing unit stress in the fasteners** (Fig. 5-8*c*):

$$f_v = P/A_v \tag{5-2}$$

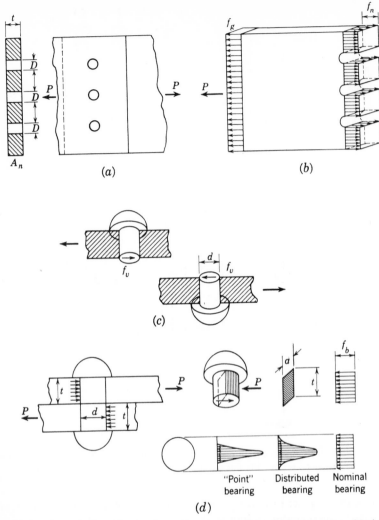

Fig. 5-8 Stresses in connections. (*a*) a riveted connection, (*b*) tensile stress, (*c*) shearing stress, and (*d*) bearing stress.

where $A_v = \Sigma \, (\pi d^2/4)$ is the total shearing area resisting load P and d is the nominal diameter of the fastener.

(*c*) **Bearing unit stress between the fastener and the plates** (Fig. 5-8*d*):

$$f_b = P/A_b \tag{5-3}$$

where $A_b = \Sigma \, dt$ is the total projected bearing area between plates and rivets resisting load P.

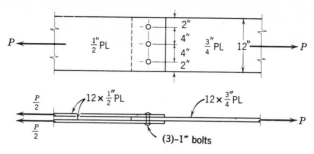

Fig. 1

Example 5.1

(*a*) For the bolted connection (Fig. 1), compute for a force of $P = 100$ kips: (1) the nominal shearing stress in the bolts, (2) the nominal bearing stress between the bolts and plates, and (3) the nominal tensile stress in the plates, assuming punched holes.

(*b*) Determine the ultimate load P for the connection if the ultimate shearing strength of the bolts is 44 ksi, the ultimate bearing strength is 95 ksi, and the ultimate tensile strength of the plates is 60 ksi.

Solution

(*a*) *Nominal Stresses.*

1. Shearing stress:

Bolt area $= 0.785$ in.^{2} (see AISC Manual, 1963, p. 4-6)

$$f_v = \frac{P}{A_v} = \frac{100}{2 \times 3 \times 0.785} = 21.20 \text{ ksi}$$

2. Bearing stress: between bolts and $\frac{3}{4}$-in. plate:

Bearing area of one bolt $= d \times t = 1 \times \frac{3}{4} = 0.750$ in.^{2}

$$f_b = \frac{P}{\sum dt} = \frac{100}{3 \times 0.750} = 44.40 \text{ ksi}$$

Between bolts and two $\frac{1}{2}$-in. plates:

$$f_b = \frac{P}{\sum dt} = \frac{100}{3 \times 2 \times 0.50} = 33.30 \text{ ksi}$$

3. Tensile stresses: [AISC Specification: bolt-hole diameter $\frac{1}{8}$ in. greater than the nominal diameter of bolt.]

For $\frac{3}{4}$-in. plate:

Net area $A_n = [12 - 3(1 + 0.125)]\frac{3}{4} = 6.46$ in.^{2}

$$f_t = \frac{P}{A_n} = \frac{100}{6.46} = 15.45 \text{ ksi}$$

For two $\frac{1}{2}$-in. plates:

Net area $A_n = 2[12 - 3(1 + 0.125)]\frac{1}{2} = 8.62$ in.^{2}

$$f_t = \frac{P}{A_n} = \frac{100}{8.62} = 11.60 \text{ ksi}$$

(b) *Ultimate Loads.*   Shearing across bolts:

$$P_v = F_{vu} \times A_v = 44 \times 2 \times 3 \times 0.785 = 207 \text{ kips}$$

Bearing on $\frac{3}{4}$-in. plate (on two $\frac{1}{2}$-in. plates, not critical):

$$P_b = F_{bu} \times \sum dt = 95 \times 3 \times 0.750 = 213 \text{ kips}$$

Tension across $\frac{3}{4}$-in. plate:

$$P_t = F_{tu} \times A_n = 60 \times 6.46 = 389 \text{ kips}$$

Conclusion: Failure will occur in shearing at a load of 207 kips.

5-6 ACTUAL STRESSES

Nominal stresses as defined by Eqs. 5-1, 5-2, and 5-3 differ from the real stresses because of the following assumptions made in the calculation of nominal stresses:

1. The frictional resistance to slip between the plates is neglected in bearing type connections.
2. Deformation of the plates under load is neglected.
3. Shearing deformation of the fasteners is assumed proportional to the average shearing stress.
4. Tensile stress concentrations due to holes in the plates are neglected.
5. Shearing stress in the fasteners is assumed to be uniformly distributed over the cross section.
6. Bearing stress between fasteners and plates is assumed to be uniformly distributed over the nominal contact surface between fasteners and plates.
7. Bending of fasteners is neglected.

These assumptions are approximately valid when the connection is subjected to static loads approaching ultimate strength.[5] The actual stresses in the connections subjected to working loads, however, are not represented by these equations. Instead, they will depend on the actual deformations of fasteners and plates, amount of friction between the plates, mechanical properties of fastener and plate materials, fastener size, plate thickness, fastener pattern, extent of hole filling by the fastener, and the type of loading.

Frictional Resistance. Tests[6] indicate that tension due to "clamping action" of the hot-driven structural-steel rivets averages about 30 ksi, and that the coefficient of friction between steel plates in a riveted connection is approximately 0.4. The frictional resistance is thus equivalent to a nominal shearing stress on the rivet of $0.4 \times 30 = 12$ ksi. Thus, when the load is less than the frictional resistance, it is transmitted entirely through friction between the plates, and there are actually no shearing or bearing stresses on the rivets.

When the load just exceeds the frictional resistance, an initial slip occurs. After this slip, further increase in load is resisted partly by friction and partly by shearing and bearing stresses on the rivets. In joints with multiple rivets, slippage may not occur simultaneously; there may be considerable load differential between the first and the last slippage.

In most cases slippage in rivets cannot be avoided. Slippage may even be beneficial in certain respects; e.g., in equalizing loads between the rivets, in damping vibrations, and in absorbing some energy due to shock loads. Under a large number of cycles of alternating loads, friction may be destroyed by loss of clamping action in the rivets. In other respects slippage may be injurious, since the deflections of a structure with riveted connections which are allowed to slip are greater than the values computed neglecting such slippage.

For high-strength bolts the resistance to slip depends upon the amount of clamping force and condition of the contact surfaces; it is independent of the design stresses in the connected parts. It has been found by tests that contact surfaces of unrusted mill scale offer the least resistance to slip of any unpainted surfaces and cleaned rusted surfaces may provide up to twice as much resistance in comparison. The recommended shear value for the high-strength bolts is based on a slip coefficient of 0.35. The factor of safety against slip, under repeated loadings, indicates the margin against the condition when a reduced fatigue strength may develop. The factor of safety n against slip may be computed from

$$n = \frac{0.35(\text{minimum bolt tension})}{(\text{allowable shear stress})(\text{nominal bolt area})}$$

The designer in some instances may wish to vary the factor of safety, depending upon the loading conditions. For example, if the satisfactory performance of a structure requires joints which do not move, the designer may elect to use a higher factor of safety. Normally the factors of safety used to define allowable stresses for high-strength bolts are the same as used to define the basic allowable stress values for other structural components. In building design a factor of $n = 1.65$ and in bridge design, $n = 1.80$, are usually used.

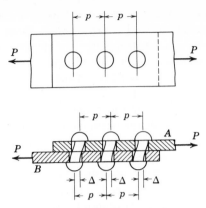

Fig. 5-9 Rigid-plate method.

Deformation of Plates and Fasteners. Assuming the plates to be perfectly rigid and the fasteners perfectly elastic permits determination of stress distribution in a multiple fastener connection as follows (Fig. 5-9). For pure translation of plate A with respect to plate B all fasteners are deformed equally by an amount Δ, and therefore develop the same unit shearing strain γ. Hence the shearing stress $f_v = G\gamma$ is the same for all fasteners, and it follows that each fastener is subjected to load $R = f_v A_v$ proportional to the shearing area A_v. Thus for the group of three rivets shown, $R = (P/3)$.

Since the plates connected by the fasteners are not absolutely rigid, their deformations must be considered. For stresses within the elastic range, the load distribution between the fasteners may be determined by the so-called *elastic-plate method*, in which the elastic deformations of the plates are considered.[7]

A joint with three rivets is shown in Fig. 5-10. Owing to symmetry, the end rivets carry the same load $R_1 = R_3 = R$. The load in each plate between rivets 1 and 2 differs from that between rivets 2 and 3, as shown. Hence the elongations in the plates will differ, the plate elongation due to R being equal to a, and that due to $(P - R)$ equal to b; the strain in the rivets will also vary, resulting in different loads on the rivets. In other words, the load transferred through the rivets will vary with the relative rigidities of the plates and the rivets. If the plates are absolutely rigid, as assumed in the rigid-plate method, then $a = b = 0$, $\Delta_1 = \Delta_2 = \Delta_3 = \Delta$, $R = P/3$, and the rivets will be equally deformed and stressed. If the plates are highly deformable compared to the rivets, practically all the load will be carried by the outside rivets, leaving the center one almost unstressed. It follows that deformability of plates always tends to increase the loads carried by the outer rivets and to decrease the loads on the inner rivets in a group. The greater the number of

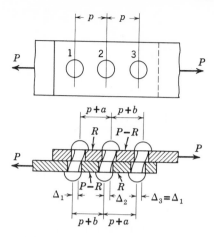

Fig. 5-10　Elastic-plate method.

rivets in a line, the greater is the difference between the loads carried by the end and the center rivets.

Thus the elastic-plate method, although seldom used in practice, should be considered as a qualitative design criterion, indicating that it is desirable to arrange a joint compactly in order to equalize the loads on the fasteners as much as possible.

The elastic-plate method is quite laborious and involves several questionable assumptions such as neglecting friction between plates, stress concentration at the holes, bending deformations of the rivets, etc. Therefore results of such an analysis are only approximate.

The distribution of load in the plastic range can be determined more readily. When the stress in the outer fasteners reaches the yield point, they will deform without taking additional loads. Hence further loading will be carried by the inner fasteners until eventually all the fasteners will be stressed to the yield point. The equalization of loads in the fasteners within the plastic range leads to the same results as those obtained in the "rigid-plate" method and is one of the main reasons for the present-day practice of using the "rigid-plate" method as the basis for analysis and design.

Effect of Stress Concentrations.　Distribution of tensile stresses across a plate with one or more circular holes may be determined by the theory of elasticity as long as the stresses do not exceed elastic limit. The stresses depend on size of hole, and spacing and arrangement of holes.

Because of high stress concentration (Fig. 5-11*a*) initial yielding may occur at working loads. This local yielding does not result in failure and, with further increase in load, the stress distribution gradually becomes more

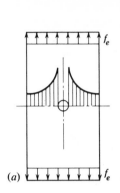

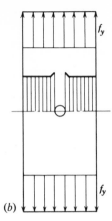

Fig. 5-11 Tension stress distribution. (*a*) Elastic stress distribution. (*b*) plastic stress distribution.

uniform across the section. In the plastic range (Fig. 5-11*b*) the stresses in the plate approach a uniform distribution, thus justifying the use of Eq. 5-1.

Shearing-Stress Distribution in the Fastener. The actual shearing-stress distribution in the fastener cannot be easily determined, but in the elastic stage of loading it certainly is not uniform. Owing to the ductility of structural steel, it may be expected that the stress distribution approaches a uniform one as the loads approach the strength of the fasteners.

Bearing-Stress Distribution. The nominal bearing-stress f_b given by Eq. 5-3 (Fig. 5-8) assumes uniform distribution of bearing stresses on half the circumference of the punched or drilled hole. Since the fastener does not fill the hole completely, the actual distribution and hence the maximum bearing stress may be quite different from the above nominal value. There are no satisfactory measurements that would define the actual stress distribution, although it is known that, in case of "point"-bearing contact, the actual stress may be very much higher than the nominal stress (Fig. 5-12).

The bearing-stress distribution along the length of the fastener depends on its bending behavior. The distribution is nearly uniform for that portion of the fastener which lies between two supporting plates (Fig. 5-12), although for the portion bearing on the side plates there may be some variation in bearing stress. This variation in bearing stress is particularly pronounced in a lap joint (Fig. 5-12). Before excessive deformation occurs, the loads on the fastener form a couple and must be resisted by an equal and opposite couple acting on the rivet heads. This partly accounts for the importance of the strength of rivet heads and the reduction in the strength of countersunk rivets. Figure 5-12 shows that, after excessive deformation occurs, the maximum bearing stress f_b can be much higher than the average bearing stress P/td. Experiments show that rivets and plates can withstand these high contact stresses without appreciable ill effects on the connection provided the nominal bearing stress $f_b \leq 2.25$ times the tensile stress allowed on the net section.

Bending in fasteners is generally disregarded because, in most cases, bending deformations are largely prevented by fit in the holes. Also the conventional flexure theory is not applicable when fastener length-to-diameter ratio is less than about 10. Bending in the fastener may be significant only if the length is sufficiently large.

The foregoing discussion leads to the following conclusions.

1. Nominal stresses defined by Eqs. 5-1, 5-2, and 5-3 are based on the behavior of mechanical connections at ultimate loads, and thus represent an application of the "ultimate strength" design concept. Validity of this ultimate design concept is largely restricted to static loads. Under a large number of cycles of alternating loads, stress concentrations may substantially reduce the strength of the connection and cause a "fatigue" failure.

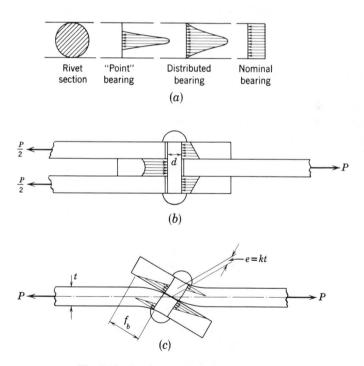

Fig. 5-12 Bearing stress distribution.

2. At low loads the stress distribution in a riveted connection is largely controlled by friction between the plates. At intermediate loads slipping occurs and shearing and bearing stresses are induced. Slip in some connections may be desirable in damping vibrations.

3. Loss of "clamping action" in the fastener results in slippage at the joints and increased deflections in the structures.

4. During the elastic deformation of the joint, the end rows tend to take more than the average load per fastener. The load distribution will depend on the relative rigidities of the plates and the fasteners.

5. As plastic yielding takes place, the load distribution among the rows tends to become more and more uniform, approaching equal loads on the fasteners.

6. During the latter part of the plastic yielding, the amount of plastic flow in the plates may be sufficient to cause detrusion of end fasteners, but load distribution among the fasteners remains nearly uniform.

Although the foregoing discussion and summary is applicable in general to all the mechanical fasteners, it is most directly descriptive of the action of rivets and high-strength bolts. There are three basic differences in the load transmission of rivets and the common (A307) structural bolt: the fit between the shank and the hole, the reduction of bolt section and stress concentration at the root of bolt thread, and the amount of initial tension in the shank determining the frictional resistance to slip.

Common Structural Bolts.   Consider two plates joined by an unfinished bolt fitting loosely in the hole with the nut just tightened without producing any appreciable clamping action on the plates (Fig. 5-13). When a shear load P is applied, the plates will slip until the bolt comes into contact with the edges of the hole and the load is transmitted by bearing on the bolt and shear in its shank (Fig. 5-13). The shearing stresses in the bolts are more or less uniformly distributed over its gross section. It should be noted that, in computing shearing stress in a properly designed and installed connection, the gross sectional area of the bolt must be used, simply because shear does not exist in the threaded portion of the bolt. Conversely, if the threaded portion of the bolt is subjected to transverse shear, the net section should be used in computing the shearing stress.

The bearing between the bolt and the plates is concentrated on the edge of the plate and produces high local stresses until plastic deformation takes place. These high local stresses may or may not be detrimental to the structure. If loading is not excessive and if it is not repeated frequently to loosen the nuts or to produce fatigue failure, then high local stresses will be of little practical significance.

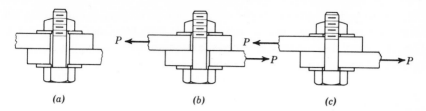

Fig. 5-13 Action of bolts under load. (*a*) unfinished bolt before load application, (*b*) unfinished bolt under load, and (*c*) turned bolt under load.

If the plates are joined by a turned structural bolt, fitting snugly into a reamed hole (Fig. 5-13), then no appreciable slip will occur and the load will be transmitted directly by bearing and shear in the bolt. Because of better fit of the bolt in the hole, the local bearing stresses in the bolt are minimized and the bearing strength of such a joint is greater than with an unfinished bolt.

The tension load on a bolted joint (Fig. 5-14) is transmitted by tension stresses in the bolt and the critical section is at the root of the thread. At this section not only is there less area, but there is also a high stress concentration, which means that, if the elastic limit of the material has not been exceeded, the actual maximum stress at this section is much higher than the average stress. At high loads, the material at the root yields and the stress distribution becomes more uniform, so that the ultimate strength of the bolt is not materially affected by the stress concentration. In a pure tension load there is no distinction between the behavior of an unfinished or a turned bolt. It must be noted that the strength of the bolt in tension may sometimes be determined by the stripping strength of bolt threads. Although there is little theoretical or experimental evidence to allow rational prediction of this strength, the method commonly used is to assume that stripping occurs on the cylindrical surface between roots of the threads and that shearing stress

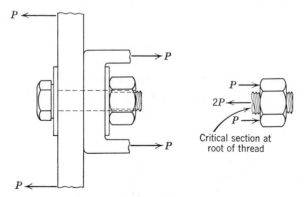

Fig. 5-14 Bolt under tension load.

is uniformly distributed over the surface within the threads engaged by the nut. A structural bolt usually has sufficient threaded length in the nut to insure the full development of its tensile strength before stripping can occur. Hence, these shearing stresses are seldom considered.

Bolts, unlike rivets, are not limited in length by problems of fabrication and high bending stresses may exist in long bolts when the grip exceeds approximately 5 diameters.

Bending stresses in short bolts are usually neglected in design; for long bolts, bending stresses must be computed in a conventional manner, although specifications make no provision for allowable stress in bending.

When considering stresses in joints using unfinished or turned bolts, it was assumed that the initial tension in them was negligible. Under actual field conditions, however, this tension may be appreciable, although its magnitude can not be accurately predicted. It is possible to tighten ordinary structural bolts so as to develop initial tension which would provide to a certain degree some of the advantages of high-strength bolts. However, because of the low strength of structural steel, only half the initial tension of high-strength bolts can be realized in structural bolts without stripping the threads or breaking the bolts. Furthermore, loss of initial tension due to creep in ordinary structural bolts seems to be more of a problem than in the high-tensile-strength bolts and for this reason initial tension in ordinary bolted connections is disregarded.

High-Strength Bolts. Consider two plates joined by a high-strength bolt, fitting loosely in the hole, as shown in Fig. 5-15. The nut is tightened to develop a clamping force on the plates which is indicated as the tensile force T in the bolt. This tension should be approximately 80 to 90% of the tensile yield strength of the bolt. When a shear load P is applied to this joint, no slip will occur until the load P exceeds the frictional resistance of the joint. When P exceeds this value, there occurs a major slip and, as load is further increased, gradual slipping brings the bolt in contact with the edges of the plate. The safe load on the bolts is defined as one slightly less than the critical load at which first major slip occurs. This load is determined by the coefficient of friction k_f and is directly proportional to the initial tension load T in the bolt. Although the value of the coefficient of friction depends on the condition of the faying surfaces, test results indicate that a value of $k_f = 0.35$ represents a conservative estimate.

Using $k_f = 0.35$ and $T = 0.8F_yA$, the initial slip will occur at $P = 0.28F_yA$. For A-325 and A-490 bolts this load corresponds to nominal shear stresses of $0.28 \times 80 = 22.4$ ksi and $0.28 \times 120 = 33.5$ ksi, respectively. The allowable shear values for "bearing-type" connection are based approximately on stresses which produce this initial slip.

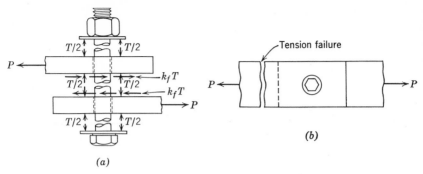

Fig. 5-15 High-strength bolt; (*a*) load transmission by friction, and (*b*) failure outside of net section.

If load P is less than $k_f T$, then slip does not occur and the entire load P is transmitted by friction on the faying surfaces of the plates. The bolt is not stressed in bearing or shear at all; the only stress in the bolt is that of initial tension. As previously mentioned, since the load P is transmitted by friction, the tension load in the plate at the hole is less than P and therefore the net section may not be critical in design. Many tests of high-strength bolt joints have shown that tensile failures occurred outside the net section (Fig. 5-15). It must be noted here that the frictional resistance similar to the shearing resistance depends on the number of contact surfaces, being double $k_f T$ for "double-shear" arrangement.

5-7 STRESSES IN PIN CONNECTIONS

Stresses in pin connections are similar to those in rivets and bolts except that there is never any direct tensile stress in pins and that bending stresses, although neglected in rivets and bolts, are generally considered in pins.

Bending Stresses.   To assure free rotation between plates and the pin, and in order to facilitate lubrication, the plates are assembled on the pin with ample clearance between them. This results in the use of relatively long pins (Fig. 5-16), and the computation of bending stresses for pins by the usual beam theory is considered to be more nearly correct than for rivets or bolts. The bending moments may be determined assuming that the load in each plate is concentrated at the center of the plate (Fig. 5-16*b*). When the main plates are reinforced for bearing or when several plates are packed together, this assumption can lead to a considerable error and an extremely conservative design. To allow for effect of load distribution, a moment diagram assuming uniform bearing distribution can be computed (Fig. 5-16*c*).

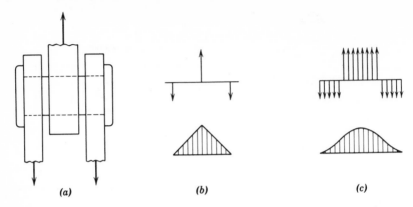

Fig. 5-16 Moments in pins; (*a*) pin connection, (*b*) moments in pin-assuming concentrated loads, and (*c*) moments in pin-assuming uniform loads.

For given bending moment M (Fig. 5-17*a*), assuming stresses within the elastic limit (Fig. 5-17*c*), the maximum fiber stress in a pin is given by Eq. 5-4:

$$f = \frac{Mc}{I} = \frac{M(d/2) \cdot}{\pi d^4/64} = \frac{10.2M}{d^3} \tag{5-4}$$

or the bending moment M in terms of maximum stress f is

$$M = \frac{f d^3}{10.2} \tag{5-5}$$

Bending moment M_y, for which the computed maximum stress f is equal to yield strength F_y, defines the yield strength of the pin in bending. When the pin is subjected to a moment greater than M_y, the stress is no longer proportional to strain and cannot be computed from Eq. 5-4. In the plastic range the ultimate bending strength M_u can be computed approximately by assuming rectangular stress distribution, with maximum stress equal to F_y

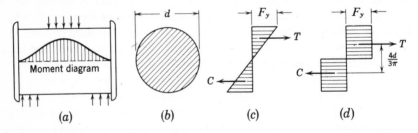

Fig. 5-17 Resisting moment of a pin; (*a*) a pin, (*b*) cross section of a pin, (*c*) elastic stress distribution, and (*d*) plastic stress distribution.

(Fig. 5-17*d*). The ultimate strength is then equal to the resisting couple defined as (stress × area × lever arm) as follows:

$$M_u = F_y \left(\frac{\pi d^2}{8} \right) \left(\frac{4d}{3\pi} \right) = \frac{F_y d^3}{6} \tag{5-6}$$

Thus, if plastic bending is permitted, the ultimate bending strength of the pin is 10.2/6 or 1.7 times greater than that indicated by the elastic theory.

Shearing Stresses. Since bending stress in a pin is computed on the basis of simple beam theory (elastic or plastic), it may seem that the shear stress in a pin should also be computed on the same basis. Assuming that bending does not exceed yield strength, the shear stress is maximum at the neutral axis of a pin and is given by the conventional elastic theory as follows:

$$f_v = \frac{VQ}{It} = \frac{V}{(\pi d^4/64)} \frac{(\pi d^2/8)(4d/6\pi)}{d} = \frac{16V}{3\pi d^2} = \frac{4}{3} \frac{V}{A} \tag{5-7}$$

The shear stresses computed by simple beam theory are in considerable error when the ratio of span to depth is as small as it usually is for pins. For this reason, the nominal shear stress based on uniform stress distribution over the section is commonly used for design of pins, in which case

$$f_v = \frac{V}{A} \tag{5-8}$$

It should be noted that the "exactness" of the nominal shear stress used for design is irrelevant, since its value is used only for comparison with an arbitrary allowable stress. It is just as reasonable to establish an allowable shear stress in pins based on average stress as it would be to establish a different allowable shear stress based on some theoretical beam shear stress, which is known to be incorrect.

Bearing Stresses. Actual bearing stresses between plates and pins depend on the accuracy of fit between them. For proper fit, uniform distribution of bearing stress may be assumed and the nominal bearing stress is

$$f_b = \frac{P}{A_b} = \frac{P}{td} \tag{5-9}$$

where *P* is the load and *t* is the total thickness of plates bearing on a pin of diameter *d*. As in riveted and bolted connections, bearing-stress concentration will be high unless a tight fit is ensured between the pin and the hole.

Combined Stresses. Since pins do not take tension, there is no combination of shearing and tensile stresses as with bolts and rivets. There is the problem

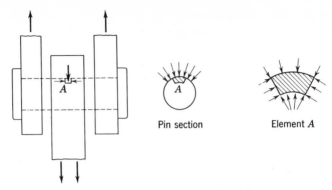

Fig. 5-18   Triaxial compression in a pin.

of combining bending stresses with bearing stresses, since the maximum compressive bending stress always occurs in a region of high bearing stress. This, however, is usually neglected, and justifiably so, because in this region the pin is subjected to triaxial compression (Fig. 5-18). Since the failure of a solid occurs when the difference in the principal stresses is large, the combined triaxial compression stresses evidently need not be investigated when the stress in each direction is kept within allowable limits.

5-8   NET SECTION—STRAIGHT AND STAGGERED PATTERNS

A connection having several rows of rivets and failing in tension will fracture along the section of maximum tensile stress (Fig. 5-19). Assuming that the shearing loads in all the rivets are equal, the proportion of total load transmitted by the plates at each section may be computed; in plates A, section 1-1 transmits the total load, sections 2-2 and 3-3 transmit only part of the total load. Theoretically it may be considered advantageous to decrease the number of rivets in section 1-1 in order to increase the net area, and to increase the number of rivets in the other rows for which less net area is required. As an example, consider the rivet arrangement shown in Fig. 5-19. In order to increase the net area on the outer rows, two alternate designs are shown and their relative efficiency will be discussed. *A theoretical efficiency may be defined as the allowable load on the connection expressed as a percentage of the allowable load on the gross area of the plate.* When the allowable load is governed by tension, this *efficiency* is equal to the ratio of average stress on the gross section to the greatest stress on the net section. Assuming that the three connections shown in Figs. 5-19a, b, and c are all critical in tension, their efficiencies are 77, 84.6, and 78%, respectively. This computation indicates that design b is theoretically better than designs a and c. However, this conclusion is not supported by test results. Numerous test results indicate that a plate with small holes, even if the holes represent only a few percent of the gross area, has a total ultimate strength generally about 20% lower than the calculated strength based on the gross area. This indicates that theoretical efficiencies in excess of 80% have little practical significance. The AISC Specification provides that net section taken through a hole shall in no case exceed 85% of the gross section.

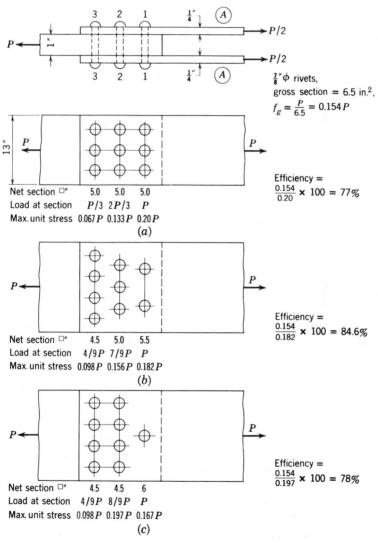

Fig. 5-19 Efficiency of rivet patterns.

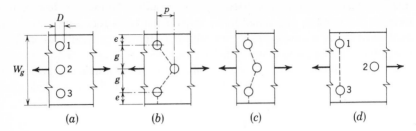

Fig. 5-20 Various proportions of rivet stagger; (*a*) straight, no stagger, (*b*) ordinary stagger, (*c*) small stagger, and (*d*) large stagger. Dotted lines show possible lines of failure.

If the space available for riveting is limited, the rivets in different rows may be staggered (Fig. 5-20). In this instance, it is necessary to determine the line of failure and the net section corresponding to the line of failure. The strength of a connection failing along a zigzag line cannot be completely determined theoretically, although some investigations have been made based on limit analysis and practical considerations.[9]

Several empirical methods for determining the strength of connection with a staggered rivet pattern have been proposed. The most commonly used method is that recommended by AREA, AISC, AASHO, and other specifications. It provides that the net width W_n of any line of failure should be computed as follows (Fig. 5-20):

$$W_n = W_g - \Sigma D + \Sigma \frac{p^2}{4g} \qquad (5\text{-}10)$$

where W_g is the gross width of the plate, ΣD is the summation of hole diameters in the line of failure 1-2-3, p and g are the pitch and gage dimensions, respectively, for the line considered. When the pitch is zero, the formula corresponds to a simple relation $W_n = W_g - \Sigma D$; when p is large, the critical net section is to be found along some other line, such as 1-3, in Fig. 5-20*d*.

A method proposed by Professor W. M. Wilson which seems to be in somewhat better agreement with test results[5] defines W_n as follows:

$$W_n = (W_g - \Sigma D)\left(1 - \frac{D}{g}\right)0.85 \qquad (5\text{-}11)$$

In Eq. 5-11, g is the transverse distance between rivets or twice the edge distance e, whichever is greater. For joints with some rivets omitted from the outer row, only rivets in the outer row are to be considered in determining ΣD and g, provided $p \geq 0.8g$. If $p < 0.8g$, all rivets in the outer row and the second row should be considered in determining ΣD and g. This method has not been accepted by any specification. The difference between Eqs. 5-10 and 5-11 is relatively small, and because of the possibility of variation in the performance of identical riveted connections, as observed in numerous tests, no particular advantage can be derived by using a more complex empirical formula, such as Eq. 5-11.

5-9 DESIGN CONSIDERATIONS

General. When the line of action of the shear load coincides with the center of gravity of the fastener areas, the conventional practice is to assume uniform distribution of shearing stress. The design of such connections is based on the nominal stresses, which must not exceed the allowable stresses.

The size of the fastener is usually dependent upon the space limitations of the connection, design load, and allowable stresses.

It is generally more economical to use a few large diameter fasteners rather than a great number of small ones which require more installation work. However, allowable load, if controlled by bearing, increases only in direct proportion to the diameter, but if it is controlled by shear, it increases as the square of the diameter. It is apparent that for any given plate thickness a greater advantage is gained by increasing the fastener size if the allowable shear controls the capacity of the fastener than when bearing controls.

Because the fastener size is generally predetermined, the thickness of plate is left to the choice of the designer. In such a case, the usual practice is to determine a minimum plate thickness to develop the full bearing value of the fastener so as to utilize its full shearing strength.

It is general practice for economy and compactness of the connection to space the fasteners as closely as possible. This also reduces the amount of additional connection material required. On the other hand, spacing should not be so close that it greatly reduces the cross section of the principal parts or produces tear-out failures between the fasteners. A minimum spacing is usually stated in specifications and is based on the clearances of the tools required to install the fastener. A maximum spacing may also be stated to insure proper action of the plates to avoid local buckling in compression members.

For economy of fabrication and standardization of connection materials, most rolled shapes have standard locations for holes referred to as gage lines. These gage lines are listed in the AISC Manual. Other dimensions may be used, but, in general, they will require retooling and therefore become more costly.

Standard terminology is used to describe dimensions in a connection, as illustrated in Fig. 5-21.

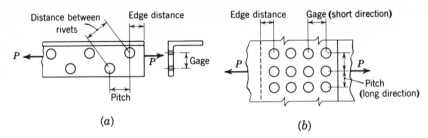

Fig. 5-21   Rivet spacing; (*a*) rolled shapes, and (*b*) wide plates.

Fastener Spacing.   In discussing fastener spacing, several definitions must be clarified. These are:

Gage.   Center-to-center spacing of fasteners measured along the width of the member or connection (Fig. 5-21).

Pitch.   Center-to-center spacing of fasteners measured along the length of the member or connection (Fig. 5-21).

Rivet Distance.   Center-to-center spacing of staggered fasteners measured obliquely on the member (Fig. 5-21).

Edge Distance.   Distance between center of fastener hole and adjacent edge of plate (Fig. 5-21).

When wide plates are connected, it is impossible to distinguish between their length and their width. In such cases, fastener spacing along the joint is known as the pitch and perpendicular to the joint as the gage (Fig. 5-21).

Design Loads.   To ensure an adequate strength of connection a design load is usually specified. The AISC Specification (Secs. 1.15.1 and 1.15.7) provides that the connection shall develop full design load, but in no case shall the connections of tension or compression truss members be designed for less than 50% of the effective strength of the member. Except for lacing in built-up members, sag bars, and girts the design load shall be not less than 6000 lb. The AASHO Specification makes similar provisions, except that 75% of the effective strength of the member is considered to be the minimum design load for connections.

Allowable Stresses for Fasteners.   The allowable stresses for fasteners are determined by applying a suitable factor of safety to their strength obtained from reliable test data on the behavior of connections.

The allowable stresses prescribed by the AISC Specification for Buildings, the AASHO Specification for Highway Bridges, and the AREA Specification for Steel Railway Bridges for rivets and bolts are summarized in Table 5-4.

Spacing of Fasteners.   The following limitations on the spacing must be observed: (*a*) net section area must not be less than required; (*b*) edge

Table 5-4 Allowable Stresses for Fasteners

Loading Conditions	A502 Grade 1† Buildings	Bridges	A502 Grade 2† Buildings	Bridges	A325 Bolts Buildings	Bridges	A490 Bolts Buildings	Bridges	A307 Bolts Buildings	Bridges	Pins Buildings	Bridges
Tension, ksi	20	–	27	–	40	36	54*	48*	14	13.5	–	–
Shear, ksi Friction type	–	–	–	–	15	13.5	20	18	–	–	$0.40F_v$	$0.40F_v$
Bearing type, threads excluded	15	13.5	20	20	22	20	32	29	10	11	–	–
Bearing type, threads not excluded	–	–	–	–	15	13.5	22.5	20	–	–	–	–
Bearing, ksi	$1.35F_y$	40	$1.35F_y$	40	$1.35F_y$	$1.22F_y$	$1.35F_y$	$1.22F_y$	$1.35F_y$	20.0	$0.9F_y$	$0.80F_y$
Bending, ksi	–	–	–	–	–	–	–	–	20	–	$0.9F_y$	$0.80F_y$
Combined‡ shear and tension, ksi }a	28	–	38	–	50	–	–	–	14	–	–	–
}b	20	–	27	–	40	–	–	–	–	–	–	–

* For static loads only.

† For highway bridges only.

‡ Allowable tension: $F_t = a - 1.6f_v \leq b$. For A325 bolts in friction-type joints the shear stress allowed shall be reduced so that $F_v \leq 15(1 - R/T)$, where R is the nominal tension load in the fastener and T is the proof load in tension. F_y—specified minimum yield point of the lowest strength connected part. The bearing stress shall not be more than the specified minimum tensile strength of the lowest strength of connected material.

135

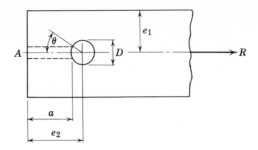

Fig. 5-22   Edge distance to prevent shear-out failures.

distances must not be less than required; (*c*) construction clearances must be observed; (*d*) spacing and edge distance must not exceed specified limits, particularly with plates in compression.

If the spacing of fasteners is small, the strength of connection may be governed by the net section of the plates; if the spacing is large, the strength of the connection will be governed by the sum of the rivet values. Hence, there must be an optimum spacing which will yield highest strength for a particular joint. Evidently, such an optimum spacing is obtained when the tensile strength of the plates equals the combined rivet shearing or bearing strength. Usually it is not possible to adopt the exact optimum spacing; however, a good design does not deviate much from this optimum.

Two types of edge distances must be considered: (*a*) perpendicular to line of stress, sometimes called "unloaded-edge" distance e_1 (Fig. 5-22) and (*b*) line of stress or "loaded-edge" distance e_2.

The minimum edge distance e_1 needed to prevent premature yielding at the "unloaded" edge B (Fig. 5-22) and consequent reduction in the effective net section width is difficult to define rationally. An empirical criterion usually employed provides that $e_1 \geq kD$, where k varies with the smoothness of the edge. AISC specifies $k = 1\frac{3}{4}$ for sheared edges and $k = 1\frac{1}{4}$ for rolled or gas-cut edges. AASHO specifies $k = 1\frac{3}{4}$ and $1\frac{1}{2}$ for sheared and for rolled or gas-cut edges, respectively.

When load R is applied to a plate through a fastener placed a distance e_2 from the edge, shearing and normal stresses are induced in the plate. If distance e_2 is inadequate, yielding may develop along lines at angle θ with the load, whereas stresses in other elements of the connection are within elastic limits. This yielding would cause excessive distortion of the hole and premature failure. The minimum edge distance (Fig. 5-22) necessary to prevent such failure is such that

$$e_2 \geq a + \tfrac{1}{2}D \cos \theta \qquad (5\text{-}12)$$

Neglecting normal stresses and using a nominal value of shearing stress on the plate equal to F_v, the minimum value of a is such that

$$a \geq \frac{R}{2F_v t} \tag{5-13}$$

where t is plate thickness. Values of θ and F_v may be reasonably taken as $40°$ and $0.95F_y$, respectively. The AISC Specification is based on $\theta = 90°$ and F_v equal to half that allowed for rivet shank material. The AISC waives this limitation on e_2 when there are more than two rivets in the line considered. However, in any case the distance e_2 shall be not less than e_1.

The minimum edge distance e_2 needed to prevent "tear-out" when edge of the plate fails in tension at point A (Fig. 5-22) is more difficult to estimate rationally. Usually edge distance in excess of two hole diameters will prevent this type of failure.

Interfastener Buckling. Thin plates subjected to compression stress may buckle unless properly stiffened. When such plates are attached by means of fasteners, the buckling may occur between the connectors if these are spaced too far apart. To prevent such instability the critical buckling stress F_{cr} should be equal to or greater than the yield point stress F_y. The critical stress may be determined from (see Sec. 9-13)

$$F_{cr} = k \frac{\pi^2 E}{12(1 - \mu^2)} \left(\frac{t}{s}\right)^2 \tag{5-14}$$

where k is a coefficient varying with boundary conditions and proportions of the plate, E and μ are elastic modulus and Poisson ratio, respectively, and s and t are dimensions shown in Fig. 5-23. For typical fastener arrangements k may vary between 0.5 and 3.0. From the condition that $F_{cr} \geq F_y$ it follows that

$$\frac{s}{t} \leq \sqrt{\frac{k\pi^2 E}{12(1 - \mu^2)F_y}} = \frac{A}{\sqrt{F_y}} \tag{5-15}$$

where A is an appropriate coefficient depending on k. The AISC Specification (see Sec. 1.9.2 and 1.18.2.3) includes the following limitations for plates in compression (Fig. 5-23)

$$s_1 \leq \left(\frac{4000}{\sqrt{F_y}}\right)t \quad \text{or} \quad 12 \text{ in.} \tag{a}$$

$$s_2 \leq \left(\frac{6000}{\sqrt{F_y}}\right)t \quad \text{or} \quad 18 \text{ in.} \tag{b} \tag{5-16}$$

$$b_1 = b_2 \leq \left(\frac{8000}{\sqrt{F_y}}\right)t \tag{c}$$

For rolled shapes the upper limit on s_1 and s_2 may be taken as 24 in.

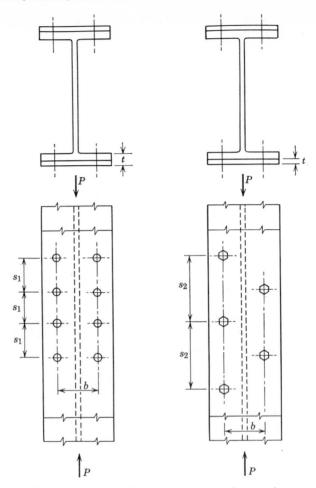

Fig. 5-23 Spacing of fasteners in compression members.

Design Procedure. The following order of procedure is suggested for the design of direct shear connections:

(a) Make a general preliminary layout of the connection and determine the loads on the connection.

(b) Determine the type of fastener to be used and estimate its size.

(c) Compute the allowable value of the fastener. The usual practice is to determine required plate thickness to develop full bearing and shearing values. If bearing is critical, the fastener size may be increased.

(d) Compute the number of fasteners required to carry the load.

(*e*) Space the fasteners properly. For economy and compactness of the connection, it is desirable to space the fasteners as close as possible. This also reduces the amount of additional connection material required.

(*f*) Check the net section on the basis of thickness determined in (*c*). If net section is not adequate, plate thickness may be increased or the fasteners rearranged so as to increase the net width of the connection.

(*g*) Check arrangement of fasteners to ensure that the spacing is not below minimum values necessary to prevent tear-out failures and to provide clearances for tools during installation. The spacing should not exceed maximum values, particularly to avoid local buckling of plates in compression members.

5-10 MOMENT CONNECTIONS—FASTENERS IN SHEAR

Consider the transmission of a pure couple through a shear connection (Fig. 5-24). The load R on each fastener varies in both magnitude and direction, and determination of the actual load distribution in such a connection becomes almost impossible if its elastic behavior and/or friction between plates are to be considered. In practice, a relatively simple solution is used. This assumes that the plates are rigid and the fasteners perfectly elastic.

Neglecting friction between plates, the applied moment will produce a slight relative rotation $d\theta$ between the plates about a center O. Assuming that the plates are perfectly rigid, all the fasteners will have a relative displacement perpendicular and proportional to their radius vector r from the center O. If the fasteners are assumed to be perfectly elastic, the unit stress on each will be proportional to its displacement $r\,d\theta$, and will be in the direction of the displacement, that is, perpendicular to its radius vector. The value of $d\theta$ is proportional to the moment $M = Pe$, and for any particular

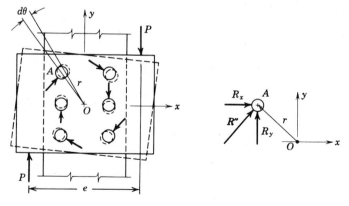

Fig. 5-24 A moment connection, rivets in shear. Dotted line shows rotated position.

value of M it is a constant for all fasteners in the connection. Thus, the unit shearing stress, $R/A \propto r(d\theta)$, and

$$\frac{R''}{A} = kr \quad \text{and} \quad R'' = krA \qquad (5\text{-}17)$$

Taking an arbitrary pair of rectangular axes through O and resolving the load R'' into x and y components, we have

$$R_x'' = R''\left(\frac{y}{r}\right) = krA\left(\frac{y}{r}\right) = kyA \qquad (a)$$

$$R_y'' = R''\left(\frac{x}{r}\right) = krA\left(\frac{x}{r}\right) = kxA \qquad (b)$$

$$(5\text{-}18)$$

Since we are considering a pure couple, the summation of loads along the x and y axes must each equal zero; hence

$$\sum R_x'' = \sum kyA = 0 \qquad \sum yA = 0 \qquad (a)$$

$$\sum R_y'' = \sum kxA = 0 \qquad \sum xA = 0 \qquad (b)$$

$$(5\text{-}19)$$

These two equations can be satisfied only if the origin O coincides with the centroid of the fastener areas A.

To determine the constant k, note that the moment to be transmitted must be balanced by resisting moments of the fasteners. From Fig. 5-24:

$$M = \sum R''r = \sum R_x''y + \sum R_y''x = \sum (kyA)y + \sum (kxA)x \qquad (a)$$

$$M = k\left(\sum Ay^2 + \sum Ax^2\right) \qquad (b)$$

or

$$k = \frac{M}{\sum Ax^2 + \sum Ay^2} = \frac{M}{\sum A(x^2 + y^2)} = \frac{M}{\sum Ar^2} \qquad (c)$$

$$(5\text{-}20)$$

Although k can be computed as $M/\sum Ar^2$, it is generally more convenient to compute k as $M/\sum A(x^2 + y^2)$, particularly if A is constant for all fasteners in the connection.

Having obtained k, we can write the following for each fastener:

$$R_x'' = kyA = \frac{M}{\sum A(x^2 + y^2)} yA \qquad (a)$$

$$R_y'' = kxA = \frac{M}{\sum A(x^2 + y^2)} xA \qquad (b)$$

$$(5\text{-}21)$$

If area A of all fasteners is the same, then

$$R_x'' = \frac{M}{\sum (x^2 + y^2)} y \qquad (a)$$

$$R_y'' = \frac{M}{\sum (x^2 + y^2)} x \qquad (b)$$

$$(5\text{-}22)$$

R'' is generally obtained as a resultant of R_x'' and R_y'':

$$R'' = \sqrt{R_x''^2 + R_y''^2} \tag{5-23}$$

R'' can also be directly obtained as follows, using the subscript i to denote the particular individual fastener:

$$R_i'' = kr_iA = \frac{M}{\sum Ar^2} r_iA \tag{5-24a}$$

For fasteners of equal area, Eq. 5-24a reduces to

$$R_i'' = \frac{M}{\sum r^2} r_i \tag{5-24b}$$

which is independent of cross-sectional area A.

It must be noted that, although the foregoing derivation is entirely logical in itself, the assumptions are not necessarily correct. The assumption of rigid plate is incompatible with the assumption of elastic fasteners; and neglecting frictional resistance is far from accurate. When the loads are small, the fasteners are elastic, but the plate deformation is usually of the same order of magnitude as fastener deformation and cannot be neglected; and the frictional resistance may be an important factor. When the loads are large, the fastener deformations are in the plastic range and the plate deformations become relatively small, which seems to justify the assumption of rigid plate; however, under large loads, the fastener stress is no longer proportional to its deformation, which invalidates the assumption of elastic fasteners. Thus the two assumptions are actually inconsistent. Although the method is not precisely correct, it is a convenient one and yields conservative results especially when fasteners' stresses are within the plastic range.

5-11 ECCENTRIC LOAD CONNECTIONS—FASTENERS IN SHEAR

Loads on fasteners in an eccentric load connection are usually determined by the method of superposition. The eccentric load (Fig. 5-25) is resolved into a direct load P through the centroid O of the fastener group plus a pure moment Pe, where e is the eccentricity of load P with respect to O. The loads on any fastener due to direct load P and due to pure moment Pe are computed separately and then added vectorially. The computation is usually simplified by resolving the load P into x and y components, P_x and P_y.

Let R' = load on fastener due to direct load P
$\quad R''$ = load on fastener due to moment Pe
$\quad R$ = resultant load on fastener

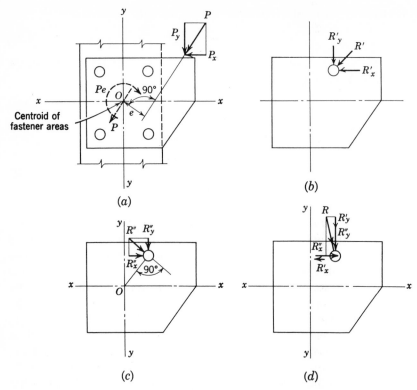

Fig. 5-25 Shear loads on fasteners due to eccentric loads; (*a*) resultant load on fastener group, (*b*) load on fastener due to direct load *P* only, (*c*) load on fastener due to moment *Pe* only, and (*d*) resultant load on fastener.

The x and y components of fastener loads are denoted by corresponding subscripts. Then for any fastener with coordinates x, y, and cross-sectional area A, the load on the fastener can be computed as follows:

$$R_x' = \frac{P_x}{\sum A} A, \qquad R_y' = \frac{P_y}{\sum A} A \tag{5-25a}$$

$$R_x'' = \frac{(Pe)yA}{\sum A(x^2 + y^2)}, \qquad R_y'' = \frac{(Pe)xA}{\sum A(x^2 + y^2)}$$

Adding these vectorially, we obtain:

$$R = \sqrt{(R_x' \pm R_x'')^2 + (R_y' \pm R_y'')^2} \tag{5-25b}$$

Instantaneous Center of Rotation. An investigation of an eccentrically loaded joint can be limited to the following: (1) the location of the most heavily loaded fastener, usually called the critical fastener; and (2) the magnitude of the load carried by this critical fastener. In simple joints the critical fastener can be located by inspection; often the following method is advantageous.

An eccentric load on a connection produces rotation about the centroid of the fastener group together with a translation of one plate with respect to the other. This rotation and translation of a rigid plate can be reduced to a pure rotation about some point which is called the "instantaneous" center (Fig. 5-26). The center is called instantaneous because it varies not only with the fastener arrangement but also with the location and direction of the eccentric load. Once this center is determined for a given loading, the relative displacement between plates at any fastener can be determined, since the magnitude of this displacement is proportional to the distance from the instantaneous center and its direction is perpendicular to the radius vector from this center. The unit shearing stress on the fasteners is proportional to the displacement and hence the most highly stressed fastener is located the greatest distance away from the instantaneous center.

The instantaneous center O_i can be located using the condition that an imaginary fastener located at O_i would carry no load (Fig. 5-26). The load on any fastener consists of components due to direct load and moment; since the load due to direct load is constant in magnitude and direction, it is evident that zero resulting load occurs at the point where the load R'' due to moment exactly counteracts the load R' due to direct load. In order that R'' shall be parallel and opposite to R', the radius vector from the centroid O of rivet pattern to O_i must be perpendicular to the load P; and O_i and P must be on the opposite sides of O. For the load R'' to exactly balance R',

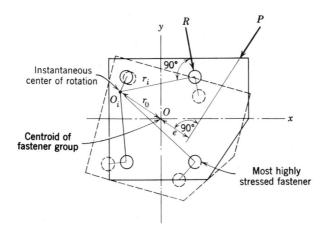

Fig. 5-26   Instantaneous center of rotation.

the center must be located at a distance r_0 from the centroid so that

$$R' = \frac{PA}{\sum A} = -R'' = -\frac{Per_0A}{\sum Ar^2} \quad \text{or} \quad r_0 = -\frac{\sum Ar^2}{e \sum A} \quad (5\text{-}26)$$

If all fasteners in the group have the same area, then $r_0 = -\sum r^2/ne$ where n is the number of rivets.

Once the instantaneous center of rotation has been found, the direction of resultant load on any rivet is located perpendicular to its radius vector r_i from that center, and the magnitude is given by

$$R = \frac{P(e + r_0)r_iA}{\sum Ar_i^2} \quad (5\text{-}27)$$

where r_i is referred to the instantaneous center and r_0 is the distance from the centroid O to the instantaneous center O_i.

The use of Eq. 5-27 in calculating fastener loads is not generally advantageous, but the concept of instantaneous center is useful in determining the fastener with the greatest unit shearing stress. Evidently the fastener with the greatest value of r_i has the greatest displacement and hence the greatest stress. If we know that the instantaneous center O_i is on the side opposite to the eccentric load P and on a perpendicular to P through the centroid O, the fastener with the greatest stress can often be determined by inspection.

Example 5-2.   A plate is connected to a column by four 1-in. fasteners as shown and carries an eccentric load of 20 kips. (*a*) Locate the instantaneous center of rotation. (*b*) Compute the resultant load on fastener A.

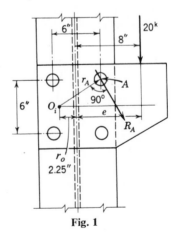

Fig. 1

Solution

The instantaneous center O_i is located at a distance r_o from the centroid.

$$r_o = -\frac{\sum Ar^2}{e \sum A}$$

or for fasteners of equal area:

$$r_o = - \frac{\sum r^2}{en}$$

$$r_o = - \frac{\sum (x^2 + y^2)}{en} = \frac{4(3^2 + 3^2)}{8(4)} = 2.25 \text{ in.}$$

The resultant load on fastener A is

$$R_A = \frac{P(e + r_o)(r_A)}{\sum r_i^2}$$

$$= \frac{20(8 + 2.25)\sqrt{(3 + 2.25)^2 + 3^2}}{2[(5.25^2 + 3^2) + (0.75^2 + 3^2)]} = 13.4 \text{ kips}$$

The direction of the load is perpendicular to the radius vector r_A.

Design of Eccentric Load Connections. The design of eccentric load connections is based on the foregoing methods of analysis. It is, of course, desirable to design the connection so that the values of the extreme fasteners will be fully utilized and hence it is often necessary to try several designs in order to reach the most economical one.

It will be helpful to keep the following relations in mind when designing eccentric load connections:

   (*a*) When the load is near the centroid of the fastener group, the effect of direct load predominates.

   (*b*) When the load is outside the fastener group, the effect of moment predominates.

   (*c*) When the load is halfway between the centroid and the edge of the fastener group, the effect of direct load and moment are about the same.

For a general case of an eccentrically loaded connection the relationship between fastener load and the applied load may be derived in the following manner. Consider m lines of n fasteners per line, all fasteners having the same cross-sectional area A and uniformly spaced at q and p inches as shown in Fig. 5-27. In order to simplify calculation of quantities $\sum Ax^2$ and $\sum Ay^2$ the fasteners can be transformed into equivalent rectangles having the same properties. For calculating $\sum Ay^2$ each fastener is transformed into a rectangle having depth p and width (A/p), with the center of the rectangle coinciding with the center of the fastener. Thus, each line of fasteners is transformed into a continuous rectangle with dimensions np and (A/p). The entire pattern is transformed into m such rectangles having the following properties:

$$\sum A = mnp \left(\frac{A}{p}\right) = mnA \tag{5-28}$$

and

$$I_x = \sum (I_x' + Ay^2) = m\frac{(A/p)(np)^3}{12} = \frac{mp^2n^3A}{12} \qquad (5\text{-}29)$$

where I_x' is moment of inertia of each equivalent fastener rectangle about its own axis, that is,

$$I_x' = \frac{(A/p)p^3}{12} \qquad \text{and} \qquad \sum I_x' = mnI_x' \qquad (5\text{-}30)$$

Then

$$\sum Ay^2 = I_x - \sum I_x' = \frac{mp^2n^3A}{12} - \frac{mnp^2A}{12} = \frac{mnp^2A}{12}(n^2 - 1) \quad (5\text{-}31)$$

When n^2 is much greater than 1, $\sum Ay^2$ is defined approximately by

$$\sum Ay^2 = \frac{mn^3p^2A}{12} \qquad (5\text{-}32)$$

Similarly it may be shown that, approximately:

$$\sum Ax^2 = \frac{nm^3q^2A}{12} \qquad (5\text{-}33)$$

For a corner fastener, which is likely to be the most highly stressed:

$$x_i = \tfrac{1}{2}q(m - 1) \qquad \text{and} \qquad y_i = \tfrac{1}{2}p(n - 1) \qquad (5\text{-}34)$$

When n and m are much greater than 1 the values of x_i and y_i are approximately:

$$x_i = \tfrac{1}{2}mq \qquad \text{and} \qquad y_i = \tfrac{1}{2}np \qquad (5\text{-}35)$$

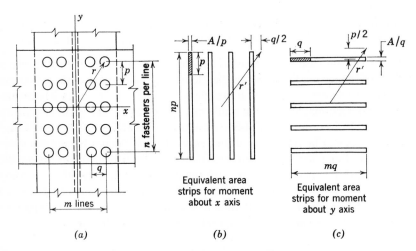

Fig. 5-27 Equivalent area of fastener group.

The load on the corner fastener can be expressed as follows:

$$R = [(R_x' + R_x'')^2 + (R_y' + R_y'')^2]^{1/2}$$

$$= \left[\left(\frac{P_x \cdot A}{\sum A} + \frac{My_i A}{\sum A(x^2 + y^2)} \right)^2 + \left(\frac{P_y \cdot A}{\sum A} + \frac{Mx_i A}{\sum A(x^2 + y^2)} \right)^2 \right]^{1/2}$$

$$= \frac{P}{mn} \left[\left(\frac{P_x}{P} + \frac{6e}{[(mq/np)^2 + 1]np} \right)^2 + \left(\frac{P_y}{P} + \frac{6e}{[(np/mq)^2 + 1]mq} \right)^2 \right]^{1/2}$$

$$= \frac{P}{C} \tag{5-36}$$

where $e = M/P$, and

$$C = \frac{mn}{\left[\left(\dfrac{P_x}{P} + \dfrac{6e}{[(mq/np)^2 + 1]np} \right)^2 + \left(\dfrac{P_y}{P} + \dfrac{6e}{[(np/mq)^2 + 1]mq} \right)^2 \right]^{1/2}} \tag{5-37}$$

Values of C can be computed and tabulated for given p, q, m, n, e, (P_x/P), and (P_y/P). Such tables have been compiled for simple patterns with common values of gage and pitch and are given in the AISC Manual, pp. 4–52 to 4–55. When such tables are available, the procedure for design is as follows:

(a) Required C is calculated as P/R_a, where R_a is the allowable value for the selected fastener.

(b) The connection having a value of C equal to or greater than the required value is selected.

5-12 BEHAVIOR OF RIVETS AND BOLTS IN TENSION

For many years engineers have been reluctant to permit the use of rivets in tension. The main argument was that it is relatively easy to chip off a rivet head, and although a chipped-off rivet is still good in shear, it cannot resist a tension load. With the use of power equipment in riveting, the quality of rivets improved considerably and defective rivets are now very rare.

Studies on rivets in tension[6,11,12] showed conclusively that rivets can be used safely in tension. Because hot-driven rivets are subjected to high tensile stresses due to cooling, it can be shown that the application of external pull on the rivets increases the tensile stresses only slightly.

To illustrate the behavior of fasteners in tension, consider an idealized case shown in Fig. 5-28a, in which a tensile load P is applied directly to the fastener head. In actual structures the tensile load can almost never be applied to the fastener head; it is convenient, however, to consider this

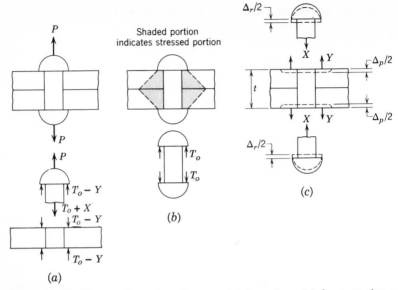

Fig. 5-28 Idealized fastener in tension—forces and deformation; (*a*) forces on plate and part of fastener, (*b*) stressed region in plates due to initial tension T_o in fastener, and (*c*) changes in deformations of fastener and plates.

idealized condition first and then evaluate the difference between the actual and the idealized cases.

Before an external load is applied each fastener head bears on the corresponding plate and the plates bear one against the other with a total compression equal to the initial tension in the rivet T_o (Fig 5-28*b*). Consider a tension load P applied to the fastener head. The initial tension T_o will increase by an amount X; the fastener will elongate slightly and thus relieve the total compression on the plates by an amount Y. If Y is less than the initial load T_o, then all the parts will still be in contact and the elongation of the rivet must equal the expansion of the plates   Figure 5-28*c* shows the free bodies and deformations of the fastener and the plate, as produced by X and Y:

For the fastener, $$\Delta_r = \frac{Xt}{(AE)_r} \qquad (a)$$

$$(5\text{-}38)$$

For the plates, $$\Delta_p = \frac{Yt}{(AE)_p} \qquad (b)$$

where t is the grip of the fastener, A is the effective area resisting the load, E is the elastic modulus of the material, and r and p are subscripts denoting fastener and plate, respectively   Consider static equilibrium of the fastener (Fig 5-28*a*):

$$X + Y = P \qquad (5\text{-}39)$$

For geometric compatability, $\Delta_r = \Delta_p$, and solving the foregoing relations for X, we obtain:

$$X = \frac{P}{1 + (AE)_p/(AE)_r} \tag{5-40a}$$

It is seen from Fig 5-28b that the effective cross-sectional area A_p is not constant throughout the thickness of the plate and its precise magnitude is indeterminate. In most practical instances, however, it can be reasonably *assumed* that $(AE)_p/(AE)_r = 10$　Thus,

$$X = \frac{P}{11} \tag{5-40b}$$

which indicates an increase in the tensile load in the fastener of only about 10% of the applied load. If the fastener is stressed beyond the elastic limit, the effective value of $(AE)_r$ becomes still smaller, as does the value of X. Thus externally applied tensile load will increase the fastener tension only slightly, under normal conditions.

In the previous paragraph it is assumed that Y is less than T_o and the plates are slightly compressed by the tension in rivets. If P is big enough to produce Y greater than T_o, that is, to cause an elongation in the fastener greater than the initial compressive deformation in the plates, then the fastener heads and the plates will not maintain contact. This means that the plates will be stress-free, and the fastener will take the entire load P. This condition will exist when $Y = T_o$ or when P is approximately $1.1T_o$ for rivets. It should be pointed out that, since the initial tensile load T_o produces stresses in the hot-driven rivets approaching the yield-point stress, the external load P should not be allowed to exceed the yield point.

For bolted joints, similar to riveted joints, the increase in the initial tensile stress is small if the external tension load does not exceed the clamping force T. If the external tension load exceeds the yield strength of the bolt shank, then the plates may separate and the bolt may elongate plastically. Then, upon removal of this high load, the initial tension in the bolt could be entirely lost or reduced materially because of the permanent set in the bolt. Questions have been raised concerning the effect of creep on the initial tension of the bolt over a long period of time, although tests seem to indicate that it is not significant.

So far only the idealized case of a fastener in tension has been considered. The difference between this and the behavior of an actual fastener is primarily due to the fact that the tension load P is not applied to the fastener head, but is usually transmitted through a connecting angle or flange of a T section, thus causing some bending deformations in the plate and changing the distributions of bearing stresses between the fastener head and the plate, shown in Fig. 5-29,

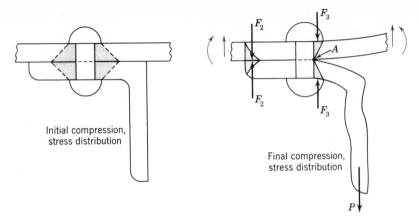

Fig. 5-29   Actual fastener in tension.

wherein it is apparent that the distribution of bearing stress is materially affected by the bending of the plate. Because of bending in the plate, forces F_2 and F_3 can never be reduced to zero and thus the initial compression in the plates can never be completely relieved. Therefore, at the time of incipient separation at point A the average expansion in the plate Δ_p will be less than that for the idealized case of Fig. 5-28 and Δ_r, the elongation in the fastener, will also be less than in the idealized case.

The design of connection with a single rivet or high-strength bolt in tension is based on a nominal tensile stress which is defined as follows:

$$f_t = \frac{P}{A_r} \tag{5-41}$$

where P is the tension load and A_r is the cross-sectional area of the fastener. This practice neglects the initial tension stress. However, as previously indicated, if P is less than the initial load then the initial tension stress on the fastener will be increased only slightly and, if P exceeds the initial tension load, then the actual tensile stress is equal to the nominal stress defined in Eq. 5-41. Test results and experience indicate that rivets and high-strength bolts can withstand static tension loads satisfactorily. When subjected to high tension load, particularly repetitive loads, rivet elongations and head deformations may cause loss of "clamping action," that is, loosening of the rivets, whereas high-tensile bolts are much better in that respect.

Example 5-3.   A 1-in. rivet is driven through three plates as shown. Assume that the rivet shrinks from a temperature of 500°F to room temperature of 70°F and that the coefficient of expansion of steel is 6.5×10^{-6} per degree Fahrenheit (Fig. 1).

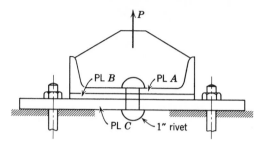

Fig. 1

(a) Neglecting compressive strain in the plates, compute the total stress in the rivet ($E = 30 \times 10^6$ psi).

(b) If a force $P = 10$ kips is applied to plate A, draw free-body diagrams of plates A, B, and C and of the rivet, showing all the forces thereon.

(c) If P increases to 30 kips, what is the resulting stress on the rivet?

Assume that the yield point of steel $= 30$ ksi. In both cases b and c, consider initial tension in the rivet.

Solution

(a) Unit shortening of the rivet produced by temperature drop:

$$\delta = (500 - 70)(6.5 \times 10^{-6}) = 2800 \times 10^{-6} \text{ in./in.}$$

Assuming this shortening is prevented from taking place and E independent of temperature, a tensile stress f_t will be produced:

$$f_t = E\delta = 30 \times 10^6 \times 2800 \times 10^{-6} = 84,000 \text{ psi}$$

This value is not realistic since it exceeds the yield point. Partly due to reduction in E at elevated temperatures and partly due to deformation in the plates, the initial tension in the rivets is about 30,000 psi and the clamping force is $30,000 \times 0.785 = 23,500$ lb.

(b) For an additional force of 10 kips applied to plate A, the force will be taken mainly by the relief of compression between the plates. The rivet, having been stressed to the yield point, can be assumed to yield slightly more without carrying additional load.

Hence the free-body diagrams will be as in Fig. 2.

(c) For $P = 30$ kips, the rivet will take the total load and the contact pressure between the plates becomes zero.

The stress in the rivet becomes

$$f = \frac{P}{A} = \frac{30,000}{0.785} = 38,200 \text{ psi}$$

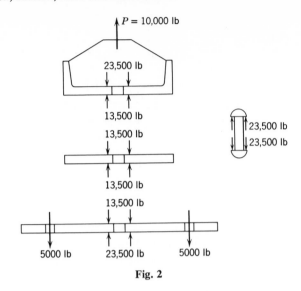

Fig. 2

5-13 FASTENER IN TENSION AND SHEAR

Rivets and bolts in shear are actually under tensile stress as well because of the existence of initial tension. These fasteners, however, are never analyzed for combined stress because experience shows that fasteners designed separately for shear and bearing will usually withstand the high combined stresses. When fasteners are subjected simultaneously to shear and external tension loads, it is then the customary practice to consider the combined effects.

In addition to tensile and shearing stresses, the fasteners may be subjected to bearing stress, although not every portion of the fastener is subjected to these three stresses at the same time. The present practice is to neglect the bearing stresses in considering the effect of combined stresses.

The actual state of stress in a fastener is rather complex. It is known, however, that, under combined loads, the external tension load will tend to decrease the frictional resistance between the plates, resulting in a loosening effect which might be serious, especially if the fasteners are subject to repeated or reversed shear loads. And, under high shear loads, the fasteners may be distorted and their capacity to carry additional tension load is thus reduced.

The present practice is to limit combined tension and shear loads by some arbitrary method. It is based on a consideration of the nominal tensile and shearing stresses only, neglecting initial tension, friction, and bearing. An empirical procedure for limiting combined stresses is based on "interaction."

The general form of the interaction equation follows:

$$A\left(\frac{f_{tu}}{F_{tu}}\right)^{x} + B\left(\frac{f_{vu}}{F_{vu}}\right)^{y} \leq 1.0 \qquad (5\text{-}42a)$$

where A, B, x, and y are empirical constants and where F_{tu} and F_{vu} are the ultimate tensile and shearing strength, respectively, when each is considered acting by itself, and f_{tu} and f_{vu} are the ultimate tensile and shearing stresses at failure. Preliminary studies[13] indicate that an equation with $A = B = 1$ and $x = y = 2$ gives reasonable agreement with test results, as in Fig. (5-30).

Transforming the foregoing ultimate-strength formula into allowable stresses, we obtain

$$\left(\frac{f_{t}}{F_{t}}\right)^{2} + \left(\frac{f_{v}}{F_{v}}\right)^{2} = 1.0 \qquad (5\text{-}42b)$$

where F_{t} and F_{v} are the allowable tensile and shearing unit stresses, respectively, when each is considered acting by itself, and f_{t} and f_{v} are the nominal unit stress components resulting from combined action. Thus the AASHO Specification has forbearing type connections.

$$\left(\frac{kf_{t}}{F_{v}}\right)^{2} + \left(\frac{f_{v}}{F_{v}}\right)^{2} = 1 \qquad (5\text{-}42c)$$

where $k =$ a constant: 0.75 for rivets and A325 bolts with threads in shear plane; 0.555 for A325 bolts with threads excluded from shear plane.

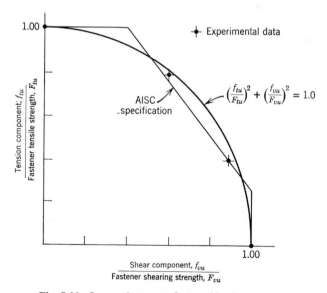

Fig. 5-30 Interaction curve for combined stresses.

AISC Specification, Sec. 1.6.3, approximates the interaction relationship by three linear functions, as shown in Fig. 5-30 and expressed by the following equations

$$\frac{f_t}{F_t} = A_1 - B_1 \frac{f_v}{F_v}, \qquad \frac{f_t}{F_t} \leq 1.0 \quad \text{and} \quad \frac{f_v}{F_v} \leq 1.0 \qquad (5\text{-}43a)$$

These equations also may be written in the following form:

$$f_t = A - Bf_v, \qquad f_t \leq F_t, \qquad f_v \leq F_v \qquad (5\text{-}43b)$$

Values of the coefficients A and B for various fasteners are specified as follows:

	A	B
A141 rivets	28,000	1.6
A195 rivets	38,000	1.6
A307 bolts	20,000	1.6
A325 bolts, bearing-type connection	50,000	1.6

For friction-type connections using A325 or A490 bolts, the following interaction equation is specified:

$$\frac{f_v}{F_v} + \frac{f_t}{F_{to}} \leq 1.0 \qquad (5\text{-}44a)$$

where F_{to} is the tension stress in the bolt due to the proof load: $F_{to} = T_b/A_b =$ proof load/bolt area; and F_v is the allowable shear stress for the fastener in the absence of external tension load. For example, this equation can be transformed for the A325 bolts into the following:

$$f_v \leq 15,000 \left(1 - \frac{f_t A_b)}{T_b}\right) \qquad (5\text{-}44b)$$

AASHO specifies the following for friction-type A325 connections:

$$F_v = 13,500 - 0.22 f_t.$$

5-14 PRYING ACTION IN CONNECTIONS WITH FASTENERS IN TENSION

A connection transmitting loads by means of fasteners in tension subjects the connecting elements to bending stresses and the fasteners to prying action (Fig. 5-29). Typical cases shown in Fig. 5-31 indicate the manner in which connecting angles, tee-shapes, or beam flanges are subjected to bending.

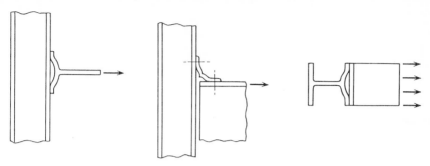

Fig. 5-31 Bending of connection elements.

In calculating the tension in the fastener the additional load resulting from the prying action must be considered. This load depends on the relative stiffness of the fastener and the connection material.

In evaluating the bending stresses in the connection material and the additional tension in the fasteners two possibilities may be considered: (*a*) fastener elongation is relatively large and the connected elements are not "clamped" by the fasteners, and (*b*) fastener elongation is relatively small and the connected elements are considered fully clamped.

In the first instance (Fig. 5-32), the tension in the fastener acts as a con-centrated load on the cantilevered portion of the connecting element. The maximum bending stress occurs at the toe of the fillet and the bending stress is

$$f_b = \frac{Mc}{I} = \frac{(Ta)(t/2)}{(bt^3/12)} = \frac{6Ta}{bt^2} \tag{5-45}$$

where b is the effective length of the connecting element subjected to load T.

In the second instance, when clamping is considered, the solution is more complicated because the problem becomes statically indeterminate. Since the relative stiffness of the fasteners and the connecting elements depend on

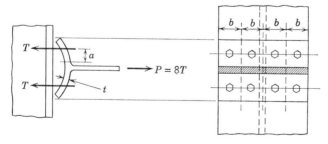

Fig. 5-32 Bending without clamping action.

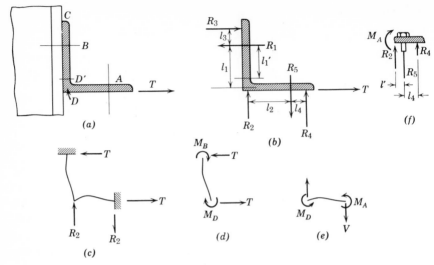

Fig. 5-33 Distortion of connection angle.

the initial flatness of the shapes and on the degree of actual clamping attained during fastener installation, only approximate solutions can be obtained. For example (Fig. 5-33), assuming that clamping is sufficient to prevent rotation at sections A and B, an elastic solution based on simple frame action of the elements leads to the following relations:

$$M_A = \frac{T l_1^2}{4 l_1 + l_2}, \qquad M_B = M_A \frac{(2 l_1 + l_2)}{l_1}, \qquad M_D = 2 M_A \qquad (5\text{-}46)$$

When $l_1 = l_2 = l$,

$$M_A = 0.2 T l, \qquad M_B = 0.6 T l, \qquad M_D = 0.4 T l \qquad (5\text{-}47)$$

The tension and shear loads in the fasteners depend on the distribution of stresses due to prying action and particularly on the magnitude and relative eccentricity of stress resultants R_3 and R_4.

Considering equilibrium of free-bodies shown in Figs. 5-33e and f the following are obtained:

$$R_2 = \frac{M_A + M_D}{l_2} \qquad \text{and} \quad R_4 = \frac{R_2 l' + M_A}{l_4}$$

Similarly

$$R_3 = \frac{T l' + M_B}{l_3}$$

Then the total loads on the fasteners are as follows:

On fastener at A: tension $R_5 = R_2 + R_4$; shear $= T$
On fastener at B: tension $R_1 = T + R_3$; shear $= 0$

A "plastic" solution, assuming that plastic hinges form at B and D' (that is, $M_B = M_{D'} = M_p$), leads to the following:

$$M_B = M_{D'} = 0.5Tl_1', \qquad M_D = M_{D'} + T(l_1 - l_1')$$

and $M_A = 0.5M_D$. The expressions for $R_1 - R_5$ are governed by statics and therefore apply to the plastic solution as well as the elastic one.

The AISC Manual (p. 4-67) suggests that the plastic method be used for evaluation of moments but uses linear elastic stress distribution due to bending with a limiting allowable bending stress of 22 ksi. Based on these assumptions values of allowable load governed by bending in the connection elements were calculated and are tabulated in the Manual.

It is interesting to compare the values of allowable load T_a recommended by AISC and the values of load T_y at initial yielding and load T_p at plastification at B.

Using AISC recommendations:

$$T_a = \frac{F_b bt^2}{6(0.5l_1')} = \frac{F_y bt^2}{1.6(3l_1')} = \frac{F_y bt^2}{4.8l_1'} \tag{5-48}$$

At initial yielding:

$$T_y = \frac{F_y bt^2}{6(0.6l_1)} = \frac{F_y bt^2}{3.6l_1} \tag{5-49}$$

At plastification at B:

$$T_p = \frac{F_y bt^2}{4(0.5l_1')} = \frac{F_y bt^2}{2l_1'} \tag{5-50}$$

It appears that the recommended AISC allowable load T_a is approximately equal to $0.42T_p$ and to $0.75\,(l_1/l_1')T_y$. Since the ratio (l_1/l_1') normally varies from 1.5 to 2.0, the value of T_a ranges from about 1.1 to 1.5 of initial yielding load T_y. However, as the allowable load still appears to be less than half of the "plastic" load T_p, the fact that T_a exceeds initial yield load should not cause distress in the structure under normal conditions.

5-15 MOMENT AND SHEAR CONNECTIONS

In design of rigid frames it is usually assumed that the connections are able to transmit both moment and shear, subjecting fasteners to combined shearing and tension stresses. Typical connections used in such frames are shown in

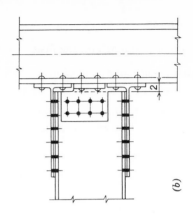

(b)

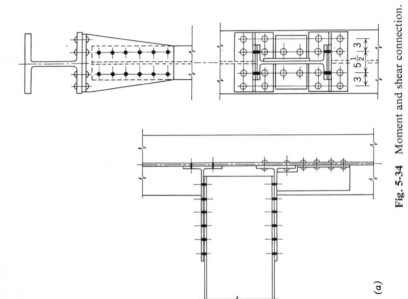

(a)

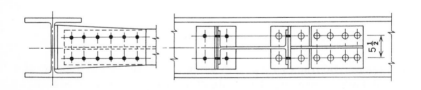

Fig. 5-34 Moment and shear connection.

158

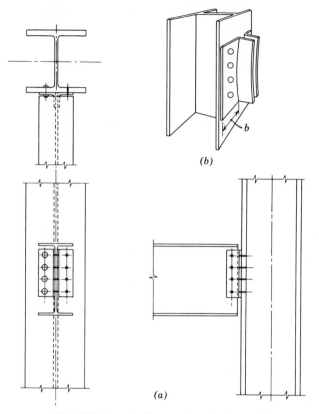

Fig. 5-35   Standard connection framing angle.

Fig. 5-34. In these connections it may be assumed that the moment is resisted by the two tee-shapes, with the tension and compression loads equal to M/h. The shear load may be assumed to be distributed equally to the fasteners.

A standard framing angle connection (Fig. 5-35a) is designed to transmit only shear. Since it is usually made up of flexible angles, it cannot transmit large moments. However, since the beam tends to rotate at the connection, the top fasteners between the angles and the column are subjected to high tension and some moment is developed[14,15,16] (Fig. 5-35b). It is common practice to neglect such moments in the design of the beam and the connections. This approximation yields a conservative design for the beam section but may not be satisfactory for the connections, especially if fatigue loadings are involved.

To compute tension loads in the fasteners as produced by the moment, three methods have been used: (a) neglecting initial tension in the fastener, (*b*) initial tension not exceeded in any fastener, (*c*) initial tension exceeded in some fasteners.

Neglecting Initial Tension in Fasteners. The moment is assumed to be transmitted through the connection by tension in the top fastener and bearing between the lower part of the angles and the column. Because initial tension is neglected, it is assumed that on the tension side the angles are separated from the column flange. The areas effective in transmitting the tensile stresses in the fastener and the bearing stresses between the angles and the column flanges are shown in Fig. 5-36. The position of the neutral axis can be determined provided that the effective width b_e of the bearing area is defined. For an approximate solution it is sufficient to assume that the legs of the angle are but little distorted in bearing. Hence, the equivalent section is defined by using fastener areas on the tension side and full bearing area on the compression side. It should be noted that the maximum tensile stress is not very sensitive to variation of b_e/b.

The fastener areas may be replaced by an equivalent rectangle, so that the width a of the equivalent area is

$$a = \frac{A}{p} m \qquad (5\text{-}51)$$

where A is fastener cross-sectional area, p rivet spacing, and m number of rivet rows. The neutral axis must pass through the centroid of this equivalent section and thus

$$\tfrac{1}{2}ac^2 = \tfrac{1}{2}b_e c_1^2 \qquad \text{or} \qquad \frac{c_1}{c} = \left(\frac{a}{b_e}\right)^{1/2} \qquad (5\text{-}52)$$

In most practical problems ratio of c_1/c varies between $\tfrac{1}{4}$ and $\tfrac{1}{8}$. Since Eq. 5-51 is only an approximation, the value of ratio c_1/c need not be computed but

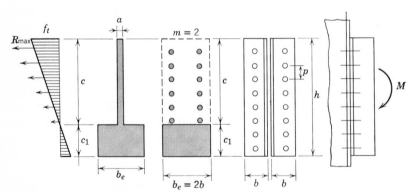

Fig. 5-36 Effect of moment on a framing angle connection neglecting initial tension.

is often assumed to be $\frac{1}{6}$. Thus, the neutral axis is located a distance $h/7$ from the compression end of the connection.

Having located the neutral axis, the maximum tensile stress f in the extreme fiber of the equivalent section can be found from Eq. 5-53.

$$f_t = \frac{Mc}{I} \tag{5-53a}$$

where

$$I = \frac{ac^3}{3} + \frac{b_e c_1{}^3}{3}$$

and

$$R_{\max} = f_t A \tag{5-53b}$$

Since the extreme fastener is located at an edge distance of approximately $p/2$ from the extreme fiber of the substitute section, distance $(c - p/2)$ should be used instead of c in Eq. 5-53a. In an approximate analysis it is hardly necessary to make this refinement unless the connection consists of only two or three fasteners in a line, where the difference might be significant. The tensile load in the fastener is defined by Eq. 5-53b and the tensile stress f_t must then be combined with the shearing stress f_v when checking the adequacy of the connection.

Initial Tension Not Exceeded in Any Fastener. When initial tension in the fasteners is considered and is not exceeded by the external load, the connecting angles will remain in complete contact with the column. The elastic properties of the connection in bearing are almost uniform and the neutral axis corresponding to points of no change in bearing stress is approximately at midheight of the connection.

Because of initial tension, the angles are clamped to the column flanges with resulting initial bearing or compression stress f_o effective over the entire contact area A_o, which may be taken as $2bh$ (Fig. 5-37). The change in the bearing stress f_o due to the applied moment M can be calculated as

$$\Delta f_o = \frac{My}{I_o} = \frac{My}{2bh^3/12} = \frac{6My}{bh^3} \tag{5-54a}$$

Effective tension load on the fastener is equal to the change in the bearing stress times the contributory contact area for the particular fastener. Thus, effective tensile load T_o on the extreme fastener, for which y is approximately equal to $h/2$, is

$$T_o = \Delta f_o bp = \frac{6Mh}{2bh^3} bp = \frac{3M}{h^2} p \tag{5-54b}$$

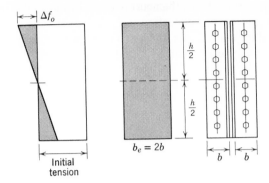

Fig. 5-37 Effect of moment on framing angle connection considering initial tension.

where p = pitch between fasteners. Hence, the nominal tension stress f_t in the extreme fastener is

$$f_t = \frac{T_o}{A_r} = \frac{3Mp}{h^2 A_r} \qquad (5\text{-}55)$$

By comparing values from Eq. 5-53a with Eq. 5-55, it can be shown that the latter yields higher computed stresses and hence is more conservative.

Initial Tension Exceeded in Some Rivets. This seldom occurs in structural design and its solution is complicated. However, if the initial tension in the fastener happens to be small, it might be exceeded in some fasteners. In such a case, the line of zero stress in the connection does not coincide with the midheight of the connection. The results obtained are somewhere between the two methods previously described. Although it is possible to derive formulas for this case, solution by means of successive approximations would be more practical.

Example 5-4. The connection shown (7-row AISC Table 1, p. 4-13), (Fig. 1) is subjected to a shear load $V = 75$ kips and a moment $M = 50$ kip-in. at the connection. Determine the combined stresses on bolt A, assuming no initial tension. A307 bolts, A36 steel.

Solution

The shear load on bolt A is

$$R_v = \frac{V}{n} = \frac{75}{14} = 5.36 \text{ kips}$$

and the shearing stress is

$$f_v = \frac{R_v}{A_v} = \frac{5.36}{0.60} = 8.93 \text{ ksi}$$

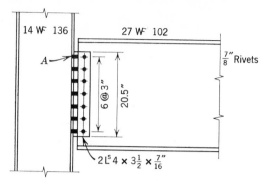

Fig. 1

The tensile load on bolt A, neglecting initial tension, and the effective areas can be determined (Fig. 2):

$$a = \left(\frac{A}{p}\right) m = \left(\frac{0.6}{3}\right) 2 = 0.4 \text{ in.}$$

$$b_e = 2b = 2(4.0) = 8.0 \text{ in.}$$

$$\frac{c_1}{c} = \frac{\sqrt{a}}{\sqrt{b_e}} = \frac{\sqrt{0.4}}{\sqrt{8.0}} = 0.22$$

$$h = \text{(approx.) } 20.5 \text{ in.}$$

$$\therefore \quad c_1 = 3.7 \text{ in.}, \qquad c = 16.8 \text{ in.}$$

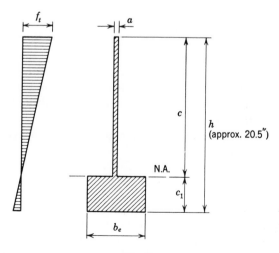

Fig. 2

$$I = \frac{ac^3}{3} + \frac{b_e c_1^3}{3}$$

$$= \frac{(0.4)(16.8)^3 + (8.0)(3.7)^3}{3}$$

$$= 767 \text{ in.}^4$$

The section modulus is

$$S = \frac{I}{c} \simeq \frac{767}{16.8} = 45.6 \text{ in.}^3$$

and the tensile stress

$$f_t = \frac{M}{S} = \frac{50}{45.6} = 1.10 \text{ ksi}$$

Discussion. Assuming initial tension, the maximum tensile stress in the extreme rivet given by Eq. 5-55:

$$f_t = \frac{3Mp}{h^2 A_r} = \frac{3 \times 50 \times 3}{20.5^2 \times 0.6} = 1.78 \text{ ksi}$$

which is higher than the previous value of 1.10 ksi.

5-16 DESIGN OF PIN CONNECTIONS

Two types of tension members that may be used with pin connections are eye-bars and conventional plates or shapes.

An economical design is frequently obtained by determining the pin size on the basis of its bending strength. This procedure may result in a pin size too small to develop the bearing strength of the members in the connection, in which case local reinforcing plates may be used. Some specifications provide that the pin diameter shall not be less than a specified fraction r of the width of the widest member in the connection; for example, AREA provides for $r = 0.8$. When the members consist of plates rather than shapes, such arbitrary restrictions of pin size may result in a very large pin and an uneconomical design.

For a given bending moment M, the required pin diameter d is obtained from Eq. 5-56 as follows:

$$d = \sqrt[3]{\frac{10.2M}{F}} \tag{5-56}$$

where F is the allowable bending stress for pins. The conventional manner of calculating the bending moment M is to assume that the loads are concentrated along the midthickness line of the members. This is evidently conservative and partially accounts for the high allowable stress in bending.

Actually the loads are distributed and it can be assumed that the distribution is more or less uniform over the thickness of the bearing plates so that the maximum moment is appreciably reduced (Fig. 5-16).

Members of a frame or truss connected by a pin do not usually lie in one plane passing through the pin axis and the determination of maximum moment on the pin must be made on the basis of the noncoplanar forces. A judicious arrangement of members on the pin may reduce the maximum bending moment effectively.

Although eye-bars are rarely used in modern steel construction, they may be economical for certain types of tension members in large frames. The proportioning of eye-bars is governed by the size of pin and the cross-sectional area of the main body of the eye-bar required to transmit the load. In addition the following relationships between the various dimension of the eye-bar (Fig. 5-38) and pin are given in the AISC Specification (see Sec. 1.14.6, p. 5-33).

1. $d_p \geq \frac{7}{8}b$
2. $b \leq 8t$
3. $t \leq \frac{1}{2}$ in.
4. $r_t \geq d_b$
5. $1.33b \leq (d_b - d_p) \leq 1.5b$

where all the dimension symbols are as shown in Fig. 5-38.

In designing pin connections the stresses in the plates around the hole must be considered. Tension plates must be so proportioned as to provide adequate net sections across and back of the pin, and in all instances sufficient thickness for bearing on the pin must be provided. In order to satisfy these requirements, it is often necessary to provide plate reinforcements, sometimes called "boss plates" or "pin plates," in order to increase bearing strength and the net section around the hole (Fig. 5-39).

Determination of stresses in the plate around the pin hole is complex. The designer should be aware of the types of stresses that are resisted by the

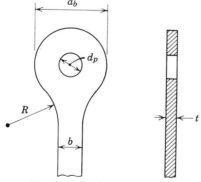

Fig. 5-38 Eyebar proportions.

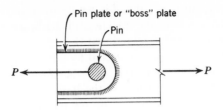

Fig. 5-39   Reinforcement around pin hole.

critical sections and must proportion the plates to minimize the stresses due to these loads. The general nature of the stresses acting on these critical sections are shown in Fig. 5-40; the magnitude of the forces, moments, and stresses will depend on the proportions of the plate and the distribution of the bearing stresses around the hole.

Although the stress distribution across net section *A-A* is not uniform, $f_{t_1} > f_{t_2}$ (Fig. 5-40), the usual procedure is to design for an average stress $f_t = P/A_n$, where A_n is the net cross-sectional area. In recognition of non-uniform distribution of stress, however, the allowable tension stress on the net section at pin holes is

$$F_t = 0.45F_y$$

a reduction of 25% from the usual $0.6F_y$ used in other connections. This reduction in allowable stress is equivalent to the requirement that the net section across the pin hole shall be not less than 1.33 times the net section of the body of the member. The net section beyond the pin hole, parallel to the axis of the member, shall not be less than two-thirds of the net section across pin hole, or about 0.9 times the net section of the body of the member. The end corners of the plate may be cut at 45° to the axis of the member, provided that the section normal to the cut is not less than that required directly beyond the pin hole.

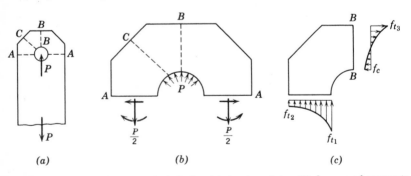

Fig. 5-40   Stresses in plate around pin hole; (*a*) tension plate, (*b*) forces and moments, and (*c*) normal stresses.

Example 5-5. An 8-in. pin in the lower chord of a highway bridge truss is acted on by forces from connecting members as shown in Fig. 1. Compute the maximum shearing, bearing, and bending stresses, and compare them with the allowable values as given by the AASHO Specification. Pin material is structural steel. $F_y = 36$ ksi.

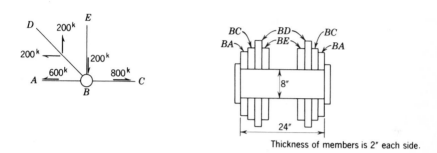

Thickness of members is 2″ each side.

Fig. 1

Solution

From Fig. 2, critical sections are:

Maximum shear between members BA and BC: $V = 300$ kips
Maximum bearing due to member BC: $P = 400$ kips
Maximum moment at load BC: $M = 600$ kip-in.

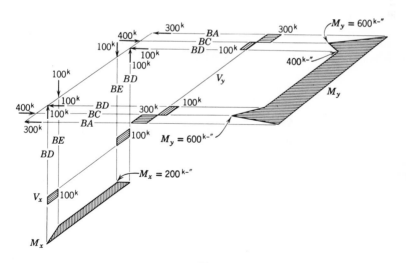

Fig. 2

For section properties of 8-in. pin, see AISC Manual, 1957, p. 269.

(a) *Shear:* $f_v = \dfrac{V}{A} = \dfrac{300}{50.3} = 5.97$ ksi,　　Allowable 14 ksi

(b) *Bearing:* $f_b = \dfrac{P}{A_b} = \dfrac{400}{16} = 25$ ksi,　　Allowable 29 ksi ⎬ AASHO

(c) *Bending:* $f = \dfrac{M}{S} = \dfrac{600}{50.3} = 11.94$ ksi,　　Allowable 29 ksi

The pin is satisfactory.

Example 5-6. The load on the end pin bearing, supporting a small bridge is 440 kips. Using the AASHO allowable stresses given in Table 5-4 for structural steel, determine the size of the pin required.

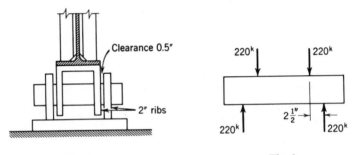

Fig. 1　　　　　　　　　　Fig. 2

Solution
(a) *Bending* $M = 220 \times 2.5 = 550$ kip-in., assuming concentrated loads.

$$F = \frac{Mc}{I} = \frac{M}{\pi r^3/4} = 27 \text{ ksi}$$

$$\therefore \quad r^3 = \frac{550}{(\pi/4)(27)} = 25.9 \text{ in.}^3, \quad r = 2.96 \text{ or } 3 \text{ in.}$$

$\therefore d = 6$ in.
(b) *Shear.* $V = 220$ kips.

$$F_v = \frac{V}{A} = \frac{V}{\pi r^2} = 13.5 \text{ ksi}$$

$$\therefore \quad r^2 = \frac{220}{(13.5)\pi} = 5.18 \text{ in.}^2, \quad r = 2.28 \text{ in.}$$

$\therefore d = 4\frac{5}{8}$ in.

(c) *Bearing.* $P = 220$ kips.

$$F_b = \frac{P}{dt} = 24 \text{ ksi}$$

$$\therefore \quad d = \frac{220}{2(24)} = 4.58 \text{ in.}$$

$\therefore d = 4\frac{5}{8}$ in.

Bending is critical. Use $d = 6$ in.

PROBLEMS

1. Two $\frac{1}{2}$-in. steel plates are spliced by two $\frac{1}{4}$-in. plates with six $\frac{7}{8}$-in. rivets in punched holes (Fig. P-1). Determine the ultimate load P as controlled by (a) shearing in the rivets at a strength of 50 ksi, (b) bearing between rivets and plates at a strength of 95 ksi, (c) tension across net section of main plates and net section of splice plates at 65 ksi. If this connection is to be designed according to the AISC Specification for Buildings, what is the allowable load P and what factors of safety are provided for the three types of failures just mentioned?

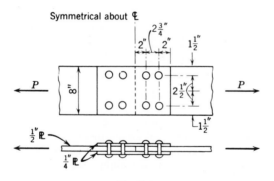

Fig. P-1

2. An 18-in. channel 45.8 lb is connected to a $\frac{5}{8}$-in. plate by $\frac{7}{8}$-in. rivets along the web of the channel, transmitting a tension load of 198 kips (Fig. P-2). (a) Check whether the rivets are sufficient. (b) What shall be the minimum spacing s to meet the net section requirement for the channel? (AISC Specification.)

3. The lower chord joint of a roof truss is shown in Fig. P-3, with a continuous chord member. Forces in two members are given and there is no load applied at the joint. Design and detail the joint using $\frac{3}{4}$-in. rivets. Note that the centroidal lines of the three members must meet at one point. (AISC Specification.)

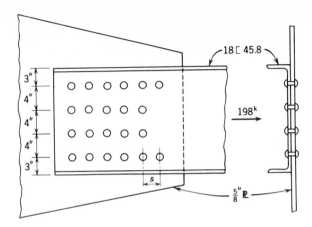

Fig. P-2

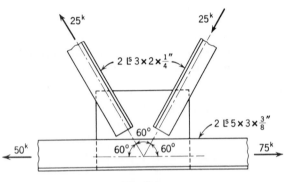

Fig. P-3

4. A connection has four rivets of the same size and carries an eccentric load *P* (Fig. P-4). Using the conventional elastic theory (assuming rigid plates and elastic rivets) compute the maximum rivet load. What is the maximum rivet load if rivet 1 is omitted from the connection resulting in a concentric load *P*?

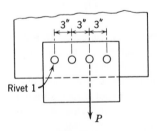

Fig. P-4

5. A bracket plate is connected to a column with four rivets as shown and carries a load of 20 kips (Fig. P-5). Compute the size of rivets required. (AISC Specification.)

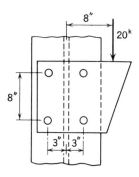

Fig. P-5

6. Two plates are connected with four rivets of the same size (Fig. P-6). The moment, shear, and direct load at the section *MM* of one plate is shown. Determine the size of rivets required for the connection. (AISC Specification.)

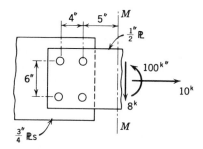

Fig. P-6

7. An ST 18 W^F 150 lb is connected to a 14 W^F 142 column by means of four 1-in. rivets (Fig. P-7). The initial tension in the rivets is known to be 25,000 psi. An external load *P* is then applied to the tee. Estimate the actual tensile stress in the rivets for (*a*) $P = 30,000$ lb, (*b*) $P = 60,000$ lb, (*c*) $P = 120,000$ lb. If the yield-point stress of the rivet steel is 33,000 psi, what load can the rivets carry at yielding? What do you think should be the permissible load?

8. A connection shown has rivets acting in both shear and tension (Fig. P-8). Determine the size of rivets required (AISC Specification) making suitable assumptions whenever necessary. For example, it may be assumed that the resisting couple is furnished by the axial force in the rivets acting with a lever arm of 4-in. between them.

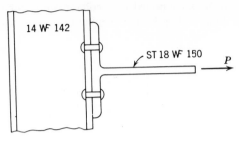

Fig. P-7

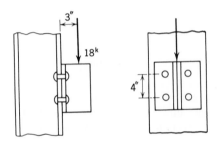

Fig. P-8

9. An 18 W̅ 50 beam is connected to a column by two split tees ST 12 W̅ 38 (Fig. P-9) to transmit a moment M. It is usually assumed that the angles carry the shear but are ineffective in resisting the moment. Compute the maximum allowable moment M as controlled by (*a*) the rivets, (*b*) the split tees—$\frac{7}{8}$-in. rivets. (AISC Specification.)

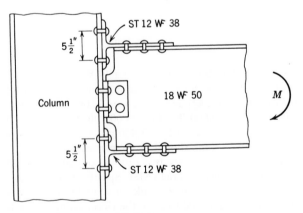

Fig. P-9

10. A 14 W^F 30 beam supports a small crane (Fig. P-10) and is connected to the supporting columns by standard 3R connections. (*a*) Compute the permissible load W as controlled by the shear capacity of the connection. (AISC Specification.) (*b*) Compute the moment capacity of the connecting angles at the start of yielding; compare this moment capacity with the simple beam midspan moment produced by the load W and the weight of the beam. (*c*) Discuss the correctness of conventional design procedure for the beam which assumes simple beam action. Discuss the possibility of fatigue failure in the connections.

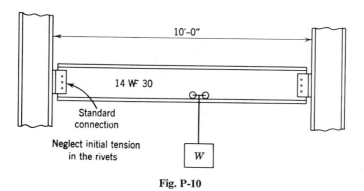

Fig. P-10

11. Compute the total allowable shear for a standard 7R connection (AISC Manual), but instead of rivets use (*a*) turned bolts, (*b*) unfinished bolts,

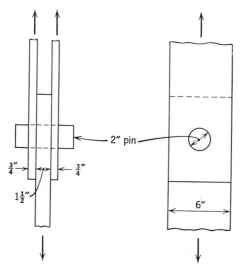

Fig. P-13

(*c*) high-tensile-strength bolts. Assume suitable allowable stresses. Refer to AISC Specification, when applicable.

12. Solve Prob. 1 and 2, assuming $\frac{7}{8}$-in. high-tensile-strength bolts instead of rivets.

13. For the pin-connected tension link shown in Fig. P-13, determine the load-carrying capacity of this link based on the following considerations: (*a*) net section strength of the link, (*b*) bearing on the pin, (*c*) shearing of the pin, (*d*) bending of the pin, assuming the loads to be concentrated at the center of the plates. (AISC Specification.)

14. A bottom-chord tension member of a bridge truss is made of two 12-in. channels 25 latticed together at 11-in. apart back to back. Design a pin connection to transmit a force of 250 kips, using reinforced plates as required. Choose a pin size, considering bearing, shearing, and bending stresses. Detail the pin plates, using $\frac{7}{8}$-in. rivets.

15. A $1\frac{3}{4}$-in. upset steel rod is to have clevises for connection (AISC Manual) to a $\frac{3}{4}$-in. gusset plate to transmit the full design load of the rod. Design a pin for the joint, considering shearing, bearing, and bending stresses. Note that the grip of the clevis equals plate thickness plus $\frac{1}{4}$-in.

REFERENCES

1. Richard de Jonge, A. E., "Riveted Joints," bibliography, American Society of Mechanical Engineers, 1945.
2. Fairbairn, William, "Experimental Inquiry into the Strength of the Wrought Iron Plates and Their Riveted Joints as Applied to Ship Building and Vessels Exposed to Severe Strain," *Phil. Trans. Roy. Soc. London*, **140**, 677 (1850).
3. Bell, M. H., "High Strength Steel Bolts in Structural Practice," *Proc. ASCE*, No. 651 (1955).
4. Munse, W. H., "High Strength Bolting," *Eng. Journal, AISC*, **4**, 1, (Jan. 1967).
5. Davis, R. E., G. B. Woodruff, and H. E. Davis, "Tension Tests of Large Riveted Joints," *Trans. ASCE* **105**, 1193 (1940).
6. Wilson, W. M., and W. A. Olive, "Tension Tests of Rivets," *Univ. Illinois Eng. Expt. Sta. Bull.* **210** (1930).
7. Hrennikoff, A., "Work of Rivets in Riveted Joints," *Trans. ASCE* **99**, 437 (1934).
8. Carter, J. W., "Stress Concentration in Built-up Structural Members," *AREA Bull.* **495** (1951).
9. Brady, W. G., and D. C. Drucker, "Investigation and Limit Analysis of Net Area in Tension," *Trans. ASCE* **120**, 1133 (1955).
10. Wilson, W. M., W. H. Munse, and M. A. Cayci, "A Study of the Practical Efficiency under Static Loading of Riveted Joints Connecting Plates," *Univ. Illinois Eng. Expt. Sta. Bull.* **402** (1952).
11. Young, C. R., and W. B. Dunbar, "Permissible Stresses on Rivets in Tension," *Univ. Toronto School Eng. Bull.* **8** (1928).
12. Munse, W. H., K. S. Peterson, and E. Chesson Jr., "Strength of Rivets and Bolts in Tension," *Proc. ASCE* **85**, ST3, part 1 (March 1959).
13. Newmark, N. M., "The Institute (AISC) Research Program—Part III," *Proc. AISC* (1952).

14. Rathbun, J. C., "Elastic Properties of Riveted Connections," *Trans. ASCE* **101,** 524 (1936).

15. Lothers, J. E., "Elastic Restraint Equations for Semi-rigid Connections," *Trans. ASCE* **116,** 481 (1951).

16. Munse, W. H., W. G. Bell, and E. Chesson Jr., "Behavior of Riveted and Bolted Beam-to-Column Connections," *Proc. ASCE* **85,** ST3, part 1 (March 1959).

6

Welded Connections

🦷🦷

6-1 INTRODUCTION

The first developments in arc welding date back to 1881; the first United States patent for a metallic arc process for welding was granted in 1889 to Charles Coffin of Detroit. Rapid development of the process came during World War I when quick repairs of armament were needed. Following the war, new techniques and equipment came at a fast pace and the acceptance of welding as a joining method became well established. The first applications were made in the shop under controlled conditions and as more experience was gained, the method was applied equally well in the field.

Today it is not uncommon to see field welding for continuous beam and girder bridges or for buildings of 50 or more stories. With proper design, choice of material, correct welding technique, and good workmanship, welding can provide reliable and economical connections. The principal advantages of welded structures are compactness of connections, economy of material, and reduced handling during fabrication in the shop.

6-2 WELDING PROCESSES

Welding is a process of connecting pieces of metal together by the application of heat with or without pressure. This broad definition applies to a large variety of processes ranging from simple soldering and brazing to underwater welding. A summary of these processes is shown in Table 6-1. The most

Table 6-1 Welding Processes

Pressure Welding
    Forge welding
    Pressure Thermit welding
    Resistance welding (a-c)
    Resistance welding (d-c)—seam and spot welding

Nonpressure (Fusion) Welding
    Metal-arc welding (a-c and d-c)—shielded, unshielded, submerged;
        manual and automatic
    Carbon-arc welding—shielded and unshielded
    Inert-gas arc welding
    Atomic-hydrogen arc welding
    Gas welding (air or oxyacetylene)
    Thermit welding

Brazing
    Electric brazing
    Furnace brazing
    Gas brazing
    Dip brazing

---

common type of welding in structural-steel work is fusion welding, which is a method of connecting pieces by molten metal. A special wire or a rod is subjected to intense heat at its tip, which melts and deposits molten metal at the point where a connection is desired (Fig. 6-1). The base metal also melts locally and unites with the deposited metal to form a welded connection. A connection may be obtained by a nonfusion process which consists of simply heating the pieces above a certain temperature and hammering them together on an anvil. The ancient process of forging can be classified as a form of nonfusion welding. In fusion welding, however, both the rod and the base metal are melted, requiring weld temperatures of approximately 2700°F. In structural steel work, metal-arc welding is used almost exclusively. For special structures, such as sheet-steel structures, resistance welding may be used; gas welding and brazing may be used for special parts and fittings requiring small welds.

In order to use welding in structural design to its fullest advantage, a thorough familiarity with the various processes is desirable. However, detailed consideration of techniques and procedures of the various methods of welding are beyond the scope of this chapter, and the following comments will describe only the essential features of the methods most commonly used

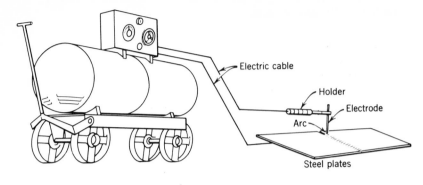

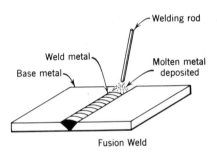

Fig. 6-1   Welding process.

in structural steel work. For further information the reader is referred to various technical publications.[1]

Metal-Arc Welding. Figure 6-1 shows a schematic arrangement of the metal-arc-welding process. The heat is generated by an electric arc formed between a steel electrode and the steel parts to be welded. The arc heat melts the base metal and the electrode simultaneously, and the electromagnetic field carries the molten metal of the welding rod (electrode) toward the base metal, while the operator moves the welding rod manually or automatically along the length of the weld with proper speed and deposits the necessary amount of weld metal. Welding is commonly done in four positions: flat, horizontal, vertical, and overhead (Fig. 6-2). Vertical and overhead welds are possible because molten metal is carried from the rod to the connection by the electromagnetic field of the arc and not by gravity. Welding position affects ease and speed of welding; thus it is of considerable practical importance in determining quality and cost of welding. Electrodes used for arc welding may be either bare steel rods (Fig. 6-3*a*) or coated with various mineral compounds (Fig. 6-3*b*). In welding with coated electrodes part of

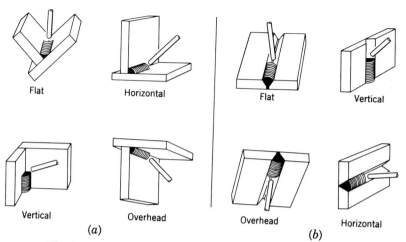

Fig. 6-2 Positions for welding: (*a*) fillet welds and (*b*) butt welds.

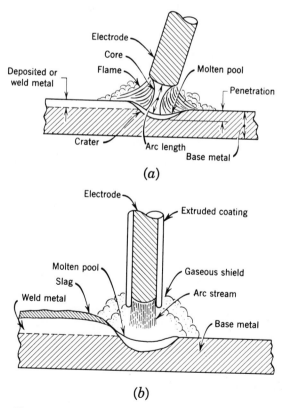

Fig. 6-3 Arc welding: (*a*) nonshielded and (*b*) shielded.

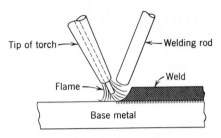

Fig. 6-4   Gas welding.

the coating melts to form a fluid slag layer and part forms a gaseous shield around the arc stream. The gaseous shield serves to stabilize the arc and to protect it from atmospheric gases. The molten slag, having lower density than the molten metal, rises to the top, retards the rate of cooling of weld metal, and also protects it from undesirable exposure to atmospheric gases. The chemical composition of weld metal may be controlled by the composition of the coating. Use of coated electrodes results in better-quality welds than can be obtained with bare electrodes and for this reason almost all modern arc welding is done with coated electrodes.

Although high quality welds are obtained in the field and in the shop using manual welding process, better and more economical results may be obtained in some applications using semi-automatic or automatic welding equipment.

Gas Welding.   In gas welding, the heat is obtained from combustion of a gas fuel; a mixture of oxygen and acetylene is commonly used and the process is then called oxyacetylene welding. The molten metal is obtained from a separate welding rod, bare or coated. Essential features of oxyacetylene welding process are shown in Fig. 6-4.

Resistance Welding.   This process is essentially a pressure welding process, which is a modern version of the ancient process of forging. The heat is generated by the electrical resistance to a current of high amperage and low voltage, passing through a small area of contact between the parts to be connected. The heat developed in the work brings the metal into a plastic state and the weld is effected by applying pressure and thus uniting the pieces locally. Various forms of resistance welding are used in the industry; spot welding and seam welding are the two most common. Essential features of this process are shown in Fig. 6-5.

6-3   TYPES OF WELDED CONNECTIONS

There are five basic types of welded joints: butt, lap, tee, edge, and corner, and four basic types of welds: groove, fillet, plug, and slot, as shown in

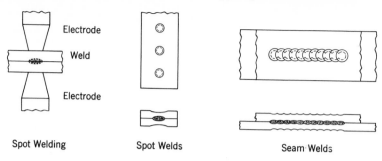

Fig. 6-5   Resistance welding.

Fig. 6-6. Groove welds are always used when the plates to be connected are aligned in the same plane; also, they may be used for a T-joint as shown. Fillet welds are used for lap, T-, and corner joints; occasionally plug or slot welds may be used for lap joints. The forms of groove welds vary depending on the manner of preparing the ends of the pieces. Common forms are shown in Fig. 6-7.

The selection of a proper type of groove weld for a butt joint is determined

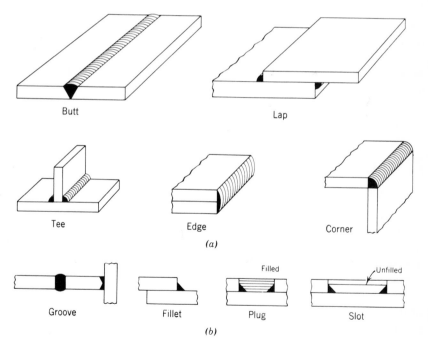

Fig. 6-6   Welded connections: (*a*) types of joints and (*b*) types of welds.

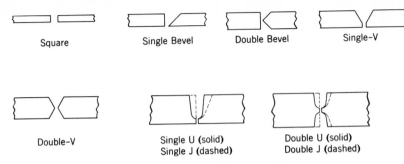

Fig. 6-7   Forms of groove welds.

by the requirement of minimum cost of preparation and welding of the connection, provided that other requirements such as strength, minimum distortion, and minimum residual stress are satisfied. The economy of a particular type of weld depends on plate thickness and the facilities of the particular welding shop for preparing and fitting the plates, and whether the joint can be welded on only one side or both sides.

The cross section of a fillet weld is characterized by its triangular shape and commonly has equal legs. Plug and slot welds are used when sufficient length of fillet welds cannot be secured or additional local connection between lap plates is desired.

In order to establish standard designations for different types of welds, the American Welding Society has specified a set of symbols which provide the means of giving complete welding information on drawings in concise form. The symbols are ideographic and indicate the type of weld required. A summary of these symbols is shown in Table 6-2.

6-4   QUALITY OF WELDS

Design of a welded connection consists of selecting the type of weld most favorable for the loads to be transmitted by the particular size and arrangement of members, and it involves the determination of arrangement and size of welds. A good design on paper cannot assure a good welded connection without good workmanship in the shop and field. The right size and chemical composition of electrode; the sequence and number of passes in welding; proper speed, voltage, and current for the particular weld; proper preparation of the connection before welding; and judicious use of jigs are some of the factors that determine the quality of workmanship and therefore the quality of the weld.

To secure good quality of welding connections requires the teamwork of

Table 6-2   Welding Symbols

Basic Weld Symbols

Type of weld									
Bead	Fillet	Plug or slot	Groove or butt						
			Square	V	Bevel	U	J	Flare V	Flare bevel
⌒	△	▽	||	V	V	U	J	⌐⌐	I⌐

Supplementary Symbols

Weld all around	Field weld	Contour	
		Flush	Convex
○	●	—	⌒

Standard Location of Elements of a Welding Symbol

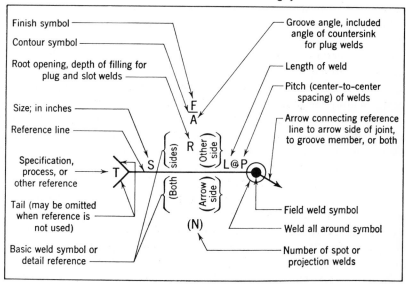

The above symbols were developed by the American Welding Society for incorporation on drawings specifying arc or gas welding. For more detailed instruction in the use of these symbols refer to "Welding Symbols and Instructions for Their Use," published by the American Welding Society.

183

competent design engineers, supervisors, and welders. However, the design of the weld originates with the engineer and, in order to specify the best weld for the particular application, the engineer must have some knowledge of the basic principles of welding metallurgy, the thermal effects of welding on base metal, and the properties of materials and techniques used in the process.

Metallurgy. Three metallurgical factors are of interest in metal-arc welding: crystalline structure, gas solubility, and oxidation.

When steel is heated to a critical temperature of approximately 1500°F, it has an almost uniform *crystalline structure* (austenite). When it is cooled slowly from this temperature, the grain structure changes to a ductile material called pearlite. When cooled very rapidly, austenite changes to a brittle material with little pearlite and mostly martensite. The critical temperatures and the rates of cooling which determine whether the steel will be ductile or brittle after welding vary with the chemical composition of steel, particularly with its carbon content. Therefore, the chemical composition of base metal and filler metal must be carefully considered in design of welded connections. The most critical factor, however, is the rate of cooling: slow cooling usually results in ductile steel; rapid cooling results in hard brittle steel.

Another metallurgical reaction of great importance in welding is *gas solubility*. Molten metal at high temperature can hold a larger amount of gases in solution than at lower temperatures. Therefore, it is important that during welding the molten metal be protected from absorbing certain gases from the atmosphere, since these gases when given off during cooling may cause formation of porous welds (gas pockets), or the gas may be retained in solution resulting in chemical and physical changes in weld metal. Also, the materials used in welding—base metal and electrode metal and coating—should *not* contain elements which may either give off gas during welding or cause an increase in the solubility of gas in molten metal. Protection of molten metal from atmospheric gases is given by coated electrodes—the coating melts, forming a protective gas shield for the arc and a slag layer for protection of molten metal. The choice of chemical properties of the electrode is partly determined by its gas solubility characteristics.

The *oxidation* reactions during welding may produce either gaseous or solid oxides. Such reactions are highly undesirable since gas oxides may produce pockets or porosity and solid oxides may produce slag inclusion or may render the weld brittle and reduce its strength.

Thermal Effects. Two effects of welding temperatures are of primary interest: one is expansion and contraction of metals with change in temperature and the other is the rate of cooling. Expansion and contraction are

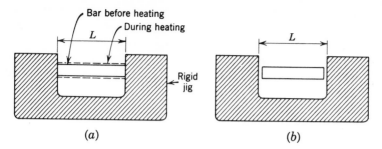

Fig. 6-8 Distortion of bar due to heating: (*a*) before and during heating and (*b*) cooled after heating.

largely responsible for the development of residual stresses and distortions in welded connections; the rate of cooling is primarily important in its effect on the crystalline structure of the welded part, as previously discussed.

Consider a steel bar restrained in a rigid jig, as shown in Fig. 6-8*a*. When this bar is heated so as to increase its initial temperature by 150°F, it tends to expand 0.001 of its initial length (coefficient of expansion of steel is 67×10^{-7} per 1°F). Since the bar is restrained and cannot expand, it must be subjected to a compression force (stress) which produces a shortening (compressive strain) equal to 0.001 of its length. The modulus of elasticity of steel is 30×10^6 psi, and the stress corresponding to this strain is 30,000 psi—nearly equal to the yield stress of the material. If the same bar is heated to a higher temperature, practically no additional stress need be applied in order to prevent it from expanding further; the piece is in a plastic state and expands laterally. At high temperature, the yield point of steel becomes lower and it is negligible before the steel melts. This means that the stress in this restrained bar will decrease with increasing temperature, the steel becoming more plastic and expanding in the radial direction. During cooling there is no restraint to shrinking of the rod and therefore it shrinks equally in all directions. Thus a completely cooled bar is shorter and has a greater diameter than it had before heating (Fig. 6-8*b*). This simple example illustrates the basic physical laws which cause residual stresses and distortion in welded parts.

For example, depositing a bead weld on one side of a plate results in

Fig. 6-9 Distortion due to a bead.

distortion, as shown in Fig. 6-9. The curvature indicates shortening of fibers in base metal adjacent to the bead. The bead tends to shorten more than the base metal, because of differences in average temperatures, but, since the strains of the bead and adjacent surface of the plate must be the same, residual tension is induced in the bead and residual bending is induced in the plate, the latter producing compression in the fibers adjacent to the bead. To minimize the residual stresses and distortions, the heat generated in welding must not be excessive.

On the other hand, slow cooling is desirable in order to minimize change in crystalline structure and ductility of steel (see preceding discussion of metallurgy in welding). Consider two plates, one thick and another thin, upon which equal bead welds are placed (Fig. 6-10). The amount of heat applied to both plates being the same, the material adjacent to the bead will cool much faster in the thick plate than in the thin one. Since little heat can escape at the surfaces (air being a poor heat conductor compared to steel), the heat has more possible paths of escape from the weld in the thick plate than from the thin one, and rapid cooling of the weld on the thick plate results. In this instance, the weld zone in the thick plate will be more brittle than in the thin one. To minimize rapid cooling and subsequent brittleness in the thick plate, application of more heat to the weld zone is desirable.

A conflict of desirable procedures to minimize thermal effects is thus apparent. To ensure slow rate of cooling, welds should develop more heat, but excessive heat results in excessive distortions, with possible higher residual stresses. A practical compromise can be achieved using proper technique, such as preheating of the base metal. Some of the design considerations based on thermal effects will be discussed in Secs. 6-5, 6-9 and 6-10.

Material Properties.   Most steel grades suitable for structural work can be welded. The relative economy and degree of ease of welding on a particular steel is generally termed "weldability." In ordinary structural steel, the carbon content is the most important factor determining weldability. Steels with carbon content below 0.1 % have a high gas absorption which is a common cause of porous welds. Steels with carbon content of 0.3 % and up become

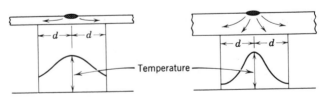

Fig. 6-10   Effect of plate thickness on rate of cooling.

brittle when cooled rapidly and for a given rate of cooling brittleness increases with increase in carbon content.

Because rapid cooling is prevalent in welding of thick parts, carbon content is particularly important for shapes and plates where thicknesses in excess of 1 or $1\frac{1}{2}$ in. occur. Two schools of thought regarding welding of thick parts exist.[2] One proposes to control chemical composition and crystallographic structure of the steel, so that ordinary welding procedures will result in satisfactory mechanical properties of welded joints. The other proposes to use common structural steel shapes and plates, and to control the mechanical properties by using special procedures and electrodes for welding thicker pieces, such as the use of preheating and low-hydrogen electrodes (AWS E-6015 or E-6026). It appears that requirement of relatively low-carbon steel provides a more positive control of the mechanical characteristics of the welded structure because it involves less hazards due to errors in workmanship.

Mechanical properties of some typical steels used in welded structures and properties of typical weld metal obtained with standard electrodes are shown in Chapter 2.

Welding Techniques. Welding technique refers to a number of details involved in the process, such as welding position, preparation of the metal before welding, fit-up of joints, electrode size and type, use of a-c or d-c equipment and proper polarity of base metal, adjustment of current and voltage for particular weld, speed of depositing filler metal, number of passes to build up a weld, maintenance of stable arc, and proper shape of weld. Because the detail requirements may be found in technical literature,[1] they will not be discussed here. Dimensional requirements of AWS qualified joints are shown in Table 6-3.

The most important defects arising from improper welding technique are undercutting, lack of fusion and penetration, slag inclusion, and porosity. Most of these defects result in stress concentrations under load and thus may reduce the strength of the weld, particularly under dynamic or repeated loads.

Undercutting is defined as burning away of base metal (Figs. 6-11a and b). The tendency for undercutting depends somewhat on electrode characteristics and welding position, frequently caused by excessive current and excessive length of arc. Undercutting, which is easily detected by visual inspection of the weld, can be corrected by depositing additional weld metal after the surface is properly cleaned.

Lack of fusion is defined as failure of base metal and weld metal to fuse at any point in the groove (other than the root, Figs. 6-11c and d). This defect is not common in arc welds unless surfaces being welded are coated with foreign material preventing fusion at that point. If surfaces are properly

Table 6-3   Welded Joints*

Joints accepted without qualification under AWS Code

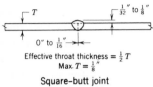

T

$\frac{1}{32}''$ to $\frac{1}{8}''$

$0''$ to $\frac{1}{16}''$

Effective throat thickness $= \frac{1}{2}\,T$
Max. $T = \frac{1}{8}''$

**Square-butt joint
welded one side**

Root need not be
chipped before welding
second side.

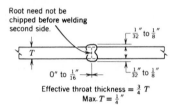

$\frac{1}{32}''$ to $\frac{1}{8}''$

T

$0''$ to $\frac{1}{16}''$

$\frac{1}{32}''$ to $\frac{1}{8}''$

Effective throat thickness $= \frac{3}{4}\,T$
Max. $T = \frac{1}{4}''$

**Square-butt joint
welded both sides**

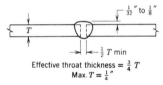

T

$\frac{1}{32}''$ to $\frac{1}{8}''$

$\frac{1}{2}\,T$ min

Effective throat thickness $= \frac{3}{4}\,T$
Max. $T = \frac{1}{4}''$

**Open square-butt joint
welded one side**

T unlimited

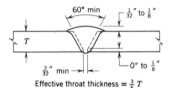

60° min

$\frac{1}{32}''$ to $\frac{1}{8}''$

T

$0''$ to $\frac{1}{8}''$

$\frac{3}{32}''$ min

Effective throat thickness $= \frac{3}{4}\,T$

**Single-V butt joint
welded one side**

T unlimited

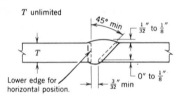

45° min

$\frac{1}{32}''$ to $\frac{1}{8}''$

T

Lower edge for
horizontal position.

$\frac{3}{32}''$ min

$0''$ to $\frac{1}{8}''$

Effective throat thickness $= \frac{3}{4}\,T$

**Single-bevel butt joint
welded one side**

T unlimited

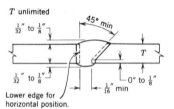

45° min

$\frac{1}{32}''$ to $\frac{1}{8}''$

$\frac{1}{32}''$ to $\frac{1}{8}''$

T

$0''$ to $\frac{1}{8}''$

$\frac{1}{16}''$ min

Lower edge for
horizontal position.

**Single-bevel butt joint
welded both sides**

T unlimited

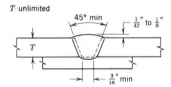

45° min

$\frac{1}{32}''$ to $\frac{1}{8}''$

T

$\frac{3}{16}''$ min

**Single-V butt joint welded
one side on backing structure**

T unlimited

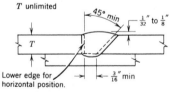

45° min

$\frac{1}{32}''$ to $\frac{1}{8}''$

T

Lower edge for
horizontal position.

$\frac{3}{16}''$ min

**Single-bevel butt joint welded
one side on backing structure**

* For butt welds, effective throat thickness $= T$ unless otherwise noted.

188

Table 6-3   (*Continued*)

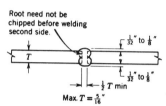

T unlimited

Double-V butt joint

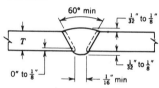

Root need not be chipped before welding second side.

Max. $T = \frac{5}{16}''$

Open square-butt joint welded both sides

T unlimited

Single-V butt joint welded both sides

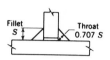

Fillet S Throat 0.707 S

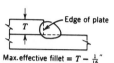

Edge of plate

Max. effective fillet $= T - \frac{1}{16}''$

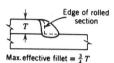

Edge of rolled section

Max. effective fillet $= \frac{3}{4} T$

0.707 T

Max. effective fillet $= T$

Edge-fillet welds
See AISC specifications sec. 24(*d*).

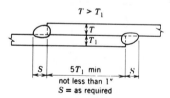

$T > T_1$

$5T_1$ min
not less than 1″
$S =$ as required

Double-fillet welded lap joint
See AISC specifications sec. 1.17.8

cleaned and the electrode size, speed, and current are properly selected, complete fusion will be assured.

Incomplete penetration is defined as the failure of the base metal and weld metal to fuse at the root (Figs. 6-11*e* and *f*). This defect may be due to faulty design of the groove, such as excessive root-face dimension, insufficient root gap or groove angle, or it may be due to faulty technique, such as use of an excessively large size electrode, excessive speed, or insufficient current.

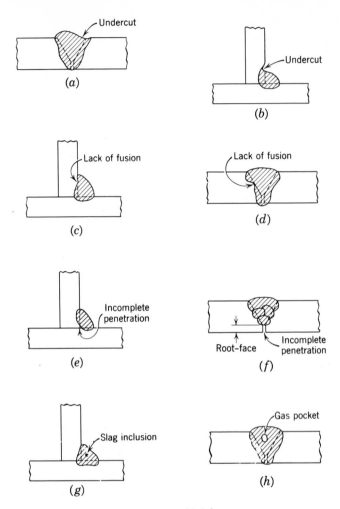

Fig. 6-11   Weld defects.

Incomplete penetration is particularly undesirable since it causes stress concentration under load and may be the cause of cracks due to shrinkage.

Slag inclusions are defined as the metallic oxides and other solid compounds sometimes found as elongated or globular inclusions (Fig. 6-11g). These oxides are the result of chemical reactions among the metal, the air, and the electrode coating during deposition and solidification of weld metal. The formation of these oxides can be largely prevented by selecting the chemical composition of electrode metal and coating which does not react with elements contained in the base metal. Since slag has lower density than the molten metal, it usually tends to rise to the surface and therefore rarely presents any difficulties in horizontal welds. Rapid chilling of weld metal and insufficient groove angle may prevent release of slag. Slag inclusions present a particular problem in vertical and overhead welding.

Porosity is defined as presence of globular voids or gas pockets in the weld metal (Fig. 6-11*h*). Gas may be trapped in the weld metal as a result of gas released because of reduced solubility with cooling, or gas formed by chemical reactions. Porosity is frequently due to the use of excessive current or excessive arc length.

Quality Control. Satisfactory welds in a structure are usually obtained when proper welding procedure is utilized and when welding is done by competent welders. The American Welding Society has developed Standard Qualification Procedures,[3] consisting of two parts: procedure qualification and operator qualification. Procedure qualification deals with properties of base and filler metal, type and size of electrodes, type of groove and position of welding, current and voltage used, and possible uses of preheating or heat treatment of parts after welding. Operator qualification requires the welding operator to produce certain test specimens which must have prescribed strength and ductility. The test welds must simulate the type and conditions of field welding and various qualification grades are provided for different types of welding. Numerous comparisons of results obtained with test specimens and actual field specimens indicate that operators who make good test specimens usually make good field welds. Nevertheless, it is not sufficient to rely on qualification tests alone, and proper inspection of all welds in the structure must be maintained in order to secure satisfactory welds. Several methods available for inspecting welds are visual, magnetic particle, dye penetrant, ultrasonic, and radiographic. All these methods require supervision by competent people who can interpret the results.

Visual. This is the simplest method and requires a competent person to observe the welder in operation when he is performing the work. This may be the most rapid and economical method.

Magnetic Particle. In this method iron filings placed on the weld are subjected to an electrical current. The patterns of the filings will indicate surface cracks to an experienced observer. In multilayer welds, each layer must be examined to inspect the weld properly.

Dye Penetrant. A dye is brushed on the surface of the weld and penetrates any cracks which may be present. The surplus is removed and a dye absorber is placed over the welds. The amount of dye oozing out of a crack will indicate its depth.

Ultrasonic. A recent development in steel making is also applicable to the inspection of welds, although expensive equipment is required. In this

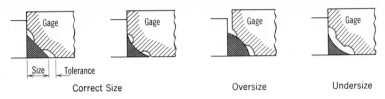

Correct Size   Oversize   Undersize

Fig. 6-12 Inspection with weld gages.

method sound waves are sent through the material and defects will affect the time interval of the sound transmission, which will identify the defects.

Radiographic. This method may use X-rays, or gamma rays, to produce the picture on the film. The technique is best applied to butt welds where the picture will show only weld material. It is not adaptable to fillet welds because the parent material will also project on the picture. Required clearances for film and equipment limit the use of this technique in the field.

These nondestructive tests may be used to supplement the visual inspection or for random checking of the welding procedures.

To determine adequacy of weld shape various gages are used; a typical one is shown in Fig. 6-12. Internal defects, such as lack of fusion or penetration, porosity, and slag inclusions, cannot be detected by visual observation of the weld but may be detected on radiographic photographs of the weld.

6-5 RESIDUAL STRESSES AND DISTORTION

Residual stresses resulting from welding are undesirable for two primary reasons: *distortions* which are usually associated with residual stresses and the possibility of *brittle fracture* if the residual stresses are high.

Distortion. Warping and distortion of welded parts may or may not be associated with residual stress. Some simple examples will be discussed. Consider a short fillet weld shown in Fig. 6-13. After the weld is cooled,

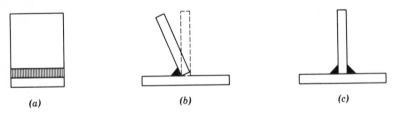

(a)   (b)   (c)

Fig. 6-13 Distortion and residual stresses in fillet welds: (*a*) side view, (*b*) distortion after weld, no residual stress, and (*c*) residual stresses induced, no distortion.

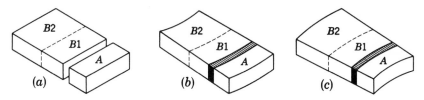

Fig. 6-14 Residual stresses in a butt weld: (*a*) before welding, (*b*) expanding state just after welding, and (*c*) contracting state after cooling.

the shrinkage is greatest along the face of the weld where large portions of the filler metal are deposited. If the vertical plate is free to move, it is pulled over by the shrinkage of the weld without residual stresses due to shrinkage, since no resistance is offered to such distortion (Fig. 6-13*b*).

A fillet weld deposited on the other side of the vertical plate (Fig. 6-13*c*) tends to pull the plate back into its original position. This shrinkage of the second weld takes place against the resistance of the first weld, high residual stresses are set up in both welds, and the plate straightens out, although it does not return fully to its original position.

Next consider a short butt weld shown in Fig. 6-14. When the molten metal is deposited, part *A*, which is small, is heated rapidly to very high temperature, allowing the steel to behave plastically. Zone *B*1, adjacent to the weld, behaves in the same manner, but zone *B*2, away from the weld, is not heated to the same extent and is assumed to behave elastically. When the weld is deposited, the heated parts tend to expand in proportion to the change in temperature. Zone *B*2, being heated less than the others, tends to expand less and therefore restrains some of the expansion of *B*1 and *A*. The distorted shape of the welded part is shown in Fig. 6-14*b*. Part *A* and zone *B*1 have been restrained from expanding while in a plastic state and therefore in cooling they tend to shrink to a length less than their original dimension. Zone *B*2, which did not undergo plastic deformation, tends to shrink to its original length and thus again will restrain *A* and *B*1 from free shrinkage. When the plates are completely cooled, part *A* will be shorter than its initial length and will present the distorted shape shown in Fig. 6-14*c*. The final residual stresses developed upon cooling are quite complex. Roughly, it may be visualized that, since *A* and *B*1 have been prevented from full shrinkage, they are subjected to some tension, whereas zone *B*2 is subjected to compression and bending resulting in essentially compression stresses at the boundary between *B*1 and *B*2, and tension stresses at the free edge of zone *B*2.

Distortions and residual stresses discussed in the preceding two examples result from asymmetry of welding sequence (Fig. 6-13) or asymmetry of thermal effects of parts joined by the weld (Fig. 6-14). If the two fillet welds

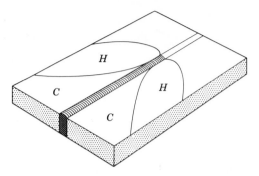

Fig. 6-15   Heating and cooling of a long weld.

in Fig. 6-13*c* could be deposited simultaneously, distortion of the vertical plate could be eliminated and residual stresses minimized. Similarly, if the plates *A* and *B* shown in Fig. 6-14 were symmetrical about the weld, the distortions and residual stresses would be reduced considerably.

The two welds considered were assumed to be of short length in order to eliminate the effect of time rate of welding. Consider a long weld (Fig. 6-15)

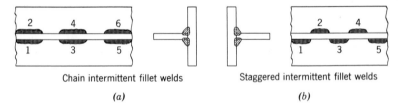

Chain intermittent fillet welds   Staggered intermittent fillet welds

(*a*)   (*b*)

Fig. 6-16   Sequences for intermittent welds: (*a*) fair, and (*b*) better.

in which one part of the plates, zone *H*, is in a "heating" or expanding state, similar to Fig. 6-14*b*, and zone *C* is in a "cooling" or contracting state, similar to Fig. 6-14*c*. Relative areas of the two zones, final distortions, and the residual stresses depend on speed of welding and on arrangement and sequence of welds designed to minimize distortions and residual stresses. Techniques for minimizing residual stresses and preventing distortion are so numerous that they cannot be discussed adequately within the scope of this book. A few of these techniques will be mentioned briefly. Figure 6-16 shows two possible sequences commonly used for staggered fillet welds. The backstep technique of sequence welding, which reduces residual stresses and distortions, is shown in Fig. 6-17. This technique may be improved by skipping some steps, for example, welding in the

Fig. 6-17   Back-step technique.

following order: 1, 3, 5, 2, 4, and 6. Other means of minimizing residual stresses include preheating the parts before welding, stress relieving by heating after welding, and peening, that is, hammering the weld to elongate it locally and to relieve the shrinkage forces. Typical preheating requirements are shown in Table 6-5, p. 213.

6-6 STRESSES IN WELDED CONNECTIONS

In actual structures, welded connections frequently consist of a group of welds rather than a single weld and the welds are usually subjected to one or more types of loading simultaneously. The variety of possible arrangements of welds in a group subjected to different combinations of loads makes it impractical, if not impossible, to obtain solutions based on elastic theory for all cases. Therefore nominal stresses are used to evaluate adequacy of welds to resist loads.

Nominal stresses in welded connections may be obtained on the basis of the following assumptions:

(*a*) The welds connecting various parts are homogeneous, isotropic, and elastic elements.

(*b*) The parts connected by the welds are rigid and their deformations are therefore neglected.

(*c*) Only nominal stresses due to external loads are considered; effects of residual stresses, stress concentrations, and shape of the welds are neglected.

These three assumptions are similar to those made in the design of riveted connections and lead to conventional formulas for axial, shear, bending, and torsional loads applied to the section constituted by the welds. The following general procedure is used for the calculation of nominal stresses in welded connections.

(*a*) Draw the effective cross section of the welded connection. The *effective dimensions* of various welds are shown in Table 6-3.

(*b*) Determine the centroid of this effective section and establish a system of orthogonal reference axes x, y, z through the centroid (Fig. 6-18). When stresses normal to the plane of the welds occur, principal axes of the section should be used.

(*c*) Determine forces and moments acting on the welded connection, defined in terms of x, y, z components.

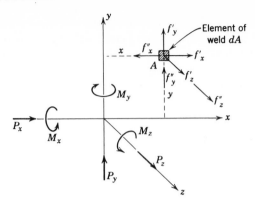

Fig. 6-18 Stress components on weld element.

(*d*) At any point of the connection, the stress on the weld due to one single component of load can be computed from conventional formulas (Eqs. 6-1, 6-2, and 6-3). In Fig. 6-18 the notation shows f_x and f_y as shearing stresses and f_z as normal stress.

Due to forces:

$$f_x' = \frac{P_x}{A}, \qquad f_y' = \frac{P_y}{A}, \qquad f_z' = \frac{P_z}{A} \qquad (6\text{-}1)$$

Due to moments:

$$f_x'' = \frac{M_z}{I_z}\, y, \qquad f_y'' = \frac{M_z}{I_z}\, x, \qquad f_z'' = \frac{M_x}{I_x}\, y + \frac{M_y}{I_y}\, x \qquad (6\text{-}2)$$

where

$$A = \int dA \qquad I_x = \int y^2\, dA \qquad I_y = \int x^2\, dA$$

and

$$I_z = \int (x^2 + y^2)\, dA = I_x + I_y$$

Resultant components of stress, with due regard to signs:

$$f_x = f_x' - f_x'', \qquad f_y = f_y' + f_y'', \qquad f_z = f_z' + f_z'' \qquad (6\text{-}3)$$

(*e*) For butt welds, both normal and shear stresses are considered in a conventional manner.

(*f*) For fillet welds, *x*, *y*, and *z* components of stress on a given leg of the weld are used to determine q_r, the maximum resultant force per unit length of weld, and the latter is arbitrarily considered a "shear" force acting on the throat section, as follows:

$$q_r = tf_r = t\sqrt{f_x^2 + f_y^2 + f_z^2} \qquad (6\text{-}4)$$

where t is the effective throat dimension.

If all welds are of the same size, a convenient procedure for computing nominal stresses in fillet welds is to determine first the resultant load q, as force per inch of weld, and then to compute nominal stress as $f = q/t$. When the size of weld is constant, it does not enter into computation of q, which depends only on the magnitude of the loads and length and arrangement of welds. This procedure is particularly advantageous in design when required size of weld is to be determined. It is important to note that all loads acting on a fillet weld are considered as shears, independent of their actual direction, and the critical section is always considered to be the throat of the weld, although other sections may actually be subjected to higher stresses.

Computation of resultant load per inch q_r by the method indicated in step f (Eq. 6-4) is entirely arbitrary, since at point A (Fig. 6-18) f_x and f_y are shear stresses and f_z is a normal stress. An alternative solution to the one indicated in step f would be to establish state-of-stress equations and to use formulas or Mohr's circle to compute maximum principal and shear stresses. Comparison of such a method with that indicated in step f has been made.[4] The results were not conclusive, although it was found that, for the cases studied, the arbitrary method of combining x, y, and z components of loads on the weld agrees with experimental results about as well as the method of determining maximum principal and shear stresses, and the arbitrary method was recommended because of its simplicity.

This procedure is illustrated in the following example.

Example 6-1. A 12-in. standard pipe is welded to its support and is loaded as shown in Fig. 1. Determine the maximum nominal stress in the weld.

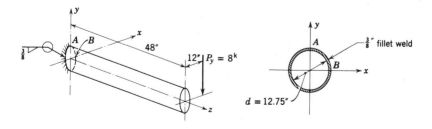

Fig. 1

Solution

Weld-Section Properties.   Assume that the mean diameter of weld section is equal to the diameter of the pipe.

Throat area:

$$A = \pi d(0.707a) = \pi \times 12.75 \times 0.707 \times 0.375 = 10.6 \text{ in.}^2$$

$$I_x = \frac{\pi d^3}{8}(0.707a) = \frac{\pi (12.75)^3}{8} \times 0.707 \times 0.375 = 215 \text{ in.}^4$$

$$I_y = I_x$$

$$I_z = I_x + I_y = 430 \text{ in.}^4$$

Forces and Moments.

$$P_x = 0, \qquad P_y = 8 \text{ kips}, \qquad P_z = 0$$

$$M_x = 48P = 384 \text{ kip-in.}, \qquad M_y = 0, \qquad M_z = 12P = 96 \text{ kip-in.}$$

Stresses due to Forces.

$$f_x' = \frac{P_x}{A} = 0, \qquad f_y' = \frac{P_y}{A} = \frac{8}{10.6} = 0.755 \text{ ksi}, \qquad f_z' = \frac{P_z}{A} = 0$$

Stresses due to Moments.   Consider points A and B.

At point A, $x = 0$, $y = 6.38$ in.,

$$f_x'' = \frac{M_z}{I_z} y = \frac{96}{430} 6.38 = 1.42 \text{ ksi}$$

$$f_y'' = \frac{M_z}{I_z} x = 0$$

$$f_z'' = \frac{M_x}{I_x} y + \frac{M_y}{I_y} x = \frac{384}{215} 6.38 = 11.42 \text{ ksi}$$

At point B, $x = 6.38$ in., $y = 0$,

$$f_x'' = \frac{M_z}{I_z} y = 0$$

$$f_y'' = \frac{M_z}{I_z} x = \frac{96}{430} 6.38 = 1.42 \text{ ksi}$$

$$f_z'' = \frac{M_x}{I_x} y + \frac{M_y}{I_y} x = 0$$

Critical stress at point A (resultant stress):

$$f_x = f_x' + f_x'' = 1.42 \text{ ksi}$$
$$f_y = f_y' + f_y'' = 0.76 \text{ ksi}$$
$$f_z = f_z' + f_z'' = 11.42 \text{ ksi}$$

$$f_r = \sqrt{f_x^2 + f_y^2 + f_z^2} = \sqrt{(1.42)^2 + (0.76)^2 + (11.42)^2} = 11.5 \text{ ksi}$$

The maximum nominal stress in the weld at A is $f_r = 23$ ksi.

6-7 STRENGTH OF BUTT WELDS

The manner in which various irregularities in the weld affect the stress distribution in butt welds can best be illustrated by the stress paths (Fig. 6-19). The stress path may be defined as a line indicating the direction of the

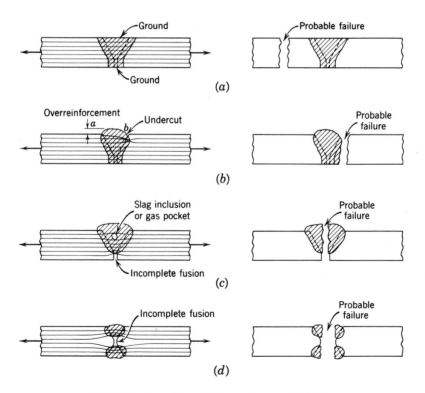

Fig. 6-19 Locations of failure of butt-welded joints.

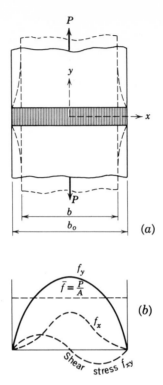

Fig. 6-20   Stress distribution assuming rigid weld.

principal stress, the spacing between these stress paths showing the magnitude of the stress. Where the lines are widely spaced the stress is low; where the lines are closely spaced the stress is high and failure is likely to start there.

In some instances, the mechanical properties of the weld metal may differ somewhat from the base metal and the stress distribution is not uniform. Consider a bar subjected to an axial tensile load P as shown in Fig. 6-20a. Under the action of this load the bar will elongate and, owing to Poisson ratio effect, its initial width b_o will decrease slightly to width b. This lateral contraction would be uniform along its length if the bar were homogeneous. At loads approaching yield the weld metal, having a yield point stress greater than the base metal, contracts laterally less than the base metal. This effect results in a varying stress distribution across the width of the welded section, as shown in Fig. 6-20b. Thus the axial tensile stress f_y at the center of the weld will be somewhat greater than the average stress, $f = P/A$. The axial, lateral, and shear stress distributions based on the theory of elasticity are shown in Fig. 6-20b. They are not exact when the load exceeds the elastic limit, and at high loads, approaching the ultimate capacity of the weld, the longitudinal stress distribution is approximately uniform.

Stress concentrations occurring at sharp re-entrant corners of butt welds connecting bars of different cross section (Fig. 6-21) reduce the strength of such connections materially. To minimize stress concentration, gradual transition from one section to the other (Fig. 6-21*b*) must be provided. Test results obtained on butt-welded specimens of variable section but with

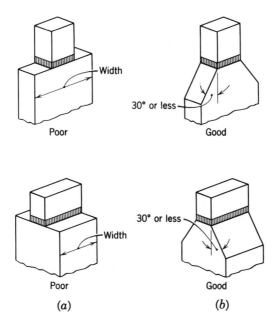

Fig. 6-21   Transition in section for butt joints.

adequate transition indicate that the tensile strength is very nearly the same as that of constant section specimens. Similarly, tests on butt-welded double-T connections (Fig. 6-22), in which bending is eliminated, indicate that the tensile strength of such connections is not materially different from straight butt-welded plates.

Tension test results of butt-welded specimens indicate that their average tensile strength is very nearly the same as the average tensile strength of the base metal. For a large number of tests the range of variation of strength values is greater for the welded specimens than for the unwelded specimens due to effects of imperfections and nonhomogeneity.

Occurrence of imperfections in some types of butt welds is so frequent that the best estimate of their strength is obtained as the product of the ultimate strength of base metal times an "effective" thickness of the weld. For most

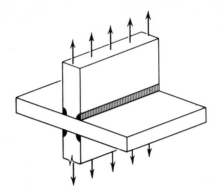

Fig. 6-22   Butt-welded double-T connection.

butt welds made on one side only, the effective thickness is as shown in Table 6-3. For butt welds made with a backing strip or for joints on both sides, no reduction is usually made.

Example 6-2.   A 6-in. standard pipe is buttwelded and is loaded as shown (Fig. 1). If the maximum allowable nominal tension stress on the weld is 20 ksi, determine the allowable load P that the pipe can carry. Use AISC specification for buildings.

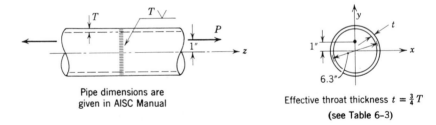

Pipe dimensions are
given in AISC Manual

Effective throat thickness $t = \frac{3}{4} T$
(see Table 6-3)

Fig. 1

Solution
Weld-Section Properties.
Throat area:

$$A = \pi \, dt = \pi \times 6.3 \times \tfrac{3}{4} \times 0.28 = 4.15 \text{ in.}^2$$

$$I_x = \frac{\pi d^3}{8} t = \frac{\pi (6.3)^3}{8} \times \frac{3}{4} \times 0.28 = 20.6 \text{ in.}^4$$

Forces and Moments.

$$P_x = 0, \qquad\qquad P_y = 0, \qquad P_z = 0,$$
$$M_x = P(1) = P \text{ kip-in.}, \qquad M_y = 0, \qquad M_z = 0$$

Stresses due to Forces and Moments.

$$f_x' = 0, \qquad f_y' = 0, \qquad f_z' = \frac{P}{A} = \frac{P}{4.15} = 0.24P \text{ ksi}$$

$$f_x'' = 0, \qquad f_y'' = 0, \qquad f_z'' = \frac{M_x}{I_x}y = \frac{P}{20.6}3.15 = 0.15P \text{ ksi}$$

Maximum nominal stress:

$$f_z = f_z' + f_z'' = 0.39P \text{ ksi}$$

Allowable load:

$$f_z = 0.39P = 20 \text{ ksi}, \qquad P = \frac{20}{0.39} = 51.2 \text{ kips}$$

Example 6-3. Two 15 ⊏ 50 channels, form a tension member in a truss. The dead load in the member is 100 kips tension and the maximum values of live load are 268 kips compression and 320 kips tension. If the number of maximum load cycles during the life of the structure is estimated as 500,000, determine whether a butt weld splice is adequate. Assume a single-bevel butt joint welded on both sides (see Table 6-3). Use AASHO Specifications for Bridges.

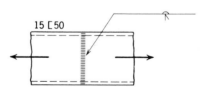

15 ⊏ 50

Fig. 1

Solution
Weld-Section Area A.
Gross section of two 15 ⊏ 50 (Fig. 1): $2 \times 14.64 = 29.3$ in.²

Maximum load $= \text{DL} + \text{LL} = 100 + 320 = 420 \text{ kips}$ (tension)
Minimum load $= \text{DL} + \text{LL} = 100 - 268 = -168 \text{ kips}$ (compression)

Weld area required:

$$A_r = \frac{P_{\max}}{F_p}$$

where F_p is the allowable stress reduced for fatigue effect corresponding to 500,000 cycles (see Fig. 3-11).

$$r = -\frac{168}{420} = -0.4; \quad F_p = 13.0 \text{ ksi,}$$

$$A_r = \frac{P_{\max}}{F_p} = \frac{420}{13.0} = 32.4 \text{ in.}^2$$

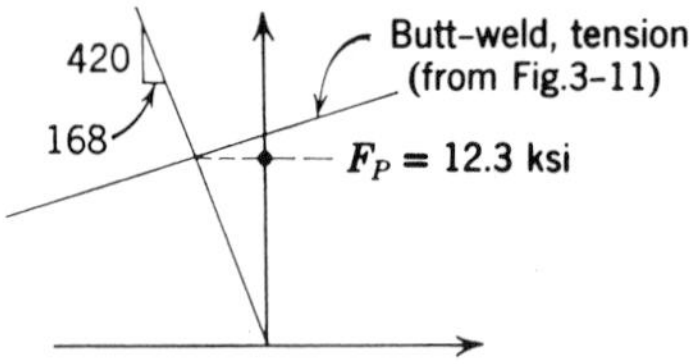

Fig. 2

Available weld area:

$$A = 29.3 \text{ in.}^2 < 32.4 \text{ in.}^2 \quad \text{(not adequate)}$$

Butt weld is not adequate; splice plates may be used or heavier shapes may be selected for butt welding.

6-8 STRENGTH OF FILLET WELDS

As shown in the previous section, butt welds form connections which are almost fully continuous with the parts they connect. Fillet welds are quite different in this respect, since they introduce eccentricities of force transmission and discontinuity of shape in the connection. A simple illustration of the different nature of the two types of welded connections is seen in Fig. 6-23, showing butt- and fillet-welded connections. Because of eccentricities and discontinuities inherent in fillet-welded connections, the actual stress distribution is extremely complex.

Although a rigorous solution for stress distribution in fillet welds cannot be obtained, valuable qualitative results can be indicated by approximate solutions based on theory of elasticity or experimental measurements on models. Consider a strap joint with side fillet welds loaded as shown in Fig. 6-24.

If the thickness of the bars connected by the welds is small compared to the length of the weld, the eccentricity of each load P with respect to the welds may be neglected and the joint is considered to be subjected to purely axial

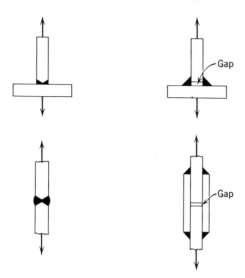

Fig. 6-23 Fillet versus butt welds.

tension load. The transfer of load from bar 1 to each weld and from the weld to bar 2 is accomplished by longitudinal shear stresses in the fillet welds (Figs. 6-24*b* and *c*). If the bars were considered to be rigid (nonelastic), whereas the welds were elastic, then because of load P the bars would not strain at all but merely displace relative to each other through a distance Δ, this being the average shearing strain in the weld (Fig. 6-24*d*). In this instance, the shearing strain in the weld is constant and therefore the shearing stress is constant throughout the length of the weld.

In actual structures the bars are never rigid and therefore some allowance must be made for their deformations. In the region of the overlap the average tensile stresses in bar 1 vary from $f' = 2P/A'$ at section a, to $f' = 0$ at section b, and, similarly, average stresses in bar 2 vary from $f'' = 0$ at a to $f'' = P/A''$ at b. If the bars are elastic, then each bar will be deformed in proportion to the stress. It is apparent that deformation of bar 1 at a is large and that of bar 2 at a is zero. This relative difference in deformation results in an increase of shearing strain in the weld at a (Fig. 6-24*c*). Similar increase in shearing strain occurs at end b and the distribution of shearing strain in the fillet weld varies as shown in Fig. 6-24*e*. Corresponding distribution of tension stresses in bar 1 at section m is shown in Fig. 6-24. The actual variation of the shear stress in the weld and of the tension stresses in the plate depends on the stiffness of the bars in tension and the rigidity of the weld in shear. When either the bars or the weld or both are stressed beyond the elastic limit, behaving more plastically than elastically, the shear in the weld

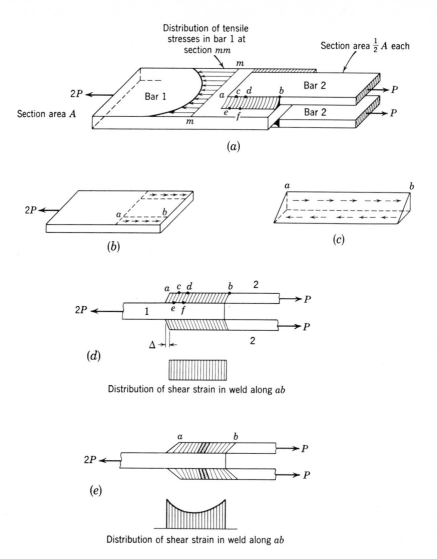

Fig. 6-24 Stress distribution in side fillet welds: (*a*) a Butt strap with side fillet welds, (*b*) free body of bar, (*c*) free body of fillet, (*d*) rigid bars and elastic welds, and (*e*) elastic bars and welds.

becomes more uniform. Under static loads at or near failure the plates and the welds are stressed beyond the elastic limit and assumption of uniform stress distribution throughout the length of weld is more nearly correct than at low or intermediate loads.

The theoretical distribution of shear stress in the fillet weld shown in Fig. 6-24 has been verified by numerous test results.[5] It has been observed in these tests that within the elastic limits the strains at the end of the fillet welds are considerably greater than at the center. Furthermore, test results show that the long side fillet welds have smaller ultimate unit strength than the short ones, although this difference is small because of redistribution of stresses in the plastic range. Tests also indicate that large-size fillet welds have smaller ultimate unit strength than small-size fillets. This is consistent with theoretical analysis, since the larger welds are more rigid and result in less uniform stress distribution and hence lower average stress.

Determination of stress distribution in end fillet welds is much more complex than that in side fillets because of the local deformations within the weld and adjacent plate material. An experimental determination of stresses in steel-plate models representing welded connections has been made[6] and the results are shown in Figs. 6-25a, b, and c. (Note that these stress distributions are valid within elastic limits of the material only.) If we compare the distribution of stresses it is apparent that:

(a) Stresses in the T joint are approximately the same as in the strap joint without friction except that f_y stresses on face CB are considerably higher for the strap joint.

(b) Friction in the strap joint reduces all stresses considerably.

In joints without friction the throat shearing stresses at the heel of the fillet weld (point B in Figs. 6-25a and b) are two to four times the value of the nominal throat shearing stress. With 40% of load transmitted by friction between the plates (Fig. 6-25c), the value of the throat shearing stress is materially reduced and is about 77% of the nominal value.

Ultimate nominal shearing strength of fillet welds is based on tests of specimens subjected to pure tension, as in Figs. 6-26a and b. For end fillet welds, nominal shearing strength is not less than seven-eighths of the minimum tensile strength of the base metal, and for side fillet welds it is not less than two-thirds of the same. Test results[5] indicate that when fillet welded connections are subjected to tension, failure usually occurs in the welds. The type of failure and strength of weld depend largely on the eccentricity of the load.

Consider connections shown in Fig. 6-26. Connection a is highly eccentric and it undergoes a large bending deformation before the weld fails along the face by prying action. Connection b, on the other hand, is symmetrical,

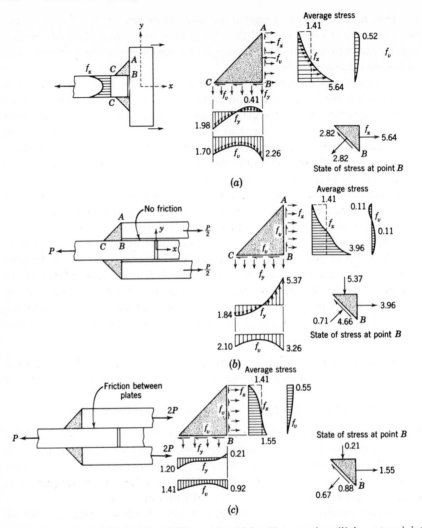

Fig. 6-25 Stress distribution in end fillet welds: (*a*) in a T connection, (*b*) in a strap joint without friction, and (*c*) in a strap joint with 40% of load transmitted by friction.

bending of the weld is prevented, and the weld fails approximately along the plane of the throat section. Connections *a* and *b* were made by means of so-called end welds. Connection *c* shows a symmetrical side weld connection, which fails approximately along the plane of the throat section, and connection *d* shows an eccentric side weld connection, which fails in about the same manner as *c*. Test results indicate that eccentricity of connection reduces the ultimate strength of end welds, such as those in Fig. 6-26*a*, about 35%, and

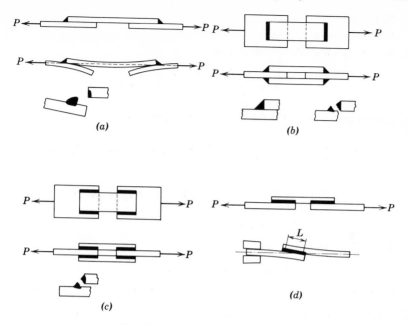

Fig. 6-26   Failure of fillet welds.

of side welds, such as those in Fig. 6-26d, about 10% when compared to similar symmetrical connections. Furthermore, test results indicate that end welds (Fig. 6-26b) are about 35% stronger than side welds (Fig. 6-26c), although side welds are generally more uniform in their strength properties.

Example 6-4.   For the fillet welded connection shown in Fig. 1 determine the allowable load P, using AISC Specification. Assume A36 steel and E-70 series electrodes.

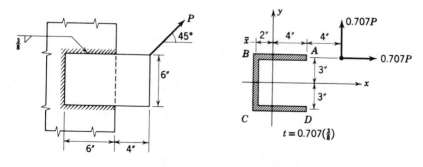

Fig. 1

Solution

Weld-Section Properties.

Section symmetrical about x axis:

$$\bar{x} = \frac{\sum Ax}{\sum A} = \frac{2 \times 6 \times 3}{3 \times 6} = 2$$

Throat area: $A = \sum Lt = 3 \times 6 \times t = 18t$ in.^{2}

$$I_x = \sum I_{ox} + \sum Ay^2 = t\frac{(6)^3}{12} + 2 \times 6 \times t(3)^2 = 126t \text{ in.}^4$$

$$I_y = \sum I_{oy} + \sum Ax^2 = 2t\frac{(6)^3}{12} + 2 \times 6t(1)^2 + 6t(2)^2 = 72t \text{ in.}^4$$

$$I_z = I_x + I_y = 198t \text{ in.}^4$$

Forces and Moments.

$$P_x = 0.707P, \qquad P_y = 0.707P, \qquad P_z = 0$$
$$M_x = 0, \qquad M_y = 0, \qquad M_z = 0.707P(8 - 3) = 3.53 \text{ kip-in.}$$

Stresses due to Forces and Moments.

$$f_x' = \frac{P_x}{A} = \frac{0.707P}{18t} = 0.039\frac{P}{t} \text{ ksi}, \qquad f_y' = \frac{P_y}{A} = 0.039\frac{P}{t} \text{ ksi}, \qquad f_z' = 0$$

$$f_x'' = \frac{M_z}{I_z}y = \frac{3.53P}{198t}y = 0.018\frac{P}{t}y \text{ ksi}, \qquad f_y'' = \frac{M_z}{I_z}x = 0.018\frac{P}{t}x \text{ ksi},$$

$$f_z'' = 0$$

Consider points A, B, C, D, Fig. 2. (Stresses shown are those resisting applied loads.) It is seen that maximum stresses occur at point D, as follows:

$$f_x'' = 0.018\frac{P}{t}(3) = 0.054\frac{P}{t} \text{ ksi}$$

$$f_y'' = 0.018\frac{P}{t}(4) = 0.072\frac{P}{t} \text{ ksi}$$

$$f_x = f_x' + f_x'' = (0.039 + 0.054)\frac{P}{t} = 0.093\frac{P}{t} \text{ ksi}$$

$$f_y = f_y' + f_y'' = (0.039 + 0.072)\frac{P}{t} = 0.111\frac{P}{t} \text{ ksi}$$

$$f_r = \sqrt{f_x^2 + f_y^2} = \frac{P}{t}\sqrt{(0.093)^2 + (0.111)^2} = 0.145\frac{P}{t} \text{ ksi}$$

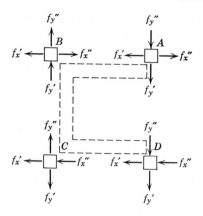

Fig. 2

Load per inch (maximum) is

$$q_r = f_r t = 0.145 \frac{P}{t} t = 0.145P \text{ kip/in.}$$

For welded connections with welds of uniform size, calculations may be simplified by considering $t = 1$ and computing q_r values directly without calculating stresses.

Allowable load per inch of $\frac{3}{8}$-in. fillet weld is (Sec. 6-9)

$$13.6 \times 0.707 \times \tfrac{3}{8} = 3.6 \text{ kip/in.}$$

Allowable load P:

$$q_r = 0.145P = 3.6 \text{ kip/in.} \qquad P = \frac{3.6}{0.145} = 24.8 \text{ kips}$$

6-9 ALLOWABLE STRESSES

American Welding Society rules have been adopted for design of most of the conventional steel structures and have been incorporated in the AISC Specification for the Design, Fabrication, and Erection of Structural Steel for Buildings and the AWS Standard Specification for Welded Highway and Railway Bridges.

The allowable stresses for butt welds subjected to static loads are usually the same as the values for base metal. Butt welds subjected to combined shearing and normal stresses shall be so proportioned that the combined stress shall not exceed the value allowed for shear. The manner in which the stresses are to be combined is not specified by codes and procedures similar to that indicated for riveted and bolted connections (Sec. 5-15) may be used.

For fillet welds the allowable stresses are taken as 65 to 72% of the allowable tensile stress for base metal, specific values depending on the type of steel and structure, as shown in Table 6-4. For example, using A36 steel for building design, the allowable fillet weld stress for E-70 series electrodes is 15.8 ksi, or 72% of 22 ksi allowed in tension. The stress in the fillet weld is considered as shear on the throat regardless of its actual direction and thus no special provision for combined stress is needed.

The design of buildings is usually based on static conditions of loading and consequently the static values in Table 6-4 will govern the design of welded connections. However, reversal of stresses must be considered for high earthquake loads or heavy moving loads, such as cranes or special heavy lifting trucks. The AISC Specification for Buildings (1963) recognizes the fatigue effects of a reversal of stress and makes provision for increasing the design loads accordingly.

The AISC Specification states that for 10,000 or fewer complete stress reversals the static values for the members, connection material, and fasteners need not be increased. For an expected repetition of loading from 10,000 to 100,000 cycles the design load to be supported at the static unit stresses is the "algebraic difference of the maximum computed stress and two-thirds of the minimum computed stress," but the stress-carrying area shall not be less than that required for either the maximum or minimum stress based upon the static allowable stress for the welds. The maximum stress is the result of the largest tension stress acting on the weld and the minimum stress is the result of the largest compression stress acting on the same weld.

Table 6-4   Allowable Static Shear Stresses for Structural Welds

	Stresses, ksi	
Type of Weld	Buildings	Bridges
Fillet plug and slot welds All Steels, E-60, SAW-1	13.6	—
A36, A242, A441		
E-70, and SAW-2	15.8	—
Fillet welds A36	—	12.4
Fillet welds A242, A441	—	14.7
Plug, All Steels	—	12.4

The specification includes provisions for greater cycles of maximum loading and the corresponding increase in the design loads in order to ensure safety of the member and its connection.

The specifications for bridge designs have included provisions for the effect of fatigue and have developed a series of expressions for various conditions of loadings. These were discussed in greater detail in Sec. 3-16.

To ensure that welds will develop their full strength, most structural specifications prescribe the type of electrode to be used and the temperature of the pieces to be connected. The advances in metallurgy and steel making usually require closer control of fabrication and material conditions. For example, the AISC Specification for Buildings requires that the base metal shall be preheated to required temperatures specified in Table 6-5 which presents minimum preheat and interpass temperatures for ordinary and low-hydrogen electrodes.

Table 6-5 Minimum Preheat and Interpass Temperatures

Thickness of Thickest Part at Point of Welding	Other Than Low-Hydrogen Welding Processes[1]		Low-Hydrogen Welding Processes[2]	
	A36 Steel	A441 Steel	A36 Steel	A441 Steel[3]
To 1 in.	None[4]	Welding with	None[4]	None[4]
Over 1 in. to 2 in.	200°F	this process not	50°F	100°F
Over 2 in.	300°F	recommended	150°F	200°F

[1] Welding with ASTM A233 E-60XX or E-70XX electrodes other than a low-hydrogen class.

[2] Welding with properly dried ASTM A233 E-XX15, 16, 18, or 28 electrodes or submerged arc welding with properly dried flux.

[3] Preheating for weldable A242 steel may need to be either higher or lower than these requirements, depending on composition of steel.

[4] Except when base metal temperature is below 32°F.

6-10 DESIGN CONSIDERATIONS—ELASTIC AND PLASTIC

General—Elastic. The principal advantages of welded structures are simplicity of design and economy of parts. A few welded connections shown in Fig. 6-27 indicate that the use of welding may result in a considerable saving of material and reduce shop handling and fabrication operations.

Butt welds are preferable to fillet welds on the basis of strength, particularly fatigue strength. On the other hand, butt welds result in higher residual stresses, require more expense in preparing edges for welding, and impose somewhat exacting limitations on the length tolerance of the parts.

A special case of interest is a butt-welded splice with and without strap plates, as shown in Fig. 6-28. Tests indicate that for structural steel, the static strength of these two connections is nearly the same as that of unwelded plates, and the fatigue strength of the unreinforced splice is considerably greater than that of the splice reinforced with straps.

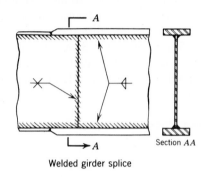

Welded girder splice

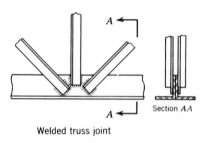

Welded truss joint

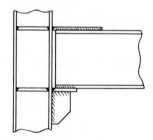

Welded frame connection

Fig. 6-27 Welded connections.

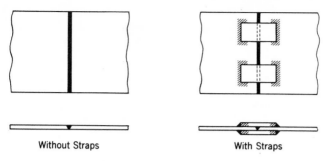

Without Straps   **With Straps**

Fig. 6-28   Butt-weld splices.

Two other factors must be considered in the choice of weld size: minimum size of weld to prevent quick cooling resulting in weld brittleness, and maximum size of weld as determined by practical limitation of obtaining proper shape of weld. The sizes of welds recommended by the AWS D2.0-63 bridges and D1.0-63 buildings are shown in Table 6-6.

Table 6-6   Recommended Fillet Weld Sizes

Weld size, in.	$\frac{3}{16}$	$\frac{1}{4}$	$\frac{5}{16}$	$\frac{3}{8}$	$\frac{1}{2}$	$\frac{5}{8}$
Maximum plate thickness, in. Buildings and Bridges	$\frac{1}{2}$	$\frac{3}{4}$	$1\frac{1}{2}$	$2\frac{1}{4}$	6	over 6

Maximum fillet weld size along edges of material is:

(*a*) For material less than $\frac{1}{4}$ in., maximum size may be equal to the thickness of the material.

(*b*) For material $\frac{1}{4}$ in. or more in thickness, maximum size shall be $\frac{1}{16}$ in. less than the thickness of the material unless otherwise specified on the drawings.

In general, the minimum practical size of weld is $\frac{3}{16}$ in., very rarely $\frac{1}{8}$ in. The most economical size of weld is around $\frac{5}{16}$ in., which is generally the maximum weld that can be obtained manually in one pass. The cost–strength ratio increases rapidly with increase in size. Consider the butt welds shown in Fig. 6-29, in which the amount of the weld is proportional to the square of the thickness. Since the cost per unit volume of weld is very nearly constant, the cost per unit load increases with the size of weld. This consideration applies to welds greater than $\frac{3}{8}$ in. in size; welds smaller than $\frac{1}{4}$ in. are generally not economical from the point of view of labor cost.

It is generally economical to use a small-size continuous weld rather than a larger discontinuous weld—both are deposited in one pass. When intermittent welds are used, it is good practice to space them apart a distance

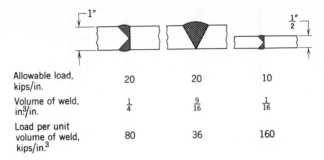

Allowable load, kips/in.	20	20	10
Volume of weld, in.³/in.	$\frac{1}{4}$	$\frac{9}{16}$	$\frac{1}{16}$
Load per unit volume of weld, kips/in.³	80	36	160

Fig. 6-29 Economy of weld size and shape.

equal to at least their own length; otherwise, a continuous weld would probably be desirable. Intermittent welds may be used to transmit calculated stress in buildings when fatigue effects are negligible. In bridge design intermittent welds should not be used to transmit calculated stress.

Various limitations on length and spacing of fillet welds are included in specifications and are shown in Table 6-7. In addition to these, if longitudinal

Table 6-7 Length and Spacing of Fillet Welds

	Buildings	Bridges
Continuous Welds		
Minimum length of weld, in.	4 × weld size	4 × weld size, $1\frac{1}{2}$ in. minimum
Intermittent Welds		not to carry calculated stress
Minimum length of weld, in.	4 × weld size, $1\frac{1}{2}$ in. minimum	Same
Maximum longitudinal clear spacing, in. (part in tension)	24 × t min.*	14 × t min.*
Same as above (parts in compression)	$\left(\dfrac{4000}{\sqrt{F_y}}\right) t$ min.*	10 × t min.*
Maximum transverse clear spacing, in.	$\left(\dfrac{8000}{\sqrt{F_y}}\right) t$	$24t$

* Not to exceed 12 in. for plates and 24 in. for rolled shapes.

fillet welds are used alone in end connections, the length of each weld shall be not less than the maximum width of the member. Also side and end fillets terminating at ends or sides shall be returned continuously around the corners for a distance not less than twice the nominal size of the weld; the length of weld return may be included in the effective length of weld. Other special provisions for length and spacing are contained in various pertinent specifications.

Design considerations for several typical welded connections will be discussed in Sec. 6-11.

Plastic Design.   All connections for a structure which is proportioned on the basis of the theory of plastic design must be capable of resisting the moments, shears, and axial loads which are acting on the connections as a result of the applied ultimate loads. Therefore, the welds must be proportioned to resist the forces produced at ultimate load using unit stresses which have been increased accordingly.

For butt-weld stresses the assumption is made that the weld material is capable of developing the tensile yield stress of the base metal on the minimum throat area.

For fillet-weld stresses the assumption is made that the weld is capable of developing at least the shearing yield stress of the weld metal on the minimum throat area. A safe design value is obtained by multiplying the elastic design allowable stress value for the weld by the ratio of F_y/F_w, where F_y is the yield strength and F_w is the allowable tensile stress of the base material.

6-11   TYPICAL WELDED CONNECTIONS FOR BUILDINGS

A designer must realize that the welding process may result essentially in a one-piece structure. Complete continuity may be achieved with welded connections and this fact must be considered in the determination of the internal forces acting on the members and their connections. Although simple beam action is possible with welded connections, the greatest advantage of welding is realized when full continuity is intentionally designed into the structure at the outset.

Welding offers many advantages to the engineer and the more important ones for design follow:

(*a*) Reduction in steel weight.
(*b*) Reduced beam depth due to continuity.
(*c*) Connection material may be eliminated.
(*d*) Smooth, uncluttered appearance is possible.
(*e*) Economical.
(*f*) More imaginative structures are possible.

A great variety of welded connections may be designed[7]; in this text only a few typical connections will be discussed and design examples based on the working load procedure will be included.

Framing Angles. These are used in building construction for beam-to-column and beam-to-girder connections (Fig. 6-30) with the angles shop-welded to the beam web and field-welded to the column or girders after erection is completed. Usually the connection is designed to transmit shear and to minimize resistance to rotation.

Considering the free body of the framing angles (Fig. 6-31), it is usually assumed that the beam weld resists a moment $M_x = \frac{1}{2}Ra_1$, and the column or girder weld resists a moment $M_y = \frac{1}{2}Ra_2$. In addition, both welds resist direct shear equal to $\frac{1}{2}R$. Actually the welds are subjected to a combination of direct shear and moments about both axes x and y. The magnitudes of

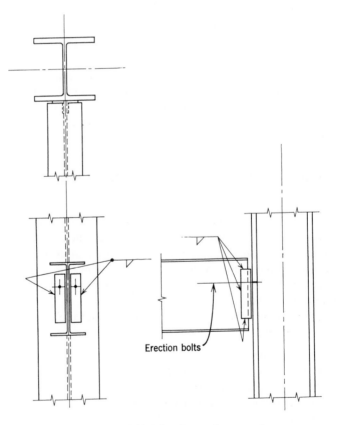

Erection bolts

Fig. 6-30 Welded framing angle connections.

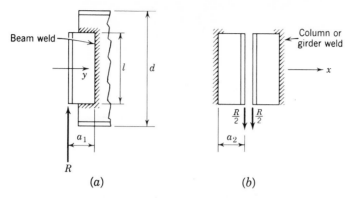

Fig. 6-31 Freebody of framing angles.

the moments are highly indeterminate, but in common practice the welds are designed only for the nominal values of moments defined previously.

In proportioning framing angle connections, the following problems arise: (a) determination of angle size, (b) column and beam weld size, (c) local shearing stresses in beam web, and (d) end rotation of the connection.

Usually a 3-in. angle leg is sufficient for connection to the beam web and a 3- or 4-in. leg may be used for the column or girder connection. The length of the angle is determined by the magnitude of shear to be carried, but it is limited by the depth of beam and suitable clearance required for welding. Thus, for a beam having over-all depth d and flange width b, the length l of the framing angle cannot exceed $(d - b/2)$ and preferably should not exceed $(d - b)$.

When the angle dimensions a_1, a_2, and l are established, the sizes of welds required to carry a given load can be determined. The thickness of the angle normally should be $\frac{1}{16}$ or $\frac{1}{8}$ in. greater than the weld size. A small thickness of angle is usually preferred in order to minimize restraint of the beam rotation at the support.

The local shearing stress f_s in the beam web at the weld is equal to $f = 2q/t_w$, where q is the load per inch of each weld and t_w is the web thickness, and this stress should not exceed 14.5 ksi allowed for shear in A36 steel beams. Normally this limitation is not critical for beams larger than 24 W^F, but it becomes important for beams shallower than 10 in.

End rotation θ of the beam (Fig. 6-32) produces displacement Δ of the top flange equal to $\theta \cdot d/2$, and displacement Δ_x of the top of the framing angle equal to $\theta \cdot l/2$. Assuming no restraint to rotation, angle θ can be determined as that for a simply supported beam under given loading conditions. For example, for a uniformly loaded beam (Fig. 6-33), $M = wL^2/8 = fI/(d/2)$

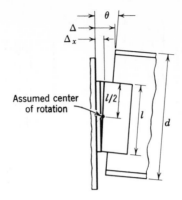

Fig. 6-32   End rotation of beam.

and $\theta = \frac{1}{2}(\frac{2}{3}ML/EI) = \frac{2}{3}fL/Ed$.   Thus, $\Delta = \theta \cdot d/2 = \frac{1}{3}fL/E$ and $\Delta_x = \theta \cdot l/2 = \frac{1}{3}(fL/E)l/d$.   It follows that for a given beam, the greater the length l, the greater is the value Δ_x; consequently, normal stresses in the weld and in the beam web may become significant and may induce appreciable restraint to rotation.

It also follows that in the value of $\Delta = \frac{1}{3}fL/E$ for a steel beam with given maximum bending stress f, Δ depends directly on L. For a given beam on 20-ft span, the movement Δ is twice that of a similar beam on a 10-ft span and therefore framing angles may not be suitable for long beams. If they are used for long beams, particular care should be exercised in their design to evaluate the effect of end restraint on the strength of the connection.

The relationship between the angle thickness t and the displacement Δ_x of the heel of the angle may be estimated as follows. Considering a strip of the angle (Fig. 6-34), $M = Ta_2 = f_a(I/c)$ and $T = M/a_2 = f_a(I/c)/a_2 = f_a t^2/6a_2$. Also $\Delta_x = Ta^3/kEI$ and, substituting values of T and I, $\Delta_x = 2f_a a_2{}^2/kEt$,

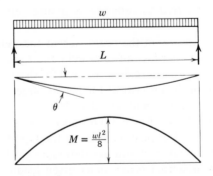

Fig. 6-33   A uniformly loaded beam.

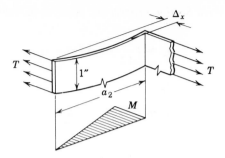

Fig. 6-34　A strip of a connection angle.

where k is a numerical coefficient varying between 2 and 3, depending on the rigidity of the heel of the angle. Using value of $k = 2$, $\Delta_x = f_a a_2^2 / Et$ and, for given t, $f_a = Et\Delta_x / a_2^2$ or, for given f_a, $t = f_a a_2^2 / E\Delta_x$. It should be noted that increase in t causes a corresponding increase in f_a, so that using more material does not necessarily give a better or stronger connection.

In general, any restraint of the movement of the heel of the angle should be avoided; thus column or girder welds should not be placed near the heel. If fitting bolts used in erection are to be left in place, they should be placed below the mid-depth of the angle.

Example 6-5.　For a simply supported 16 W̄ 36 beam carrying a uniformly distributed total (dead and live) load, $w = 2.9$ kip/ft, design a framing angle of welded connection, using AISC Specification (Fig. 1). Assume A36 steel and E-70 electrodes.

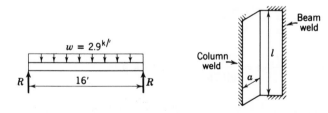

Fig. 1

Solution

$$R = \frac{wl}{2} = 2.9 \times 8 = 23.2 \text{ kips}$$

$$l = (d - b) = 16 - 7 = 9 \text{ in.}$$

Try 3 × 3 angles.

Column Weld. Weld Section Properties.   Length of weld $l = 9$ in., Fig. 2,

$$I_x = t\frac{9^3}{12} = 60.7t$$

$$I_y \cong 0$$

$$I_z = I_x + I_y = 60.7t$$

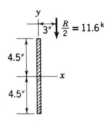

Fig. 2

Forces and Moments.

$$P_x = 0, \qquad P_y = \frac{R}{2} = 11.6 \text{ kips}, \qquad P_z = 0$$

$$M_x \cong 0, \qquad M_y \cong 0, \qquad M_z = \frac{R}{2}\,a = 11.6 \times 3 = 34.8 \text{ kip-in.}$$

Load per inch:

$$q_y' = \frac{P_y}{L} = \frac{11.6}{9} = 1.29 \text{ kip/in.}$$

$$q_x'' = \frac{M_z}{I_z}\,yt = \frac{34.8}{60.7t}\,4.5t = 2.57 \text{ kip/in.}$$

$$q_r = \sqrt{q_y{}^2 + q_x{}^2} = \sqrt{(1.29)^2 + (2.57)^2} = 2.87 \text{ kip/in.}$$

The allowable load per inch for a $\frac{5}{16}$-in. weld is $15.8 \times 0.707 \times \frac{5}{16} = 3.4$ kip/in. Use $\frac{5}{16}$-in. weld, E-70 electrodes.
    Angle thickness:

$$\text{Weld size} + \tfrac{1}{8} \text{ in.} = \tfrac{5}{16} + \tfrac{1}{8} = \tfrac{7}{16} \text{ in. thick}$$

($\frac{3}{8}$ in. may be used; see Table 6-4.)

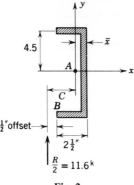

Fig. 3

Beam Weld. Weld Section Properties.   Symmetrical about x axis Fig. 3:

$$\bar{x} = \frac{\sum Ax}{\sum A} = \frac{2 \times 2.5 \times 1.25}{2 \times 2.5 \times 9} = 0.446 \text{ in.}$$

$$I_x = \sum I_{ox} + \sum Ay^2 = 2 \times 2.5(4.5)^2 t + \frac{(9)^3}{12} t = 162.0t \text{ in.}^4$$

$$I_y = \sum I_{oy} + \sum Ax^2 = 2 \times \frac{(2.5)^3}{12} t + 2 \times 2.5(1.25 - 0.45)^2 t + 9t(0.45)^2$$

$$= 7.6t \text{ in.}^4$$

$$I_z = I_x + I_y = 169.6t \text{ in.}^4$$

Forces and Moments.

$$\text{P}_y = 11.6 \text{ kips,} \qquad M_z = \frac{R}{2} C = 11.6(3 - 0.45) = 29.6 \text{ kip-in.}$$

Maximum load per inch (critical point B):

$$q_y' = \frac{P_y}{L} = \frac{11.6}{14} = 0.83 \text{ kip/in.}$$

$$q_y'' = \frac{M_z}{I_z} \times t = \frac{29.6}{169.6t} 2.05t = 0.36 \text{ kip/in.}$$

$$q_x'' = \frac{M_z}{I_z} \times t = \frac{29.6}{169.6t} 4.5t = 0.79 \text{ kip/in.}$$

$$q_r = \sqrt{q_y^2 - q_x^2} = \sqrt{(0.83 + 0.36)^2 + (0.79)^2} = 1.43 \text{ kip/in.}$$

Using $\frac{3}{16}$-in. weld, $q = 1.8$ kip/in. ($\frac{1}{4}$-in. weld may be used.) The reader should verify selection of column and beam welds using tables in the 1963 AISC Manual, p. 4-23.

Check rotation and angle thickness. See Fig. 6-32.

$$\theta = \frac{2}{3}\frac{fL}{Ed} = \frac{2}{3} \cdot \frac{19.8 \times 16 \times 12}{30 \times 10^3 + 15.85} = 5.35 \times 10^{-3}\,\text{radian}$$

For $l = 9$ in.,

$$\Delta_x = \theta \times \frac{l}{2} = 5.35 \times 10^{-3} \times 4.5 = 0.024 \text{ in.}$$

Using $t = \frac{7}{16}$-in. angle,

$$f_a = \frac{Et\Delta_x}{a^2} = \frac{30 \times 10^3 \times 7 \times 0.024}{16 \times (3)^2} = 35 \text{ ksi}$$

Using $t = \frac{3}{8}$-in. angle,

$$f_a = \tfrac{6}{7}(35) = 30 \text{ ksi}$$

Both values of t indicate high bending stresses in the angle. But such high stress is not objectionable unless many cycles of repeated loadings are expected. Use $t = \frac{3}{8}$ in. as minimum permissible to accommodate $\frac{5}{16}$-in. column weld.

Deflection of beam, top flange $\Delta = 5.35 \times 10^3 \times 15.85/2 = 0.0424$ in.
AISC, 1963, Sec. 1.15.4 limits deflection of top flange of beam to

$\Delta = 0.007d$ for uniform load and live load deflection of $\frac{1}{360}$ span

$\therefore\ \Delta = 0.007(15.85) = 0.111$ in. > 0.0424

Therefore connection angles are satisfactory.

Beam Seat.   Beam-seat connections (Fig. 6-35) facilitate ease of erection and provide adequate connections for moderate end reactions. The beam seat is usually shop-welded to the column and the beam is field-welded to the seat after erection is complete. The top angle is used for lateral bracing of the beam and is also field-welded. Usually it is not designed to carry any portion of the end shear.

The beam seat (Fig. 6-36) must be designed to provide adequate bearing length for the beam, to resist safely the bending at the heel of the angle, and to provide a weld safely resisting the direct shear and the moment produced by the reaction. The size of outstanding leg a is determined by beam bearing and usually a 3- or 4-in. leg is adequate. The thickness t of the angle is determined by stresses at the section m–m of critical moment just inside the

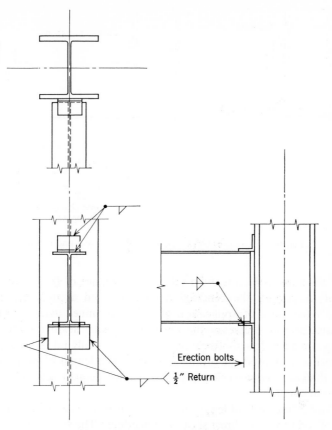

Fig. 6-35 Beam-seat connection.

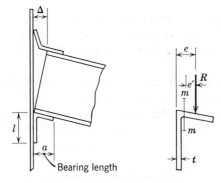

Fig. 6-36 Beam-seat design.

225

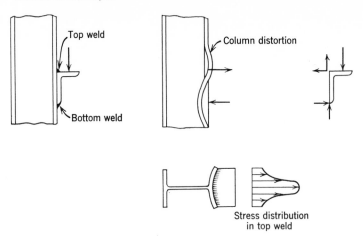

Fig. 6-37   Undesirable welds between beam seat and column.

heel of the angle, which is often taken at a distance of $(t + \frac{3}{8})$ in. from the back of the angle; the bending moment resisted at this section is Re'. Normally the point of application of R is determined assuming uniformly distributed bearing stresses on the angle. For greater flexibility, the allowable bending stress for design of beam-seat angle is usually taken as 20% greater than normally allowed for ordinary beams. The length of beam-seat angle and its thickness may be selected to provide an adequate section to resist the bending moment.

The size of the vertical leg of the beam seat is determined by the required length of weld to resist direct shear and bending. The size of this leg and the size of the weld may be varied to provide an adequate and an economical design.

As in framing-angle connection, the beam is assumed to rotate freely at the end and requires relatively free movement at the top of the beam. The top angle therefore should be flexible and usually a $4 \times 4 \times \frac{1}{4}$-in. angle is selected.

For flexible connection, the beam-seat attachment to the column should be made along the vertical sides of the seat angle (Fig. 6-35) and not along top and bottom (Fig. 6-37). The top-bottom connection tends to distort the column flanges and produces great stress concentration and possible failure in the top welds near the web of the column (Fig. 6-37).

Example 6-6.   For the beam of Example 6-5, proportion the beam seat and the welds in the connection as shown in Fig. 1. AISC Specification. A36 steel and E-70 electrodes.

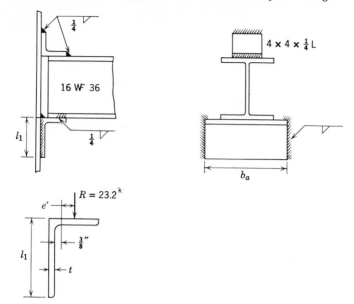

Fig. 1

Solution

Try: outstanding leg of seat angle 4 in., angle thickness $t_a = \frac{3}{4}$ in.
Then: e' to critical section for bending in angle:

$$e' = (\tfrac{4}{2} - t_a - \tfrac{3}{8}) = 0.87 \text{ in.}$$

For maximum bending stress in angle, $f_b = 22 \times 1.2 = 26.4$ ksi, determine b_a:

$$f = \frac{Mc}{I} = \frac{(Re')6}{b_a(t_a)^2} = \frac{23.2 \times 0.87 \times 6}{b_a(\frac{3}{4})^2} = 26.4 \text{ ksi}, \qquad b_a = 8.125 \text{ in.}$$

Use $b_a = 10$ in. (providing approximately 1 in. on each side of the beam flange). With $b_a = 10$ in., we may use angle thickness

$$t_a = \sqrt{\frac{8.125}{10}} \left(\frac{3}{4}\right) = 0.675 \approx \frac{3}{4} \text{ in.}$$

Assume $l_1 = 6$ in., weld throat $= t$.
Weld-Section Properties (Fig. 2).

$$L = 2l_1 = 2 \times 6 = 12 \text{ in.}$$

$$I_x = 2t \frac{(6)^3}{12} = 36t$$

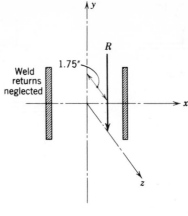

Fig. 2

Forces and Moments.

$$P_y = R = 23 \text{ kips}, \qquad M_x = R \times (\tfrac{3}{8} + \tfrac{3}{8} + \tfrac{7}{8}) = 46 \text{ kip-in.}$$

Stresses (Load per Inch).

$$q_y{}' = \frac{P_y}{L} = \frac{23.2}{12} = 1.93 \text{ kip/in.}$$

$$q_z{}'' = \frac{M_x}{I_x} \times t = \frac{46.0}{36t} \, 3t = 3.83 \text{ kip/in.}$$

Resultant:

$$q_r = \sqrt{q_y{}^2 + q_z{}^2} = \sqrt{(1.93)^2 + (3.83)^2} = 4.29 \text{ kip/in.}$$

Use $\tfrac{7}{16}$-in. fillet weld, E-70 electrodes.

$$q_a = 15.8 \times 0.707 \times \tfrac{7}{16} = 4.9 \text{ kip/in.}$$

Use $4 \times 6 \times \tfrac{3}{4}$-in. angle $\times$ 10 in. long.

The reader should verify selection of angle and weld size, using tables in the AISC Manual.

Stiffened Seat. For heavy reactions, say in excess of 40 kips, adequate beam-seat angles become excessively thick and a stiffened seat is more economical. The stiffened seat is usually made up of a standard T section, although sometimes two welded plates may be utilized. A typical beam-to-column-web connection is shown in Fig. 6-38.

When a stiffened seat is used, the reaction is concentrated near the part of the seat away from the support because of the tendency of the beam to rotate.

Thus, the moment on the welds attaching the stiffened seat to the support is greater than for ordinary beam seats.

The stiffened seat must be designed to provide adequate bearing length for the beam, prevent distortion or buckling of the stiffener, and resist safely the direct shear and bending moments on the weld attaching the seat to the support.

For a typical connection the point of application of the load is located at a distance of $\frac{2}{3}a$ from the support, Fig. 6-39. The stiffener is designed to resist direct compression and bending. The combined axial and bending stresses on the critical section should not exceed the allowable value, normally taken as 22 ksi. For the usual proportions of the stiffener, buckling is not critical, but in any case the thickness of the stiffener should not be less than that of the beam web.

The flange of the T-shaped seat should be wider than the flange of the beam and should accommodate a field-welded connection. The stem of the seat should be determined by the required length of weld to resist direct shear and bending. The length of stem and the size of weld should be selected to provide an economical design. The vertical stem of the stiffened seat when attached to a plate, such as column

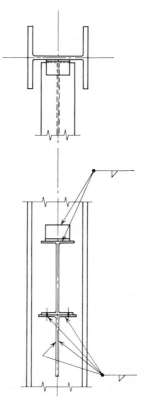

Fig. 6-38 Stiffened beam seat connection.

or beam web, produces bending in the plate, and if this web plate is relatively thin, it may be overstressed. For T-shaped stiffened seats the flange of the T stiffens the plate and tends to reduce the bending stresses in the web. The allowable moment $M_a = Ve$ (Fig. 6-39) on the seat may be limited by the stresses in the web plate, nevertheless. If the web thickness is t, and length of the stiffener is L, then M_a can be taken as

$$M_a = kLt^2$$

where k depends on bracket shape and web plate edge conditions. For T-shaped seat coefficient k may be taken approximately 20 to 24.[8]

Angle Connection. Angles are frequently used for tension or compression members in light and medium trusses. Welded connections of an axially loaded angle to a gusset plate or to another member are shown in Fig. 6-40.

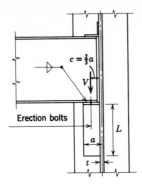

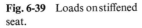

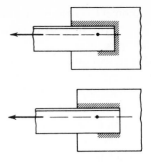

Fig. 6-39 Loads on stiffened seat.

Fig. 6-40 Design of angle connection to eliminate eccentricity.

Properly designed connections are proportioned to eliminate eccentricity in the plane of the welds by varying lengths of weld on each side of the leg. The criterion for design requires the resultant of uniformly stressed welds to be collinear with the external force. The proportioning of the welds is further subject to restrictions on length and spacing of welds previously discussed in this section.

Example 6-7.   Determine the length of welds required for the connection in Fig. 1. Welds are $\frac{1}{4}$ in. Use AISC Specification and A36 steel.

Solution

The eccentricity of the load out of plane of welds is neglected.

For $\frac{1}{4}$-in. welds, $q_a = 2.8$ kip/in. for A36 steel, with E-70 electrodes.

For equilibrium,

$$\sum F = 0, \quad 2.8(L_1) + 2.8(L_2) = 32 \text{ kips} \tag{1}$$

$$\sum M = 0, \quad (2.8L_1)2.86 = (2.8L_2)1.14 \tag{2}$$

From Eq. 2,

$$L_1 = \frac{1.14}{2.8} L_2 = 0.398L_2$$

Substitute in Eq. 1:

$$2.8(0.398L_2) + 2.8L_2 = 32 \text{ kips}$$
$$L_2 = 8.17 \text{ in.}$$
$$L_1 = 0.398L_2 = 3.25 \text{ in.}$$

Let $L_1 = 4$ in. and use $L_2 = 9$ in.

Although this example illustrates an exact procedure for calculating the size and length of weld to attach an angle to a gusset plate, practical considerations of behavior simplify the connection. The AISC Specification,

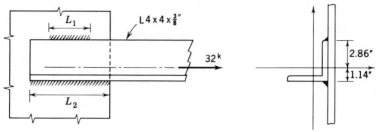

Fig. 1

Sec. 1.15.3, permits equal length welds sufficient to carry the entire load. The necessity to balance welds according to the forces is not required for end connections of single-angle, double-angle, and similar-type members.

Example 6-8.   Column Base.   A typical column base to transfer large moments is easily fabricated with angles or plates welded to the column flange.   These are indicated in Fig. 1.

Design a column base plate attachment for a 14 W 87 lb column to withstand a moment of 690 kip-in. (Fig. 1).   AISC specification.

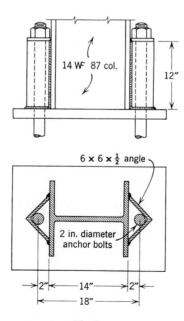

Fig. 1

Anchor Bolts.   Assume direct load to be small and design bolts for this moment.

Load on bolt

$$F = \frac{M}{d} = \frac{690}{18} = 38.4 \text{ kips}$$

Required bolt area (A36 steel) for bolts $= \dfrac{38.4}{22}$

$$= 1.75 \text{ sq in.}$$

Use a 2-in.-diameter bolt; area at root of thread $= 2.30$ sq in.

Determine amount of welding; treat weld as a line.
Section modulus of weld:

$$S_w = \frac{2L^2}{6} = \frac{L^2}{3}$$

Bending force on weld:

$$f_b = \frac{M}{S_w} = \frac{Fe}{\dfrac{L^2}{3}}$$

Vertical shear force on weld:

$$f_v = \frac{F}{2L}$$

Resultant force on weld:

$$f_r = \sqrt{f_v^{\,2} + f_h^{\,2}} = \sqrt{\frac{F^2}{4L^2} + \frac{9F^2 e^2}{L^4}}$$

$$= \frac{F}{2L^2} \sqrt{L^2 + 36e^2}$$

and letting $w =$ fillet leg size

$$f_r = 15.8 \times 0.707 \times w = 11.2w$$

or

$$w = \frac{F}{22.4L^2} \sqrt{L^2 + 36e^2}$$

Solving directly for L, this becomes

$$L^2 = \frac{1}{2}\left(\frac{F}{22.4w}\right)\left[\left(\frac{F}{22.4w}\right) + \sqrt{\left(\frac{F}{22.4w}\right)^2 + 144e^2}\,\right]$$

Here, assume $w = \frac{1}{4}$-in. fillet

$$\left(\frac{F}{22.4w}\right) = \frac{38.4}{(22.4)(\frac{1}{4})} = 6.85$$

$$e = 2.0 \text{ in.}$$

and

$$L^2 = \tfrac{1}{2}(6.85)[6.85 + \sqrt{6.85^2 + 144(2)^2}] = 109.0$$

or

$$L = 10.44 \text{ in.} \quad \text{or} \quad \text{use } L = 11.0 \text{ in.}$$

If a direct axial load is acting in combination with a moment, the net tensile force acting on the bolt must be determined before performing the foregoing analysis.

Example 6-9. Beam to Column Connection. Design a fully welded beam to column connection for a 14 W 30 lb beam to an 8 W 40 lb column to transfer an end moment of 740 kip-in. and a shear of 16.0 kips. This example will be considered with two variations although others are possible.

14 in. W 30 lb beam properties:

$$S = 41.8 \text{ in.}^3$$

$$t_w = 0.270 \text{ in.}$$

$$t_f = 0.383 \text{ in.}$$

$$b_f = 6.733 \text{ in.}$$

First Method, Fig. 1. If the full section of the beam is welded to the column, the bending stress in the outer fiber may be computed as follows: (AISC Sec. 1.5.1.4)

$$f_b = \frac{M}{S} = \frac{740}{41.8}$$

$$= 17,700 \text{ psi} < 22,000 \text{ psi (A36 steel)}$$

Second Method, Fig. 2. Another method would be to separate this connection into the two basic welds, assuming the flange weld to transfer all the bending forces and the weld on the web to transfer only the shear forces.

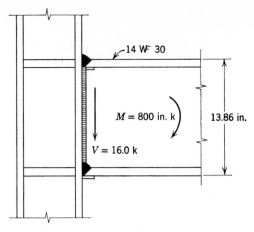

Fig. 1

The bending force in the flange may be found by dividing the bending moment by the distance between the centers of gravity of the top and bottom flange.

$$F_b = \frac{M}{d} = \frac{740}{13.475} = 54.8 \text{ kips}$$

$$A_f = (6.733)(0.383) = 2.58 \text{ sq in.}$$

$$f_b = \frac{F}{A_f} = \frac{54.8}{2.58}$$

$$= 21,200 \text{ psi} < 22,000 \text{ psi (A36 steel)}$$

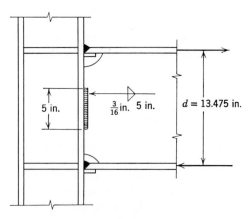

Fig. 2

Welding on web: for design purposes, assume leg size of fillet weld is $\frac{2}{3}t_w =$ 0.180 in. Use $\frac{3}{16}$ in. fillet weld, E-70 electrodes.

$$\text{weld allowable} = q_w = 0.707 \times 15.8w \text{ kip/in.}$$
$$= (11.2)(0.1875) \text{ kip/in.}$$
$$\text{length of weld} = L = \frac{V}{f_w} = \frac{16.0}{(11.2)(0.1875)}$$
$$= 7.9 \text{ in. (total)}$$

Minimum of 4 in. on each side ($\frac{3}{16}$-in. fillet leg); use 5 inches.

The need for stiffeners in the column should be investigated to prevent crippling of the column web.

6-12 COMBINED WELDED AND BOLTED CONNECTIONS

There may be occasions when it is necessary to use bolts acting in conjunction with welds, such as for repairs to old structures or in new work when clearances and working space are at a minimum.

In making alterations by welding, existing rivets and properly tightened high-strength bolts may be assumed to carry the stresses from the existing dead loads and the new welded connection may be assumed to carry all the additional stresses.

In new work which uses rivets, A307 bolts, or high-strength bolts used in bearing-type connections, no combination of load-carrying capacity may be made with welds. The welds are more rigid and hence prevent any load going to the bolts until the welds have yielded. Therefore, if welds are used, they must be designed to carry the entire stress in the connection. In new work, when high-strength bolts are properly installed as a friction-type connection before welding, they may be considered to share the load with the welds. In this instance, the high-strength bolts provide a sufficient resistance to slip to be considered as acting simultaneously with the welds.

6-13 WELDING DIFFERENT STEEL GRADES

The acceptance of higher-strength steels for construction purposes has prompted the AISC to include some of these steels in the 1963 Specification.

The higher-strength steels require low-hydrogen electrodes to be used in the dry condition. Only E-70 low-hydrogen electrodes are approved for manual arc welding or Grade SAW-2 for submerged arc welding on A441 or weldable A242 steels. Fillet welds on high-strength steels may be made by E-60 series low-hydrogen electrodes or Grade SAW-1 submerged arc process.

When using several steel strength levels in the same component, such as a plate girder, the steels may be welded together to act as a unit. For example, a web plate of A36 steel may be butt welded to a plate of A441 steel by using the proper electrode (E-70 low-hydrogen) and proper procedure for making the welds. Any steel grade may be welded to another grade to develop the necessary strength to transfer the stresses.

It is not always necessary to use the same grade of steel throughout the structure and economies may be gained by using low-priced steels for some components. When a beam is designed for a high-strength steel, such as A441, the framing angles may be made of A36 steel provided they can transfer the required loads. However, a beam of A441 steel should never be welded directly to a column of A36 steel because the flange of the column with its lower yield stress will not be able to accept the full transfer load from the beam.

PROBLEMS

Use A36 steel unless otherwise noted.

1. A steel plate subjected to an axial tension load is to be spliced by using a square butt joint welded on one side. If the plate is 10 in. wide, determine the required thickness for the following values of tensile load P: (*a*) $P = 12$ kips, (*b*) $P = 35$ kips. (AISC Specification.)

2. A steel plate subjected to an axial tension load is to be spliced by using a double-bevel butt joint. If the plate is 10 in. wide, determine the required thickness for the following values of tensile load P: (*a*) $P = 60$ kips, (*b*) $P = 160$ kips. (AISC Specification.) A441 Steel.

3. Determine the plate thickness and the size of the welds indicated in Fig. P6-3. (AISC Specification.)

4. For the welded connections shown in Fig. P-4, determine the maximum stress f in the welds in terms of weld size s, eccentricity e, load P, and weld length L.

5. A 6-in.-diameter standard steel pipe is loaded as shown in Fig. P-5. For a single-bevel butt joint welded from one side determine maximum allowable load P. (AISC Specification.)

6. The pipe of Prob. 5 is subjected to a load $P = 20$ kips. If the end welds are to be made using fillet welds, determine the required size of weld. (AISC Specification.)

7. Two plates are welded together as shown in Fig. P-7. Check whether the connection is overstressed according to the AISC Specification. What is the minimum required size of weld permissible? (AISC Specification.)

8. Two $\frac{5}{8}$-in. plates are to be connected by fillet welds as shown in Fig. P-8. Determine the minimum size of welds required. (AISC Specification.)

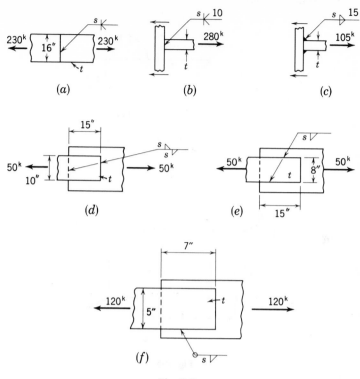

(a)   (b)   (c)

(d)   (e)

(f)

Fig. P-3

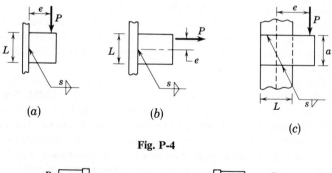

(a)   (b)   (c)

Fig. P-4

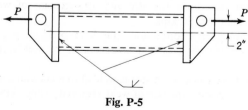

Fig. P-5

237

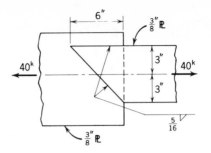

Fig. P-7

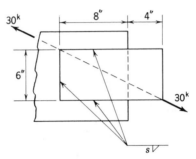

Fig. P-8

9. Welds shown in Fig. P-9 are proportioned to eliminate effects of eccentricity in the plane of the weld. Out-of-plane eccentricity is neglected. Determine lengths L_1 and L_2, using $\frac{3}{8}$-in. welds. (AISC Specifications.)

10. The connection of two 2 in. × 2 in. × $\frac{1}{4}$ in. angles attached to ST 8 W^F 29 as shown in Fig. P-10 is to develop load P equal to the full axial tension capacity of the angles. Determine size and arrangement of welds. (AISC Specification.) (A441 steel.)

11. The connection of Prob. 10 is to develop a maximum axial tension P equal to 10 kips. Determine the size and arrangement of welds. (AISC Specification.)

12. A seat angle welded to a column is shown in Fig. P-12. Assuming that the top angle does not transmit any load to the column, determine the size of weld required to support a beam reaction of 20 kips. (AISC Specification.)

13. A seat angle used to support an offset beam is shown in Fig. P-13. Using AISC Specification, determine the size and arrangement of welds required to support a beam reaction of 20 kips.

14. For the framing angle connection using $3\frac{1}{2}$ in. × $2\frac{1}{2}$ in. × $\frac{3}{8}$ in. angles shown in Fig. P-14, determine the size and arrangement of welds required. (AISC Specification.)

15. A T bracket used as a beam seat is welded to a column and carries a 50-kip load (Fig. P-15). Determine the size of weld required, using AISC Specification.

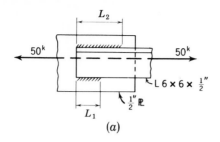

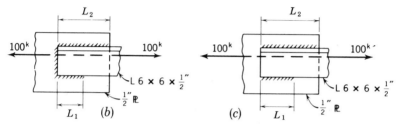

Fig. P-9

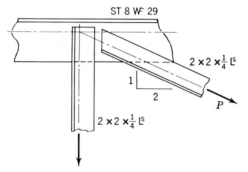

Fig. P-10

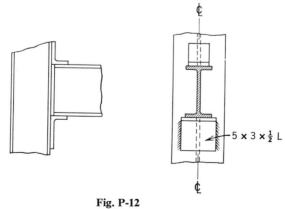

Fig. P-12

239

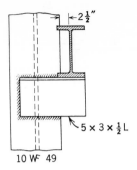

Fig. P-13

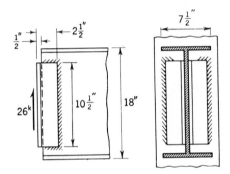

Fig. P-14

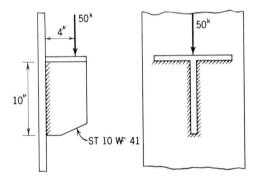

Fig. P-15

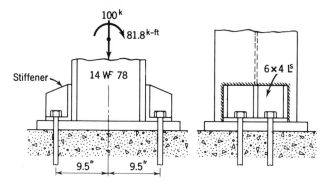

Fig. P-16

16. A steel column of a building frame is designed to resist an axial load of 100 kips combined with a bending moment of 81.8 kip-ft (Fig. P-16). The column base connection is made, using clip angles and anchor bolts as shown. Assume that the load is resisted by uplift on the anchor bolts and bearing on the opposite column flange. Determine the uplift load on anchor bolts and the size and arrangement of welds required for connection of the angle to the column flange. (Use AISC Specification.)

17. A special rig used for lifting large assemblies with a single crane is shown in Fig. P-17. The horizontal spreader strut is made up of two 8 in. × 4 in. × $\frac{1}{2}$ in. angles back to back and is welded to the end plate as shown. Assuming that the maximum design hook load is 40 tons, select the size of fillet weld and determine dimensions L_1 and L_2. (AISC Specification.) (A441 steel.)

18. A typical end connection in a welded roof truss is shown in Fig. P-18. The maximum load in the top chord is 43 kips compression. Determine the maximum load in the double-angle diagonal and determine the size of weld and arrangement and length of welds attaching the diagonal to the top chord. (AISC Specification.)

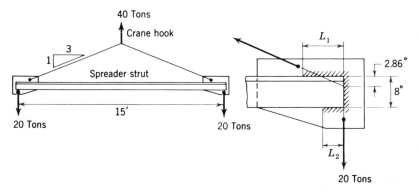

Fig. P-17

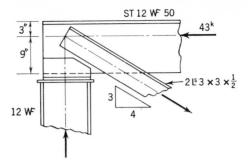

Fig. P-18

19. A tension member in a highway bridge truss is made up of two 15-in. channels 50 lb connected by lacing (Fig. P-19). The dead load force in the member is 100 kips tension; the live load varies from 268 compression to 320 kips tension. We want to splice the member. Neglecting the effects of fatigue: (*a*) check the splice using full-penetration butt welds in channels, (*b*) design an alternate splice, using plates fillet welded to web of channels and $\frac{5}{8}$-in. plates welded to channel flanges in place of lacing. (AASHO Specification.)

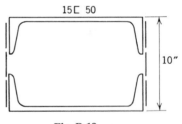

Fig. P-19

20. Repeat Prob. 19*b*, considering effect of fatigue if the number of load cycles during the life of the structure is estimated at 600,000 cycles. (AASHO Specification.)

REFERENCES

1. *Welding Handbook*, American Welding Society; *Procedure Handbook*, Lincoln Electric Co.; *Manual of Design for Arc Welded Steel Structures*, Air Reduction Co.; *Welding J.*, American Welding Society.

2. Greenberg, S. A., "Steel for Welded Bridges and Building," *Proc. ASCE* No. 344 (1954).

3. Standard Qualification Procedure, American Welding Society.

4. Jensen, C. J., "Combined Stresses in Fillet Welds," *Welding J.* **12,** No. 2, 17–21 (1934).

5. Report of the Structural Steel Welding Committee, American Bureau of Welding, 1931.

6. Bierret, G., and G. Grunning, "Stress Distribution in and the Strength of Joints with End Fillet Welds," *Stahlbau* **6,** No. 22, 169–75 (1933).

7. Blodgett, Omer W., and John B. Scalzi, "Design of Welded Structural Connections," James F. Lincoln Arc Welding Foundation, Cleveland, Ohio.

8. Abolitz, A. L., and M. E. Warner, "Bending Under Seated Connections," *Engineering Journal AISC* **2,** 1 (1965).

7

Tension Members

◆◆◆

7-1 INTRODUCTION

A simple tension member is a straight member subjected to two pulling forces applied to its ends. It is an efficient and economical member because it uses the full area of the material effectively at the maximum uniform stress permitted by the designer. It is usually easy to fabricate, ship, and erect into position in a structure. There is an increasing interest in the use of tension members in building designs such as hangers for floors and cables for roofs. Many roof forms can be developed which span large areas and present pleasing esthetic lines.

In general, there are four groups of tension members: wires and cables, rods and bars, single structural shapes and plates, and built-up members. These will be discussed separately in the following sections.

7-2 TYPES OF TENSION MEMBERS

Wires and Cables. The mechanical properties of wires and cables are discussed in Sec. 2-6. Wire rope with fiber cores are used almost exclusively for hoisting purposes. Wire ropes with strand cores or independent wire rope cores are used for "standing lines" or hoisting ropes. The manufacturing practices for these ropes vary depending upon the intended use. Wire ropes are used for hoists, derricks, rigging slings, guy wires, and hangers for suspension bridges. Main cables for suspension bridges are made up of untwisted, parallel wires placed on the job site by special devices. Strands are sometimes used for smaller bridges. The advantages of wire rope and strand are flexibility and strength. Special fittings are required to provide proper end connections for wire rope and strand. Several of these fittings are shown in Fig. 7-1.

Although wire rope is occasionally used for bracing members, its application is limited because of its inability to resist compression, the need for special connecting devices, and its excessive elongation when the strength is fully utilized. In some special structures, such as steel towers with guy lines,[1,2] wires and ropes can be designed with considerable initial tension so as to prestress the frame in order to increase its effectiveness in resisting external loads. Prestressed wires and ropes can carry compression resulting from external loads, provided compression does not exceed the initial tension.

Rods and Bars. Small tension members are often made from hot-rolled square or round rods, or flat bars. The tensile strength of such members depends on the type and grade of steel; structural carbon steel is the most commonly used material for them. Because of the slenderness of these members, their compression strength is negligible.

Rods and bars are employed as tension members in bracing systems, such as diagonal bracings or sag rods, or as main members in very light structures, such as radio towers. When bars are used, it is desirable to place the larger dimension in the vertical plane in order to reduce sagging. Welded end connections for rods and bars are relatively simple, since no special fabrication is required (Fig. 7-2a). Otherwise, rods can be threaded and bolted, employing various details for connections (Fig. 7-2b). Threading the ends reduces the net section of the rod and thereby its strength, but it does not appreciably affect the rigidity of the member. When rod sizes are chosen for rigidity rather than for strength, this loss of area at the threads is often unimportant. If it is desired to maintain the strength of the main section, the rods can be upset at the ends and then threaded (Fig. 7-2b). Upset rods are costly because of the additional work required for forging the ends and they may prove economical only if a sufficient quantity of them is ordered. Rods can also be connected with a clevis, or forged into a loop (Fig. 7-2c). Flat bars may be welded, riveted, or bolted to an adjoining part, or forged into loop or eyebar ends and connected to a pin; an eyebar is shown in Fig. 7-2d. In current practice eyebars are rarely used and have to be ordered from the mill as a special item.

The main disadvantage of rods and bars is inadequate stiffness, resulting in noticeable sag under their own weight, especially during erection. In addition, it is almost impossible to fabricate rods and bars to fit perfectly in the structure. If too long, they will buckle when forced into position; if too short, they will have to be pulled into place and may produce undesirable

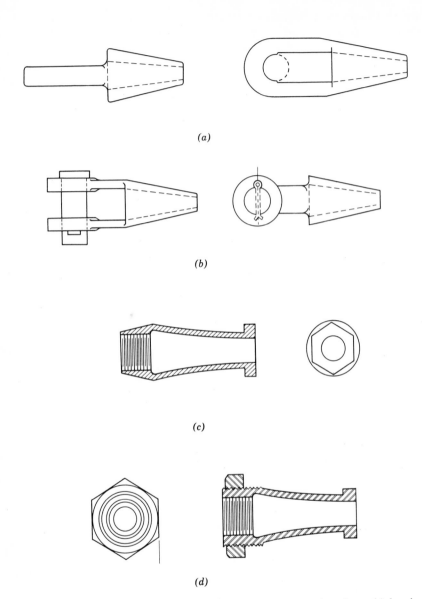

Fig. 7-1 Cable fittings. (a) closed rope sockets, (b) open strand sockets, (c) bearing type socket—internal threads, and (d) tensile-type socket—internal and external threads.

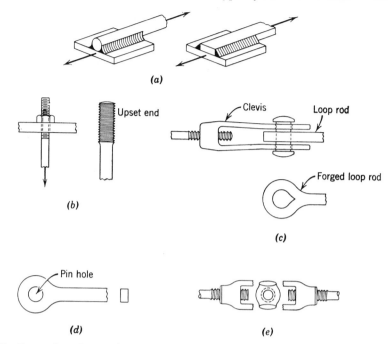

Fig. 7-2　Connections for tension members. (*a*) welded connections, (*b*) threaded and bolted connection, (*c*) clevis with loop rod, (*d*) eyebar end, and (*e*) turnbuckle.

initial stresses in the rods and the structure. For this reason, turnbuckles or sleeve nuts (Fig. 7-2*e*) are often required to adjust for the variation in bar length.

Single Structural Shapes.　When some amount of rigidity is required or when load reversals may subject the tension member to some compression, wire ropes, rods, and bars will not usually serve the purpose. In such cases, structural shapes, single or built-up, must be employed. The simplest and most common rolled shape used as a tension member is the angle. One serious objection to the single angle is the presence of eccentricity in the connection. Figures 7-3*a*, *b*, and *c* show riveted, welded, and lug-angle connections used with an angle. The riveted connections are eccentric in both planes, whereas welded connections may be so designed as to produce concentric connection in one plane, with eccentricity only in the other plane. The lug-angle connection may be designed to transmit the stress in the outstanding leg directly to the lug; in this manner a stiffer connection with more rivets is provided.

Angles are considerably more rigid than wire ropes, rods, and flat bars, but they may still be too flexible if the members are very long. Therefore, single

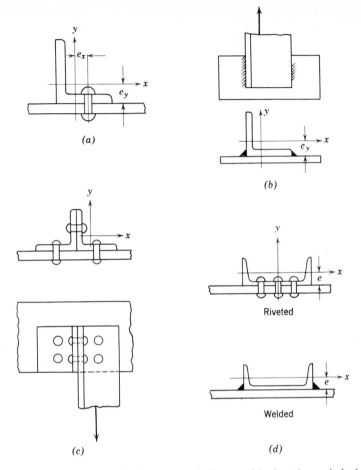

Fig. 7-3 End connections for single structural shapes. (*a*) riveted- or bolted-angle connection, (*b*) welded-angle connection, (*c*) lug-angle connection, and (*d*) channel connection.

angles are used mainly for bracing, for light-truss tension members, and where the length of members is not excessive.

Single channels can sometimes be effectively employed as tension members. For the same cross-sectional area as provided by an angle, the channel has less eccentricity and can be conveniently riveted, bolted, or welded (Fig. 7-3*d*). The rigidity of a single channel is high in the web direction but low in the flange direction; unless the channel is braced in the weak direction at intermediate points it cannot be used for long members.

Standard I and W sections are used occasionally as tension members. Although for the same area, the W sections are more rigid than the American

Standard I sections, they are often inconvenient in connections because each variation of nominal size shape has a different over-all depth. The standard I's have several sections for each depth and thus can be more conveniently fitted into a structure, but there is not a sufficient variety of sections to make an economical choice. Single rolled sections are usually more economical than built-up sections and should be used, provided adequate strength, rigidity, and suitable connections can be obtained.

Built-up Members. Built-up members are obtained by connecting two or more plates or shapes which then act as a single member (Fig. 7-4). Such members may be made necessary by the requirement of area, which cannot be provided by a single rolled shape; or by the requirement of rigidity because for the same area, much greater moment of inertia can be obtained with built-up sections than with single rolled shapes; or by the requirement of suitable connection, where the width or depth of member necessary for proper connection cannot be obtained in a standard rolled section. Another advantage of built-up members is that they can be made sufficiently stiff to

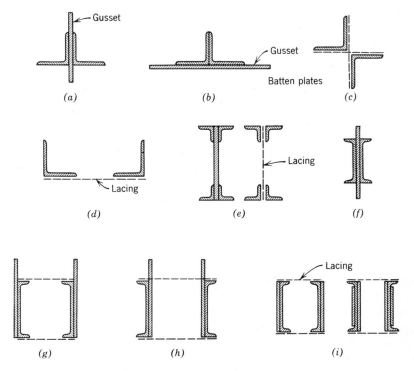

Fig. 7-4 Built-up angle or channel members.

carry compression as well as tension and hence they are desirable when stress reversals might occur.

A common built-up member is a two-angle section. Two angles can be placed back to back on two sides of a connection gusset (Fig. 7-4a). When they are connected to the same face of the gusset (Fig. 7-4b), eccentricity exists in one plane, subjecting the angles to tension and bending simultaneously. A star arrangement (Fig. 7-4c) provides a symmetrical and concentric connection and also slightly greater rigidity. The starred angles are best connected by batten plates spaced at distances approximately equal to 50 times the leg width (for A36 steel), the batten plates being alternately placed in the two perpendicular directions. In a two-plane truss, where two parallel gussets are provided at each connection, the two angles will have to be arranged as shown in Fig. 7-4d and laced together—an arrangement often used for bracing. When bigger area and symmetry of members are desired, for main members, four angles can be built into one member (Fig. 7-4e). These angles are tied together with lacing and/or tie plates, or with a solid plate when such additional area is required to carry the stress. One advantage of the shapes shown in Figs. 7-4d and e is that the distance back to back of angles can be adjusted to any required value.

Two channels placed back to back with a gusset plate in between can be used for medium loads in a single-plane truss (Fig. 7-4f). Such a member is not common because loads on a single-plane truss are usually light and two channels are seldom required. In a double-plane truss two channels (Fig. 7-4g) are frequently used, with the flanges turned inward in order to simplify transverse connections and minimize lacing. Occasionally, the flanges are turned outward to provide greater lateral rigidity (Fig. 7-4h), which is a common arrangement for compression members. When greater area is required, plates are added to the webs, as in Fig. 7-4i.

When heavier members are required, as in medium and heavy bridge trusses, plate and angle sections are frequently employed. These start from a minimum of four angles (Fig. 7-5a) and may require many additional plates and angles for very heavy members (Figs. 7-5b, c, d, and e). The flexibility of these built-up members in adapting to any required depth, width, and cross-sectional area, the simplicity of splices and connections to gusset plates, the ability to carry compression in case of stress reversal, and the high bending strength make these sections very advantageous in medium and heavy construction.

Besides load, rigidity, and connection requirements, the choice of a proper type of member is further guided by considerations of economy, simplicity, and ease of fabrication. Latticed construction (using lacing) is often desirable for long members transmitting relatively small loads, whereas short members transmitting heavy loads should be built up with plates. Members that have

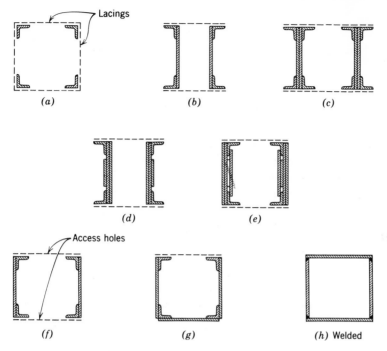

Fig. 7-5   Built-up plate and angle members.

small radii of gyration in one plane should be avoided because they are easily bent and difficult to handle. Small box members are efficient but difficult to rivet or weld and require access holes in the plates (Fig. 7-5*f*). The use of one latticed face is an improvement and is often acceptable (Fig. 7-5*g*), although this destroys the symmetry of the member. More often, latticing on two sides is used (Fig. 7-5*e*). Current practice is to use four plates without hand holes welded at the corners (Fig. 7-5*h*).

7-3   STRESSES DUE TO AXIAL LOADS

If the load in the tension member is axial, that is, coincident with the longitudinal centroidal axis of the member (Fig. 7-6), then the stress distribution in the member can be assumed to be uniform and is defined by the familiar formula,

$$f = \frac{P}{A} \qquad (7\text{-}1)$$

where f is unit stress, P is total load, and A is cross-sectional area.

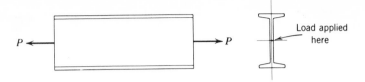

Fig. 7-6   Axial load in a tension member.

The assumption of uniform stress distribution is justified only if a plane section remains plane, that is, no distortion of the section takes place. However, it is not restricted to the elastic range of the material. For axial loading, stress distribution in a steel member remains approximately uniform in the plastic range all the way up to failure. If end forces acting on a tension member are concentrated over a small part of the sectional area, uniform stress distribution cannot be assumed at that section. Under such conditions, distortion of the section will occur, resulting in high local stress. At a section some distance away from the applied load a plane section remains plane, as shown by laboratory and field measurements, and the assumption of uniform stress distribution is then justified. Possible variation of stress distributions at various sections of a tension member is shown in Fig. 7-7.

Stress concentrations in a tension member may also be due to holes in the member or to abrupt changes in the shape or elastic properties of the member. Theoretical stress distributions in members with holes or other discontinuities can be obtained mathematically. For a general case such computations require a great deal of numerical calculations and at best lead to a somewhat approximate answer. The theoretical stress distribution in a member computed on the basis of elastic theory may be quite different from that at ultimate load, particularly if the material is ductile and failure is governed by its plastic behavior. For members of structural carbon steel, the redistribution of stresses after yielding is so great that the elastic stresses may have very little significance.

The theoretical elastic stress distribution may be significant for more brittle

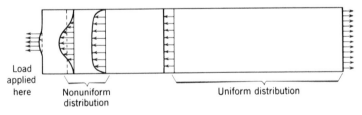

Fig. 7-7   Stress distributions in a tension member with concentrated load applied at end.

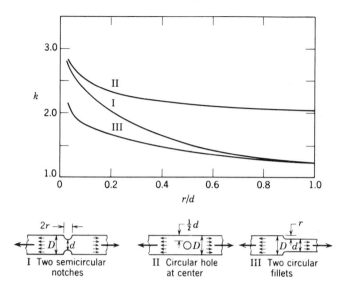

Fig. 7-8 Stress concentration factor $k =$ (maximum unit stress)/(average stress on net section).

materials which are nearly elastic up to failure, or even for ductile materials if the members are subjected to a large number of cycles of pulsating loads which might produce fatigue failures. In such cases, it may be necessary to compute the maximum stress based on the elastic theory. The maximum tensile stress which exists at the section through the hole or notch can be defined in terms of stress concentration factor k and the average stress of either the net or the gross section thus: maximum stress $f_t = kP/A$. Values of k based on the net section for various conditions producing stress concentrations are given in Fig. 7-8.

7-4 COMBINED AXIAL TENSION AND BENDING

Elastic Stress Distribution. Pure axial load in tension members can seldom be obtained in actual structures. The connections may not be concentric; the member itself may not be straight, resulting in eccentric load at some sections; the member may not be vertical and thus may be subjected to bending due to its own weight; there may be wind, vibration, or earthquake forces producing bending in the member. If at any section the axial load and the bending moment due to combined effects of eccentricity and transverse forces are known, then the stress distribution at that section may be defined

by the equation

$$f = \frac{P}{A} + \frac{My}{I} \qquad (7\text{-}2)$$

where f = unit stress at a point
  M = bending moment in a principal plane
  y = distance of point from centroidal axis
  I = moment of inertia about this centroidal axis
and P and A are as defined in Eq. 7-1.

This formula is valid only when bending takes place about a principal axis and the stresses are within the elastic limit.

If bending takes place about some plane other than one through a principal axis, Eq. 7-2 must be modified as follows:

$$f = \frac{P}{A} \pm \frac{M_x y}{I_x} \pm \frac{M_y x}{I_y} \qquad (7\text{-}3)$$

where M_x and M_y are components of M about x and y principal axes, respectively, and I_x and I_y are the respective moments of inertia about these axes.

In all cases, when a section has an axis of symmetry, that axis is one of the principal axes; the other principal axis is a centroidal axis perpendicular to the axis of symmetry. The general case of unsymmetrical bending will be discussed more fully in Chapter 8 and it suffices to state here that neglecting effects of dissymmetry may result in significant errors.

Accurate determination of stress distribution under combined axial load and bending in the elastic range is complicated by the effect of the deflection of the structure and the axial load P on the magnitude of the bending moment M.

The effect of tension load is always to decrease the primary bending moment. In conventional structures, the effect of deflection is small and is usually neglected. For long slender members, such as long-span suspension bridges, the effect of cable deflections on bending moments in stiffening trusses becomes extremely important (Fig. 7-9) and may reduce the required truss depth by 75% or more. For ordinary tension members, however, the effect

Fig. 7-9 Design of long-span suspension bridges. (*a*) by elastic theory and (*b*) by deflection theory.

of deflection is usually small. A method for evaluating this effect more precisely is discussed in detail in Chapter 8.

When design is based strictly on elastic considerations, the maximum tension stress due to combined axial tension and bending is given by Eqs. 7-2 and 7-3 and its magnitude must be limited to a value which would provide the desired factor of safety against yielding. In such designs usually a single-limit stress value is specified for tension due to axial loads and that due to bending.

The design requirement may be expressed as follows:

$$f = \frac{P}{A} + \frac{Mc}{I} = \frac{P}{A} + \frac{Mc}{Ar^2} \leq F \tag{7-4}$$

Therefore, required sectional area A is

$$A \geq \frac{P}{F} + \frac{Mc}{Fr^2} = \frac{(P + BM)}{F} \tag{7-5}$$

where $B = c/r^2$ is known as the "bending factor," which reduces the effect of bending moment M to an equivalent axial load BM. For selected W^F sections, particularly suitable for combined action of axial load and bending, the values of B_x corresponding to bending about the strong axis, and of B_y, for bending about the weak axis, are shown in the AISC Manual, pp. 3-13 to 3-25. A graph of approximate values of B_x and B_y plotted as a function of depth or width dimension of W^F section is shown in Fig. 7-10.

When inelastic behavior is considered, the maximum stress due to combined axial tension and bending cannot be evaluated from Eqs. 7-2 and 7-3 since the principle of superposition is no longer valid. Also, because of nonlinear distribution of bending stresses in the inelastic range, the limit stresses for pure tension and pure bending are different. In such instances, "interaction" equations are used to evaluate the capacity of the member to resist combined loadings. This method is discussed in more detail in Chapter 8. Here it will be used without derivation.

The design requirement based on the elastic theory may be expressed as follows:

$$\frac{P/A}{F_t} + \frac{Mc/I}{F_b} \leq 1.0 \tag{7-6}$$

This can be transformed to

$$\frac{P}{AF_t} + \frac{Mc}{Ar^2 F_b} \leq 1.0 \tag{7-7}$$

from which

$$A \geq \frac{P}{F_t} + \frac{BM}{F_b} \tag{7-8}$$

where $B = c/r^2$ is the bending factor, as before.

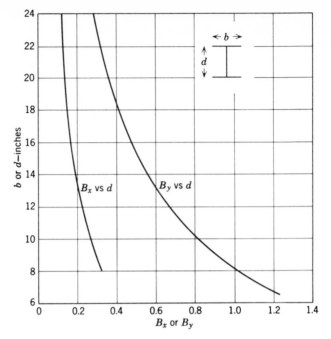

Fig. 7-10 Approximate values of bending factors.

If P/F or P/F_t is defined as A_1, the area required to carry load P safely without bending moment M, and if BM/F or BM/F_b is defined as A_2, the area required to carry the moment M safely without axial load P, then the required area A can be defined as

$$A \geq A_1 + A_2 \qquad (7\text{-}9)$$

It follows from Eq. 7-9 that the total area required is approximately equal to the sum of the areas required when axial load and bending are acting singly. Accurate determination of required area A is again a process of trial and error, and the number of trials depends largely on the experience and judgment of the designer.

7-5 ALLOWABLE STRESSES

Axial Load Only. To simplify design calculations, the size of a tension member is determined by comparing a computed nominal stress with a specified allowable stress.

The nominal stress is computed as

$$f = \frac{P}{A} \qquad (7\text{-}10)$$

The allowable stress is usually given by specifications which govern the design of the structure. For example, the allowable stresses for buildings are given by the AISC Specification for the Design, Fabrication, and Erection of Structural Steel for Buildings. The AASHO and AREA Specifications prescribe the allowable values to be used for highway and railway bridges, respectively. The factor of safety against yielding for tension members for buildings is 1.65 and for bridges it is 1.80. These factors of safety have been established for a great number of years and take into account many of the variables of design and construction which do not lend themselves to calculation.

In most specifications tabulated values for allowable stresses are presented and are usually of a nature which indicates that they have been "rounded off" to convenient design values. Therefore, they may present a slightly different factor of safety, but the variation is usually of a negligible amount.

For design convenience, the AISC Specification has given the allowable stress in tension in terms of the yield stress of the steel, thus enabling all grades of steel to be considered and evaluated for design purposes, thus,

$$F_t = 0.60 F_y \qquad (7\text{-}11)$$

on the net section, except in pin holes, and

$$F_t = 0.45 F_y \qquad (7\text{-}12)$$

on the net section at pin holes in eyebars, pin-connected plates, or built-up members. The reduced value for pin holes takes into account stress concentrations and is based on research conducted on pin-connected plate links.[3]

The allowable stresses for a selected group of steels is listed in Table 7-1.

Table 7-1   Allowable Axial Tension Stress

Type of Steel	F_y, Yield Stress, ksi	F_t, Allowable Stress, ksi AISC	F_t, Allowable Stress, ksi AASHO
A36	36	22	20
A441, A440, A242	42	25	22
A441, A440, A242	46	27.5	24
A441, A440, A242	50	30	27
A514	100	60	–

The allowable stresses just discussed are for static loads and must be reduced for repeating and alternating loads as outlined in Chapter 3.

As noted in Chapter 2, wires and cables do not have a definite yield point stress and as a result the ultimate strength is used as the basis of design. Most experience to date in the design of cable-supported structures has been with a factor of safety of three against the minimum breaking strength. However, as more experience is gained on the use and behavior of cable structures the factor of safety may change. The safety factor to be used is usually determined by the engineer based on a careful consideration of factors such as exposure to weather, dynamic forces, and other special conditions.[1,2] For other applications, such as rigging, where greater uncertainty is involved, the factor of safety may range from 3 to 5.

Axial Load and Bending. Many tension members are subjected to combined axial and bending stresses, and because many specifications give different allowable stresses for axial loads and bending, acting independently, a question may arise with respect to the proper allowable stress which should be used for the combined stresses, defined by Eqs. 7-2 or 7-3.

A rational solution for this situation would be to verify the safety of the member by several evaluations.

One evaluation is to determine accurately the actual maximum tension stress from all sources, including secondary effects, and to compare this design stress with the allowable stress in tension only or against the yield point with an appropriate factor of safety. In those specifications in which the allowable stress in tension and bending is the same, the sum of the two load effects on the member must be less than the common allowable stress.

The AISC Specification, (Sec. 1.6.2, p. 5-21), states that members subject to both axial and bending stresses shall be proportioned according to an interaction relationship expressed as

$$\frac{f_a}{0.6F_y} + \frac{f_b}{F_b} \leq 1.0 \tag{7-13}$$

where f_a = computed axial stress

f_b = computed bending tensile stress

F_b = allowable bending tensile stress

F_y = yield stress of steel

In addition, as a precaution against lateral buckling, the AISC Specification requires that the computed bending compressive stress, taken alone, shall not exceed the value which will prevent lateral buckling of the member, when subjected to bending forces only.

7-6 ECONOMY OF STEELS IN TENSION

The selection of the most economical strength grade of steel to be used in a particular application depends to a great extent on the type of structure under consideration. For example, for a suspended roof only cables are considered. In some instances a wide selection of steels is available and a choice may be possible simply on the basis of economy alone. However, steel material prices alone may not be sufficient to arrive at a decision.

Because a tension load is a direct load independent of cross-sectional area, an economic comparison of various grades of steel may be made easily. The price per pound of steel may be obtained from steel producers at any given time and relative prices* may be easily established for the various steels under consideration. The required area of steel for a given tension load is dependent only upon the allowable design stress and is essentially independent of the shape of the cross section. If the factor of safety against yielding remains constant, as in most specifications, the area is inversely proportional to the allowable stress or to the yield strength of the steel.

A typical problem may be to select the most economical steel for a tension member of 100 in. in length required to support a direct load of 22 kips.

Table 7-2   Relative Economy of Various Steels in Tension

	Steel	A36	A572-42	A242	A441	A572-50	A 514 Grade B	Grade F
Yield strength, psi		36,000	42,000	50,000	50,000	50,000	100,000	100,000
Allowable stress,[1] psi		22,000	25,000	30,000	30,000	30,000	60,000	60,000
Area, sq in.		1.000	0.880	0.734	0.734	0.734	0.367	0.367
Relative weight		1.000	0.880	0.734	0.734	0.734	0.367	0.367
Elongation, in.[2]		0.076	0.086	0.104	0.104	0.104	0.207	0.207
Relative elongation		1.000	1.135	1.364	1.364	1.364	2.723	2.723
Relative price[3]		1.000	0.921	1.022	0.862	0.840	0.808	0.953

[1] Allowable tensile stress = 0.60 F_y.
[2] Modulus of elasticity = 29,000,000 psi.
[3] Based on average net mill prices (cents per lb), 1966.
  Based on $\frac{3}{4}$-in. thickness for all steels.

* Relative price is the ratio of the price of the higher-strength steel weight required for the load to the price of the A36 steel required to carry the same load.

Based on an allowable stress of 0.6 of the yield point or yield strength, the required area may be determined and evaluated as in Table 7-2. The table indicates that the higher-strength steels require less area and therefore less weight, produce a slightly greater elongation under load, but may be more economical in price. Assuming that weight and elongations under load are satisfactory for all steels, then the choice of a $\frac{3}{4}$-in. plate is one of price only and the relative price row indicates the economy to be as follows, listed in the order of maximum economy: A514 Grade B, A572-50, A441, A572-42, A514 Grade F, A36, and A242. Similar comparisons for several different types of applications will indicate a trend which the designers may use in future considerations.

7-7 DESIGN OF CABLES

The design of cables as tension members is often a simple problem. If the cable forms a straight tension member and does not bend around a pulley sheave or drum, the size is determined simply by the tensile strength of the cable. The total load which the cable must support is computed from the analysis of forces acting on the structure. The allowable load which the cable can support is determined by the engineer either from cable manufacturer's recommendations or his own evaluation of the specific conditions.

Although a cable size is first determined by stress requirements to support the load with safety; flexibility or excessive elongation may require that a larger size be used.

Elastic elongation of a cable Δ due to tensile load P can be determined by the conventional formula

$$\Delta = \frac{PL}{AE} \tag{7-14}$$

where L = length of wire rope or cable, A = cross-sectional area of the metallic part of the rope, and E = the effective modulus of elasticity of the rope. The cross-sectional area of the metallic part of the wire rope depends on its construction and generally varies from $0.35D^2$ to $0.6D^2$, where D is the nominal rope diameter in inches. The effective modulus of elasticity E also depends on the type of rope and varies from 9×10^6 to 24×10^6 psi. In addition to the elastic elongation, the cable usually undergoes some permanent stretch due to adjustment of individual wires in the strand and proper seating of the strands on the core of the rope. This permanent stretch occurs during the initial period of loading of the wire and varies from $\frac{1}{4}$ to 1 % of the length of the rope. If the wire rope is prestretched, some permanent deformation will take place and the effective modulus of elasticity will be relatively high.

If the cable is to be bent around sheaves or drums, then the type of cable and its size determine the permissible radius of bend. For a small radius of bend, great flexibility is required, necessitating a larger number of wires and strands in its construction. The usual practice is shown in Table 7-3.

Table 7-3   Permissible bend Diameter

Rope Construction	Sheave or Drum Diameters
6 × 7	72 × rope diameter
6 × 19	45 × rope diameter
8 × 19	31 × rope diameter
6 × 37	27 × rope diameter

Cables can be used effectively in prestressed structures, that is, structures where initial stresses have been introduced during fabrication or erection in order to stiffen or to increase the strength of the structure under various applied loads. Cables are also used effectively as guys to stabilize towers and stacks. The problems of determining stresses in the guys and in the segments of the tower are outside the scope of this book. The designer confronted by these problems will find the references listed at the end of the chapter helpful.[1,2]

Example 7-1. The floorbeam suspender of a suspension bridge is to be made of bridge strands, with two strands at a panel point, carrying a dead load of 110 kips and a maximum live and impact load of 68 kips (Fig. 1). Design the member and investigate the elongation (assuming a length of 150 ft) $E = 24,000$ ksi; use table of minimum breaking strengths and a factor of safety of 3.

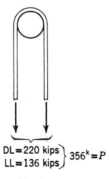

$$\left.\begin{array}{l}\text{DL}=220\text{ kips}\\\text{LL}=136\text{ kips}\end{array}\right\} 356^{k}=P$$

Fig. 1

Solution

Total load on cable $= 110 + 68 = 178$ kips
    Cable ultimate load required $= (3)(178) = 534$ kips
    From Table 2-7 2–$2\frac{1}{8}$-in. ϕ Class A coating $= 554$ kips
Elongation due to live load:

Metallic area for $2\frac{1}{8}$-in. strand $= 2.71$ in.^{2}

$$\Delta = \frac{(P/2)L}{AE} = \frac{(68/2)(150 \times 12)}{(2.71)(24{,}000)} = 0.94 \text{ in.}$$

Discussion

The total elongation due to full dead load and live load would be $\frac{89}{34} \times 0.94 = 2.46$ in. For high live and impact loads, the allowable stress should be decreased because the elongation due to live load and impact could be excessive. Detail of the connection is shown in Fig. 2.

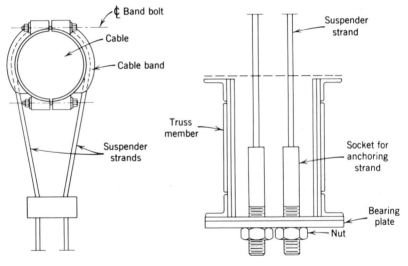

Fig. 2

7-8 DESIGN OF RODS AND BARS

Determining the choice of round or square rods, and flat bars for use as tension members has been discussed in Sec. 7-2. These members are generally slender and light and their end connections are usually nearly concentric. The net cross-sectional area therefore is determined on the basis of load and allowable tensile stress, as follows:

$$A_n = \frac{P}{F_t} \qquad (7\text{-}15)$$

The size of rod or bar corresponding to the net section A_n determined from Eq. 7-15 may be too small for practical reasons. For example, with a load $P = 4$ kips, allowable stress $F_t = 20$ ksi, the required net area $A_n = \frac{4}{20} = 0.2$ in.², and if a round bar is used, a $\frac{1}{2}$-in. diameter would be sufficient. The radius of gyration of this bar is 0.125 in. and, if the tension member is 15 ft long, the resulting L/r is $15 \times 12/0.125 = 1440$. It has been found from experience that members with such high slenderness ratios, L/r, are difficult to handle and that accidental loads imposed on them under service conditions may cause damage or failure. These accidental loads cannot be taken directly into account in design. For example, they may occur when maintenance men want to use the tie rod as a prop, or a crew of workmen use the member to support some construction equipment. Because it is not feasible to limit the slenderness ratio of rods and bars to some definite minimum amount, the designer must use his judgment in selecting the proper size. For example, in the example just mentioned, with $P = 4$ kips and $L = 15$ ft, a $\frac{3}{4}$-in.-diameter size with $L/r = 960$ would be much better than the $\frac{1}{2}$-in. size.

Flat bars are sometimes used as tension members because of the ease of making the connection and the usual practice is to place the width in the vertical plane to reduce sagging. The design of flat bars with welded, riveted, or bolted connections is quite simple. Usually eccentricity of connection and weight of the member can be neglected, although in wide flats the effect of eccentricity may be important. In studying the slenderness of the member, the radius of gyration in the weak direction must be considered. The flexibility of the flat bar as well as difficulty in adjusting its length are the primary objections to its use.

Design of eyebars with pinned connections is no longer common, as a result of the development of riveting, welding, and bolting which often provide more economical methods of connecting the more usual types of tension members.

7-9 DESIGN OF SINGLE STRUCTURAL SHAPES

When a single angle is used as a tension member with eccentric end connections, large secondary bending stresses may develop. If the axial load and the eccentricity are known, the bending moment can be determined and the maximum stress can be computed by the combined stress formula (Eq. 7-2 or 7-3). It should be noted that usually the bending moment is not in the principal plane and the simplified equation $f = Mc/I$ does not apply.

Some specifications (AISC Sec. 1.15.3) permit the eccentricity in the end connection to be neglected. This means that a nominal stress in the angle may be computed by assuming uniform distribution of stress over the full area. This specification may appear to be unreasonable and possibly unsafe as judged by the elastic theory. In practice it may be justified because the combined stress formula based on elastic theory gives a higher calculated stress than actually would exist in the plastic range and, therefore, the actual factor of safety is considerably greater than would be indicated by the elastic theory.

Example 7-2. Each end of a 6 × 6 × $\frac{1}{2}$-in. angle member is connected along one leg to a $\frac{3}{4}$-in. gusset plate, using one line of $\frac{7}{8}$-in. rivets. Determine allowable tension load for the member based on net section A_n of the angle and allowable stress $F_a = 20$ ksi.

Solution

Net section: $A_n = A_g - \sum dt = 5.75 - 1 \times (\frac{7}{8} + \frac{1}{8})0.50 = 5.25$ in.²

$$P = A_n F_a = 5.25 \times 20 = 105 \text{ kips}$$

A single channel is often used as a tension member and may be conveniently welded, riveted, or bolted. For the same area, a channel has less eccentricity than a single angle, as illustrated in Fig. 7-3d.

Although structural shapes are considerably more rigid than rods and bars, their size may sometimes be determined by requirements of rigidity and not of strength. As indicated before, a certain amount of rigidity is necessary to insure ease of handling and safety under accidental loads. Although no definite limit on L/r has been specified for rods and bars, various specifications provide that tension members other than bars and rods shall not exceed the following limits:

	Main Members	Secondary Members
AISC	240	300
AASHO	200	240
AREA	200	200

In using structural shapes as horizontal tension members, bending due to dead weight is usually neglected, except for long and heavy members. The definition of "long and heavy" members depends on how much bending stress can be neglected. The following approximate analysis indicates how the magnitude of the bending stress due to weight can be quickly estimated.

Consider a horizontal member in tension, hinged at both ends. The maximum bending moment due to dead weight w per unit length is equal to $wl^2/8$, where l is the length and w is equal to the product of the density of material (0.28 lb per cu in. for steel) and cross-sectional area A. The maximum fiber stress due to bending f_b is therefore equal to

$$f_b = \frac{Mc}{I} = \left(\frac{wl^2}{8}\right)\left(\frac{c}{I}\right) = \left(\frac{0.28Al^2}{8}\right)\left(\frac{c}{Ar^2}\right) = 0.035c\left(\frac{l}{r}\right)^2 \qquad (7\text{-}16)$$

The distance between the extreme fiber and neutral axis c generally varies between $0.5d$ and $0.7d$, where d is the depth of the member, and the radius of gyration r generally varies between $0.3d$ and $0.4d$. For maximum f_b, assuming that $c = 0.7d$, and $r = 0.35d$,

$$f_b = 0.035(0.7)d\left(\frac{l}{0.35d}\right)^2 = 0.21\left(\frac{l}{d}\right) \qquad (7\text{-}17)$$

where l and d are in inches and f_b is in pounds per square inch.

When the l/d ratio is large, the member is long and flexible and the stress may be reduced because of the effect of axial tension on bending moment (see Sec. 7-3). If this effect is neglected, f_b may be estimated from Eq. 7-17. Thus, for $l/d = 40$ and $l = 25$ ft (300 in.), $f_b = 2400$ psi and may or may not be considered excessive in design.

When bending stresses are not negligible, they must be combined with the stresses due to direct tension.

Example 7-3. A tower is framed with main verticals, 8 × 8-in., and horizontal members, 3-in. × 3-in. angles (Fig. 1). Using AISC Specification,

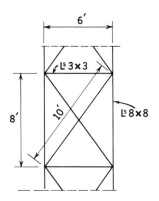

Fig. 1

A36 steel, design the diagonals for a tension load of 8 kips. Consider two designs:
   (*a*) Single-angle diagonals bolted to the vertical members.
   (*b*) Tie rods with clevis attachments (see AISC Manual).

Solution

(a) Single-Angle Diagonals Bolted to the Vertical Members.

Unsupported length, $L = 10 \text{ ft} = 120$ in. From Sec. 1.8.4 AISC Manual, p. 5-23: Maximum $L/r = 240$; then minimum $r = L/240 = 120/240 = 0.5$ in. If diagonals are connected to each other at point of intersection: $L = 5$ ft and $r = 0.25$. Usually this connection is neglected and full unsupported length is considered.

Area required:

$$A_n = \frac{P}{F} = \frac{8}{22} = 0.36 \text{ in.}^2$$

$A \geq 0.36$ in.^{2} }from AISC Manual. Use $2\frac{1}{2}$-in. $\times$ $2\frac{1}{2}$-in. angle.

$r \geq 0.5$ in. }$(r = 0.49$ in., considered satisfactory).

Minimum thickness: Use $\frac{1}{4}$ in. for exterior bracing members. Try $2\frac{1}{2}$ in. $\times$ $2\frac{1}{2}$ in. $\times$ $\frac{1}{4}$ in. angle, and $\frac{3}{4}$ in. diameter bolts.

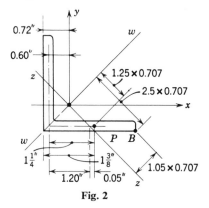

Fig. 2

Note: In these calculations, as is usual in practice, eccentricity of loading on the angle has been neglected. Considering the eccentricity of loading (Fig. 2), the stresses are

$$f = \frac{P}{A} + \frac{Mc}{I}$$

e_z = eccentricity = $(1.25 \times 0.707) = 0.88$ in.

$M_w = Pe_z = 8 \times 0.88 = 7.04$ kip-in.

I_{ww} = gross section = $I_x + I_y - I_z = I_x + I_y - r^2A$
     $= 0.70 + 0.70 - (0.49)^2(1.19) = 1.11$ in.^{4}

$e_w = 0.05 \times 0.707 = 0.036$ in. (negligible)

$$A_n = A_g - \left(d + \frac{1}{8}\right)t = 1.19 - \left(\frac{7}{8}\right)\frac{1}{4} = 0.97 \text{ in.}^2$$

$c_z = 1.76$ in.

$$f_b = \frac{P}{A} + \frac{M_w}{I_w}c_z = \frac{8}{0.97} + \frac{(7.04)(1.76)}{1.11} = 8.25 + 11.17 = 19.42 \text{ ksi}$$

The exact theoretical stresses will be different because the zz and ww axes as well as I_{zz} and I_{ww} will be different for the net section.

Bolted connections: It is good practice to use no fewer than two bolts. Load per bolt:

$$\frac{P}{n} = \frac{8}{2} = 4 \text{ kips}$$

Allowable shear loads in common bolts, AISC Manual, p. 4-4, $\frac{3}{4}$-in. bolts (A307), without washers, allowable $P = 4.42$ kips.

Edge distance for $\frac{3}{4}$-in. bolts in $2\frac{1}{2}$-in. angle:

AISC, p. 4-82: Edge distance $= 2\frac{1}{2} - 1\frac{3}{8} = 1\frac{1}{8}$ in.

AISC, p. 5-37: Minimum edge distance $= 1$ in.

Minimum spacing distance, AISC Sec. 1.16.4:

$$s = 3d = 3 \times \tfrac{3}{4} = 2\tfrac{1}{4} \text{ in.}$$

Minimum edge distance, AISC Sec. 1.16.6:

$$s = \frac{1.5A}{t} = 1.5\frac{(0.44)}{0.25} = 2.64$$

This distance may be reduced as follows:

$$s_n = \frac{R}{R_a}s = \frac{4}{4.42} \times 2.64 = 2.39 \text{ in.}$$

From Table 1.16.5, AISC, minimum edge distance should be $1\frac{1}{4}$ in. Use $2\frac{1}{2}$ in. See connection detail (Fig. 3).

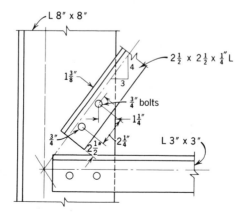

Fig. 3

Check net section:

$$A_n = 0.97 \text{ in.}^2 > 0.4 \text{ in.}^2$$

(*b*) *Tie Rod with Clevis.* No limitations on L/r. Net section required:

$$A_n = \frac{P}{F} = \frac{8}{20} = 0.4 \text{ in.}^2$$

AISC, p. 4-91: net area of threaded rod (no upset), use $\frac{7}{8}$-in. diameter rod.

$$A_n = 0.419 \text{ in.}^2$$

Size of clevis required: AISC, p. 4-94: for $\frac{7}{8}$-in. rod, diameter of tap $\frac{7}{8}$ in., use clevis no. $2\frac{1}{2}$ with $\frac{7}{8}$-in. hole. (Fig. 4)

Size of pin required: double shear. Bending negligible. Using common bolt, AISC, p. 4-4: $\frac{7}{8}$-in. bolt, allowable $P = 12.03$ kips; bearing on main angle 8-in. × 8 in. is not critical (minimum $t = \frac{1}{2}$ in.).

Note: For a $\frac{3}{4}$-in. bolt, allowable $P = 8.84$ kips, but this bolt is undesirable; for clevis no. $2\frac{1}{2}$ with a $\frac{7}{8}$-in. hole, a $\frac{7}{8}$-in. bolt is preferred.

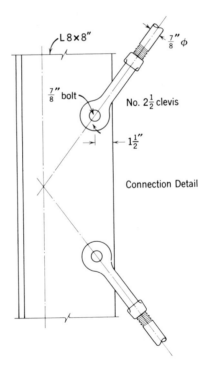

Fig. 4

7-10 DESIGN OF BUILT-UP MEMBERS

Built-up tension members are designed for requirements of strength, size, and economy. The principal strength requirements are: (*a*) the computed nominal stress must be within the allowable value, (*b*) the slenderness ratio L/r must be within the allowable value, and (*c*) tie plates and lacing or batten plates must satisfy specification requirements in order to assure composite action of the elements of the section. Dimensional limitations of the structure require consideration of necessary clearances and ease of connection, and in all cases economy of material, fabrication, and maintenance will govern the final design.

The first step in the design of a built-up member is to select the general type of section on the basis of size and economy requirements, as discussed in Art. 7-2. In considering the limitations of the slenderness ratio for a given member, the value of the radius of gyration for a given over-all size and type of built-up member often has to be estimated. For conventional arrangements approximate values of ratios of radius of gyration to width and depth of members are listed in Table 7-4. If the member is subjected to axial load only, the allowable stress is known and the required net cross-sectional area can be easily computed from Eq. 7-1. If the member is subjected to an eccentric load, a trial-and-error procedure must be used.

Design of the plates and lattice or batten-plate systems is much less critical for tension members than it is for compression members. The segments of a built-up member must be well tied together at the ends of the member and also braced together at intermediate points so that each element considered as an independent tension member will satisfy the strength and rigidity requirements.

Horizontal and inclined tension members are subjected to transverse loads due to their own weight. The shears due to these loads must be carried by the lacing system in the members. Also, if bending caused by transverse loads, eccentricity, or any other effect is appreciable, then the axial tension load may have a shear component perpendicular to the elastic curve of the member and the lacing system must be designed for such loads. The magnitude of these loads is generally small; however, the specifications (such as AISC, AASHO) are more than sufficient to ensure adequate design for these effects.

Built-up members are not limited to rolled structural shapes. The use of welding enables the fabrication of many shapes from steel plates. This is illustrated by Example 7-4.

Table 7-4 Approximate Values of the Radii of Gyration for Typical Structural Sections
(When two figures are given, they are the maximum and minimum values)

Section	r_x/h	r_y/b	Section	r_x/h	r_y/b	Section	r_x/h	r_y/b
(solid circle)	0.250	0.250	(section)	0.40 0.45	0.19 0.22	(section)	0.40 0.44	0.38 0.42
(hollow circle, $b=h$)	0.350	0.350	(section)	0.40 0.45	0.40 0.45	(section)	0.40 0.44	0.50 0.54
(solid square)	0.294	0.294	(section)	0.25 0.30	0.19 0.22	(section)	0.37 0.40	0.48 0.52
(hollow rectangle)	$0.5\sqrt{\dfrac{3b+h}{3(b+h)}}$	$0.5\sqrt{\dfrac{3h+b}{3(h+b)}}$	(section)	0.27 0.30	0.35 0.38	(section)	0.40 0.44	0.25 0.30
(section)	0.38 0.42	0.22 0.25	(section)	0.20 0.25	0.20 0.25	(section)	0.39 0.44	0.30 0.33

Section	r_x/h	r_y/b	Section	r_x/h	r_y/b	Section	r_x/h	r_y/b
$h=b$	0.20	0.40		0.38 0.40	0.19 0.23		0.33 0.36	0.52 0.54
$h=b$	0.30 0.32	0.30 0.32		0.38 0.43	0.19 0.22		0.42 0.46	0.27 0.29
$b>h$	0.28 0.31	0.31 0.33		0.40 0.45	0.20 0.25		0.49 0.52	0.30 0.33
$2h=b$	0.30 0.32	0.21 0.22		0.35 0.35	0.20 0.25		0.49 0.52	0.27 0.29
$2h>b$	0.31 0.33	0.19 0.21		0.35 0.37	0.42 0.46		0.46 0.49	0.27 0.30
$2h<b$	0.28 0.30	0.22 0.25		0.35 0.37	0.54 0.56		0.36 0.40	0.52 0.56
$h=b$	0.21 0.22	0.21 0.22		0.33 0.36	0.52 0.54		0.36 0.40	0.23 0.26

Note: These figures are valid only for conventional proportions.

271

Example 7-4. A tension member in a bridge truss is a welded H shape 43 ft long and 16 in. deep over-all. The maximum load on the member is 352 kips tension and the minimum load is 58 kips compression. Select proportions for the plates forming the member using A441 steel. (AASHO Specification.)

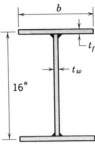

Fig. 1

Solution

Design load $= P_{max} + \frac{1}{2}P_{min} = 352 + \frac{1}{2}(58) = 381$ kips.

Allowable tensile stress $F_t = 27$ ksi.

$$\text{Required area } A = \frac{P}{F_t} = \frac{381}{27} = 14.1 \text{ in.}^2$$

Assume compression is not critical.

Minimum width of flange governed by slenderness ratio (L/r) where $r = 0.25b$, approximately.

$$\frac{L}{r} = 200 \qquad r = \frac{L}{200} = \frac{43 \times 12}{200} = 2.58 \text{ in.} = 0.25b$$

$$b = \frac{2.58}{0.25} = 10.3 \text{ in. minimum.} \qquad \text{Try } b = 11 \text{ in.}$$

Using the same plate thickness t for web and flanges (see Fig. 1):

$A = (2 \times 11 + 16 - 2t)t = (38 - 2t)t = 14.1 \text{ in.}^2 \qquad \therefore \quad t = 0.38 \text{ in.}$

Try varying plate thickness t_f and t_w. Use flange plates 0.5 in.

$$A_f = 2 \times 11 \times 0.5 = 11 \text{ in.}^2$$

$$A_w = A - A_f = 14.1 - 11.0 = 3.1 \text{ in.}^2$$

$$A_w = [16 - 2(\tfrac{1}{2})]t_w = 15t_w = 3.1 \qquad t_w = 0.207$$

Use $t_w = \frac{1}{4}$-in. plate.

Conclusion. Use

$$t_f = \tfrac{1}{2} \text{ in.} \qquad b = 11 \text{ in.} \qquad t_w = \tfrac{1}{4} \text{ in.}$$

7-11 CONNECTIONS AND SPLICES

Connections. Design of typical end connections for wires, cables, rods, bars, and other structural shapes has been discussed briefly in Secs. 7-1 and 7-5. It must be emphasized that design of terminal connections and splices for structural members deserves considerable attention since the structure is no stronger than the weakest point and, if the connections or splices are inadequate to resist the loads, the entire structure will collapse, regardless of the reserve strength of the members themselves. Particular attention must be given to the "stress paths," that is, the mechanism of load transmission in the end connections and splices. Accurate determination of stress distribution is often impossible, and in such cases approximate calculations must be made, assuming a mechanism of load transmission that takes into account the elastic deformations of members and connections.

A "simple" problem of connecting a single angle tension member to a gusset plate (Fig. 7-11) illustrates some of the difficulties that may arise in design of connections. If the angle, rivets, and gusset plate were all perfectly rigid, then tension in the angle could be transmitted to the connections in an infinite number of ways, since conditions of equilibrium of forces and

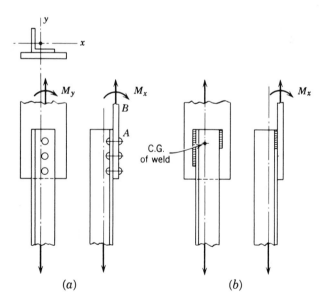

(a)   (b)

Fig. 7-11 Single angle connection—tension coinciding with centroid of angle. (a) riveted or bolted and welded.

moments are not sufficient to define a particular stress distribution in the angle and rivets, and the gusset plate.

One possibility is shown in Fig. 7-11a, where the resultant tension coincides with the centroidal axis of the angle. This load is eccentric with respect to both rivets and gusset plate, and results in transverse shear and tension loads on rivets at A, as well as bending of the gusset plate at B. Using welded connections in place of rivets, it is possible to eliminate eccentricity of the load with respect to one of the axes of the welds (Fig. 7-11b), but eccentricity with respect to the plane of the welds and to the gusset plate still remains. However, provided connections are not subjected to fatigue, many specifications permit equal length welds for end connections of single or double angles and similar flexible members.

Another possibility is shown in Fig. 7-12a in which the resultant tension coincides with the centroid axis of the gusset plate and is eccentric with respect to both axes of the angle. The instances just considered neglected deformations of the angle, rivets, and gusset plate. For a given connection, however, the elastic deformations determine the stress distribution in the connection. For example, if the gusset plate is considered to be much more flexible than the angle in bending about the x axis, but much more rigid than the angle in bending about the y axis, then the connection will deform in such a manner that the tension load will produce little or no bending about

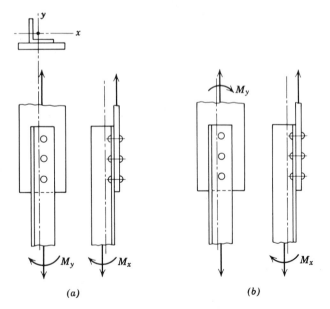

(a)   (b)

Fig. 7-12 Single angle connection.

the x axis in the gusset plate and little or no bending about the y axis in the angle, as shown in Fig. 7-12*b*.

Exact analysis of the highly indeterminate problem of load transmission in this seemingly simple connection is practically impossible. In conventional design all effects of eccentricities are often neglected; the tension member, the rivets in the connection, and the gusset plate are designed as if all are subjected to purely axial loads. Although such an assumption conflicts with the basic condition of equilibrium, it is believed that in many cases effects of eccentricity are secondary in magnitude and no serious overstressing would occur under actual conditions. However, the designer should evaluate the validity of such an assumption and, in the design of connections for single angles, eccentricities preferrably should be taken into consideration.

Lug angles are often used to connect the outstanding leg of an angle to a gusset plate (Fig. 7-13). The lug angles reduce secondary bending stresses in the angle and also provide additional connecting rivets, which may permit the reduction of gusset-plate dimensions. Sometimes the addition of lug angles is considered to increase the stiffness of the whole connection, although in most instances this increase is quite negligible.

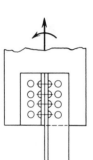

When the size of tension members is determined by consideration of deflection, fatigue, stiffness, minimum dimensions, corrosion protection, appearance, etc., the allowable loads they can support based on allowable stresses in the member may exceed the computed design loads. This raises the question of whether the connections and splices should be designed to carry the computed design load only, or to develop the full *allowable* load on the member. Many engineers consider it good practice to design for the allowable load of the member regardless of the computed design loads.

Fig. 7-13

This increased strength of the connection is not always obtained at negligible cost; for example, if the number of rivets or length of welds in the connection is increased, the cost of additional riveting or welding may be appreciable and may not be justified.

Some specifications, such as AISC and AASHO, set minimum strength requirements for connections, such as prescribed minimum load capacity or a prescribed percentage (50–75%) of the effective capacity of the material connected.

Splices. Steel rods used as tension members are usually not very long and seldom require splices. When splices are required, turnbuckles or sleeve nuts

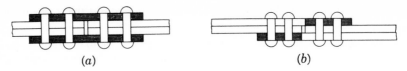

(a)   (b)

Fig. 7-14   Riveted splices.

can be used advantageously, since they also serve to adjust the length or to produce some initial tension if necessary. For rolled shapes and built-up sections two types of riveted splices are used: butt splices and lap (or shingle) splices. In butt splices, the member is entirely cut at one section and the connection is made by additional splice material (Fig. 7-14*a*). In lap (shingle) splices, the different parts of the member are cut at successive sections and the connection is made by overlapping of the various parts of the member (Fig. 7-14*b*). Comparing the two splices, it is obvious that the additional material (cross-hatched) required for splicing is smaller in a shingle than in a butt splice. On the other hand, the butt splice is easier to assemble, especially in the field.

When computing the rivets required in a splice, distinction should be made between direct and indirect riveting (Fig. 7-15). In Fig. 7-15*a*, the load in the angles is transmitted directly to the vertical splice plate, whereas the load in the main plate is indirectly transmitted through a filler. In Fig. 7-15*b*, the load transmission from the angles to the vertical splice is indirect, whereas that from the main plate to the splice plate is direct. Indirect splicing usually requires additional rivets. AISC provides that fillers shall be extended beyond the splice and secured by sufficient rivets to develop the stress in the filler. For example, in Fig. 7-15*a*, the additional rivets connecting the filler ahead of the splice must be designed for the following load:

$$P_f = \frac{P}{A + A_f} A_f \qquad (7\text{-}18)$$

where A is the area of the two angles and the main plate, A_f the area of the filler plate, and P the total load on the member.

For Fig. 7-15*b*, the additional rivets due to indirect splicing cannot be defined in the same manner because no filler plate is involved. AASHO provides that, as an alternate, if splice plates are not in direct contact with parts that they connect, the number of rivets on each side of the joint required for direct-contact splice shall be increased by two transverse lines of rivets for each intervening plate.

In some butt connections one of the splice plates may be overdimensioned, without redistributing the rivets according to the area of splice plates. This practice is sometimes criticized because the heavy plate may overload its

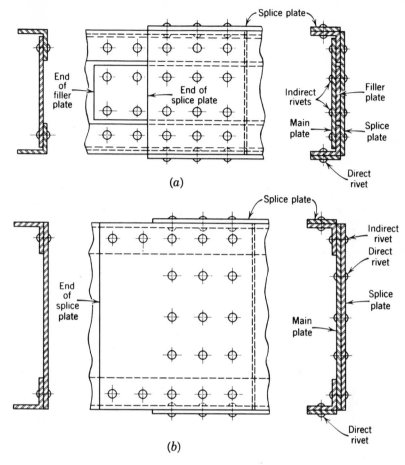

Fig. 7-15 Direct and indirect riveting.

rivets on the basis of elastic stress distribution. Actually, the rivets in the overdimensioned plate may yield prematurely, but no failure will occur until the load is redistributed in proportion to the rivet strength. This case is similar to that of a long splice in which end rivets, although overstressed in the elastic range, will be almost equally stressed in the plastic range. Thus no valid objection can be made to slight over-dimensioning of one plate, provided the other components of the splice are adequately designed.

A simple unsymmetrical lap splice (Fig. 7-16) is subject to high bending stresses due to eccentricity of tensile load and therefore is seldom used unless the couple due to this eccentricity can be adequately resisted by lateral support.

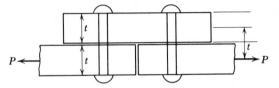

Fig. 7-16   Stress distribution in an eccentric splice.

A combination of butt and lap splices is shown in Fig. 7-17, in which two plates are cut at different sections, partly splicing each other and partly spliced by the additional plates S_1 and S_2. Various assumptions can be made in the design of such splices. For example, it may be assumed that plate M_2 transmits its entire load to M_3 in the region between sections B and C, and that load in plate M_1 is transmitted to splice plates S_1 and S_2 in the region between sections A and B. This load is distributed in inverse proportion to the distance of these splices from M_1. In this instance, if thickness of splice plates is denoted by x, then the load in S_1 is given by

$$P_s = P_1 \frac{(3t + x)}{(2t + x)2} = P_1 \frac{3t + x}{4t + 2x} \tag{7-19}$$

If the stress in S_1 at B is the same as the stress in M_1 at A, then $P_1/t = P_s/x$, and $P_s = P_1(x/t)$. Substituting in Eq. 7-19, we get

$$P_1 \frac{3t + x}{4t + 2x} = P_1 \left(\frac{x}{t}\right) \tag{7-20}$$

$$x = 0.69t \tag{7-21}$$

This result is valid for the case of two main plates only, the splice plate thickness increasing with the number of plates.

Another design may be made on the assumption that the total load in plates M_1 and M_2 produces uniform stress distribution at sections B and C.

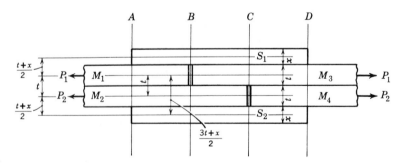

Fig. 7-17   Load distribution in a splice.

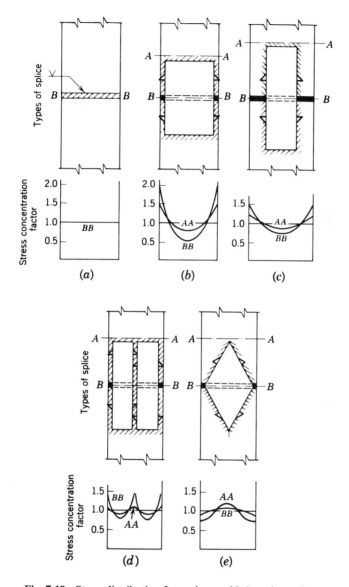

Fig. 7-18 Stress distribution for various welded tension splices.

This assumption implies that the couple due to eccentricity of the load on section B or C is resisted by lateral support of the splice plates, similar to the I-beam cover-plate splices considered previously. In this instance, the load P_s transmitted by each splice plate is

$$\text{at } B \quad P_{sB} = \frac{P_1 + P_2}{A_2 + 2A_s} A_s \quad \text{and} \quad \text{at } C \quad P_{sC} = \frac{P_1 + P_2}{A_1 + 2A_s} A_s \quad (7\text{-}22)$$

where P_1, P_2 are the loads in plates M_1 and M_2, respectively; A_1 A_2 are the areas of plates M_1 and M_2, respectively; and A_s is the area of each splice plate, S_1 and S_2. In this case, there is no load transmission to or from splice plates between sections B and C, whereas the load transmitted between plates M_1 and M_2 in this region is $(P_1 + P_2) - 2P_s$.

Design of Welded Splices. The general principles of design of welded connections and splices have been discussed in Chapter 6. Butt welds designed to develop the full applied load usually form an excellent connection, provided full-penetration butt welds are used. The fatigue unit strength of butt welds is considerably greater than that of fillet welds, although it is not as great as the strength of the unwelded member. Thus, if the connection is subjected to fatigue, a butt weld may not be sufficient to develop the strength of the member. The addition of fillet-welded cover straps to reinforce a butt weld is generally not satisfactory for a connection subjected to alternating loads because they cause stress concentrations which reduce the fatigue strength of the connection. Under static loading, butt welds in structural carbon steel should develop the full strength of base metal and reinforcing straps are thus unnecessary.

Various types of fillet-welded butt-strap splices can be designed. The results of numerous tests on butt-strap splices show that, for optimum stress distribution, a gradual transfer from main to splice plates is desirable. Figure 7-18 shows some typical splices and stress distribution observed in them under axial tensile load. The advantage of butt-welded splice is clearly evident in that minimal stress concentration is developed.

PROBLEMS

Use A36 Steel unless otherwise indicated

1. A tension member made of two 12-in. channels 20.7 lb is connected to two $\frac{3}{8}$-in. gusset plates (Fig. P-1), using seven $\frac{7}{8}$-in. rivets for each channel. Compute the allowable load on the section, considering the net section of the channels, as well as the strength of the connections. (AISC Specification.)

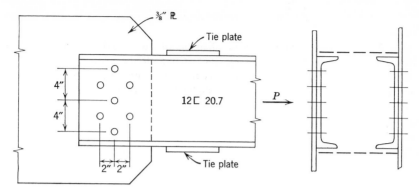

Fig. P-1

2. A 6-in. × 6-in. angle is to be connected to a $\frac{1}{2}$-in. plate on one leg only, to transmit a tension load of 95 kips. Compute the thickness of the angle required: (*a*) using fillet welds for connection, (*b*) using 1-in. rivets. Detail both connections. (AISC Specification.) For approximation, neglect eccentricity, but assume that one-half of the outstanding leg is ineffective in carrying the load. Discuss the correctness or the degree of accuarcy of such an approximation.

3. A tension member made up of two angles 5-in. × 5-in. × $\frac{1}{2}$-in. is connected to a $\frac{1}{2}$-in. plate (Fig. P-3) to transmit an axial load of 160 kips. Design and detail the connection: (*a*) using $\frac{7}{8}$-in. rivets, (*b*) using fillet welds. (AISC Specification.) Would the welded connection permit smaller angles for the member?

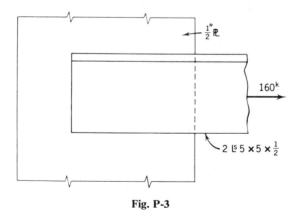

Fig. P-3

4. The web member of a small bridge truss is to be made of two equal-leg angles, connected to a $\frac{1}{2}$-in. gusset plate. The member has a length of 12 ft. It carries a maximum tension of 100 kips and a minimum tension of 20 kips. Design the member and detail the connection, using fillet welds. Assume 2×10^6 cycles of loading. (AASHO Specification.)

5. Solve Prob. 4 if $\frac{7}{8}$-in. high-tensile-strength bolts are used instead of fillet welds.

6. A 16 W℉ 142 is used as a tension diagonal of a truss and is connected to two 1-in. gusset plates by 1-in. rivets (Fig. P-6). Compute the allowable load on the section, assuming that tension in the gusset plates is not critical but considering the carrying capacity of the rivets. AASHO Specification. *Note:* Without the addition of the splice and filler plates, the connection would require much larger gusset plates.

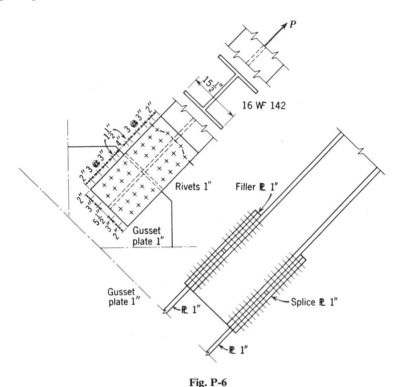

Fig. P-6

7. Determine the plastic strength P of a 6-in. × 6-in. × $\frac{1}{2}$-in. angle in tension, (Fig. P-7*a*). Assume stress distribution as shown in Fig. P-7*b* and yield point stress $f_y = 33$ ksi. *Note:* This stress distribution neglects eccentricity in the horizontal plane.

8. The bottom chord of a highway bridge truss is made up of two 15-in. channels 40 lb (Fig. P-8). It is to be spliced away from the joints, transmitting a total load of 300 kips. Use splice plates outside the webs and add plates inside the webs if required, but do not splice the flanges. Design the connection, using $\frac{7}{8}$-in. rivets.

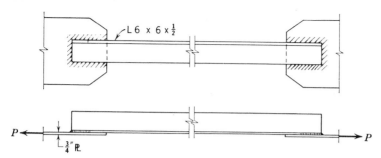

L 6 x 6 x $\frac{1}{2}$

$P \leftarrow$ ... $\rightarrow P$

$\frac{3}{4}''$ PL

Fig. P-7a

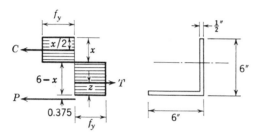

Fig. P-7b

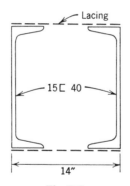

Lacing

15 ⊏ 40

14″

Fig. P-8

9. A roof truss is to be designed for the loading shown in Fig. P-9 using one $\frac{3}{8}$-in. gusset plate at each joint and $\frac{3}{4}$-in. rivets. Design the member AB of the bottom chord, using a two-angle section. Detail its connection to the gussets. Note that, by adding the hanger CD to minimize the bending of member AB, only the direct tension must be considered for AB.

10. Tension members of a heavy building truss carry total dead and live loads varying from 250 to 700 kips per member. The members are connected using two gusset plates at each joint spaced a constant distance apart (which may be chosen anywhere between 12 and 18 in.). Choose one type of member make-up

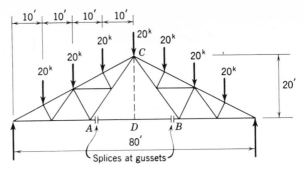

Fig. P-9

suitable for these members. First consider riveted connections and then welded connections. Make preliminary computations and show typical connections.

11. A hanger in a building is to be made of a round steel rod with threaded ends screwed to a steel plate bearing on concrete (Fig. P-11). The hanger is 30 ft long and carries a total design load of 90 kips. Design the rod to meet the AISC Specification and to limit the total elongation to within $\frac{1}{4}$ in.

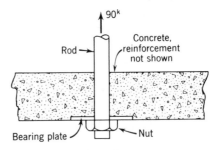

Fig. P-11

12. A round rod with upset ends is spliced with a turnbuckle at midlength. The rod carries no dead-load stress but is subject to live-load stresses varying between 10 kips compression and 20 kips tension. By means of the turnbuckle, an initial tension is to be introduced into the rod so that a residual tension of at least 2000 psi will remain under the design compressive load. What size of rod is required and what should be the initial tension?

REFERENCES

1. Cohen, E., and H. Perrin, "Design of Multi-Level Guyed Towers: Wind Loading," *Proc. ASCE J. Structural Div.* (September 1957); discussion (March 1958).

2. Rowe, R. S., "Amplification of Stress and Displacement in Guyed Towers," *J. Structural Div. Proc. ASCE* (Oct. 1958); discussion (May 1959).

3. Johnston, B. G., "Pin-Connected Plate Links," *Trans. ASCE* **104,** 2023 (1939).

8

Bending and Torsion of Beams

ଡ଼

8-1 INTRODUCTION

Structural steel shapes loaded transversely are a common form of structural member in buildings, bridges, and other structures. In most instances, the beams have their loads applied in the plane of the web, thus producing bending about the strong axis. In some applications, the loads are applied in a direction perpendicular to the web and produce bending about the weak axis of the cross section. In these applications, the load is considered to be passing through the shear center of the cross section and therefore is producing simple bending about either or both axes. The determination of these stresses for various cross sections is discussed in this chapter.

When the loads do not pass through the shear center, a torsional moment is produced which develops twisting of the member and additional stresses. The combination of bending and torsional stresses is also discussed.

When structural members are subjected to combined bending and axial load, the deformation of the member may have an important influence on the internal moments resisted by the member. The effect is particularly important if the axial load is compressive. The influence of axial loads, both tensile and compressive, on the internal bending moments and on the capacity of a section to resist these moments is considered.

8-2 SIMPLE BENDING

Stresses and deformations in members subjected to flexure within the elastic range can be determined with a good degree of accuracy. Experiments

285

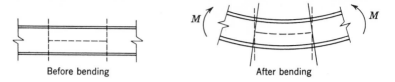

Before bending   After bending

Fig. 8-1 Plane section remains plane.

indicate that a plane section before bending remains very nearly plane after bending (Fig. 8-1) and that within elastic range stress is proportional to strain. Using these observed phenomena and the equations of equilibrium, simple equations for normal and shearing stresses in a beam (Fig. 8-2) are obtained when bending takes place about a principal axis and there is no twisting of the section:[1]

$$f_b = \frac{M_x}{I_x} y \tag{8-1}$$

$$f_v = \frac{V_y Q_x}{I_x t} \tag{8-2}$$

Normal stress f_b acts on a fiber at a distance y from the neutral axis xx; M_x is the bending moment about the neutral axis xx at the section where f_b is acting; and I_x is the moment of inertia of the cross section taken about the xx neutral axis. Shearing stress f_v (Fig. 8-2b) is due to the shearing force V_y. The location of the element subjected to shearing stress f_v is defined by the quantity $Q_x = A\bar{y}$, where A is the area of a part of cross section measured between the element and the free edge or the edges of the cross section, and $\bar{y}$ is the distance from the centroid of this area A to the neutral axis. The thickness t of the material is measured as the true thickness and is not necessarily measured parallel to the neutral axis.

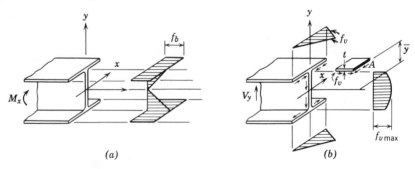

(a)   (b)

Fig. 8-2 Normal and shearing stresses in a beam. (a) normal stresses and (b) shearing stresses.

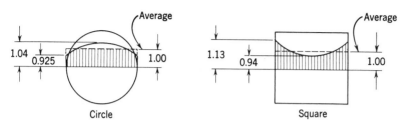

Fig. 8-3 Shearing stress variation across thick beams.

The value of the shearing stress defined by Eq. 8-2 is only an average stress across the thickness t and therefore it is accurate only when t is small compared to the depth of the beam. When the thickness of the beam is large, the shearing stress is not distributed uniformly across the thickness. For simple sections of regular shape, the actual distributions of shearing stress across the width of the section are shown in Fig. 8-3 in terms of an average shearing stress of unity.[2] It should be noted that Eq. 8-2 defines shearing stress only for open sections for which the static moment Q is computed for a portion of the section between a given fiber and a free edge.

The simple flexure formulas (Eqs. 8-1 and 8-2) have the following limitations: (*a*) the material is elastic, with stress directly proportional to strain and with equal moduli in tension and compression, (*b*) the stresses do not exceed the elastic limit, (*c*) the deflections are small compared to beam depth, (*d*) all loads are parallel to a principal plane and the sections rotate about a neutral principal axis perpendicular to the plane of loading, (*e*) the sections do not twist, (*f*) the stress is taken at a section away from the point of application of a concentrated load, at a distance equal to or greater than the depth of the beam, (*g*) the beam cross section is constant along the span, (*h*) the beam is initially straight or very nearly so, and (*i*) the span is at least four or five times as great as the depth of the beam. Limitations *a* and *b* are well known from standard textbooks on strength of materials and will not be discussed here. The other limitations are important in special instances and are generally not treated in elementary textbooks. Therefore some of these cases will be briefly reviewed here.

8-3 UNSYMMETRICAL BENDING

A section of arbitrary shape (Fig. 8-4) has two principal axes, u and w, both passing through the centroid and perpendicular to each other. Any axis of symmetry is a principal axis; for sections having no axis of symmetry, the principal axes correspond to the maximum and minimum moment of

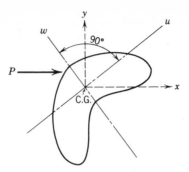

Fig. 8-4   Unsymmetrical bending, *u* and *w*—principal axes.

inertia axes and are defined by the condition that the product of inertia I_{uw} about the principal axes is equal to zero; that is,

$$\int_A uw \, dA = 0$$

When the applied loads are not in a principal plane, the stresses in the beam can be found by superposition. In this instance, the loads are resolved into components parallel to the principal axes, and then

$$f_b = \frac{M_u}{I_u} w + \frac{M_w}{I_w} u \tag{8-3}$$

$$f_v = \frac{V_w Q_u}{I_u t} + \frac{V_u Q_w}{I_w t} \tag{8-4}$$

where the notation is similar to that used in Eqs. 8-1 and 8-2. Numerical values and directions of stresses in Eqs. 8-3 and 8-4 are obtained by super-position of the effects of bending about the principal axes.

In practical cases when the section has no axis of symmetry, determination of principal axes and of section properties with respect to them is tedious. Expressions for bending and shearing stresses in such sections can be written with reference to any arbitrary set of centroidal axes x and y (Fig. 8-4) which are perpendicular to each other. This form usually simplifies the numerical calculations of the stresses, provided that a consistent sign convention is followed.

$$f_b = \frac{M_y I_x - M_x I_{xy}}{I_x I_y - I_{xy}^{\,2}} x + \frac{M_x I_y - M_y I_{xy}}{I_x I_y - I_{xy}^{\,2}} y \tag{8-5}$$

$$f_v = \frac{V_x I_x - V_y I_{xy}}{I_x I_y - I_{xy}^{\,2}} \frac{Q_y}{t} + \frac{V_y I_y - V_x I_{xy}}{I_x I_y - I_{xy}^{\,2}} \frac{Q_x}{t} \tag{8-6}$$

The sign convention is such that moment M is positive when producing compression in the first quadrant (x, y positive) and f_b is positive when the fiber is in compression. After the direction of the normal stresses is established, the direction of the shearing stresses can be obtained by drawing a free-body sketch and using conditions of equilibrium, as in Example 8-1. Note that the equations for bending stress are applicable to both open and closed sections, but the equations for shearing stress are applicable to open sections only.

Example 8-1. For the 6-in. × 4-in. × $\frac{1}{2}$-in. angle loaded as shown (Fig. 1), determine (*a*) section properties A, I_x, I_y, I_{xy}; principal axes u, w; and moments of inertia I_u, I_w; (*b*) bending stresses at points A, B, and C, using generalized Eq. 8-5; (*c*) location of neutral axis, using graphical method; (*d*) maximum shearing stress in the angle; and (*e*) deflection at mid span.

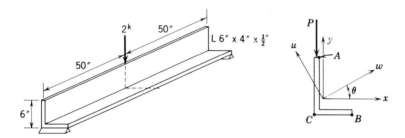

Fig. 1

Solution

(*a*) From AISC Manual, p. 1-31, properties of sections:

$$I_x = 17.4 \text{ in.}^4 \quad I_y = 6.3 \text{ in.}^4 \quad A = 4.75 \text{ in.}^2 \quad r_u = 0.87 \text{ in.} \quad \tan\theta = 0.440$$
$$\therefore \theta = 23° 45'$$

$$I_{xy} = -\frac{I_x - I_y}{2} \tan 2\theta = -\frac{17.4 - 6.3}{2}(1.091) = -6.1 \text{ in.}^4$$

$$I_u = A r_u{}^2 = (4.75)(0.87)^2 = 3.6 \text{ in.}^4$$

$$I_u + I_w = I_x + I_y = 17.4 + 6.3 = 23.7 \text{ in.}^4$$

$$\therefore \quad I_w = (I_x + I_y) - I_u = 23.7 - 3.6 = 20.1 \text{ in.}^4$$

(*b*) The general equation is

$$f_b = \frac{M_x I_y - M_y I_{xy}}{I_x I_y - I_{xy}{}^2} y + \frac{M_y I_x - M_x I_{xy}}{I_x I_y - I_{xy}{}^2} x = (\alpha)y + (\beta)x$$

For this beam:

$$M_x = \frac{PL}{4} = \frac{2 \times 100}{4} = 50 \text{ kip-in.} \qquad M_y = 0$$

$$\alpha = \frac{M_x I_y - M_y I_{xy}}{I_x I_y - I_{xy}^2} = \frac{50(6.3) - 0}{(17.4)(6.3) - (-6.1)^2} = +4.32 \text{ kip/in.}^3$$

$$\beta = \frac{M_y I_x - M_x I_{xy}}{I_x I_y - I_{xy}^2} = \frac{0 - 50(-6.1)}{(17.4)(6.3) - (-6.1)^2} = +4.18 \text{ kip/in.}^3$$

The bending stresses are: $f_b = 4.32y + 4.18x$

Point A (-0.49, $+4.01$):

$$f_b = (4.32)(4.01) + 4.18(-0.49) = 15.28 \text{ ksi} \quad \text{(compression)}$$

Point B ($+3.01$, -1.99):

$$f_b = (4.32)(-1.99) + 4.18(+3.01) = +3.99 \text{ ksi} \quad \text{(compression)}$$

Point C (-0.99, -1.99):

$$f_b = (4.32)(-1.99) + 4.18(-0.99) = -12.74 \text{ ksi} \quad \text{(tension)}$$

(*c*) Locate the neutral axis (Fig. 2).

Plot stress diagrams to scale; establish points D, E, corresponding to zero stress; line DE is the neutral axis. Check; it must pass through C.G.

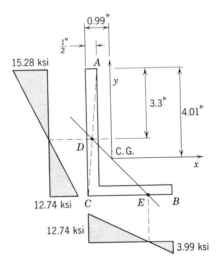

Fig. 2

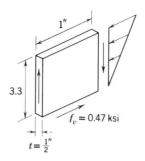

Fig. 3

(*d*) Maximum shear stress will occur at the neutral axis at either point *D* or *E*. By observation, point *D* is critical (Fig. 3). Assume that variation of stress across the thickness of the angle leg is small. Then

$$f_v = \frac{V_x I_x - V_y I_{xy}}{I_x I_y - I_{xy}^2} \frac{Q_y}{t} + \frac{V_y I_y - V_x I_{xy}}{I_x I_y - I_{xy}^2} \frac{Q_x}{t}$$

$$V_x = 0, \quad V_y = 1 \text{ kip}$$

$$Q_x = \tfrac{1}{2}(3.3)\left(4.01 - \frac{3.3}{2}\right) = 3.89 \text{ in.}^3$$

$$Q_y = \tfrac{1}{2}(3.3)\left(-0.99 + \frac{0.5}{2}\right) = -1.22 \text{ in.}^3$$

$$f_v = \frac{-1(-6.1)}{72.9} \frac{(-1.22)}{0.5} + \frac{1(6.3)}{72.9} \frac{(3.89)}{0.5} = 0.467 \text{ ksi}$$

(*e*) For a simple beam with a concentrated load *P* at the midspan,

$$\Delta = \frac{PL^3}{48EI} \qquad E = 30 \times 10^6 \text{ psi}$$

Consider the components of load and deflection along the principal axes (Fig. 4):

$$P_u = P \cos \theta = 2(0.915) = 1.83 \text{ kips}$$

$$P_w = P \sin \theta = 2(0.403) = 0.806 \text{ kips}$$

$$\Delta_u = \frac{P_u L^3}{48EI_w} = \frac{1.83(100)^3 \times 1000}{48 \times 30 \times 10^6 (20.1)} = 0.063 \text{ in.}$$

$$\Delta_w = \frac{P_w L^3}{48EI_u} = \frac{0.81(100)^3 \times 1000}{48 \times 30 \times 10^6 (3.6)} = 0.156 \text{ in.}$$

$$\Delta_R = \sqrt{\Delta_u^2 + \Delta_w^2} = 0.168 \text{ in.}$$

Note: The resultant deflection is perpendicular to the neutral axis.

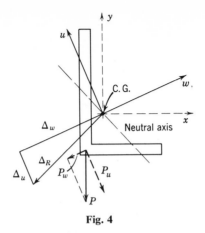

Fig. 4

8-4 SHEAR CENTER OF OPEN SECTIONS

Equations 8-2 and 8-4 define shearing stresses that are valid only if the beam does not twist: that is, the shear force passes through a definite point, called the center of twist or the shear center. For beams of open cross section (Fig. 8-5) having relatively thin webs and flanges in which shearing deformations may be neglected, the location of the shear center may be obtained from the condition that the shearing stress distribution defined by Eq. 8-2 or 8-4 is in equilibrium with the shearing force passing through the shear center. For the beam shown in Fig. 8-5, equilibrium conditions stating that $\Sigma F_x = 0$ and $\Sigma F_y = 0$ are automatically satisfied by the conventional

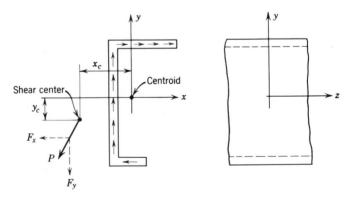

Fig. 8-5 Shear center location for unsymmetrical open section.

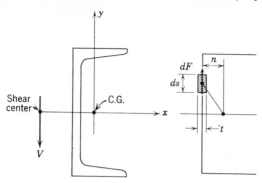

Fig. 8-6   Shear center location for open section with one axis of symmetry.

shearing-stress distribution, regardless of the line of action of the shearing force. The equilibrium condition that $\Sigma\, M_z = 0$ is satisfied only when the shearing force passes through a particular point, defined as the shear center. The location of this point (x_c, y_c) is obtained by solving equation $\Sigma\, M_z = 0$ for two independent shearing forces which do not act simultaneously. For sections having one axis of symmetry (Fig. 8-6), the shear center lies on the axis of symmetry at a distance x_c from the centroid, where x_c is obtained from the following equation:

$$V x_c = \int dM_z = \int n\, dF = \int n f_v t\, ds$$

or

$$x_c = \frac{\displaystyle \int n f_v t\, ds}{V} \qquad (8\text{-}7)$$

In Eq. 8-7 V is an arbitrary assumed force perpendicular to the plane of symmetry, dF is a shearing force on a differential element ds, n is the lever arm of force dF taken with respect to the centroid of the section, and f_v is the shearing stress defined by Eq. 8-2 or 8-4. The numerical work of solving Eq. 8-7 can be simplified by using quantity $(f_v t) = $ (shearing stress $\times$ thickness) as a shearing force dF per unit length ds. This quantity dF/ds, sometimes called "shear flow" and denoted by a symbol q, is defined from Eq. 8-2 as $q = VQ/I$. For determination of the location of the shear center, the magnitude of shear force V is arbitrary and may be taken numerically equal to I, so that V/I is 1, and $q = Q = A\bar{y}$. Use of this simplified method of locating the shear center is illustrated in Example 8-2.

Example 8-2.   Determine shear center for a 6 ⊏ 8.2 lb.

Solution

Section properties (Fig. 1):

$I_x = 13.0 \text{ in.}^4$; $A = 2.39 \text{ in.}^2$; $A_{\text{web}} = 1.06 \text{ in.}^2$; $A_{\text{flanges}} = 1.32 \text{ in.}^2$;
i.e. 0.66 in.² each.

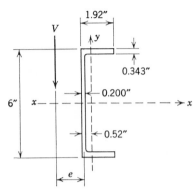

Fig. 1

Determine shear flow (Fig. 2):

$$q = \frac{VA\bar{y}}{I_x} \quad \text{Set} \quad V = I = 13 \text{ kips} \quad \therefore \quad \frac{V}{I} = 1 \text{ kip/in.}^4 \quad \text{and} \quad q = A\bar{y} \text{ kip/in.}$$

$$q_1 = \frac{VA_f}{I}\frac{h}{2} = (0.66)(2.83) = 1.87 \text{ kip/in.}$$

$$q_2 = q_1 + \frac{V\frac{1}{2}A_{\bar{w}}\bar{y}}{I_x} = 1.87 + \frac{1}{2}(1.06)(1.33)$$

$$= 1.87 + 0.70 = 2.57 \text{ kip/in.}$$

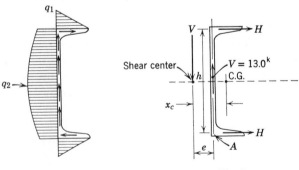

Fig. 2   **Fig. 3**

Determine shear center (Fig. 3):

$$\sum M_A = 0 = Hh - Ve$$

$$H = \frac{(1.87)(1.92)}{2} = 1.79 \text{ kips}$$

$$e = \frac{Hh}{V} = \frac{(1.79)(5.66)}{13} = 0.78 \text{ in.}$$

$$x_c = e + 0.52 - 0.10 = 1.2 \text{ in.}$$

For unsymmetrical cross sections (Fig. 8-5), the coordinates of the shear center x_c, y_c, taken with respect to the center of gravity, can be found from the following equations:

$$x_c = \frac{\int (nf_v t)_y \, ds}{V_y} = \frac{\int (nq)_y \, ds}{V_y}, \qquad y_c = \frac{\int (nf_v t)_x \, ds}{V_x} = \frac{\int (nq)_x \, ds}{V_x} \quad (8\text{-}8)$$

In these expressions, V_x and V_y are forces parallel to the principal axes x and y, and q_x and q_y are shear flows corresponding to the shears V_x and V_y. The tedium of locating the principal axes may be avoided by assuming an arbitrary bending-stress distribution, finding corresponding shearing-stress and shear-flow distributions, and corresponding components F_x and F_y of the shear force V, where x and y are arbitrary orthogonal centroidal axes. Then the equation for moment equilibrium can be written:

$$F_x y_c + F_y x_c = \int n \, dF = \int nq \, ds \quad (8\text{-}9)$$

This equation involves two unknowns, x_c and y_c, and cannot be solved directly. A similar equation can be written, however, if another arbitrary bending-stress distribution is assumed and the process repeated. Solving two such equations simultaneously yields values of x_c and y_c, as illustrated in Example 8-3.

Example 8-3. Locate the shear center of the section shown in Fig. 1. Neglect bending stresses in plate elements. Flange cross-sectional area = 1.81 in.[2]

Solution

Since the bending stresses in plate elements are to be neglected, the section may be idealized as one consisting of concentrated flange areas carrying normal forces only, connected by webs carrying shear forces only, as shown in Fig. 2.

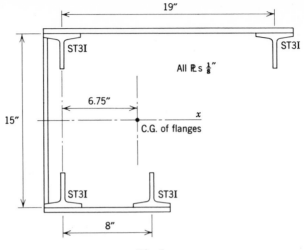

Fig. 1

Assume arbitrary direction of shear so that it will cause a horizontal neutral plane and bending-stress distribution as shown, and flange stress increment per unit length top and bottom: $\Delta f/\Delta L = 1$ ksi per inch. Then flange load increment per unit length

$$\frac{\Delta P}{\Delta L} = \frac{\Delta f}{\Delta L} A_f = A_f = 1.81 \text{ kip/in.}$$

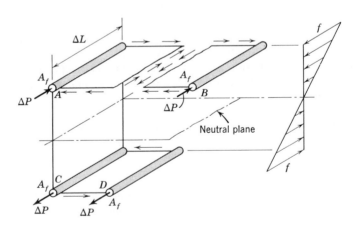

Fig. 2

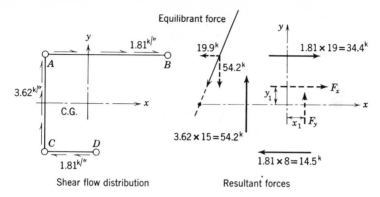

$$\text{Fig. 3}$$

Hence shear flows on the section can be determined as follows (Fig. 3):
For equilibrium:

$$\sum F_x = 0 \qquad F_x + 34.4 - 14.5 = 0 \qquad F_x = -19.9 \text{ kips}$$

$$\sum F_y = 0 \qquad F_y + 54.2 = 0 \qquad F_y = -54.2 \text{ kips}$$

$$\sum M = 0 \qquad F_x(y_1) - F_y(x_1) + 34.4(7.5) + 54.2(6.75) + 14.5(7.5) = 0$$

$$(-19.9)y_1 - (-54.2)x_1 + 733 = 0$$

$$x_1 - 0.367y_1 + 13.5 = 0$$

(line of resultant force corresponding to assumed stress distribution, no twist).

Since no twist is caused by the resultant force just shown, the shear center is located at some point on this line. To locate this point, it is necessary to assume another stress distribution, to find another line of force causing no twist. Intersection of these two lines will define the shear center.

Assume the direction of shear so that it will cause a vertical neutral plane (Fig. 4) and linear distribution of bending-stress increment per unit length.

$$\text{At } A \text{ and } C, \qquad \frac{\Delta f}{\Delta L} = 1 \text{ ksi/in.}$$

and

$$\text{at } B, \qquad \frac{\Delta f}{\Delta L} = \left(\frac{12.25}{6.75}\right)(1) = 1.815 \text{ ksi/in.}$$

$$\text{at } D, \qquad \frac{\Delta f}{\Delta L} = \left(\frac{1.25}{6.75}\right)(1) = 0.185 \text{ ksi/in.}$$

The values of $\Delta P/\Delta L = (\Delta f/\Delta L)A_f$, are:

$$\text{at } A \text{ and } C, \quad \frac{\Delta P}{\Delta L} = \frac{\Delta f}{\Delta L} A_f = 1.81 \text{ kip/in.}$$

$$\text{at } B, \quad (1.815)(1.81) = 3.29 \text{ kip/in.}$$
$$\text{at } D, \quad 0.185(1.81) = 0.335 \text{ kip/in.}$$

The shear flows are then as shown in Fig. 4.

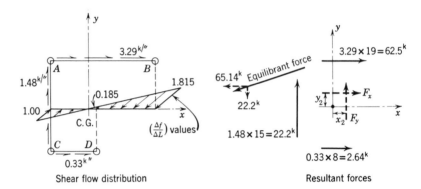

Shear flow distribution   Resultant forces

Fig. 4

For equilibrium:

$$\sum F_x = 0 \quad F_x + 62.5 + 2.64 = 0 \quad F_x = -65.14 \text{ kips}$$
$$\sum F_y = 0 \quad F_y + 22.2 = 0 \quad F_y = -22.2 \text{ kips}$$
$$\sum M = 0$$
$$(-65.14)y_2 - (-22.2)x_2 + 62.5(7.5) + 22.2(6.75) - 2.64(7.5) = 0$$
$$(-65.14)y_2 + (22.2)(x_2) + 599.2 = 0$$
$$x_2 - 2.94y_2 + 27.0 = 0 \quad \text{(line of resultant, no twist)}$$

For location of shear center, find the intersection of two forces graphically or solve two equations simultaneously.

$$x - 0.367y + 13.5 = 0$$
$$\underline{x - 2.94y + 27.0 = 0}$$
$$2.57y - 13.5 = 0 \quad y = \frac{13.5}{2.57} = 5.25 \text{ in.}$$

$$x = -13.5 + 0.367y = -13.5 + (0.367)(5.25) = -11.6 \text{ in.}$$

Shear center location is shown in Fig. 5.

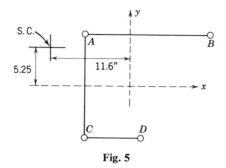

Fig. 5

8-5 TORSION OF CYLINDRICAL SECTIONS

The stresses produced by torsional moments acting on elastic cylindrical members of circular cross section may be obtained readily.

Tests have indicated that there is no distortion (warping) of the circular cross section in the plane of the section or normal to it, that is, there are no relative displacements in the radial direction or along the axis of the cylinder. Therefore the only stress is the shearing stress which acts tangentially and is proportional to the distance from the centroid (Fig. 8-7). For circular cross sections the maximum shear stress may be expressed as

$$f_v = \frac{Tr}{J} \tag{8-10}$$

where r is the radius, J is the polar moment of inertia, and T is the torque acting on the section. The cylinder is considered to be in a state of pure shear due to torsion.

In a member of arbitrary cross section the torsional shear stress is not proportional to the radius vector and cannot be obtained by means of a simple equation as in the circular member.

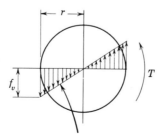

Fig. 8-7 Torsional shearing stress distribution in a circular section.

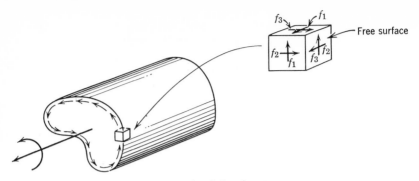

Fig. 8-8 Torsional shearing stresses.

Consider a prismatic member of arbitrary cross section subjected to torsion without warping (Fig. 8-8). In order to satisfy the boundary conditions, the stresses f_2 and f_3 in the free surface must be zero. Therefore at the boundary the only possible shear stress is f_1, which is parallel to the boundary. Whenever the boundary has a corner (Fig. 8-9) the shear stress must be zero. The resultant of all the shearing stresses on the cross section must be equal to the applied torque.

The solution of torsion on prismatical bars subjected to couples at the ends was given by Saint-Venant and these stresses are often referred to as pure or Saint-Venant torsional stresses. The basic assumption is that the member is free to warp during the application of the torsional forces.

The solution for free torsion of a rectangular bar[3] indicates that the torque T and angle of twist per unit length $d\phi/dz$ are related as follows:

$$\frac{d\phi}{dz} = \phi' = \frac{T}{GK_t} \qquad (8\text{-}11a)$$

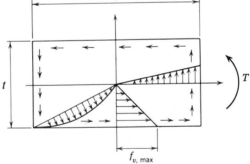

Fig. 8-9 Shearing stress distribution in a rectangular section.

where K_t is the torsional rigidity of the bar and G is the shearing modulus of the material. The maximum stress occurs in the middle of the longer side b (Fig. 8-9) and is given by

$$f_{v,\max} = Gt\phi' \tag{8-11b}$$

where t is the thickness of the bar. For thin rectangular plates the ratio of b/t is greater than 3 and K_t can be approximated by

$$K_t = \frac{1}{3}\left(1 - 0.63\frac{t}{b}\right)bt^3 \tag{8-12}$$

When b/t is greater than ten, K_t is very nearly equal to $\frac{1}{3}bt^3$.

8-6 FREE TORSION OF OPEN SECTIONS

A type of member that is often used in structures is a narrow rectangular cross section, either alone or in various shapes, such as a wide-flange rolled beam or a channel.

The maximum shearing stress and angle of twist due to free torsion of rolled shapes, such as angles, channels, wide-flange, and I-shaped sections, may be determined approximately by considering the shapes as a series of connected rectangles.

In order to evaluate the torsion stresses in a wide-flange shape, let us consider the behavior of a beam subjected to a torque T on each end, as indicated in Fig. 8-10. The member is assumed to be freely acted upon by the twisting moments and has no other supports or restraints.

The analysis is based on the assumptions that the stresses are within the elastic limit, the angle of twist is small, and the cross section does not change its shape. Thicker the section, the better the correlation with the third assumption. It is further assumed that the web distortion (Fig. 8-11) is small compared to the other deformations and can be neglected.[4,5]

The beam rotates through an angle $d\phi$ per unit length dz (Fig. 8-10). This rotation causes a lateral displacement du of the top fiber and a warping of the entire cross section measured by the angle ω. Then

$$du = d\phi\frac{h}{2} \quad \text{and} \quad \omega = \frac{2du}{dz} = h\frac{d\phi}{dz} \tag{8-13}$$

If all the cross sections along the span are free to warp and the torque is constant along the span, the beam is subjected to shearing stresses only. This type of action on a wide-flange beam is referred to as "free" or "unrestrained" torsion and is further classified as pure or simple torsion or merely as St. Venant's torsion.

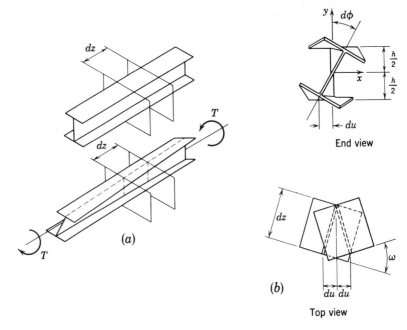

(a)

End view

(b)

Top view

Fig. 8-10   A W beam subjected to twisting moment.

These simple torsional stresses for an open section may be computed approximately on the basis of a series of rectangular elements and the torsional rigidity for the entire cross section is considered the sum of the individual rigidity of each rectangle. Thus, for a symmetrical shape (Fig. 8-12), the torsional stresses may be determined.

For pure torsion of such a section, Eq. 8-11 applies. Assuming that the torsional rigidity of the section K_t can be represented by the sum of the rigidities of component rectangles and assuming that each rectangle has a ratio of (b/t) greater than 10, the torsional rigidity is

$$K_t = \tfrac{1}{3}(2b_f t_f^{\,3} + d_w t_w^{\,3})$$ (8-14)

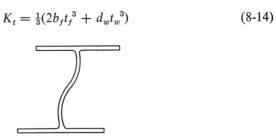

Fig. 8-11   Web distortion of beam subjected to twisting moment.

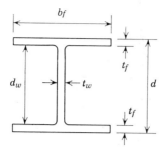

Fig. 8-12   Symmetrical open section.

and

$$\frac{d\phi}{dz} = \phi' = \frac{T}{\frac{1}{3}(2b_f t_f^3 + d_w t_w^3)G} \tag{8-15}$$

The maximum torsional shear stress $f_{v,\max}$ is then

$$f_{v,\max} = Gt\left(\frac{d\phi}{dz}\right) = \frac{Tt}{\frac{1}{3}(2b_f t_f^3 + d_w t_w^3)} \tag{8-16}$$

where t is either flange or web thickness, depending on the location where $f_{v,\max}$ is desired.

The value of K_t defined by Eq. 8-14 is approximate because the rectangular elements are assumed to be thin and the contribution of the thicker portion at the flange-to-web junction is neglected. For standard W shapes values of K_t which account for the actual dimensions have been calculated and are tabulated in Table B-1 in Appendix B.

8-7 RESTRAINED TORSION OF OPEN SECTIONS

In actual structures, most members subjected to torsion are not free to warp. Warping is restrained at some section, such as the fixed end of a cantilever beam (Fig. 8-13), which produces shear and bending of the flanges in the plane of the flanges.

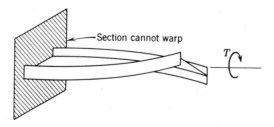

Fig. 8-13   Restrained torsion of beam.

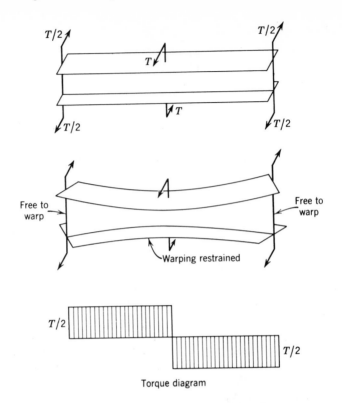

Fig. 8-14   Warping restraint in a beam subjected to torque.

It may also be reasoned that changes in torque along the span of a beam (Fig. 8-14) will result in an internal restraint of warping.

It can be seen that the cross section at the center of the beam where the torque is applied is restrained in the same manner as that of a fixed-end beam (Fig. 8-14). Therefore, in most structures torsion of beams is of the "restrained" or "partially restrained" type rather than "free." This action is also evident even when the ends of the beam are free to warp, as for simple beam shear connections. It becomes evident that warping produces additional shear and bending stresses in the flanges which must be added to the pure torsional shear and simple bending stresses in order to determine the maximum resultant stresses in the beam section.

Restrained torsion introduces two problems in design: (*a*) determination of deformations and of resisting torques caused by the applied loads and (*b*) determination of stresses in the member subjected to combined bending and torsional forces. The determination of deformations and resisting torques are problems which have been included in textbooks on the theory of structures and will not be discussed further here. However, the determination of internal stresses in a beam will be amplified here.

Consider a wide-flange beam subjected to a pure torsional load such as the cantilever indicated in Fig. 8-15.

Section *mm* twists through an angle ϕ and the flanges have a lateral deflection u at the section where $u = (h/2)\phi$.

The total torque T at any section is balanced by the torsional resistance (St. Venant) T_t of the cross section and the torsion-bending resistance T_b of the flanges, so that

$$T = T_t + T_b \tag{8-17}$$

The torsional resistance T_t depends on the angle of twist ϕ and from Eq. 8-11a is

$$T_t = K_t G\phi' \quad \text{(St. Venant's torsion)} \tag{8-18}$$

The torsion-bending resistance T_b, neglecting web resistance to distortion, is

$$T_b = Fh \tag{8-19}$$

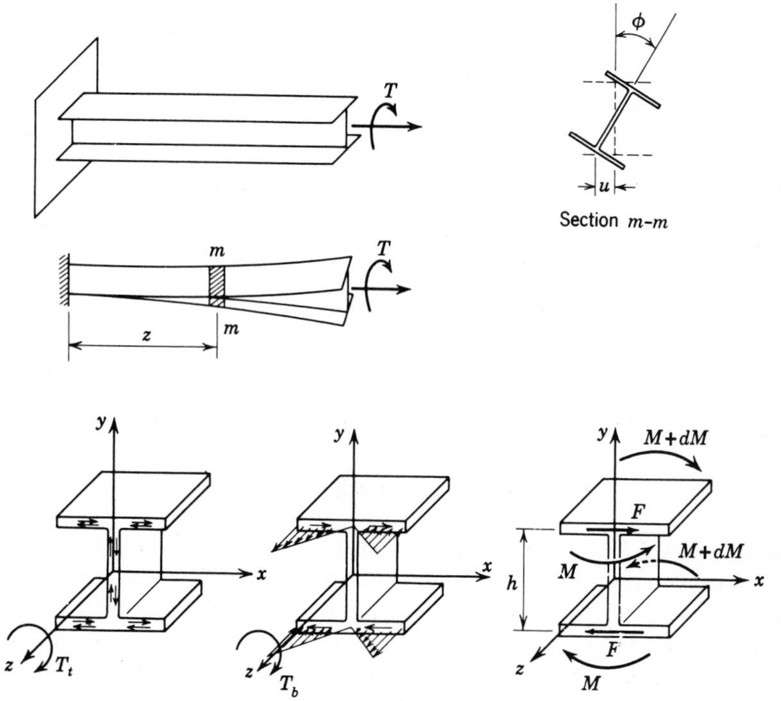

Fig. 8-15 W beam under pure torsional loading.

where F is the shearing force in the flanges. This force can be found if the variation of the bending moment M in the flange is known. This bending moment M is related to the lateral deflection u as:

$$M = -EI_f \frac{d^2u}{dz^2} \tag{8-20}$$

The shearing force F is

$$F = \frac{dM}{dz} \tag{8-21}$$

and substituting:

$$F = -EI_f \frac{d^3u}{dz^3} = -EI_f\left(\frac{h}{2}\right)\left(\frac{d^3\phi}{dz^3}\right) = -EI_f\left(\frac{h}{2}\right)\phi''' \tag{8-22}$$

where I_f is the moment of inertia of the flange about the y axis.

The torsion bending resistance T_b is

$$T_b = Fh = -EI_f\left(\frac{h^2}{2}\right)\left(\frac{d^3\phi}{dz^3}\right) = -EI_f\left(\frac{h^2}{2}\right)\phi''' \tag{8-23a}$$

The quantity $I_f h^2/2$ is a measure of the torsional resistance of a wide-flange shape due to bending of the flanges in their respective planes. This resistance may be determined for other than W and I-shapes as a numerical coefficient K_b characteristic of the shape so that

$$T_b = -K_b E\phi''' \tag{8-23b}$$

and values of K_b for various shapes are shown in Table B-2, Appendix B.

Then for torsional equilibrium at a section,

$$T = T_t + T_b = K_t G\phi' - K_b E\phi''' \tag{8-24a}$$

For a variable torque $T(z)$ the differential equation becomes

$$\frac{dT}{dz} = K_t G\phi'' - K_b E\phi'''' \tag{8-24b}$$

The general form of the solution for this differential equation is

$$\phi = A \sinh\frac{z}{a} + B \cosh\frac{z}{a} + C + D(z) \tag{8-25}$$

where $a = \sqrt{EK_b/GK_t}$, coefficients A, B, C, and $D(z)$—usually a polynomial in z—are determined by the beam geometry and the loading and boundary conditions. The coefficients A, B, C, and $D(z)$ have been evaluated for a variety of loadings and boundary conditions and are included in *Torsion Analysis of Rolled Steel Sections*, by C. P. Heins, Jr., and P. A. Seaburg, published by Bethlehem Steel Corporation. This publication also includes tabulations for torsional properties K_t and a, of structural-steel shapes, and

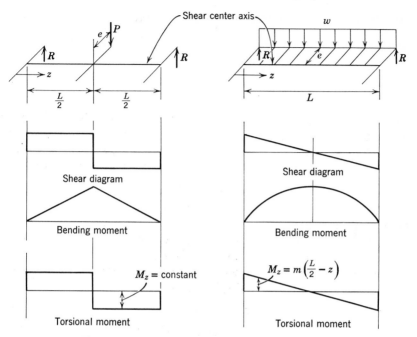

Fig. 8-16 Combined bending and torsional loading.

charts for the twist $\phi(z)$ and its derivatives which simplify considerably the calculation of torsional stresses. Torsional properties for some of the structural shapes are given in Tables B-1 and B-2 and the coefficients for the solution of Eq. 8-25 for specific cases are included in Table B-3, Appendix B.

In evaluating the coefficients for the solution of a specific case the torsional loading and boundary conditions must be defined. Combined bending and torsional loading occurs when the load is eccentric with respect to the shear center of the cross section. The effects of bending and torsion can be evaluated separately and the resulting stresses and deformations can be obtained by superposition, as long as the stresses do not exceed elastic limit.[6] It is assumed that the problem is statically determinate and the equilibrium problem for the beam can be solved without introduction of compatibility conditions. Thus, for a given beam, shear, bending moment and torsional moment diagrams can be constructed (Fig. 8-16).

In addition to the loading conditions it is necessary to define the torsional boundary conditions. Considering twisting rotation and warping restraint, three idealized boundary conditions are possible:

(*a*) "Free" end—free to twist and to warp:

$$\phi \neq 0, \qquad \phi' \neq 0, \qquad \phi'' = 0 \qquad \text{(resists torque } T_t)$$

(*b*) "Pinned" end—not free to twist but free to warp:

$$\phi = 0, \qquad \phi' \neq 0, \qquad \phi'' = 0 \qquad \text{(resists torque } T_t)$$

(*c*) "Fixed" end—not free to twist, not free to warp:

$$\phi = 0, \qquad \phi' = 0, \qquad \phi'' \neq 0 \qquad \text{(resists torque } T_b; \ T_t = 0)$$

With the loading and boundary conditions established, the coefficients in the general solution can be evaluated and the torsional deformations are defined by Eq. 8-25. The torsional stresses—both shearing and normal—can be expressed in terms of twisting deformation and its derivatives.

These stresses can be classified into three categories (Fig. 8-17).

(*a*) Shearing stresses due to "pure" torsion, designated by a prime superscript, f_v' (Fig. 8-17).

(*b*) Shearing and normal stresses due to restrained torsion, designated by a double-prime superscript, f_v'' and f_b'' (Fig. 8-17).

(*c*) Shearing and normal stresses due to usual flexure, designated by a triple-prime superscript, f_v''' and f_b''' (Fig. 8-17).

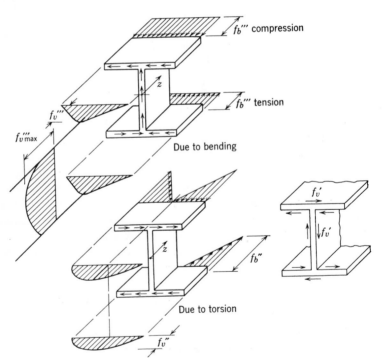

Fig. 8-17 Stresses due to torsion bending.

A second subscript is added to denote the location in the section where stresses are considered; $f_{b,1}'''$ denotes the normal stress due to usual flexure at point 1 (Fig. 8-17). Deformations and stresses in W^F shapes are defined by equations given in Table B-4.

A condition that is often present in steel frame buildings is the problem of a spandrel beam loaded eccentrically by the exterior wall. The designer will generally select the beam size based on the complete wall in place; however, the erection sequence may be such that a greater torque is produced and a larger beam size may be required.

Example 8-4. Determine the maximum stresses in the spandrel beam indicated in Fig. 1 for the masonry wall shown. The beam is simply supported

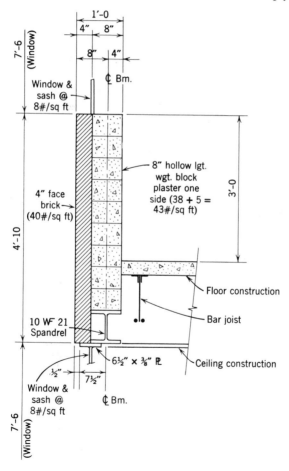

Fig. 1

at the columns. Span is 16 feet, A36 steel, section is 10 W⁼ 21 with shelf plate for face brick.

Solution

Check 10 W⁼ 21 spandrel beam for torsion. Assume torsionally pinned end conditions provided by web angle connections. Neglect $6\frac{1}{2}'' \times \frac{3}{8}''$ shelf plate in evaluating properties of beam.

Loads	Load/foot			Torque/foot
4-in. face brick	$40 \times 4.83 = 193$ lb/ft	$\times 6$	in. =	1158 in.-lb/ft
8-in. hollow block, plaster	$43 \times 4.0 = 172$ lb/ft	0	in. =	0 in.-lb/ft
Window and sash	$7.5 \times 8.0 = 60$ lb/ft	4	in. =	240 in.-lb/ft
10 W⁼ 21	$= 21$ lb/ft	0	in. =	0 in.-lb/ft
$6\frac{1}{2}'' \times \frac{3}{8}''$ shelf plate	$= 9$ lb/ft	4.25	in. =	38 in.-lb/ft
	455 lb/ft			1436 in.-lb/ft

$$\text{Eccentricity } e = (1436/455) = 3.16 \text{ in.}$$

Angle of Twist. An idealized model of the spandrel beam is shown in Fig. 2. This corresponds to case 4, Table B-3, Appendix B, where the angle of twist ϕ is defined as follows:

$$\phi = \frac{ma^2}{GK_t}\left[-\tanh\frac{L}{2a}\sinh\frac{z}{a} + \cosh\frac{z}{a} - \frac{z^2}{2a^2} + \frac{zL}{2a^2} - 1\right]$$

The following section and material properties for the 10 W⁼ 21 beam are obtained from Table B-1 and AISC Manual.

$K_t = 0.23$ in.⁴, $a = 50.26$ in.   (from Table B-1),

$b = 5.75$ in., $t_f = 0.34$ in., $t_w = 0.24$ in., $d = 9.90$ in.,

$I_x = 106.3$ in.⁴, $S_x = 21.5$ in.³, $Q_f = 4.55$ in.³, $Q_w = 12.04$ in.³,

$E = 29.5 \times 10^6$ psi, $G = 11.2 \times 10^6$ psi.

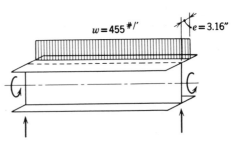

Fig. 2   Idealized model of spandrel beam.

Plane Bending and Shear Stresses.
Moment at $z = L/2$

$$M = \frac{wL^2}{8} = \frac{0.455 \times 16^2 \times 12}{8} = 174.72 \text{ kip-in.}$$

Normal stress at $z = L/2$:

$$f_{b,1}''' = \frac{M}{S_x} = \frac{174.72}{21.5} = 8.13 \text{ ksi} \qquad \text{(see Fig 8-17)}$$

Shear at $z = 0$:

$$V = \frac{wL}{2} = \frac{0.455 \times 16}{2} = 3.64 \text{ kips}$$

Shearing stress at $z = 0$:

$$\text{web: } f_{v,3}''' = \frac{VQ_w}{I_x t_w} = \frac{3.64 \times 12.04}{106.3 \times 0.24} = 1.72 \text{ ksi}$$

$$\text{flange: } f_{v,1}''' = \frac{VQ_f}{I_x t_f} = \frac{3.64 \times 4.55}{106.3 \times 0.34} = 0.46 \text{ ksi}$$

Normal stresses at support and shearing stress at midspan due to plane bending only are zero.

Restrained Torsion—Normal and Shearing Stresses

$$m = w \cdot e = 0.455 \times 3.156 = 1.435 \text{ kip-in. per ft}$$

$$\frac{ma^2}{GK_t} = \frac{1.435(50.26)^2}{11.2 \times 10^3(0.23)12} = 0.1175$$

$$\frac{L}{a} = \frac{16 \times 12}{50.26} = 3.82, \quad \frac{L}{2a} = \frac{3.82}{2} = 1.91, \quad \sinh\frac{L}{2a} = 3.31, \quad \cosh\frac{L}{2a} = 3.44$$

$$\tanh\frac{L}{2a} = \tanh 1.91 = 0.956 \text{ (from Table B-5)}$$

Then

$$\phi = 0.1175\left[-0.956\sinh\frac{z}{a} + \cosh\frac{z}{a} - \frac{z^2}{2a^2} + \frac{zL}{2a^2} - 1\right]$$

Derivatives of ϕ:

$$\phi' = 0.1175\left[-\frac{0.956}{50.26}\cosh\frac{z}{a} + \frac{1}{50.26}\sinh\frac{z}{a} - \frac{z}{(50.26)^2} + \frac{16 \times 12}{2(50.26)^2}\right]$$

$$= 10^{-3}\left[-2.24\cosh\frac{z}{a} + 2.34\sinh\frac{z}{a} - 0.0466z + 4.46\right]$$

$$\phi'' = \left[-\frac{2.24}{50.26}\sinh\frac{z}{a} + \frac{2.34}{50.26}\cosh\frac{z}{a} - 0.0466\right] \times 10^{-3}$$

$$= \left[-4.46\sinh\frac{z}{a} + 4.66\cosh\frac{z}{a} - 4.66\right] \times 10^{-5}$$

$$\phi''' = \left[-\frac{4.46}{50.26}\cosh\frac{z}{a} + \frac{4.66}{50.26}\sinh\frac{z}{a}\right] \times 10^{-5}$$

$$= \left[-8.9\cosh\frac{z}{a} + 9.28\sinh\frac{z}{a}\right] \times 10^{-7}$$

then

	ϕ	ϕ'	ϕ''	ϕ'''
At $z = 0$	0	$+2.24 \times 10^{-3}$	0	-8.9×10^{-7}
At $z = L/2$	0.127	0	-3.38×10^{-5}	0

Stresses (from Table B-4, Appendix B)
Torsional shear (at $z = 0$):

Flange $f'_{v,1} = Gt_f\phi' = 11.2 \times 10^3 \times 0.34 \times 2.24 \times 10^{-3} = 8.5$ ksi

web $f'_{v,1} = Gt_w\phi' = 11.2 \times 10^3 \times 0.24 \times 2.24 \times 10^{-3} = 6.0$ ksi

Warping shear (at $z = 0$):

Flange $f''_{v,1} = E\frac{b^2d}{16}\phi''' = 29.5 \times 10^3 \times \frac{(5.75)^2 9.9}{16}(-8.9 \times 10^{-7})$

$$= -0.53 \text{ ksi}$$

Web $f''_{v,2} = f''_{v,3} = 0$

All normal stresses at the support ($z = 0$) are 0. Therefore combined torsional and bending shear stresses:

$$\text{Flange} f_{v,1} = f'_{v,1} - f''_{v,1} + f'''_{v,1} = 8.5 + 0.53 + 0.46 = 9.5 \text{ ksi}$$

$$\text{Web} f_{v,3} = f'_{v,3} + f''_{v,3} + f'''_{v,3} = 6.0 + 0 + 1.72 = 7.7.\text{ksi}$$

and combined normal stresses are 0.

Torsional shear (at $z = L/2$)

$$f'_v = Gt\phi' = 0$$

Warping shear (at $z = L/2$)

$$f''_v = E\frac{b^2 d}{16}\phi''' = 0$$

All combined shear stresses at midspan are 0. Warping normal stress (at $z = L/2$)

$$f''_{b,2} = E\frac{bd}{4}\phi'' = 29.5 \times 10^3 \frac{5.75 \times 9.9}{4.0}(-3.38) \times 10^{-5} = -14.2 \text{ ksi}$$

Therefore combined torsional and bending normal stresses are:

$$f_{b,2} = f'_{b,2} - f''_{b,2} = 8.13 + 14.2 = 22.3 \text{ ksi}$$

Twisting of a channel (Fig. 8-18a) occurs about its shear center and produces bending in the web. The vertical bending deflection of the web is $\Delta = e\phi$, where e is the distance from the shear center to the web. Here both the flanges and the web bend, and the determination of stresses, are complicated by the conditions of compatibility of stresses at the web-flange

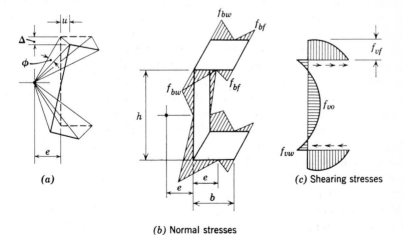

(a)

(b) Normal stresses

(c) Shearing stresses

Fig. 8-18   Twisting of a channel.

junctions. The following expressions for normal and shearing stresses in a channel (Fig. 8-18b, c) are given without derivation.[7]

$$f_{vw} = -E(eh/2)(d^2\phi/dz^2)$$
$$f_{bf} = -E(h/2)(b - e)(d^2\phi/dz^2)$$
$$f_{vf} = -E(h/4)(b - e)^2(d^3\phi/dz^3) \tag{8-26}$$
$$f_{vw} = -E(bh/2)(b/2 - e)(d^3\phi/dz^3)$$
$$f_{vo} = -E(h/2)(eh/4 + be - b^2/2)(d^3\phi/dz^3)$$

When twisting and bending deformations of the beam are appreciable, their effects on the moments and on the torque should not be neglected, since such deformations produce secondary bending stresses and increase the torque. As in the case of beam columns these secondary effects may be critical and produce flexural–torsional buckling. The problem has been solved mathematically in general terms, including effects of axial thrust, and forms the basis for the general theory of flexural–torsional buckling of beams and columns.[8]

8-8 TORSION AND BENDING OF TUBULAR SECTIONS

Stress distribution in and deformation of beams with tubular sections (Fig. 8-19a) differ from those of "open" sections (Fig. 8-19b) because of the torsional rigidity of the tubular or "boxed" sections. Compare the twisting of a split thin-wall circular tube with that of a continuous tube (Figs. 8-19a and b). The end section of the split tube is greatly warped and the section twists through an appreciable angle; the end of the continuous tube is not warped at all and the angle of twist is very small. The shearing-stress distribution in the split tube varies across the thickness (Fig. 8-19c)—it is zero at the midthickness line; it also varies along the perimeter of the tube, although in a thin plate this variation is not significant. In the closed tube subjected to pure torsion (Fig. 8-19d) the shearing stress varies only slightly across the thickness and along the perimeter of the section. If the thickness of the wall varies but remains small, the shearing-stress distribution across the thickness may be considered uniform and acting tangentially to the boundary, and, when the sections are free to warp, there are no longitudinal stresses due to the applied torque. From the equilibrium condition $\Sigma F_z = 0$ (Fig.8-19e) for an element of tube wall it follows that $f_{v1}t_1 = f_{v2}t_2$, or in terms of shear flow, $q_1 = q_2$. This condition states that for a "boxed" section with relatively thin walls the shearing stress may vary with the thickness, but the shear flow remains constant. The magnitude of this shear flow q, for free torsion of a boxed section, can be obtained from the equilibrium condition $\Sigma M_z = 0$

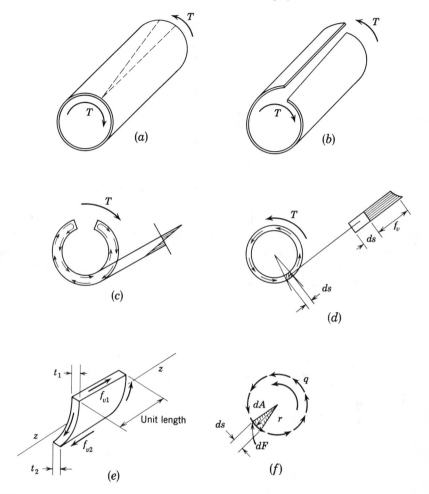

Fig. 8-19 Tubular sections under torsion: (*a*) closed section, (*b*) open section, (*c*) stress distribution for *b*, (*d*) stress distribution for *a*, (*e*) an element of wall, and (*f*) shear flow in element.

for the section. It follows that (Fig. 8-19*f*)

$$T = \int r \, dF = \int rq \, ds = q \int r \, ds = q \int 2 \, dA = 2Aq \qquad (8\text{-}27a)$$

or

$$q = \frac{T}{2A} \qquad (8\text{-}27b)$$

where A is the area enclosed by the wall center line of the tubular section.

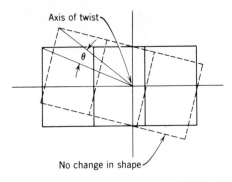

Fig. 8-20   Twisting of section with multiple cells.

Equation 8-27 is applicable to single-cell boxed beams only. For multiple cells (Fig. 8-20), determination of shear-stress distribution becomes statically indeterminate and the twisting deformation of each cell must be considered. For any boxed section subjected to bending and torsion free from warping, the angle of twist $d\phi$ per unit length dz has been shown[1] to be equal to

$$\frac{d\phi}{dz} = \frac{\displaystyle\int q\,\frac{ds}{t}}{2AG} \tag{8-28}$$

For a single-cell section the values of q are statically determinate and the angle of twist may be readily calculated as illustrated in Example 8-5.

Example 8-5.   Determine the torsional rigidities $d\phi/dz$ of open and tubular sections in Fig. 1.

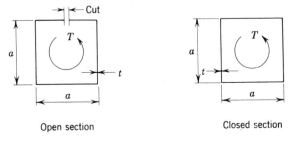

Open section   Closed section

Fig. 1

Solution

Open section:

$$\frac{d\phi}{dz} = \frac{T}{KG}, \qquad K \simeq 4(0.333)at^3, \qquad \frac{d\phi}{dz} = \frac{T}{1.33at^3G}$$

Closed section:

$$\frac{d\phi}{dz} = \frac{\int q \frac{ds}{t}}{2AG}, \qquad q = \frac{T}{2A} = \frac{T}{2a^2}, \qquad \frac{d\phi}{dz} = \frac{T(4a)}{2a^2t(2a^2G)} = \frac{T}{ta^3G}$$

Comparison of rotations under a given torque:

$$\frac{\text{Open section}}{\text{Closed section}} = \frac{T/1.33at^3G}{T/ta^3G} = \frac{a^2}{1.33t^2}$$

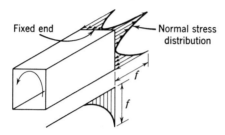

Fixed end

Normal stress distribution

f

f

Fig. 8-21   Cantilever box beam subjected to torque (from Ref. 9).

If $t = 0.1$ in., $a = 10$ in.

$$\frac{a^2}{1.33t^2} = 7500$$

that is, open section rotation is 7500 times closed section rotation.

Restrained torsion of tubular sections induces normal stresses which are of a local nature.[9] Consider a cantilever box beam subjected to a torque T. The normal stress distribution at the fixed end is shown in Fig. 8-21. The magnitude of these stresses decreases rapidly away from the fixed end and, for practical purposes, they are negligible at a section located a distance equal to twice the depth of the beam from the fixed end. The maximum normal stress f at the fixed end depends on the shape of the box section; it is zero for a circular or square box and increases as the box becomes flatter (Fig. 8-22).[9]

Comparison of normal stresses due to restrained torsion of a box beam with those of an I-beam having approximately the same proportions and weight is of interest. Consider cantilever beams shown in Fig. 8-23. The normal stresses (see Example 8-6) at the fixed end are 280 and 9350 psi for the box and I-beams, respectively. The normal stresses at the midspan are approximately zero and 2800 psi for the box and I-beam, respectively. It is clearly seen that the effect of restrained torsion on boxed shapes is considerably less important than that on open shapes. Therefore, in many practical instances it is permissible to neglect the normal stresses due to restrained torsion of boxed beams and consider that the shearing stresses derived for

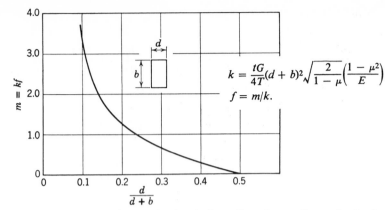

Fig. 8-22 Maximum normal stress at fixed end of cantilever box section under torsion (from Ref. 9).

free torsion (Eq. 8-27) also apply to the restrained torsion. This approximation is quite good where normal stresses are zero, that is, throughout most of the span, except locally at points of restraint or application of the torque.

Example 8-6. For the sections shown in Fig. 8-23 compute stresses at the fixed ends and at midspan.

Solution

(a) *Box-beam*

$$d = 6 \text{ in.} \qquad E = 30 \times 10^6 \text{ psi}$$
$$b = 12 \text{ in.} \qquad G = 0.375E$$
$$t = 0.25 \text{ in.} \qquad \nu = 0.33$$
$$T = 6000 \text{ in.-lb} \quad [d/(d+b)] = \tfrac{6}{18} = 0.33$$

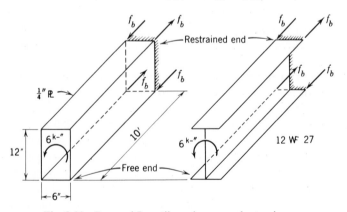

Fig. 8-23 Box and I cantilever beams under torsion.

From Fig. 8-22:

$$m \cong 0.54 \quad \text{at} \quad z = 0$$

$$k = \frac{tG(d + b)^2}{4T} \sqrt{\frac{2}{1 - \nu}} \left(\frac{1 - \nu^2}{E} \right)$$

$$\therefore \ k = \frac{(0.25)(0.375E)(18)^2 \sqrt{3} \ (0.891)}{4(6000)E} = 1.95 \times 10^{-3}$$

$$f_{max} = \frac{m}{k} = \frac{0.54}{1.95 \times 10^{-3}} = 277 \text{ psi}$$

At $z = L/2$,

$$f_{max} \cong 0 \quad \text{(see Fig. 8-21)}$$

(b) *12 W 31*

$$T = 6000 \text{ in.-lb}, \qquad I_y = 91.8 \text{ in.}^4 = 2I_f$$

From Table B-1, Appendix B

$$a = 54.58 \text{ in.}$$

From Table B-4, Appendix B

$$M_{max} = EI_y \frac{h}{4} \frac{d^2\phi}{dz^2}$$

For Case 2, Table B-3, Appendix B

$$\phi = \frac{T}{GK_t} \left(-a \sinh \frac{z}{a} + a \tanh \frac{L}{a} \cosh \frac{z}{a} + z - a \tanh \frac{L}{a} \right)$$

At $z = 0$:

$$\frac{d^2\phi}{dz^2} \bigg]_{z=0} = \frac{-T}{GK_t a} \tanh \frac{L}{a}$$

$$M^*_{max} = \frac{EI_y}{GK_t} \frac{hT}{4a} \tanh \frac{L}{a}$$

$$f_{max} = \frac{M}{I_f} \frac{b}{2} = \frac{Ehb}{GK_t} \frac{T \tanh (L/a)}{4a}$$

$$L = 120 \text{ in.}, \qquad h = 12.09 - 0.465 = 11.625 \text{ in.}$$

$$b = 6.525 \text{ in.}, \qquad t_w = 0.265, \qquad t_f = 0.465$$

$$K_t = 0.58 \quad \text{(see Table 8-1)}$$

$$f_{max} = \frac{E(11.625)(6.525)(6000)(0.976)}{(0.375E)(0.58)(4)(54.58)} = 9350 \text{ psi}$$

* Sometimes given as $M_{max} = \dfrac{Ta}{h} \tanh \dfrac{L}{a}$, which is equivalent.

At $z = L/2$,

$$\frac{d^2\phi}{dz^2} = \frac{T}{GK_t}\left(\frac{1}{a}\sinh\frac{z}{a} - \frac{1}{a}\tanh\frac{L}{a}\cosh\frac{z}{a}\right)$$

$$f = \frac{Ehb}{4}\frac{T}{GK_t}\left(\frac{1}{a}\sinh\frac{z}{a} - \frac{1}{a}\tanh\frac{L}{a}\cosh\frac{z}{a}\right)$$

$$\sinh\frac{z}{a}\Bigg]_{z=L/2} = \sinh\frac{L}{2a} = 1.336, \qquad \cosh\frac{L}{2a} = 1.669$$

$$f_{z=L/2} = \frac{E(11.625)(6.525)(6000)[1.336 - 0.976(1.669)]}{4(0.375E)(0.58)(54.58)} = 2800 \text{ psi}$$

A summary of stresses follows.

	Box beam psi	12W 31 psi
$z = 0$	277	9350
$z = \dfrac{L}{2}$	0	2800

Shear-stress distribution caused by transverse loads on a beam of tubular section cannot be determined from Eqs. 8-2, 8-4, or 8-6 because the static moments Q_x and Q_y cannot be clearly defined. For open sections Q_x and Q_y were defined as the static moments of area for the part of section between the free edge and a given fiber (Fig. 8-2b). There are no free edges in a boxed section, although points of zero shear stress may be considered equivalent to a free edge in defining values of Q. If the location of the fiber having zero shear stress in a boxed section were known, the shearing-stress distribution could be determined from Eqs. 8-2, 8-4, or 8-6. For beams of symmetrical cross section, carrying a load in the plane of symmetry, the fibers of zero shearing stress lie on the axis of symmetry parallel to the load and therefore the location of the zero shearing stress is known by observation. If the location of zero shearing-stress fiber is not known, an arbitrary fiber A (Fig. 8-24) actually subjected to a shearing stress f_{va} or shear flow q_a can be taken as a reference zero stress fiber and the distribution of shearing stress corresponding to this assumption can be calculated. Consider a portion of the section between fibers A and B. The change in shear flows between these fibers is due to the change in the normal forces acting on this portion of the section. This change in normal forces per unit length is equal to the shear-flow change $q_n = q_b - q_a$ and is

$$q_n = q_b - q_a = \Delta F = \frac{V_y Q_x}{I_x} + \frac{V_x Q_y}{I_y} \qquad (8\text{-}29a)$$

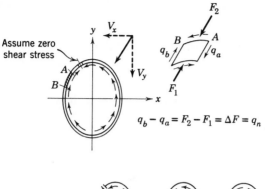

Fig. 8-24 Shearing stress computation for closed section.

and the shear flow at fiber B is

$$q_b = q_a + q_n = q_a + \frac{V_y Q_x}{I_x} + \frac{V_x Q_y}{I_y} \qquad (8\text{-}29b)$$

where static moments Q are taken about the principal axes xx, yy, assuming fiber A to be a "zero" stress fiber or a "free" edge. The value of q_a can be found from the condition of equilibrium $\Sigma T = 0$; that is, the twisting moment of applied shears V_x and V_y taken about the axis must be in equilibrium with the moment of the resisting shear flows taken about the same axis. It is convenient to divide the torsional moment of the resisting shear flows in two parts: T_1 due to a constant unknown shear flow q_a, and T_2 due to known shear flows q_n (Eq. 8-28). Then

$$T_1 = \int r \, dF = \int r q_a \, ds = q_a \int r \, ds = 2A q_a \qquad (a)$$

$$\qquad\qquad\qquad\qquad\qquad\qquad\qquad\qquad\qquad (8\text{-}30)$$

$$T_2 = \int r \, dF = \int r q_n \, ds = \int r \left(\frac{V_y Q_x}{I_x} + \frac{V_x Q_y}{I_y} \right) ds \qquad (b)$$

and, from $\Sigma T = 0$,

$$T_1 + T_2 = V_x y + V_y x \qquad (8\text{-}31)$$

Assuming the location of a convenient fiber A and computing shear flow q_n and moments T_2, $V_y x$ about any convenient axis z, the value of T_1 is determined as follows:

$$T_1 = (V_x y + V_y x) - T_2 \qquad (8\text{-}32)$$

and, from Eq. 8-30a,

$$q_a = \frac{T_1}{2A} \tag{8-33}$$

where A is the area enclosed by the tubular boundary of the cross section. With q_a determined, shear flow at any other fiber can be obtained from Eq. 8-29.

Example 8-7. Calculate and plot the shear-flow distribution (neglecting the bending stresses in the $\frac{1}{8}$-in. plates) and determine the maximum shear stress (Fig. 1).

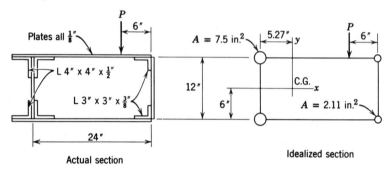

Fig. 1

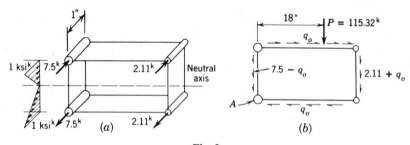

Fig. 2

Solution

Assume unit change in bending stress per unit length (Fig. 2a) and assume shear flow in top web q_o (Fig. 2b). Calculate the resultant shear-flow distribution in terms of q_o and determine the corresponding P from ΣV and q_o from ΣM.

$$\Sigma V = P = (7.5 - q_o)12 + (2.11 + q_o)12 = 115.32 \text{ kips}$$
$$\Sigma M = 18P = q_o(24)(12) + (2.11 + q_o)(24)(12) = 576q_o + 608$$
$$q_o = \frac{18(115.32) - 608}{576} = 2.55 \text{ kip/in.} \quad \text{(See Fig. 3)}$$

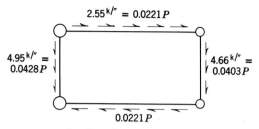

Resultant shear-flow distribution
($P = 115.32$ kips)

Fig. 3

Maximum shear flow, $q_{max} = (7.5 - 2.55) = 4.95$ kip/in. for $P = 115.32$ kips.

$$q_{max} = 0.0428P$$

$$f_v = \frac{q}{t} = 8q$$

$$\therefore \quad \max f_v = 8(0.0428P) = 0.343P \text{ ksi} \quad (P \text{ in kips})$$

Equation 8-28 may also be used to determine the location of the shear center—center of twist—of a box section, where the shear center is defined as the point through which the shear force must pass if the section is to bend without twisting. Example 8-8 describes such a solution.

Example 8-8. Determine the shear center for section shown in Fig. 1 (neglecting bending stresses in the plates).

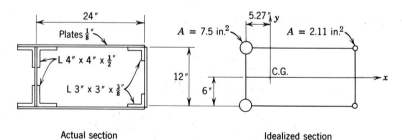

Actual section   Idealized section

Fig. 1

Solution

$$\frac{d\phi}{dz} = \frac{\int \frac{q}{t}\,ds}{2AG} = 0 \qquad \therefore \int \frac{q}{t}\,ds = 0, \quad t \text{ constant} \qquad \therefore \int q\,ds = 0$$

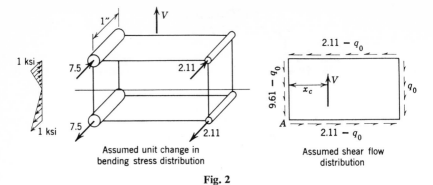

Assumed unit change in
bending stress distribution

Assumed shear flow
distribution

Fig. 2

Assume unit change in bending stress distribution (Fig. 2). Then, for

$$\frac{d\phi}{dz} = 0 \qquad \int q \, ds = (2.11 - q_o)(48) + (9.61 - q_o)(12) - 12q_o = 0$$

$$q_o = \frac{216.7}{72} = 3.0 \text{ kip/in.}$$

For equilibrium,

$$\sum F_y = 0, \quad V = 9.61 \times 12 = 115.32 \text{ kips}$$

$$\sum M = 0, \quad 3.0(12)(24) + (0.89)(24)(12) - Vx_c = 0$$

Solving for x_c,

$$x_c = \frac{1120}{115.32} = 9.7 \text{ in.}$$

Force V produces no twist, hence it passes through shear center (Fig. 3).

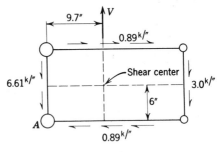

Fig. 3

The bending and torsional stresses in box-shaped sections defined in the preceding sections are valid provided the distortion of the cross section is negligible and the plate elements of the box section are subjected to longitudinal normal and shearing stresses only. These conditions are satisfied when sufficient diaphragms are incorporated into the structural system which preserve the over-all shape of the section, and the transverse loads on the plates produce essentially negligible bending stresses in them. In such instances the transverse loads may be assumed to be concentrated at the diaphragms and the plates between them act only as membranes. When these conditions are not satisfied, a more detailed analysis must be carried out which considers both the membrane and bending action of the plate elements. When the plates are reinforced in one direction by stiffeners, the analysis must account for the orthotropic nature of the plates.

Greater emphasis on esthetics and economy resulted in the increasing use of box sections for bridges and buildings. When the plates are not stiffened, a simplified analysis for torsion and bending stresses may be used. Application of such an analysis is illustrated in Example 8-9.

Example 8-9. Determine shear stress distribution in multicell box beam shown in Fig. 1. $T = 200$ in.-kips.

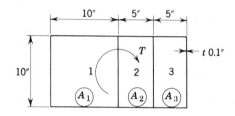

Fig. 1

Solution

$$A_1 = 100 \text{ in.}^2, \qquad A_2 = A_3 = 50 \text{ in.}^2, \qquad t = 0.10 \text{ in.}$$

Angle of twist for each box θ; $\quad 2G\theta = \dfrac{1}{A} \displaystyle\int_A \dfrac{q\,ds}{t}$

Line integrals: $a = \displaystyle\oint \dfrac{ds}{t}$

Subscripts denote regions shown in Fig. 1. The region outside the box is denoted by 0.

$$a_{10} = \frac{3(10)}{0.1} = 300 \qquad a_{12} = \frac{10}{0.1} = 100 \qquad a_{20} = \frac{2(5)}{0.1} = 100$$

$$a_{23} = \frac{10}{0.1} = 100 \qquad a_{30} = \frac{5 + 5 + 10}{0.1} = 200$$

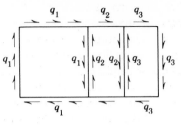

Fig. 2

The angles of twist are all equal. Then for cells 1, 2, and 3, Fig. 2

(1) $\quad 2G\theta = \dfrac{1}{A_1}[q_1 a_{10} + (q_1 - q_2)a_{12}] = 0.01(300q_1 + 100q_1 - 100q_2)$

$\quad\quad 2G\theta = 4q_1 - q_2$

(2) $\quad 2G\theta = \dfrac{1}{A_2}[(q_2 - q_1)a_{12} + q_2 a_{20} + (q_2 - q_3)a_{23}]$

$\quad\quad\quad\quad = 0.02(100q_2 - 100q_1 + 100q_2 + 100q_2 - 100q_3)$

$\quad\quad 2G\theta = -2q_1 + 6q_2 - 2q_3$

(3) $\quad 2G\theta = \dfrac{1}{A_3}[(q_3 - q_2)a_{23} + q_3 a_{30}] = 0.02(100q_3 - 100q_2 + 200q_3)$

$\quad\quad 2G\theta = -2q_2 + 6q_3$

(4) For equilibrium: Σ torque $= 200$ in.-kips $= 2qA_1 + 2q_2 A_2 + 2q_3 A_3$

$\quad\quad\quad 100q_1 + 50q_2 + 50q_3 = 100$

$\quad\quad\quad\quad 2q_1 + q_2 + q_3 = 2$

Solving these four equations simultaneously:

$\quad q_1 = 0.505$ kips/in., $\quad q_2 = 0.560$ kip/in., $\quad q_3 = 0.430$ kip/in.

Shear flows as in Fig. 3.

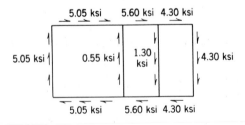

Fig. 3

8-9   TAPERED BEAMS

The application of simple formulas established for stresses in beams of constant cross section to tapered beams may indicate values appreciably different from actual stresses. It is not possible to establish simple stress-distribution formulas for the general case of tapered beams, although more or less approximate solutions of special cases can be indicated.

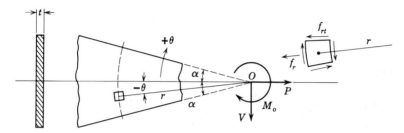

Fig. 8-25   Stresses in a tapered beam of rectangular section.

Precise definitions of normal and shearing stresses—f_r and f_{rt}, respectively—in a tapered beam (Fig. 8-25) must be made on a section normal to a radial fiber. For this reason it is often convenient to define the position of the load with reference to the apex of the beam. For a beam of rectangular cross section tapered in depth the radial and tangential stresses f_r and f_{rt} (Fig. 8-25) are defined as follows:[13]

Load at Apex
(*a*) Axial load P

$$f_r = \frac{P \cos \theta}{tr(\alpha + \tfrac{1}{2} \sin 2\alpha)}, \qquad f_{rt} = 0 \qquad (a)$$

(*b*) Shear V $\hspace{9cm}$ (8.34)

$$f_r = \frac{V \sin \theta}{tr(\alpha - \tfrac{1}{2} \sin 2\alpha)}, \qquad f_{rt} = 0 \qquad (b)$$

(*c*) Bending moment M_o

$$f_r = \frac{2M_o \sin 2\theta}{tr^2(\sin 2\alpha - 2\alpha \cos 2\alpha)}$$

$$f_{rt} = \frac{M_o(\cos 2\alpha - \cos 2\theta)}{tr^2(\sin 2\alpha - 2\alpha \cos 2\alpha)}$$

$\hspace{9cm}$ (c)

It seems paradoxical that for apex load V the tangential shearing stress f_{rt} is equal to zero. This results because the radial forces are concurrent with the load V, so that each pair of radial fibers is in equilibrium with a part of the vertex load independently of the adjacent fibers, therefore there is no tendency for slip (shear) between the fibers. In the case of moment M_o, the radial fiber forces pass through the apex and cannot produce a resisting moment about the apex. Therefore this loading condition requires tangential stresses defined by Eq. 8-34c for equilibrium.

The stresses defined by Eqs. 8-34 do not result in plane deformations of the section. These stresses are valid only when the stress distribution at the supports is identical with that just defined. If the type of support requires plane sections to remain plane, then the stress distribution at the support differs from that just defined and the solution is valid only for sections at some distance from the supports.

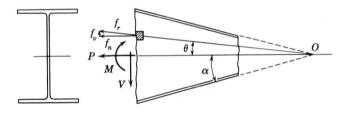

Fig. 8-26   Tapered beam of nonrectangular section.

An approximate method for determining stresses in tapered beams, not restricted to rectangular cross sections, is based on a modification of the conventional formulas. The essential feature of this method is that normal stresses f_n are computed for a section perpendicular to a line bisecting the tapered beam (Fig. 8-26) and these stresses are considered normal components of the radial stress. The shearing components of the radial stress and the resultant radial stress are then defined as

$$f_v' = f_n \tan \theta \qquad (a)$$

$$f_r = \frac{f_n}{\cos \theta} \qquad (b) \qquad (8\text{-}35)$$

where the normal stress f_n is

$$f_n = \frac{P}{A} \pm \frac{My}{I} \qquad (8\text{-}36)$$

and M and I are taken at the section considered. The shearing components of the radial stress derived from loads P and M result in a resisting force $V' = \int f_v' \, dA$, which does not necessarily balance the total shear V. Therefore V' resists only a part of the total shear V, and the balance of the shear, $V'' = V' - V$, must be resisted essentially by tangential shearing stresses. Their distribution on the section may be approximated by $f_v'' = V''Q/It$. This method gives results that are satisfactory provided the taper angle 2α is less than 45°, which covers most practical instances of structural design.

8-10 CURVED BEAMS WITH IN-PLANE LOADING

The conventional formula for calculating bending stress $f_b = M_y/I$ does not apply to curved beams if the initial radius of curvature is small, say less than six times the depth of the beam. Curved elements subjected to bending occur in chain links, crane hooks, and ends of eye-bars (Fig. 8-27). Consider an

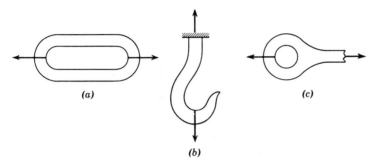

Fig. 8-27   Examples of curved beams.

element of a curved beam subjected to moment M and axial force P, where M is taken with respect to the centroidal axis (Fig. 8-28). Experimental evidence indicates that in a curved beam, as in a straight beam, a section that is plane before bending remains plane after bending. Therefore the total strains are proportional to the distance from the neutral axis $x'x'$; that is, $\delta = ky'$. The location of the neutral axis, however, may not coincide with the centroidal axis xx, so that, in general,

$$\delta = ky' = \delta_o + (d\phi)y \qquad (8\text{-}37)$$

where δ_o is the strain at the centroidal axis and $d\phi$ is the angle of rotation due to bending. If the neutral axis coincides with the centroidal axis, then δ_o is zero. The unit strain is obtained by dividing the total strain by the length of the fiber ds, where ds is $(R + y)\,d\theta$. The normal stress f_n at any point defined by the ordinate y is equal to the modulus of elasticity times the unit strain, or

$$f_n = E\,\frac{\delta}{ds} = E\,\frac{\delta_o + (d\phi)y}{(R + y)\,d\theta} = E\,\frac{\delta_o + y\,d\phi + R\,d\phi - R\,d\phi}{(R + y)\,d\theta}$$

$$= E\left[\frac{d\phi}{d\theta} + \frac{\delta_o - R\,d\phi}{(R + y)\,d\theta}\right] \qquad (8\text{-}38)$$

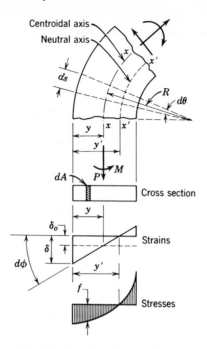

Fig. 8-28   Bending of curved beams.

from which

$$f_n = a + \frac{b}{R + y} \tag{8-39}$$

where a and b are numerical coefficients, defined as follows:

$$a = E \frac{d\phi}{d\theta}, \qquad b = E \frac{\delta_o - R\, d\phi}{d\theta} \tag{8-40}$$

It can be seen from Eq. 8-39 that the stress distribution is nonlinear, with maximum stress on the inside fiber, that is, where y is negative, as shown in Fig. 8-28. Coefficients a and b may be determined from two equations of equilibrium:

$$\int f_n \, dA = P \quad \text{and} \quad \int f_n y \, dA = M \tag{8-41}$$

From these equations:

$$b = \frac{M}{A - R \int \dfrac{dA}{R + y}} = \frac{M}{A - RZ}, \qquad Z = \int \frac{dA}{R + y} \tag{8-42}$$

$$a = \frac{P - bZ}{A} \tag{8-43}$$

In these equations A is the area of the cross section and Z is determined as a property of cross section, similar to moment of inertia. The value of Z can be determined using either calculus or numerical integration. The sign convention is such that M is taken positive when it tends to increase the initial curvature of the beam (Fig. 8-28), P is taken positive when it tends to produce tension on the section, and y is positive away from center of curvature. With a and b computed as before, the stress f_n may be determined from Eq. 8-39.

In addition to the tangential normal stress f_n, radial normal stress f_r may be of importance in curved members. Consider an element of a beam shown in Fig. 8-29. The tangential stresses in the fiber at a and c have components

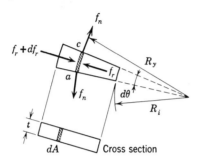

Fig. 8-29 Element of a curved beam.

that produce radial stresses within ac. From equilibrium conditions, the change in radial stress component in any fiber is

$$df_r t(R_y\, d\theta) = f_n\, dA\, d\theta \quad \text{or} \quad df_r = \frac{1}{R_y t} f_n\, dA \qquad (8\text{-}44)$$

The value of radial stress f_r is zero on the inner and the outer surfaces if there are no radial external forces acting on these surfaces. In this instance, the radial stress on any fiber is

$$f_r = \int_{R_i}^{R_y} df_r = \frac{1}{R_y t} \int_{R_i}^{R_y} f_n\, dA \qquad (8\text{-}45)$$

It can be seen from Eq. 8-45 that the radial stress may become important when $R_y t$ is small.

In shapes with abrupt changes in thickness, such as H or T shapes, because of stress concentration effects the radial stress at the re-entrant corners is appreciably greater than the average stress defined by Eq. 8-45. For very sharp corners the stress concentration factor is approximately 2.0, whereas there is practically no stress concentration for a filleted corner with radius of fillet equal to twice the web thickness.[11] Variation of tangential and radial stresses for an H-shape with small radius of curvature is shown in Fig. 8-30.

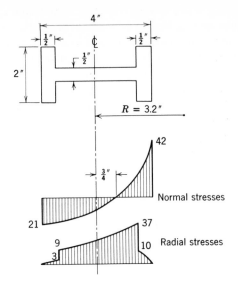

Fig. 8-30   Stress variations for H-shape with sharp corners.

The distribution of shearing stresses in curved beams of various shapes is not well known. Usually it is considered adequate to approximate shearing stress by the conventional VQ/It value. No adequate solution exists for the problem of torsion, and torsion combined with bending, for a curved beam with a small radius of curvature. Solution of this problem may become important in ring girders subjected to a complex combined loading.

8-11   CONCENTRATED LOADS

Theoretically, under the point of application of a concentrated load the fiber stress is infinite. In actual structures there are no concentrated loads; the loads are distributed over a very small length along the span (Fig. 8-31*a*). These loads produce high local bearing stresses,[12] and the distribution of the local bending and shearing stresses differs considerably from the conventional Mc/I and VQ/It stresses. For a rectangular beam the stress distribution corresponds to a localized load P as shown in Fig. 8-31*b*. It can be seen that the stresses at a section adjacent to the point of load application differ considerably from the conventional stress distribution; the stresses at a distance approximately equal to the depth of the beam agree very well with the conventional theory. The stress concentration due to local loads in I-shaped beams can be decreased materially by using load-bearing stiffeners

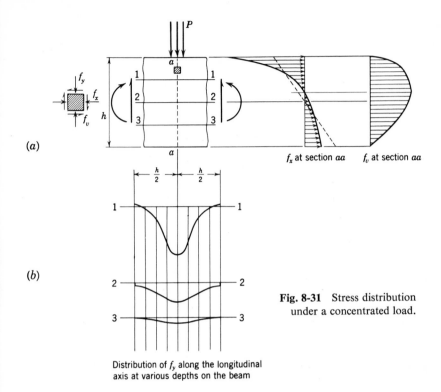

Fig. 8-31 Stress distribution under a concentrated load.

which transmit the concentrated load to the web by shears distributed along the web depth.

Combinations of shearing, bending, and bearing stress which occur at points adjacent to concentrated loads may be approximated as shown in Fig. 8-32. A factor of safety n against local yielding of a plate element subjected to normal and shearing stresses may be defined with reasonable accuracy by the Hencky-Von Mises criterion:

$$(f_x^2 + f_y^2 - f_x f_y + 3f_v^2)^{1/2} = \frac{f_y}{n}$$

where f_y is the tensile yield strength of the plate. The bending, bearing, and shearing stresses in the web element vary within the beam or girder, and depend on the type of loading. Several typical cases are shown in Fig. 8-32. The magnitudes of the normal and shearing stresses cannot be accurately determined in all instances because of the unknown effects of load concentration and the post-yielding or post-buckling stress redistribution. Therefore,

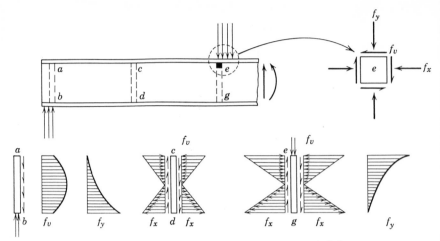

Fig. 8-32 Combinations of stresses in beams and girders.

only approximate values of these stresses can be used for design purposes with a suitable factor of safety. In many instances a factor of safety of 1.5 may be sufficient, and smaller values may be accepted when plasticity reduces local stress concentrations without permanent damage.

8-12 COMBINED BENDING AND AXIAL LOAD—ELASTIC RANGE

Accurate determination of stress distribution under combined bending and axial load in the elastic range is complicated by the effect of the deflection of the structure and the axial load P on the magnitude of the bending moment M. In Fig. 8-33, let M_o be the bending moment at any point along the member due to the external loads neglecting the effect of the deflection of the member. If y is the eventual deflection of the member at any point due to the combined effect of the axial and bending loads, and P is the magnitude of tension load (Fig. 8-33a), then the actual bending moment M at any section is

$$M = M_o - Py \qquad (8\text{-}46a)$$

and if P is the magnitude of compression load (Fig. 8-33b), then the actual bending moment M at any section is

$$M = M_o + Py \qquad (8\text{-}46b)$$

It can be seen from this equation and Fig. 8-33 that the effect of tension load is always to decrease the initial bending moment M_o and the effect of compression load is to increase the initial bending moment M_o. Therein lies

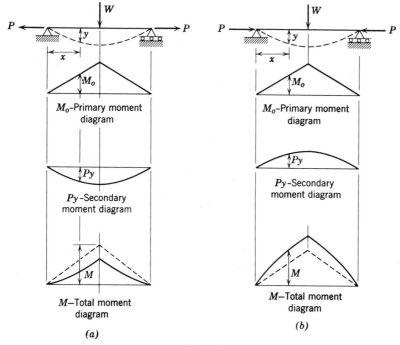

Fig. 8-33

the basic difference between tension and compression members: tension load tends to decrease the bending in the member and hence the effect of deflection can be neglected on the side of safety, whereas compression load tends to increase the bending, and the effect of deflection may often be the critical factor which determines the strength of the member.

Moment M_o is easily computed when the loading is known, but the value Py cannot be easily determined because deflection y is dependent on the moment M, which in turn depends on deflection y.

Two different methods can be used for solving this problem. One is the numerical method or the method of successive approximations; the other method involves the solution of a differential equation where one variable can be eliminated. The numerical procedure for determining moments and deflections[13] follows:

1. Assume a shape of the deflected member.
2. Substitute the assumed values of y in Eq. 8-46, obtaining the first trial values of M.
3. For a number of points along the member, compute values of y based on the first trial values of M determined in step 2. Conventional methods

such as moment area, double integration, or "conjugate beam" can be used for this computation.

4. Compare the values of y computed in step 3 with those assumed in step 1. If these values agree closely, then the approximation is considered satisfactory. If the agreement is not satisfactory, the process must be repeated. It is assumed that the axial load P is less than the critical value P_{cr} so that the process of successive approximations will converge.

In simple cases moments and deflections can be obtained by solving a differential equation derived from either Eq. 8-46a or Eq. 8-46b, depending on the direction of axial load—tension or compression. For example, when P is compressive, differentiating M in Eq. 8-46b twice with respect to x results in

$$\frac{d^2M}{dx^2} = \frac{d^2M_o}{dx^2} + P\frac{d^2y}{dx^2} = f_1(w) + P\frac{d^2y}{dx^2} \qquad (8\text{-}47)$$

where $f_1(w)$ is a function of the loading and equals d^2M_o/dx^2. In order to eliminate y from Eq. 8-47 the relationship between y and M based on conventional flexure theory can be used:

$$\frac{d^2y}{dx^2} = -\frac{M}{EI} \qquad (8\text{-}48)$$

By substituting Eq. 8-48 into Eq. 8-47, the following differential equation is obtained

$$\frac{d^2M}{dx^2} + \frac{P}{EI}M = \frac{d^2M}{dx^2} + \frac{1}{j^2}M = f_1(w) \qquad (8\text{-}49)$$

where $j = \sqrt{EI/P}$. The solution of this differential equation results in the following expression:

$$M = C_1 \sin\frac{x}{j} + C_2 \cos\frac{x}{j} + f(w) \qquad (8\text{-}50)$$

where $f(w)$ is such a function that $d^2f(w)/dx^2 = f_1(w)$, and C_1 and C_2 are numerical constants depending on the boundary conditions, that is, type of loading and type of end supports.

It can be shown that when P is tensile, solution of the differential equation results in the following expression:

$$M = C_1' \sinh\frac{x}{j} + C_2' \cosh\frac{x}{j} + f(w) \qquad (8\text{-}51)$$

where $f(w)$ is the same as just defined and C_1' and C_2' are numerical constants depending on loading, beam characteristics, and boundary conditions. Equations 8-50 and 8-51 are valid only for members of constant cross section.

For members with variable cross section, solution of the differential equation becomes difficult and often impossible and the method of successive approximations must be used to obtain a solution.

The deflection of the beam can be easily determined from Eq. 8-52:

$$y = \frac{M - M_o}{P} \qquad (8\text{-}52)$$

If both the M diagram and the M_o diagram are drawn, then the difference between these two curves is a measure of the deflection. The maximum deflection y_{max} will occur where this difference is greatest and its location often can be determined by observation. If the location of y_{max} is not obvious, it can be found analytically by differentiating y with respect to x and equating to zero; thus

$$\frac{dy}{dx} = \frac{1}{P}\left(\frac{dM}{dx} - \frac{dM_o}{dx}\right) = \frac{1}{P}(V - V_o) = 0 \qquad (8\text{-}53)$$

This indicates that* $V = V_o$ at the point of maximum deflection; thus the point of maximum deflection can be located if the V and V_o diagrams are drawn.

If at any section the axial load and the bending moment due to combined effects of eccentricity and transverse forces are known, then the stress distribution at that section may be defined by the equation

$$f = \frac{P}{A} + \frac{My}{I} \qquad (8\text{-}54)$$

This formula is valid only under the following conditions:

(a) Bending takes place about a principal axis.
(b) A plane section remains plane.
(c) Stresses are within the elastic limit.

For given loading conditions the general expression for bending moment M is given by Eq. 8-50 or 8-51. The maximum moment M_{max} can be obtained by plotting a bending moment diagram for M; that is, evaluating boundary and loading conditions and substituting these values of C_1, C_2, and $f(w)$ into Eq. 8-50, computing values of M corresponding to values of x, and plotting these values. In order to facilitate solution of simple problems, values of M_{max} for both cases when P is tensile or compressive are given in Table 8-1. For several other instances of loading with compressive axial load the expressions for C_1, C_2, and $f(w)$ are given in Table 8-2.

* For definition of V, see Eq. 8-58.

Table 8-1 Maximum Moments—Combined Axial Load and Bending

L = span
P = axial load—tensile or compressive
I = moment of inertia (constant)
$z = (PL^2/4EI)^{1/2}$
W = total transverse load on the beam
M_o = applied end couple
M_e = moment in the beam at the support
M_s = moment in the beam at midspan

Loading Condition	Tension	Compression
1. Equal and opposite end couples M_o.	$M_e = M_o$ $M_s = \dfrac{M_o}{\cosh z}$	$M_e = M_o$ $M_s = \dfrac{M_o}{\cos z}$
2. Transverse concentrated load W at the center, beam on simple supports.	$M_e = 0$ $M_e = \dfrac{WL}{4}\dfrac{\tanh z}{z}$	$M_e = 0$ $M_s = \dfrac{WL}{4}\dfrac{\tan z}{z}$
3. Transverse concentrated load W at the center, beam with fixed ends.	$M_e = \dfrac{WL}{4}\dfrac{(\cosh z - 1)}{z \sinh z}$ $M_s = \dfrac{WL}{4}\left(\dfrac{\tanh z}{z}\right.$ $\left. -\dfrac{\cosh z - 1}{z \sinh z \cosh z}\right)$	$M_e = \dfrac{WL}{4}\dfrac{(1 - \cos z)}{z \sin z}$ $M_s = \dfrac{WL}{4}\left(\dfrac{\tan z}{z}\right.$ $\left. -\dfrac{1 - \cos z}{z \sin z \cos z}\right)$
4. Transverse uniformly distributed load W, beam on simple supports	$M_e = 0$ $M_s = \dfrac{WL}{4}\dfrac{(\cosh z - 1)}{z^2 \cosh z}$	$M_e = 0$ $M_s = \dfrac{WL}{4}\dfrac{(1 - \cos z)}{z^2 \cos z}$
5. Transverse uniformly distributed load W, beam with fixed ends.	$M_e = \dfrac{WL}{4}\dfrac{(z - \tanh z)}{z^2 \tanh z}$ $M_s = \dfrac{WL}{4}\dfrac{(\sinh z - z)}{z^2 \sinh z}$	$M_e = \dfrac{WL}{4}\dfrac{(\tan z - z)}{z^2 \tan z}$ $M_s = \dfrac{WL}{4}\dfrac{(z - \sin z)}{z^2 \sin z}$

Table 8-2   Moment Coefficients—Combined Axial Load and Bending

$$M = C_1 \sin (x/j) + C_2 \cos (x/j) + f(w)$$

$$j = \sqrt{EI/P}$$

Loading	C_1	C_2	$f(w)$
M_1, M_2 beam	$\dfrac{M_2 - M_1 \cos (L/j)}{\sin (L/j)}$	M_1	0
w_0 distributed	$\dfrac{w_0 j^2 [1 - \cos (L/j)]}{\sin (L/j)}$	$w_0 j^2$	$-w_0 j^2$
W point load	$x < a$: $\dfrac{Wj \sin (b/j)}{\sin (L/j)}$	0	0
	$x > a$: $\dfrac{Wj \sin (a/j)}{\tan (L/j)}$	$Wj \sin \dfrac{a}{j}$	0
M_a applied moment	$x < a$: $-\dfrac{M_a \cos (b/j)}{\sin (L/j)}$	0	0
	$x > a$: $-\dfrac{M_a \cos (a/j)}{\tan (L/j)}$	$M_a \cos \dfrac{a}{j}$	0

If M defined by Eq. 8-50 is a continuous function, then the point of maximum moment may be found from the condition that $(dM/dx) = 0$:

$$\frac{dM}{dx} = \frac{1}{j}\left(C_1 \cos \frac{x}{j} - C_2 \sin \frac{x}{j}\right) + \frac{d}{dx} f(w) = 0 \qquad (8\text{-}55)$$

From Table 8-2 it follows that for most loading conditions $df(w)/dx = 0$, and for such conditions $M_{\max}$ occurs at point x_m such that

$$\tan \frac{x_m}{j} = \frac{C_1}{C_2} \qquad (8\text{-}56)$$

Solving for $\sin (x_m/j)$ and $\cos (x_m/j)$ corresponding to Eq. 8-56 and substituting these values into Eq. 8-50, the following equation is obtained:

$$M_{\max} = \sqrt{C_1{}^2 + C_2{}^2} \qquad (8\text{-}57a)$$

which is easily computed with values of C_1 and C_2 given in Table 8-2. Substituting the values of $M_{\max}$ in Eq. 8-54, the maximum stress $f_{\max}$ can be computed as

$$f_{max} = \frac{P}{A} + \frac{M_{max}}{I} c \qquad (8\text{-}57b)$$

Usually the shears in beams do not govern the design of the members. They can be computed, however, in the following manner. Just as the total bending-moment M differs from the primary moment M_o, the actual total shear V differs from the primary shear V_o. If the shear is defined as the

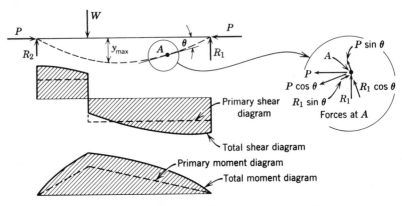

Fig. 8-34   Shear and moment diagrams for a beam column.

total force in the plane of the section normal to the elastic curve (Fig. 8-34), then, for compressive axial load, differentiating Eq. 8-46b, we obtain

$$V = \frac{dM}{dx} = \frac{dM_o}{dx} + P\frac{dy}{dx} = V_o + P\frac{dy}{dx} \qquad (8\text{-}58)$$

This equation indicates that the total shear is greater than the primary shear V_o by the amount of shearing component produced by P. To be exact, the actual V_o is slightly less than the shear R_1 that would be computed by conventional methods, as $V_o = R_1 \cos \theta$. When θ is small, $\cos \theta$ is nearly unity, and V_o is assumed to be equal to R_1, whereas $P(dy/dx) = P \sin \theta$ cannot be neglected when P is large. Differentiating Eq. 8-50, we obtain

$$V = \frac{dM}{dx} = \frac{1}{j}\left(C_1 \cos \frac{x}{j} - C_2 \sin \frac{x}{j}\right) + \frac{df(w)}{dx} \qquad (8\text{-}59)$$

and, by plotting V against x in this equation, the total shear diagram for the beam column can be obtained (Fig. 8-34). The shearing stress at any point in the beam column can then be computed in the conventional manner as follows:

$$f_v = \frac{VQ}{Ib} \qquad (8\text{-}60)$$

The conventional solution of the beam-column problem given above neglects deformations due to shear in the same manner as the conventional flexure theory neglects shear deformations. If the shear deformations are considered, the solution of this problem becomes more complex. Usually, however, the effect of shear deformations is not important.

The shears, moments, deflections, and stresses just defined are not linear functions of the transverse load since they depend on the axial load in a nonlinear manner. This is apparent from the solutions of the differential equations which involve trigonometric or hyperbolic functions (Eq. 8-50 or 8-51).

In a statically determinate system for a given axial load, the response is linear with transverse bending load. Therefore for a combination of transverse loads the principle of superposition can be used in determining values of M and V, and hence values of deflections and stresses, provided that in adding the effects of each transverse load, the full axial load is included in the evaluation of the effect of each individual transverse load.

This solution is not too difficult, but it does require tedious computations; whereas the approximate solution presented in the following requires much less computation and is in good agreement with the theoretical solution, provided that the deflected shape of the beam can be approximated by a single sine wave or a parabola. The derivation of this approximate equation is given in textbooks[1] and will not be included here. It shows that the deflection of a beam column at any point in the beam is given approximately by the formula

$$y = y_o \frac{1}{1 \pm P/P_{cr}} \qquad (8\text{-}61)$$

where y_o is the deflection of the same beam at the same point computed by conventional elementary methods neglecting the effect of axial load P, and P_{cr} is the critical buckling load of the column defined by the Euler formula

$$P_{cr} = \frac{\pi^2 EI}{L^2} \qquad (8\text{-}62)$$

In this equation the plus sign is used when the axial load P is tensile and the minus sign when P is compressive.

8-13 COMBINED BENDING AND AXIAL LOAD—PLASTIC RANGE

Experiments show that in the plastic range a plane section still remains plane, but stress is no longer proportional to strain and the bending stress distribution is no longer proportional to the distance from the neutral axis.

Hence Eq. 8-48 as well as Eqs. 8-50 and 8-51 are no longer valid and the solutions of differential equations described in the preceding section do not apply for beams in the plastic range. The numerical method[13] outlined on p. 335 is applicable with the modification that the relationship between given moment and curvature used in step 2 to evaluate the deflection includes the effect of inelastic behavior of the material and of the axial load. This condition makes it practically impossible to obtain a solution for the plastic capacity of a section subjected to combined bending and axial load by this method without the use of a digital computer.

A convenient and powerful method for determining the strength of such members is that of interaction.[14] The general criterion for failure is expressed by a functional relationship in terms of ratios of actual load to the strength of a member under pure axial or pure bending load as follows:

$$\frac{P}{P_u} = f_1 \frac{M}{M_u} \quad \text{or} \quad \frac{M}{M_u} = f_2 \frac{P}{P_u} \qquad (8\text{-}63)$$

where $P =$ actual axial load, $M =$ maximum bending moment acting simultaneously with P, $P_u =$ strength of the particular member when subjected to pure axial load, and $M_u =$ strength of the particular member when subjected to pure flexure.

The derivation of this type of equation requires knowledge of stress-strain relationship (or its idealized approximation) for the material, definition of capacity in terms of limiting stress or strain, and solution of two equilibrium conditions for the given cross section, that is,

$$\int f \, dA = P \quad \text{and} \quad \int f y \, dA = M \qquad (8\text{-}64)$$

The derivations of interaction equations will be illustrated for several simple cases in the plastic range and also the general elastic case.

1. General Elastic Case. Given a section subjected to axial load P and bending moment M about axis of symmetry. If the stress-strain relation is linear,

$$f_m = \frac{P}{A} + \frac{M}{S} \qquad (8\text{-}65)$$

where f_m is maximum (limit) stress, and A and S are cross-sectional area and section-modulus, respectively. This equation can be rewritten as follows:

$$1 = \frac{P}{A f_m} + \frac{M}{S f_m}$$

or

$$\frac{P}{P_u} + \frac{M}{M_u} = 1 \qquad (8\text{-}66)$$

where $P_u = Af_m$ and $M_u = Sf_m$ are "ultimate" or limit values of axial load and bending moment, respectively, when acting separately.

2. *Rectangular Section—Ideal Plastic Case.* Consider a rectangular section subjected to loads P and M, as shown in Fig. 8-35. Idealizing the stress-strain relationship as rigid-plastic (ideally plastic) the stress distribution will be that shown in the figure, with maximum stress $f_m = f_y$. Then, from equilibrium equations,

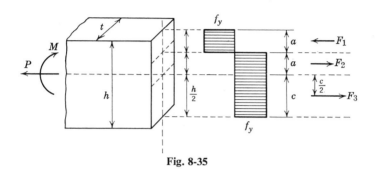

Fig. 8-35

$\int f\, dA = P$:

$$F_1 - F_2 - F_3 + P = 0 \tag{8-67}$$

or, since $F_1 = F_2 = f_y at$

$$P = F_3 = f_y ct \tag{8-68}$$

$\int fy\, dA = M$:

$$F_1 \cdot a + F_3\left(\frac{h}{2} - \frac{c}{2}\right) - M = 0 \tag{8-69}$$

or, since $\frac{1}{2}(h - c) = a$

$$f_y a^2 t + f_y cta = M = f_y at(a + c) \tag{8-70}$$

For pure bending $M_u = f_y(th^2/4)$ and for pure axial load $P_u = f_y ht$. Then

$$\frac{P}{P_u} = \left(\frac{f_y \cdot ct}{f_y ht}\right) = \frac{c}{h} = \frac{h - 2a}{h} = 1 - 2\frac{a}{h} \tag{8-71a}$$

or

$$\frac{a}{h} = \frac{1}{2}\left(1 - \frac{P}{P_u}\right) \tag{8-71b}$$

Furthermore,

$$\frac{M}{M_u} = \frac{f_y at(a+c)}{f_y th^2(\frac{1}{4})} = \frac{4a(a+c)}{h^2} \tag{8-72a}$$

or

$$\frac{M}{M_u} = 4\left(\frac{a}{h}\right)\left(\frac{h-a}{h}\right) = 4\left(\frac{a}{h}\right)\left(1 - \frac{a}{h}\right) \tag{8-72b}$$

Substituting for a/h, we obtain

$$\frac{M}{M_u} = 2\left(1 - \frac{P}{P_u}\right)\left(1 - \frac{1}{2} + \frac{1}{2}\frac{P}{P_u}\right) = 1 - \left(\frac{P}{P_u}\right)^2 \tag{8-73a}$$

or

$$\frac{M}{M_u} + \left(\frac{P}{P_u}\right)^2 = 1 \tag{8-73b}$$

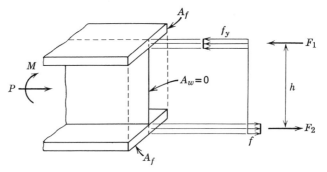

Fig. 8-36

3. *Ideal W Section—Ideal Plastic Case.* An "ideal" W section considered here is one in which the web area is negligible when compared with flange areas. This section is not necessarily ideal from a practical point of view; it is only ideal because it simplifies the solution of the problem considered here and so serves to illustrate the influence of the shape of cross section on the form of interaction equation. Consider such a section shown in Fig. 8-36 and subjected to bending moment M and axial load P. Idealizing the stress-strain relationship as rigid plastic and noting that flange areas A_f are equal and web area $A_w = 0$, it can be shown that f must be tensile and must be less than f_y. Then from equilibrium conditions:

$\int f \, dA = P$:

$$F_1 - F_2 - P = 0 \tag{8-74a}$$

or

$$P = A_f(f_y - f) \tag{8-74b}$$

$\int fy\, dA = M$:

$$F_1 \cdot \frac{h}{2} + F_2 \cdot \frac{h}{2} - M = 0 \qquad (8\text{-}75a)$$

or

$$M = \frac{h}{2} A_f(f_y + f) \qquad (8\text{-}75b)$$

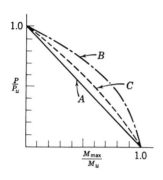

(A) Straight-line equation.
$1 = P/P_u + M_{max}/M_u$
(B) Ideally plastic equation.
$1 = (P/P_u)^2 + M_{max}/M_u$
(C) I shape with thin web.

Fig. 8-37 Interaction curves, bending and axial loads.

For pure bending $M_u = f_y A_f \cdot h$, and for pure axial load $P_u = 2A_f f_y$. Then,

$$\frac{P}{P_u} = \frac{A_f(f_y - f)}{2A_f f_y} = \frac{1}{2}\left(1 - \frac{f}{f_y}\right) \qquad (8\text{-}76)$$

And,

$$\frac{M}{M_u} = \frac{(h/2)A_f(f_y + f)}{hA_f f_y} = \frac{1}{2}\left(1 + \frac{f}{f_y}\right) \qquad (8\text{-}77)$$

Adding,

$$\frac{P}{P_u} + \frac{M}{M_u} = \frac{1}{2}\left(1 - \frac{f}{f_y} + 1 + \frac{f}{f_y}\right) = 1 \qquad (8\text{-}78a)$$

or

$$\frac{M}{M_u} + \frac{P}{P_u} = 1.0 \qquad (8\text{-}78b)$$

Because the real W^F shape in which $A_w \neq 0$ falls somewhere between a rectangular shape and the ideal W^F shape in this derivation, an interaction curve for such a real shape will fall between those for a rectangular and ideal W^F shapes.

The interaction equations for the foregoing three cases are plotted in Fig. 8-37.

PROBLEMS

Use A36 steel unless otherwise specified.

1. A $6 \times 3\frac{1}{2}Z$ 15.7 carries a concentrated load P at midpoint of span L. Determine (a) section properties A, I_x, I_y, principal axes u, w, and moments of inertia I_u and I_w, (b) bending stresses at critical points, (c) location of neutral axes, (d) maximum shearing stress, and (e) deflection at midspan.

2. Repeat Prob. 1 for the section shown in Fig. P-2.

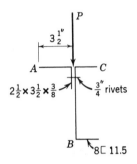

Fig. P-2

3. Determine the spacing of $\frac{3}{4}$-in. diameter rivets required to connect the two elements of the section in Prob. 2.

4. Determine the shear center location for the sections shown in Fig. P-4, neglecting bending stresses in plate elements.

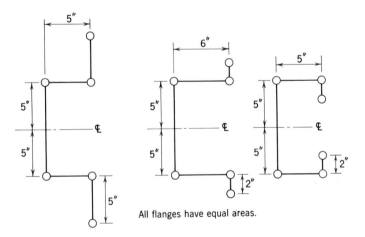

All flanges have equal areas.

Fig. P-4

5. Determine the shear center for the sections shown in Fig. P-5.

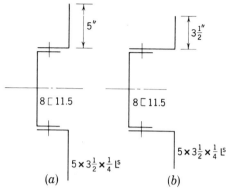

Fig. P-5

6. Determine the shear center for the section of Prob. 2.

7. For the beam of Example 8-1 determine the additional shearing stress and the angle of twist at midspan if the load is applied at the c.g. instead of at the heel. Assume no twist at the supports and ends free to warp.

8. A 12 WF 27 is simply supported with ends free to warp. It is loaded with a concentrated load P at midspan applied with 6-in. eccentricity with respect to the web. Determine the stress distribution at the end sections and at midspan due to bending and torsion. Consider the following values of span L: (*a*) $L = 8$ ft, (*b*) $L = 12$ ft, (*c*) $L = 18$ ft, and (*d*) $L = 24$ ft.

9. For Prob. 8 determine the angle of twist at midspan assuming no twist at the ends.

10. The beam shown in Fig. P-10 carries a crane rail. Maximum crane wheel loads are applied as shown. Determine maximum normal and shearing

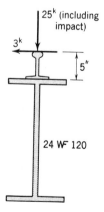

Fig. P-10

stresses, considering the effects of torsion. The beam is simply supported with span $L = 25$ ft. The ends are free to warp.

11. Repeat Prob. 10 for the beam shown in Fig. P-11. A441 steel.

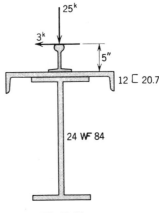

Fig. P-11

12. Determine the shearing stress distribution due to torsion in the section shown in Fig. P-12. Neglect stresses induced by torque.

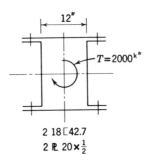

Fig. P-12

13. For the section of Prob. 12 determine the angle of twist per foot of beam length.

14. Determine the bending and shearing stresses in a boxed girder shown in Fig. P-14. Neglect bending stresses induced by torque. Load P is applied at midpoint of the 88-ft span.

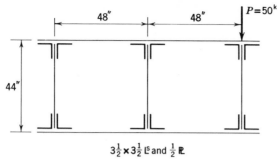

$3\frac{1}{2} \times 3\frac{1}{2}$ ℄ and $\frac{1}{2}$ ℞

Fig. P-14

15. For the girder of Prob. 14 determine angle of twist per foot of girder length.

16. For the beams shown in Fig. P-16, neglecting bending resistance of the web, determine (*a*) the maximum normal stress in the top and bottom flanges,

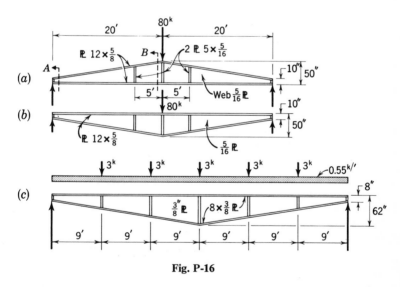

Fig. P-16

(*b*) the shearing stress in the web at the support (section *A*) and at midspan (section *B*), (*c*) the load in the stiffener at midspan, and (*d*) the variation in shearing stress in the web between sections *A* and *B* (plot diagram).

17. Repeat Prob. 16, parts (*a*), (*b*), and (*c*) only, considering the bending resistance of the web.

18. For the "butterfly" beam shown in Fig. P-18, carrying shear and moment due to lateral forces, determine the maximum normal stresses in the flanges and the web. Assume the beam is pinned at midspan and fixed at the supports.

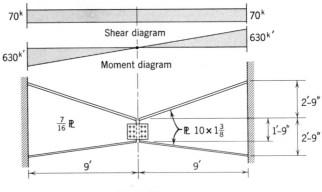

Fig. P-18

19. Determine the flange and web stresses for the beam of Prob. 18, when it carries a vertical concentrated load $P = 80$ kips at midspan.

REFERENCES

1. Popov, E. P., *Introduction to Mechanics of Solids*, Prentice-Hall, Englewood Cliffs, N.J. 1967.
2. Timoshenko, S., *Theory of Elasticity*, McGraw-Hill Book Co., New York, pp. 290–296, 1934.
3. Timoshenko, S., *Theory of Elasticity*, McGraw-Gill Book Co., p. 245, 1934.
4. Kubo, G. G., B. G. Johnson, and W. J. Eney, "Non-uniform Torsion of Plate Girders," *Trans. ASCE* **121**, 759 (1956).
5. Goldberg, J. E., "Torsion of I-Type and H-Type Beams," *Trans. ACSE* **118**, 771 (1953).
6. Sourochnikoff, B., "Strength of I-Beams in Combined Bending and Torsion," *Trans. ASCE* **116** (1951).
7. Timoshenko, S., *Strength of Materials*, Part II, D. Van Nostrand Co., Princeton, N.J., pp. 288–292, 1941.
8. Timoshenko, S., "Theory of Bending Torsion, and Buckling of Thin-Walled Members of Open Cross-Section," *J. Franklin Inst.* **239**, Nos. 3, 4, 5 (March–April–May 1945).
9. von Kármán, T., and Chien Wei Zang, "Torsion with Variable Twist," *J. Aeronaut. Sci.* **13**, No. 10 (October 1946).
10. Moorman, R. B., "Wedge Shaped Structural Members under Direct Stress and Bending," *Trans. ASCE* **110** (1945).
11. Seely, F. B., and R. V. James, "The Plaster-Model Method of Determining Stresses Applied to Curved Beams," *Univ. Illinois Eng. Expt. Sta. Bull.* **195** (1929).
12. Timoshenko, S., *Theory of Elasticity*, McGraw-Hill Book Co., New York, p. 98, 1934.
13. Newmark, N. M., "Numerical Procedures for Computing Deflections, Moments and Buckling Loads," *Trans. ASCE* **108** (1943).
14. Shanley, F. R., *Strength of Materials*, McGraw-Hill Book Co., New York, 1957.

9

Buckling of Prismatic Members, Frames, and Plates

 GGGGGGGGGGGGGGGGGGGGGGGGGGGGGGGGGGGGGGGGGGGGGGGGGGGGGGGGGGG

9-1 INTRODUCTION

Slender compression members fail by instability (buckling) when the load reaches a critical value. A structural system or member is stable when it returns to its original state after a small disturbance (force or displacement) is removed. Under certain conditions the system cannot attain an equilibrium state and the disturbance causes a deformation of an indeterminate magnitude. Such a condition corresponds to a critical (buckling) state of the system.

Consider a slender pin-ended compression member subjected to an axial load P and a transverse load W at midlength (Fig. 9-1). The behavior of this beam column in the elastic range is described by the following equilibrium and geometric conditions, Eqs. 9-1 and 9-2.

$$M_x = \frac{W}{2} x + Py \tag{9-1}$$

and

$$M_x = -\frac{1}{\rho} EI = -EI \frac{d^2y}{dx^2} \tag{9-2}$$

The resulting differential equation

$$EI \frac{d^2y}{dx^2} + Py + \frac{W}{2} x = 0 \tag{9-3}$$

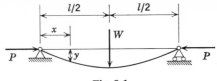

Fig. 9-1

has the following solution:

$$y = \frac{1}{2} \frac{W}{P} x \left[\frac{\sin \alpha x}{\alpha x \cos (\alpha L/2)} - 1 \right]$$ (9-4)

where $\alpha = (P/EI)^{1/2}$, E is the elastic modulus, and I is the moment of inertia of the cross section.

It is apparent from Eq. 9-4 that when cos $(\alpha L/2)$ is zero, the deflection y is infinite, even when W is infinitesimally small. This condition corresponds to

$$\frac{\alpha L}{2} = \frac{\pi}{2}, \quad \text{or} \quad \alpha = \left(\frac{P}{EI} \right)^{1/2} = \frac{\pi}{L}, \quad \text{or} \quad P = P_{\text{cr}} = \frac{\pi^2 EI}{L^2}$$ (9-5)

Thus the critical buckling load P_{cr} for an axially loaded slender compression member is obtained. This simple example may be used to illustrate several fundamental concepts about buckling or instability.

From Eq. 9-4 it is seen that when $P < P_{\text{cr}}$, deflection y vanishes as W becomes zero. Only when $P = P_{\text{cr}}$ deflection y does not vanish; it becomes indeterminate as W vanishes. This is characteristic of instability.

In addition from Eq. 9-4 it is seen that the relationship between y and W is not linear, since y depends on both W and P. For a given value of $P \neq P_{\text{cr}}$, y varies linearly with W; but since y is a trigonometric function of P, it rapidly approaches infinity as P approaches P_{cr}. This nonlinearity is another characteristic of instability and is contrary to the normally assumed linear behavior of structural systems. Usually it is assumed that a slight change in loading condition produces a proportionately slight change in stress or displacement.

The instability results from the fact that change in the geometry of the structure (deformation) influences the equilibrium conditions (Eq. 9-1). In conventional structural analysis the deformation is neglected in considering equilibrium conditions. Consideration of the change in geometry does not always lead to instability. Some systems are inherently stable; for example, axial tension tends to straighten out a member subjected to bending and therefore tension does not lead to instability. On the other hand, compression tends to increase the curvature of a member subjected to bending and therefore may lead to instability.

Various forms of buckling (instability) occur in structural systems. A slender column, such as the one just examined, failed by translational buckling, that is, by translation of cross section without change of its shape. Some shapes of cross section could fail by torsional buckling, when the cross section rotates as well as translates, or by local buckling, when a portion of the cross section—usually a thin plate—fails by local plate instability before the entire column section buckles by translation or rotation. Framed systems also can fail by over-all instability and this mode of failure is called "general frame instability." In the following sections the buckling loads corresponding to some modes of instability are considered. A more detailed treatment of buckling of steel members and structures is given in other sources.[1,2,3]

9-2 ELASTIC BUCKLING OF AXIALLY LOADED PRISMATIC MEMBERS

The critical buckling load defined by Eq. 9-5 is valid for a straight prismatic member idealized by the following assumptions:

(*a*) The material is linearly elastic and proportional limit stress is nowhere exceeded.

(*b*) The elastic moduli in tension and compression are equal.

(*c*) The material is perfectly homogeneous and isotropic.

(*d*) The member is perfectly straight initially and the load is perfectly concentric with the centroid of the section.

(*e*) The ends of the member are perfect frictionless hinges which are so supported that axial shortening is not restrained.

(*f*) The section of the member does not twist and its elements do not undergo local buckling.

(*g*) The member is free from residual stresses.

(*h*) Small-deflection approximation may be used in defining geometric curvature of the deformed shape.

For such an ideal member instability is characterized by zero deflection y with load P increasing up to the critical value P_{cr}, and a bifurcation at the critical load, with either zero y or indeterminate y satisfying the mathematical solution. This relationship is shown in Fig. 9-2 by the solid line OAB. Actually, because conditions (*c*) and (*d*) cannot be fully satisfied even with extreme care and precision, the lateral deflection is more nearly progressive but with sudden increase at loads closely approaching P_{cr}, $OA'B'$, (Fig. 9-2).

For an ideal column, satisfying assumptions (*a*) to (*h*) the stress is uniformly distributed over the cross section for all loads up to the critical

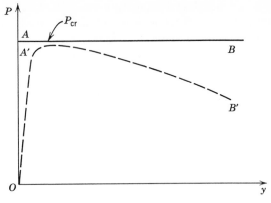

Fig. 9-2

and therefore the critical stress f_{cr} can be defined as follows:

$$f_{cr} = \frac{P_{cr}}{A} = \frac{\pi^2 EI}{L^2 A} = \frac{\pi^2 E}{(L/r)^2} \tag{9-6a}$$

where $r = \sqrt{I/A}$ is the least radius of gyration of the cross section.

It is sometimes convenient to express this equation in nondimensional terms. Dividing both sides by yield strength F_y leads to the following:

$$\frac{f_{cr}}{F_y} = \frac{1}{\lambda^2} \tag{9-6b}$$

where $\lambda = (L/r)(F_y/\pi^2 E)^{1/2}$.

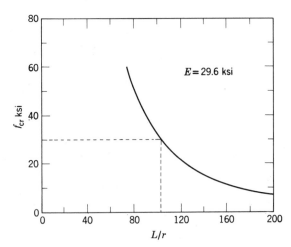

Fig. 9-3

Equation 9-6a, usually called Euler formula in honor of the Swiss mathematician Euler, who derived the expression for critical load P_{cr} in 1757, defines the critical stress as a function of elastic modulus E and slenderness ratio L/r—a nondimensional geometric characteristic of an ideal column. For steel columns the variation of f_{cr} with L/r, defined by Eq. 9-6, is shown in Fig. 9-3.

Experiments show that for slender steel columns with L/r values greater than 100–120, test results agree closely with the values obtained from Eq. 9-6. For shorter, less slender columns, experimental results deviate from the ideal elastic buckling values,[1] primarily because local stresses exceed the proportional limit.

9-3 INELASTIC BUCKLING OF AXIALLY LOADED PRISMATIC MEMBERS

The assumption of linearly elastic material behavior for the ideal column is valid only as long as the critical stress f_{cr} does not exceed f_p, proportional limit. In actual columns the material has a stress-strain diagram with a curved portion above the proportional limit, as in Fig. 9-4. At some stress $f > f_p$ the slope of the stress-strain curve is defined by the tangent modulus E_t, which is smaller than the initial modulus E based on linear f-ϵ relation. Therefore the critical load P_{cr} based on the assumption of linear elastic behavior is no longer valid when the critical stress exceeds the proportional limit.

Engesser (1889) suggested that the critical load of an axially loaded column in the inelastic range can be defined by the Euler equation provided that the modulus E is replaced by tangent modulus E_t which corresponds to the

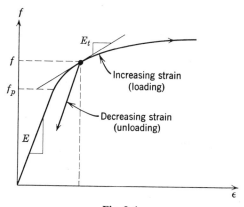

Fig. 9-4

critical stress, that is,

$$f_{\text{cr}} = \frac{\pi^2 E_t}{(L/r)^2} \qquad (9\text{-}7)$$

This relationship is based on the assumption that the deformation of all fibers of the cross section is controlled by the law $(df/d\epsilon) = E_t$, that is, no unloading of fibers takes place. However, if the column is slightly curved, an increase in curvature corresponds to increase in compression on the concave side and decrease in compression on the convex side. The unloading on the convex side will follow a linear stress-strain law, whereas the increase in compression will follow the nonlinear law $(df/de) = E_t$. This concept proposed by Considere and developed by Karman leads to the so-called reduced modulus theory, where $E_t < E_r < E$, and the critical stress f_{cr} is

$$f_{\text{cr}} = \frac{\pi^2 E_r}{(L/r)^2} \qquad (9\text{-}8)$$

Shanley has shown that critical stress depends on the conditions preceding buckling and the tangent modulus approach is the lower bound solution for this critical stress. Therefore Eq. 9-7 is now generally accepted as the appropriate solution for the critical stress in the inelastic range.

The Engesser equation (see Eq. 9-7) cannot be solved directly because E_t and $f_t = f_{\text{cr}}$ are interdependent and f_{cr} must be known before E_t can be determined. Trial calculations may be made until the calculated f_{cr} and E_t values are consistent with each other, based on specific material stress-strain curve. To assist in this calculation it may be expedient to plot the values of E_t versus f, as shown in Fig. 9-5.

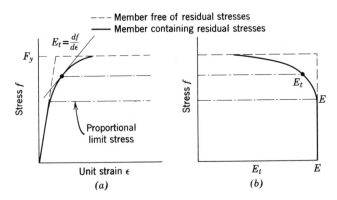

Fig. 9-5   Tangent modulus for mild steel.

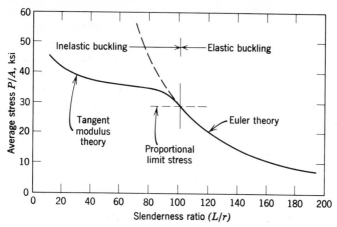

Fig. 9-6 Column curve for structural carbon steel.

By substituting the corresponding values of E_t and f_t into Eq. 9-7, values of L/r may be determined as

$$\frac{L}{r} = \left(\frac{\pi^2 E_t}{f_t}\right)^{\frac{1}{2}} \qquad (9\text{-}9)$$

and the tangent-modulus curve for critical stress in the inelastic range can be plotted as an extension of the Euler curve for the elastic range, as in Fig. 9-6.

9-4 BUCKLING OF ECCENTRICALLY LOADED PRISMATIC MEMBERS

The nature of loading is a major factor in determining the strength of a compression member. The applied loads are seldom concentric with the centroid and in addition to eccentricities introduced by the geometric configuration of the structure, other so-called accidental eccentricities introduce bending into the member. In this section an eccentrically loaded member is considered in which the plane of the bending moment due to eccentricity coincides with the plane of buckling, as in Fig. 9-7.

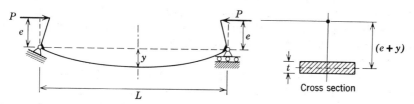

Fig. 9-7 Buckling under eccentric load.

Although for an ideal axially loaded member buckling could be characterized by a sudden lateral deflection occurring when the load reaches a
critical value, for an eccentrically loaded column the deflection increases
gradually as the load increases and the definition of the critical load must be
amplified.

Consider a slender, eccentrically loaded, pin-ended compression member
(Fig. 9-7). The behavior of this beam column in the elastic range is described
by the following:

$$\text{equilibrium:} \quad M_x = Pe + Py \tag{9-10}$$

$$\text{geometry:} \quad M_x = -EI\frac{d^2y}{dx^2} \tag{9-11}$$

$$\text{therefore:} \quad EI\frac{d^2y}{dx^2} + Py + Pe = 0 \tag{9-12}$$

$$\text{solution;} \quad y = e\left[\frac{\cos(\alpha L/2) - \alpha x}{\cos(\alpha L/2)} - 1\right] \tag{9-13}$$

where $\alpha = (P/EI)^{\frac{1}{2}}$. The form of Eq. 9-13 is similar to Eq. 9-4 and so it
may be concluded that the critical buckling load corresponding to infinite
deflection can be obtained from $\cos(\alpha L/2) = 0$ which leads to $P = P_{cr} = \pi^2EI/L^2$, as in Eq. 9-5. This solution is correct if the maximum stress f_m due
to combined effect of axial load P, eccentricity e, and maximum deflection
y_m (at $x = L/2$) is less than the proportional limit stress f_p. This stress f_m is

$$f_m = \frac{P}{A} + \frac{P(e + y_m)c}{I} \tag{9-14}$$

where c is the distance from the centroidal axis to the extreme fiber. Substituting from Eq. 9-13 into Eq. 9-14,

$$f_m = \frac{P}{A}\left(1 + \frac{ec}{r^2}\sec\frac{L}{2r}\sqrt{\frac{P}{AE}}\right) \tag{9-15}$$

As P approaches the Euler load π^2EI/L^2, the contribution of the bending
stress to the value of f_m increases rapidly and therefore f_m may reach the
proportional limit value at a relatively low value of average stress P/A. Thus
the validity Eq. 9-13 for evaluation of the critical (buckling) load for an
eccentrically loaded member is much more limited than it is for axially
loaded members.

To evaluate properly the buckling load for an eccentrically loaded member
it is necessary to consider inelastic behavior at stresses beyond the proportional limit. This problem, formulated by Karman, requires determination

of the relationship between the load P and the deflection y in the inelastic range and is sensitive to the stress-strain relationship, slenderness of the member, and geometry of cross section. In general, it requires a numerical solution, since closed-form mathematical solution is not feasible. Such solutions are described in detail elsewhere.[1,2] To demonstrate the significance of such solutions consider the curve of P versus maximum deflection y_m

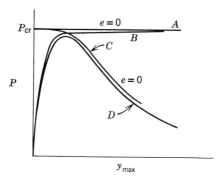

Fig. 9-8　Buckling characteristics: *A*—elastic, $e = 0$; *B*—elastic, $e > 0$; *C*—inelastic, $e = 0$; and *D*—inelastic, $e > 0$.

shown in Fig. 9-8. For a member with given length, cross section, and stress-strain characteristics of the steel, calculated load-deflection relationships are shown assuming elastic and inelastic behavior. Curve *A* corresponds to elastic response for axial loading and curve *B* to elastic response for eccentric loading, both approaching the same value of critical buckling (Euler) load. Curve *C* corresponds to calculated inelastic response for axial loading, showing the unstable nature of this response, that is, decreasing load P with increasing deflection y. The point corresponding to buckling load is given by a modified Euler load based on tangent modulus. Curve *D* corresponds to calculated inelastic response for an eccentrically loaded column. The ascending branch of this curve represents a stable equilibrium condition and the descending branch represents the unstable equilibrium. The critical load corresponds to the peak load value between the stable and the unstable response.

In defining the critical load for eccentrically loaded columns it has been suggested that the loading which first produces external fiber stress equal to yield strength and calculated on the basis of linear elastic theory be taken as the critical value. This load can be obtained from Eq. 9-15 by equating f_m to F_y and solving for P. A closed-form solution is not possible because P occurs as an argument in the trigonometric term of the equation. However, a graphical solution is possible. The average stress P/A corresponding to

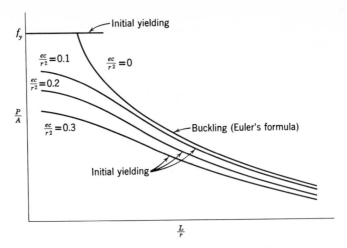

Fig. 9-9   Yielding under eccentric compression.

this criterion of limit load is defined Eq. 9-16 and shown in Fig. 9-9.

$$\frac{P}{A} = \frac{F_y}{1 + \dfrac{ec}{r^2} \sec \dfrac{L}{2r} \sqrt{\dfrac{P}{AE}}} \tag{9-16}$$

The limit load P defined by Eq. 9-16 and the critical buckling load P_{cr} for an eccentrically loaded column, obtained by appropriate analysis based on inelastic behavior, represent two basically different design criteria. The difference between these values ranges widely: it may be small or large, depending on shape of cross section, slenderness, relative eccentricity, and stress-strain characteristics.

Although Eq. 9-16—the so-called secant formula—may be useful in formulating design criteria for compression members, it must be realized that the loads obtained from this criterion are not directly related to the actual buckling load.

9-5 INFLUENCE OF RESIDUAL STRESSES

The final stage during the manufacture of rolled-steel shapes is that of cooling to room temperature from the high temperatures required for rolling the billet into the desired shape. Because cooling rate depends on thickness, the thinner sections cool first and as a result thin portions of a structural shape of nonuniform thickness develop internal tension stresses, whereas the thicker parts which cool last develop internal compressive stresses. These cooling stresses are referred to as residual stresses and vary approximately as indicated in Fig. 9-10a. Residual stresses are also introduced into the steel shapes by fabrication operations such as straightening and welding.

The net effect of the residual stresses is to alter the stress-strain diagram for the shape as compared to the ideal material specimen. If instead of the coupon test, a stub section of the column is used to determine the stress-strain diagram, a realistic curve is obtained which includes the effect of the residual stresses. A plot will be obtained which may be considered an average stress-strain curve and the tangent modulus to this curve will reflect the presence of residual stresses and the variation of the yield strength over the cross section. This effect can be illustrated by the stress-strain curve in Fig. 9-10b, which indicates the behavior of the usual small test coupon as a dotted curve and the actual behavior of a stub column as a solid line. The appropriate value for E_t, the tangent modulus of the entire cross section, may be obtained from this curve. If an actual stub test is not available, an idealized distribution of residual stresses may be assumed, as indicated in Fig. 9-10a. A review of the research on residual stresses in rolled shapes[4] indicates that the average value of the maximum residual stress in compression is approximately $f_{rc} = 0.3F_y$.

Results of one investigation[4] are shown in Fig. 9-11 in which column capacity curves are plotted for concentrically loaded steel W shapes. The curves are plotted in nondimensional form (f_{cr}/F_y) versus λ for three cases. Case A corresponds to elastic buckling without the effect of residual stress. Case B corresponds to inelastic buckling about the strong axis, with residual stresses as shown used for evaluating the inelastic behavior. Case C corresponds to inelastic buckling about the weak axis.

A curve, designated D in Fig. 9-11, represents an empirically defined "basic column strength." This curve is a parabola tangent to the Euler curve at $f_{cr} = \frac{1}{2}F_y$ and has a horizontal tangent at $\lambda = 0$.

The basic curve is conservative for bending about the strong axis and slightly unconservative for the weak axis.

9-6 TORSIONAL BUCKLING

The primary buckling considered previously is due to bending without twisting, that is, the sections displace from their original position by translation without rotation. Thin-wall members with open cross-sectional shape, unlike boxed or thick-wall sections, are sometimes weak in torsion and hence may buckle by twisting rather than bending. Torsional buckling occurs when the torsional rigidity of the member is appreciably smaller than its bending rigidity.[5]

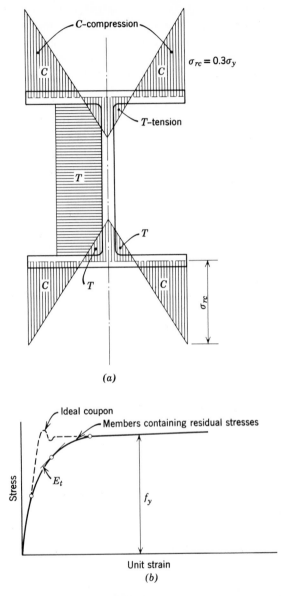

(a)

(b)

Fig. 9-10 Residual stresses in rolled shapes. (*a*) Assumed residual stress distribution in a rolled shape. (*b*) Influence of residual stress on the stress-strain diagram.

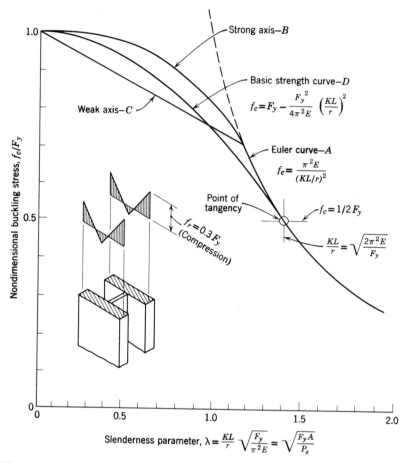

Fig. 9-11 Effect of residual stress on column strength of wide-flange shape. (Ref. 4.)

In addition to bending and torsional modes of buckling, some sections fail in a combined torsion–bending mode of buckling. Consider a T-section column shown in Fig. 9-12 loaded through the centroid of the section and assumed to buckle laterally in bending. Because of the bent shape of the column axis, the vertical load is no longer normal to the cross-section plane but has a normal and a shearing component. This centroidal shearing component does not pass through the shear center and hence causes twisting of the column. It can be seen that pure lateral buckling of this T section is impossible; that is, lateral bending is accompanied by twisting resulting in a torsion–bending mode of buckling. The critical load P_{cr} for such a combined mode of buckling is smaller than the Euler load $P_{x,cr}$ for pure lateral

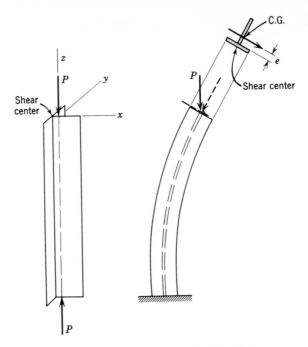

Fig. 9-12 Combined torsion-bending buckling.

buckling about the xx axis. The difference between P_{cr} and $P_{x,cr}$ may be small for long columns with adequate torsional rigidity, but for torsionally weak columns of intermediate length this difference is appreciable.

The magnitude of the critical load $P_{z,cr}$ for pure torsional buckling can be determined from consideration of restrained torsion combined with compression, just as the Euler load can be determined from consideration of bending combined with compression. The mathematical derivation of $P_{z,cr}$, omitted here, results in the following expression.[5]

$$P_{z,cr} = \frac{1}{r_z^2}\left(\frac{\pi^2 E K_b}{L^2} + G K_t\right) = \frac{G K_t}{r_z^2}\left(1 + \pi^2\frac{a^2}{L^2}\right) \qquad (9\text{-}17)$$

where r_z = polar radius of gyration about centroidal z axis

E, G = Young's modulus and shearing modulus, respectively

K_b = torsion–bending constant—property of cross section (see Sec. 8-7)

K_t = torsional rigidity constant (see Sec. 8-6)

$a = (EK_b/GK_t)^{1/2}$ (see Sec. 8-7)

The mode of buckling, that is, pure bending, pure torsion, or combined torsion–bending, depends on eccentricity of the load, location of the shear

center, and symmetry of cross section. Neglecting effects of local yielding
or buckling, a pin-ended member subjected to eccentric load will buckle at
a critical buckling load P_{cr} because of either bending or torsion, or combined
torsion–bending. The magnitude of this load is given by the smallest of the
roots of Eq. 9-18.

$$(P_{cr} - P_{x,cr})(P_{cr} - P_{y,cr})(\alpha P_{cr} - P_{z,cr})$$
$$- P_{cr}{}^2[\beta_y(P_{cr} - P_{x,cr}) + \beta_x(P_{cr} - P_{y,cr})] = 0 \quad (9\text{-}18)$$

where $P_{x,cr}$ and $P_{y,cr}$ are the Euler critical loads for buckling about x and y
principal axes, respectively; $P_{z,cr}$ is the critical load for pure torsional
buckling (see Eq. 9-17); and α, β_x, β_y are coefficients depending on the geometrical properties of cross section, to be defined.

$$\alpha = 1 - \frac{e_x x_o - e_y y_o}{r_z{}^2 + x_c{}^2 + y_c{}^2}$$

$$\beta_x = \frac{(x_c - e_x)^2}{r_z{}^2 + x_c{}^2 + y_c{}^2} \quad (9\text{-}19)$$

$$\beta_y = \frac{(y_c - e_y)^2}{r_z{}^2 + x_c{}^2 + y_c{}^2}$$

where e_x, e_y are the eccentricities of load with respect to the y and x principal
axes, respectively; x_c, y_c are the coordinates of shear center with respect to
the x and y principal axes; and x_o, y_o are coordinates of a point characteristic
of cross-section shape, to be defined.

$$x_o = 2x_c - \int_A \frac{x(x^2 + y^2)\,dA}{I_y}$$

$$y_o = 2y_c - \int_A \frac{y(x^2 + y^2)\,dA}{I_x} \quad (9\text{-}20)$$

The significance of load eccentricity, location of shear center, and symmetry of cross section can be seen from a consideration of the following
special cases:

**Case 1. Section with Two Axes of Symmetry, Shear Center Coincides with
the Centroid**

$$x_c = y_c = 0, \qquad x_o = y_o = 0, \qquad \alpha = 1, \qquad \beta_x = \left(\frac{e_x}{r_z}\right)^2, \qquad \beta_y = \left(\frac{e_y}{r_z}\right)^2$$

If load P is concentric, that is, $e_x = e_y = 0$, then, from Eq. 9-18,

$$(P_{cr} - P_{x,cr})(P_{cr} - P_{y,cr})(P_{cr} - P_{z,cr}) = 0 \quad (9\text{-}21)$$

and the critical load P_{cr} is the smallest value of the three roots $P_{x,\mathrm{cr}}$, $P_{y,\mathrm{cr}}$, $P_{z,\mathrm{cr}}$; the column fails either by pure lateral buckling defined by Euler load or by pure twisting (Eq. 9-17).

Case 2. Section with One Axis of Symmetry, Symmetrical about x Axis.

$$y_c = y_o = 0, \qquad \alpha = 1 - \frac{e_x x_o}{r_z^2 + x_c^2}, \qquad \beta_x = \frac{(x_c - e_x)^2}{r_z^2 + x_c^2},$$

$$\beta_y = \frac{e_y^2}{r_z^2 + x_c^2}$$

If load P passes through the centroid, $e_x = e_y = 0$, then

$$\alpha = 1, \qquad \beta_y = 0, \qquad \beta_x = \frac{x_c^2}{r_z^2 + x_c^2}$$

In this case it can be shown that the smallest root of Eq. 9-18 is either $P_{\mathrm{cr}} = P_{y,\mathrm{cr}}$, or smaller than either $P_{x,\mathrm{cr}}$ or $P_{z,\mathrm{cr}}$.

If the load passes through the shear center, $e_x = x_c$, $e_y = y_c$, then

$$\beta_x = \beta_y = 0, \qquad \alpha = 1 - \frac{e_x x_o}{r_z^2 + x_c^2}$$

and the critical load P_{cr} is the smallest value of the three roots $P_{x,\mathrm{cr}}$, $P_{y,\mathrm{cr}}$, and $P_{z,\mathrm{cr}}/\alpha$. It should be noted that the load $P_{x,\mathrm{cr}}$ or $P_{y,\mathrm{cr}}$ is a critical buckling load only when it is less than $P_{z,\mathrm{cr}}/\alpha$ and when either the shear center coincides with the centroid or the load is applied at the shear center. In the latter instance, if x is the axis of symmetry, the load applied at the shear center is eccentric with respect to y axis and, if $P_{y,\mathrm{cr}}$ is less than $P_{x,\mathrm{cr}}$ then buckling is not of a sudden type, but occurs as bending deflections increase, approximately hyperbolically, with increase in load P.

9-7 EFFECTIVE LENGTH OF COMPRESSION MEMBERS

In the derivation of the secant and Euler's formulas, it was assumed that the ends of the member are free to rotate, as shown in Fig. 9-13a. In actual structures, such an ideal condition is the exception rather than the rule, since the ends are usually riveted or welded to other members and thus are restrained against rotation. Even if they are pin-connected, there is only freedom to rotate about one axis; and there is pressure and friction between the pins and the member, preventing completely free rotation. The amount of this end restraint varies greatly with different structures. If the compression

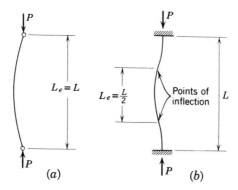

Fig. 9-13 Effective lengths of columns: (*a*) pin-ended column and (*b*) fixed-ended column.

member is rigidly attached to relatively rigid members, then the end conditions approach complete fixity (Fig. 9-13*b*). If it is pin-ended at one end and rigidly fixed at the other end, it will deform as shown in Fig. 9-14*a*. Sometimes, a compression member may have a completely free end, that is, not only free to rotate but also free to translate, as in Fig. 9-14*b*. In all instances, the strength of a compression member of actual length *L* with any degree of restraint at the ends can be compared to an equivalent pin-ended member of length L_e, so that the equivalent member has the same strength as the actual member.

The physical significance of the equivalent length becomes apparent if we consider the shape of the buckled member (buckling mode). As an example, consider a pin-ended member with length L_e and another of length $L = \frac{1}{2}L_e$, which is fixed at one end but free at the other (Fig. 9-14*b*). It is obvious that the deflected shape of the two members is the same and therefore the critical buckling load is the same. In other words, the "effective" length L_e of

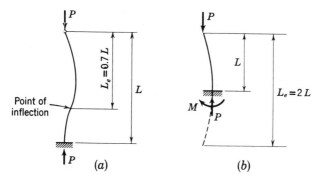

Fig. 9-14 (*a*) one end pinned, other fixed and (*b*) one end free, other fixed.

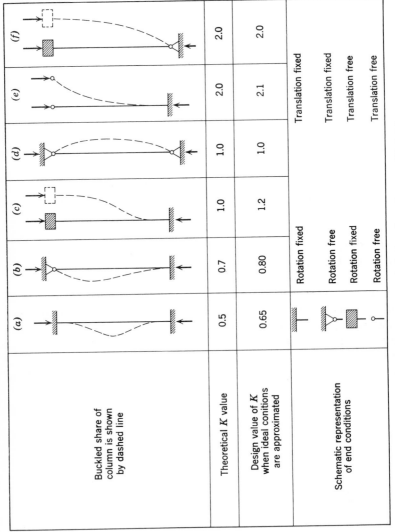

	(a)	(b)	(c)	(d)	(e)	(f)
Buckled share of column is shown by dashed line						
Theoretical K value	0.5	0.7	1.0	1.0	2.0	2.0
Design value of K when ideal conitions are approximated	0.65	0.80	1.2	1.0	2.1	2.0
Schematic representation of end conditions		Rotation fixed	Rotation free	Rotation fixed	Rotation free	Translation fixed	Translation fixed	Translation free	Translation free

Fig. 9-15 Effective length factors for columns. (Ref. 3.)

column in Fig. 9-14*b* is $L_e = 2L$ and the critical buckling load is

$$P_{cr} = \frac{\pi^2 EI}{L_e^2} = \frac{\pi^2 EI}{4L^2} \tag{9-22}$$

"Effective" length is sometimes called "unsupported" length. This terminology may be confusing since the effective length may be either greater or less than the distance between supports, which is the true unsupported length. By the proper use of this "effective" length, most formulas derived for pinended members can be applied to columns of other end conditions.

The concept of "effective" length is based largely on its use in the Euler formula and its use in other types of formulas may or may not be correct. Defining the effective length of a column in terms of its full unsupported length and the end fixity coefficient C, the critical Euler load for a member with end restraints is given by the following equation:

$$P_{cr} = \frac{C\pi^2 EI}{L^2} = \frac{\pi^2 EI}{L^2/C} \tag{9-23}$$

If the effective length L_e is defined as $L/\sqrt{C} = KL$, Eq. 9-23 is simplified as follows:

$$P_{cr} = \frac{\pi^2 EI}{L_e^2} \tag{9-24}$$

Approximations of actual end conditions may be described by idealized end restraints, which are considered as independent of adjacent members. Four basic conditions are defined: (*a*) the end is free to rotate but is fixed against translation ("pinned" or "hinged"); (*b*) the end is fixed against both rotation and translation; (*c*) the end is fixed against rotation but is free to translate; and (*d*) the end is free to rotate and to translate. Figure 9-15 shows these common types of end conditions with the theoretical K value for each case and also a recommended[3] design value of K which takes into account the expected deviations of practical structures from the theoretical condition.

9-8 MEMBERS OF VARIABLE CROSS SECTIONS

The buckling criteria presented in the preceding articles are valid only for members of constant cross section. The problem of determining the critical load for a member with variable cross section is often encountered in the design of towers, crane booms, and similar structures. Usually one or both ends of the column are tapered in section; sometimes the column is made up of several segments. The critical buckling load in such members can be

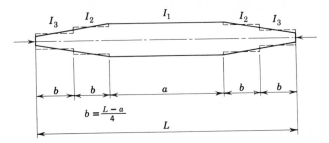

Fig. 9-16   A tapered column approximated by an equivalent stepped column.

determined by using numerical methods of successive approximations.[2,6] From a practical viewpoint, the solution of this problem is complicated by a number of factors, not the least of which is the tediousness of such computations.

For practical design purposes a tapered member can be approximated by an equivalent stepped column[7] (Fig. 9-16). Then the critical load for such member with pin ends can be expressed as an Euler load for an equivalent column having a constant section I_1 and an effective length L_e, so that

$$P_{cr} = \frac{\pi^2 E I_1}{L_e{}^2} \tag{9-25}$$

where the effective length L_e depends on ratios I_2/I_1, I_3/I_1, and a/L. The determination of this effective length is greatly simplified by the use of charts shown in Fig. 9-17.

9-9   INTERMEDIATE LOADS ON COMPRESSION MEMBERS

Occasionally compression members are made as continuous members of two or three spans, with compression loads varying in these spans (Fig. 9-18). The most efficient design of such a continuous column will result if all spans will buckle simultaneously as pin-ended Euler columns and exert no restraint on each other. This condition will occur only if the load in each span is equal to the Euler load for the span; that is,

$$P_1 = \frac{\pi^2 E I_1}{L_1{}^2} \quad \text{and} \quad P_2 = \frac{\pi^2 E I_2}{L_2{}^2} \tag{9-26}$$

This cannot always be achieved because values of I_1 and I_2 may be fixed by some consideration other than satisfying these equations. In this instance, buckling will occur when the load in one span is smaller than its critical load, whereas in the other span the applied load is greater than its critical load, and it can be said that one span provides end restraint for the other span.

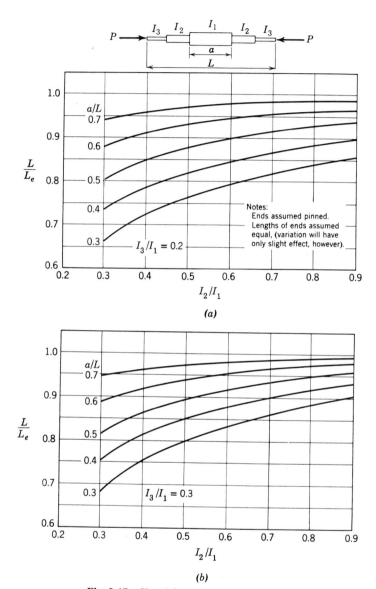

Fig. 9-17 Charts for columns of variable section.

371

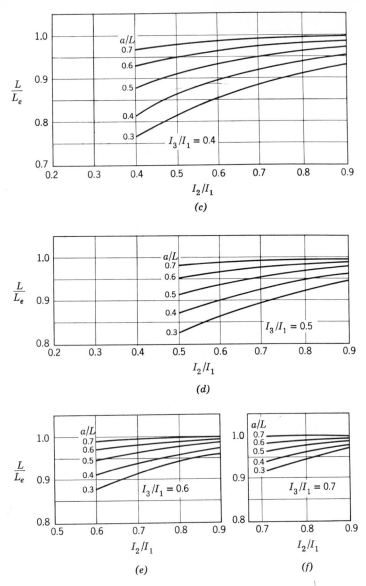

Fig. 9-17 Charts for columns of variable section (*continued*).

The critical load for a span of a continuous member can be defined in terms of an effective length L_e for that span. The value of this effective length depends on the relative stiffnesses EI/L of adjacent spans and relative magnitudes of the loads P. The stiffness value of EI/L must be calculated with the modulus of elasticity corresponding to the actual stresses. The mathematical formulation of L_e for a general case is rather complex.[2] Several simple cases have been solved and the results are shown in Fig. 9-19. Furthermore, it is good design practice to reinforce the column below the points of application of intermediate loads. Thus a mathematical solution is complicated by the variation of moment of inertia along the length of the member.

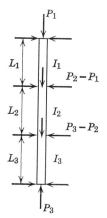

Fig. 9-18 Columns with lateral supports and intermediate loads.

An approximate solution of this problem is possible by defining an average load and an average moment of inertia for a pin-ended column subjected to several loads and reinforced in steps, so that the I of the section is approximately proportional to the total load at the section (Fig. 9-20). These average values are

$$P_a = \frac{P_1 \Delta L_1 + (P_1 + P_2)\Delta L_2 + \cdots + (P_1 + \cdots + P_n)\Delta L_n}{\sum_1^n \Delta L} \quad (9\text{-}27)$$

$$I_a = \frac{I_1 \Delta L_1 + I_2 \Delta L_2 + \cdots + I_n \Delta L_n}{L} \quad (9\text{-}28)$$

where $L = \Sigma_1^n \Delta L$. Then the critical average load for the column can be written as

$$(P_a)_{\mathrm{cr}} = \frac{\pi^2 E I_a}{L^2} \quad (9\text{-}29)$$

9-10 MEMBERS WITH ELASTIC LATERAL SUPPORTS

In order to decrease the effective length of a compression member, lateral bracing between end supports can be used very effectively. In most ordinary structures, the bracing is stiff enough to prevent almost completely any lateral movement at the intermediate supports. Sometimes, however, the lateral support is elastic—it will displace an appreciable amount and hence cannot be considered a rigid support. A general theoretical analysis of the problem is rather complicated,[1,2] although for special cases charts for the effective length of a column with elastic lateral supports can be used.[8]

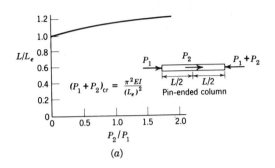

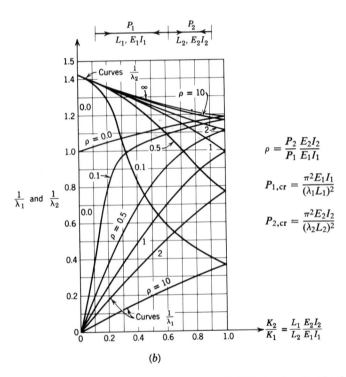

Fig. 9-19 Intermediate loads on compression members. (*a*) Effective length of column loaded at both ends and at midpoint. (*b*) Buckling of a two-span column, with unequal spans, moments of inertia, and loads. Hinged end supports. (*c*) Buckling of a three-span symmetrical column. Same conditions as (*b*).

374

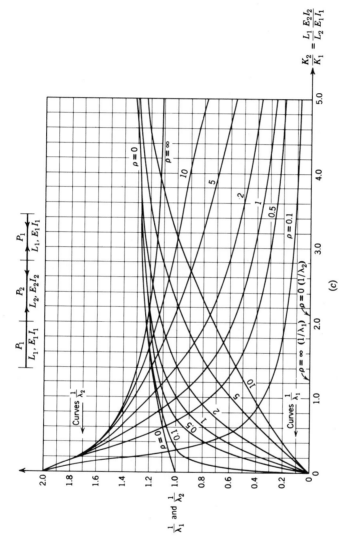

Fig. 9-19 Intermediate loads on compression members (*continued*).

375

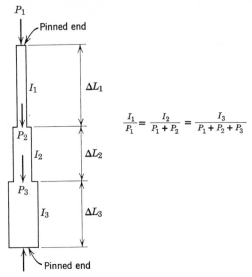

$$\frac{I_1}{P_1} = \frac{I_2}{P_1 + P_2} = \frac{I_3}{P_1 + P_2 + P_3}$$

Fig. 9-20 Stepped column with intermediate loads.

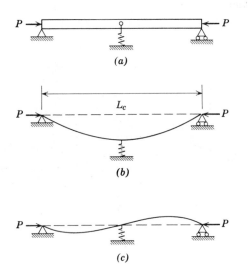

Fig. 9-21 Column with elastic lateral support.

Consider a pin-ended member supported laterally at its midlength by an elastic spring having a stiffness value k, as shown in Fig. 9-21a. The behavior of the member subjected to axial load P will depend largely on the stiffness of this spring as compared to the lateral stiffness of the column. If the spring is relatively weak, it can offer little resistance to lateral buckling and the column will buckle in one wave (Fig. 9-21b) with $L_e = L$,

$$P_{cr} = \frac{\pi^2 EI}{L_e^2} \tag{9-30}$$

On the other hand, if the spring is very stiff, it will prevent entirely lateral buckling at the support and the column will buckle in two waves (Fig. 9-21c), $L_e = \frac{1}{2}L$, and

$$P_{cr} = \frac{\pi^2 EI}{(L/2)^2} = \frac{4\pi^2 EI}{L^2} \tag{9-31}$$

which is four times as great as that obtained with a weak spring.

Therefore, it is apparent that, for any stiffness of elastic support, the critical load can be expressed by Eq. 9-30. Figure 9-22a gives the values of L_e to be used for one intermediate support. Similar reasoning applies to multiple supports. If the supports are all of the same elastic properties and equally spaced, the effective lengths for two supports are given in Fig. 9-22b and those for an infinite number of supports are given in Fig. 9-22c.

9-11 BUCKLING OF COMPRESSION MEMBERS IN FRAMES

The buckling strength of isolated compression members subjected to specified loading has been discussed in the preceding sections of this chapter. For simple idealized instances, the critical load can be determined using the effective length of member.

The determination of a realistic critical loading which will produce general instability in frames is much more difficult because of the following complications: (a) deformations of the individual members of the frame influence the distribution of the loads which in turn influences the stiffness of these members; (b) inelastic behavior associated with yielding or strain hardening; (c) in statically indeterminate systems buckling of an individual member does not lead to general instability. Nevertheless, theoretical formulation of the solution for the problem of general frame instability is possible[9,10] although actual solution depends on the availability of a sufficiently large computer to carry out the necessary numerical operations. In the future such general computer methods may become practicable, but meanwhile somewhat

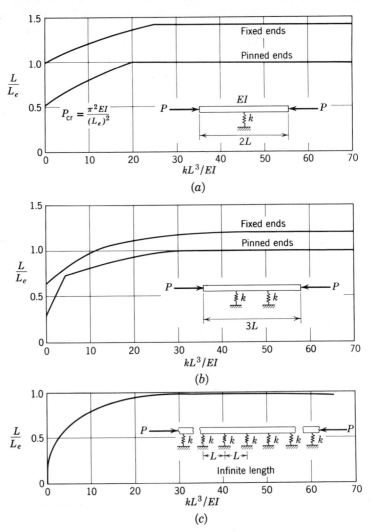

Fig. 9-22 Effective length of column on elastic supports.

approximate methods must be used which obtain varying degrees of accuracy depending on the simplicity of the model used for analysis.

When all columns in the portion of the framework considered reach their individual critical loads simultaneously, the critical load may be defined using an effective length $L_e = KL$ in the Euler equation. In a rectangular frame consisting of girders and columns, if the adjacent girders are rigidly attached to column ends A and B and if the equivalent stiffness of these girders are known, then the effective length factor $K = L_e/L$ can be determined.

An approximate method for determining the value of K for columns of constant section with ends partially restrained against rotation and fully restrained against translation has been proposed,[11] as follows:

$$K^2 = \frac{(\pi^2 m_A + 2)(\pi^2 m_B + 2)}{(\pi^2 m_A + 4)(\pi^2 m_B + 4)} \tag{9-31}$$

where m_A and m_B are dimensionless parameters of stiffness, defined by the equations

$$m_A = \left(\frac{EI}{L}\right)\left(\frac{1}{k_A}\right), \qquad m_B = \left(\frac{EI}{L}\right)\left(\frac{1}{k_B}\right) \tag{9-32}$$

Here k_A and k_B are the stiffnesses of the ends of column AB. If a moment M_{BA} is applied at the end B of column AB producing a rotation ϕ_A at A and inducing a moment M_{AB} at A, then the stiffness at A is defined as

$$k_A = \frac{M_{AB}}{\phi_A} \tag{9-33a}$$

Similarly, stiffness at B is defined as

$$k_B = \frac{M_{BA}'}{\phi_B'} \tag{9-33b}$$

when a moment M_{AB}' is applied at the end A and produces moment M_{BA}' and rotation ϕ_B' at B. Equation 9-31 has been shown to be valid for all values of m_A and m_B with a maximum error of about 4%.

An alternative solution can be obtained from a solution[3] of a transcendental equation in terms of K and relative stiffnesses G_A and G_B, which can be solved with the aid of alignment charts shown in Fig. 9-23. The value of G at given end A or B is defined as

$$G = \frac{\sum (I/L)_c}{\sum (I/L)_g} \tag{9-34}$$

in which Σ indicates a summation for all members rigidly connected to that joint and lying in the plane in which buckling of the column is being considered, I is the moment of inertia of the cross section taken about an axis perpendicular to the plane of buckling, L is the unsupported length of the member, and c and g are subscripts denoting column and girder, respectively.

To obtain a definitely conservative estimate of K, refinements in girder stiffness $(I/L)_g$ should be made when conditions at the far end of the particular

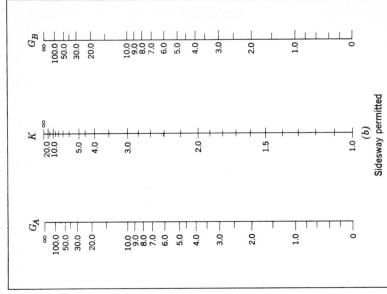

Sidesway prevented

(a)

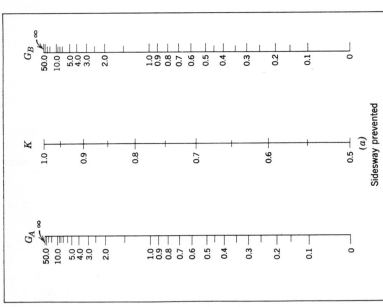

Sidesway permitted

(b)

Fig. 9-23 Charts for effective length of columns in continuous frames.

girder are known. These values are
multiplied by the following:

2.0—no sidesway, far end of girder
fixed against rotation.

1.5—no sidesway, far end of girder
hinged.

0.5—with sidesway, far end of girder
hinged.

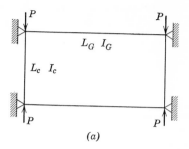

(a)

With values of G_A and G_B calculated,
the value of K can be obtained from the
charts (Fig. 9-23) and the effective length
L_e of the column thus determined.

For the column end not rigidly connected to the foundation, as in a "pinned"
end, the theoretical value of G is infinite.
However, unless actually designed and
constructed as a frictionless pin, the relative stiffness of the column end, for
practical purposes, may be taken as $G =$
10. If the column end is rigidly fixed to a
rigid foundation the theoretical value of G
is zero. In most practical instances, how-

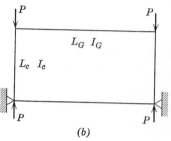

(b)

Fig. 9-24 Frame member buckling.
(a) Frame restrained against sidesway
(b) Frame unrestrained against sidesway.

ever, the relative stiffness of such a column
end may be taken as $G = 1.0$, although smaller values may be justified
occasionally.

The use of the charts is illustrated in the following simple examples.
Consider a rigid frame shown in Fig. 9-24*a*, which is restrained against sidesway and will buckle into the shape represented by the solid lines. If $I_g = 2I_c$
and $L_g = 2L_c$, then $G_A = 1.0$. If the column base is rigidly connected to a
substantial footing which will prevent any significant rotation of the foundation, then $G_B = 1.0$. From the chart (Fig. 9-23), $K = 0.78$ and $L_e = 0.78L_c$.
If the same rigid frame is not prevented from sidesway, that is, it is dependent
upon its own bending stiffness for stability against sidesway, it will buckle
into the shape represented by solid lines in Fig. 9-24*b*. For $G_A = G_B = 1.0$,
from the chart (Fig. 9-23), $K = 1.31$ and $L_e = 1.31L$. Another illustration
is given in Example 9-1.

Example 9-1. For the steel frame shown in Fig. 1, determine the effective
length of columns C_2, C_3, and C_6. All columns are 8 W̄ 31 and all girders
are 16 W̄ 58 of A36 steel. All members are used with webs parallel to the
plane of the drawing and all columns are assumed to be adequately braced

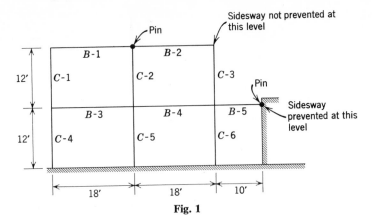

Fig. 1

in the weak direction. Girders B-2 and B-5 have flexible connections at one end, which may be considered hinged, as shown in the figure.

Solution

(I/L) Values.

Girders: B-1 = B-2 = B-3 = B-4

$$\frac{I}{L} = \frac{746.4}{216} = 3.45$$

$$B\text{-}5 \qquad \frac{I}{L} = \frac{746.4}{120} = 6.22$$

Columns: C-1 = C-2 = C-3 = C-4 = C-5 = C-6

$$\frac{I}{L} = \frac{109.7}{144} = 0.76$$

Note: u and l subscripts denote upper and lower ends of members.

Column C-2:

$$G_u = \frac{\sum (I/L)_c}{\sum (I/L)_g} = \frac{0.76}{3.45} = 0.22$$

$$G_l = \frac{\sum (I/L)_c}{\sum (I/L)_g} = \frac{0.76 + 0.76}{3.45 + 3.45} = 0.22$$

From the figure

$$K = 1.06, \qquad L_e = KL = 1.06 \times 120 = 127.2 \text{ in.}$$

Column C-3:

$$G_u = \frac{\sum (I/L)_c}{\sum (I/L)_g} = \frac{0.76}{0.5 \times 3.45} = 0.44$$

$$G_l = \frac{\sum (I/L)_c}{\sum (I/L)_g} = \frac{0.76 + 0.76}{3.45 + 1.5(6.22)} = 0.19$$

From the figure:

$$K = 1.1, \qquad L_e = KL = 1.1 \times 120 = 132 \text{ in.}$$

Column C-6:

$$G_u = \frac{\sum (I/L)_c}{\sum (I/L)_g} = \frac{0.76 + 0.76}{3.45 + 1.5(6.22)} = 0.19$$

$G_l = 1.0$ (column rigidly attached to a properly designed footing)

From the figure:

$$K = 0.675, \qquad L_e = KL = 0.675 \times 120 = 81 \text{ in.}$$

Depending on the relative stiffness of the girder to the columns and the type of base support, the effective length of column may exceed two times its actual length. However, if the frame were braced against sidesway by a system of diagonals, the effective length would be less than the actual length because the girder and bracing members would provide resistance to joint rotation, thus producing some degree of fixity at the ends of the columns. Although the effective length may be less than the actual length in theory, it is recommended as good design practice to use no less than the actual length in the column formulas.

The type of column base connection to the foundation will influence the degree of restraint against rotation. Although a simple base connection may be assumed to have pin action, the flat end of the column under vertical load does exert some restraining effect.

Consideration should also be given to the types of soil on which the footing rests and the likelihood of differential soil settlement under the footing producing a pin effect at the base of the column.

In the past, in multistory frame structures the existence of heavy masonry walls combined with similar interior partitions and the standard moment connections of beams to columns provided sufficient lateral support for tier buildings to prevent lateral movement (sidesway). The current use of light curtain walls, the omission of interior partitions, and wider column spacings produce a situation where the lateral movement must be resisted by the bending stiffness of the frame itself.

Entering the appropriate chart for sidesway, with the quantities of G for the top and bottom of the column, the effective length may be read directly. As in all solutions by trial, a second design may have to be verified in a similar manner using the newly determined sections for columns and girders.

This procedure is applicable to frames without large end moments acting on the columns. When the design of a building frame is based upon the effect of a large lateral loading, or drift, the effective column length may be taken as the actual length between floors. This is possible because the lateral loading causes end moments in the columns which produce a reverse curvature to the column and thus an inherent rotation at the floors which tends to restrict the build-up of a large effective length.

In truss frames made up of triangulated panels, loads are usually applied at the panel points or joints of the members and axial stresses are induced in all members. Because most trusses are welded or bolted at the joints, instead of being pinned, secondary bending is also induced at the joints. These bending distortions are usually small and do not affect the buckling analysis. Therefore the problem of buckling reduces to one of a compression member partially restrained against rotation at both ends. Usually, lateral bracing frames are provided in the plane of the compression chord, and therefore transverse movements at the ends are negligible. When such bracing is not provided, the stiffness of transverse restraint at the end must be considered.

If the truss has been so designed that the compression and tension members will reach their maximum capacities at about the same loading (for fixed static loads), the compression members should be designed with an effective length equal to the distance between the panel points. Where continuity of the compression chord of constant cross section exists over the entire span of the truss, a reduced length may be used and K may be taken as 0.9. For a member adjacent to the panel point where the chord stresses change from compression to tension, K may be taken as 0.85 provided the adjacent tension and compression members are continuous and their length and cross sections are similar.

In some instances, the magnitude of the load changes at a subpanel point, that is, at a joint which is not laterally braced. Examples are shown in Fig. 9-25. In such cases,[3] effective length factor may be taken as

$$K = 0.75 + 0.25 \frac{P_2}{P_1} \qquad (9\text{-}35)$$

where P_2 is less than P_1 if both are compressive, and P_2 is negative when tensile.

Web compression members are designed with $K = 1$ if the truss is subjected to fixed loads. When these members are designed on the basis of moving live loads and the critical loading for a particular web member does not produce maximum stress intensities in adjacent members, the effective length may be reduced using $K = 0.85$.

When lateral bracing frames in the plane of compression members are omitted and bracing is provided by flexible cross frames, the influence of such elastic supports on the buckling of the compression member must be considered. The approximate values of effective length given in Sec. 9-10 may be used if the stiffness of lateral supports is known.

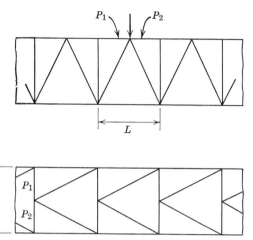

Fig. 9-25   Truss frames with subdivided panels.

9-12   LATERAL BUCKLING OF BEAMS

The compression flange of a beam subjected to bending may be considered as a column subjected to an axial force, which, depending on the type of loading, may be constant or varying along the span (Fig. 9-26). The tension flange of the beam tends to remain straight and restrains the compression flange from buckling transversely as a free column. When the critical value of the bending moment is reached, however, the compression flange will buckle laterally, causing lateral bending and twisting of the beam (Fig. 9-26). Lateral buckling is accompanied by twisting of the section and it is apparent that the torsionally weak shapes in which the flexural rigidity of the beam in the plane of bending is significantly greater than its lateral rigidity will buckle more readily than torsionally rigid ones. Thus boxed shapes do not usually buckle laterally before full plastic bending strength is realized.

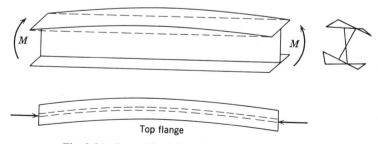

Top flange

Fig. 9-26   Lateral buckling of beam: constant load.

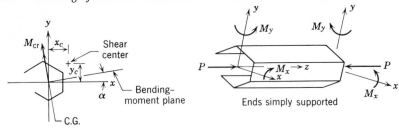

Fig. 9-27 Thin-walled section under bending.

The value of the critical bending moment M_{cr} depends on the properties of material, shape and dimensions of the beam, length of span, support conditions, and type of loading. The determination of this critical moment for the general case of arbitrary beam cross section, type of loading, and condition of support is mathematically complex. The problem is simplified if it is restricted to pure bending of simply supported beams of relatively thin-walled open, that is, not boxed, shapes.

The theoretical expression for the critical moment M_{cr} for lateral buckling of a simply supported thin-walled open section subjected to pure bending (Fig. 9-27) has been derived[5] on the basis of the following assumptions:

(a) The cross section and the bending moment are constant along the length of the beam.

(b) The stresses nowhere exceed the proportional limit.

(c) The external loads, hence also the plane of bending, remain parallel to the original direction when the points of application of these loads are displaced.

(d) The distortion of cross section is neglected.

The value of M_{cr} can be obtained as a root of the following equation:

$$(P_x \sin^2 \alpha + P_y \cos^2 \alpha)M_{cr}^2 + P_x P_y(y_0 \sin \alpha + x_0 \cos \alpha)M_{cr}$$
$$- P_x P_y P_z r_z^2 = 0 \quad (9\text{-}36)$$

In Eq. 9-36, x and y denote the principal centroidal axes of the section, z is the longitudinal centroidal axis, α is the inclination of the bending-moment plane with respect to the x axis so that $M_x = M_{cr} \sin \alpha$ and $M_y = M_{cr} \cos \alpha$, I_x and I_y are the principal moments of inertia of the cross section, $r_z = \sqrt{I_z/A}$ is the radius of gyration of the polar moment of inertia, $I_z = I_x + I_y$, x_c and y_c are the coordinates of the section's shear center with due regard to signs, and x_0 and y_0 are the coordinates of a point characteristic of the cross-section shape, defined in Eq. 9-20 as follows:

$$x_0 = -\int_A \frac{x(x^2 + y^2)\,dA}{I_y} + 2x_c \quad \text{and} \quad y_0 = -\int_A \frac{y(x^2 + y^2)\,dA}{I_x} + 2y_c$$

P_x and P_y are critical Euler loads, that is, $P_x = \pi^2 EI_x/L^2$ and $P_y = \pi^2 EI_y/L^2$; and P_z is the critical torsional buckling load, defined in Eq. 9-17, as follows:

$$P_z = \frac{1}{r_z^2}\left(\frac{\pi^2 E K_b}{L^2} + GK_t\right) = \frac{GK_t}{r_z^2}\left(1 + \pi^2 \cdot \frac{a^2}{L^2}\right)$$

E is Young's modulus; L is the effective span length distance between supports for a simply supported beam; G is the shearing modulus, $G = \frac{1}{2}E/(1 + \mu)$, μ is the Poisson's ratio; K_t is the torsional constant, that is, $K_t = \frac{1}{3}\Sigma\, bt^3(1 - 0.63t/b)$; and K_b and a are the torsion-bending constants defined in Sec. 8-7. For special cases Eq. 9-36 can be greatly simplified and several of them follow.

Case 1. Symmetrical Section, Two Axes of Symmetry. A typical example of such a section is a W beam shown in Fig. 9-28a. In this case $x_0 = y_0 = 0$, and Eq. 9-36 becomes

$$(P_x \sin^2 \alpha + P_y \cos^2 \alpha)M_{cr}^2 - P_x P_y P_z r_z^2 = 0 \tag{9-37}$$

(a) *Bending in One Principal Plane Only:* $\alpha = \pi/2$, $M_y = 0$, $M_x = M_{cr}$.

$$M_{cr}^2 - P_y P_z r_z^2 = 0$$
$$M_{cr} = \pm(P_y P_z r_z^2)^{1/2} \tag{9-38}$$

(b) *Bending due to Moment M in a Plane Inclined at an Angle α with the x Axis.*

$$M_{cr} = \pm\left(\frac{P_x P_y P_z r_z^2}{P_x \sin^2 \alpha + P_y \cos^2 \alpha}\right) \tag{9-39}$$

Case 2. Antisymmetrical Section. An antisymmetrical section is one in which every element of area has an equal element at an equal distance on the opposite end of the diameter passing through the centroid of the section, for example, a Z section with equal flanges (Fig. 9-28b). For this section $x_0 = y_0 = 0$ and Eqs. 9-37, 9-38, and 9-39 apply.

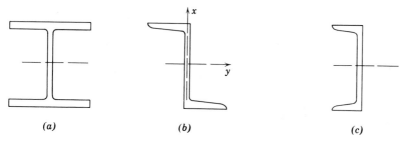

(a)   (b)   (c)

Fig. 9-28

Case 3. Symmetrical Section, One Axis of Symmetry. A typical example of such a section is a standard channel (Fig. 9-28c) symmetrical about x axis, $y_0 = 0$.

(a) *Bending about Axis of Symmetry:* $\alpha = \pi/2$. $\sin \alpha = 1.0$, $\cos \alpha = 0$, $M_y = 0$, $M_x = M_{cr}$. Equation 9-38 applies.

(b) *Bending about Principal Axis Perpendicular to Axis of Symmetry:* $\alpha = 0$.

$$\sin \alpha = 0, \qquad \cos \alpha = 1.0$$

$$M_{cr} = \tfrac{1}{2}[P_x x_0 \pm (P_x^2 x_0^2 + 4P_x P_z r_z^2)^{\frac{1}{2}}] \tag{9-40}$$

In this case two critical moments are obtained, one positive and one negative, different in magnitude, depending on the sign used with the second term in the brackets.

For *a narrow rectangular beam*, $K_b = 0$, and Eq. 9-7 can be simplified as follows:

$$M_{cr} = (P_x P_z r_z^2)^{\frac{1}{2}} = \frac{\pi}{L}(EI_y GK_t)^{\frac{1}{2}} \tag{9-41}$$

For *an I beam*, similarly:

$$M_{cr} = \frac{\pi^2 E}{L}\left[I_y\left(\frac{K_b}{L^2} + \frac{GK_t}{\pi^2 E}\right)\right]^{\frac{1}{2}} = \frac{\pi}{L}\left[EI_y GK_t\left(1 + \pi^2 \frac{a^2}{L^2}\right)\right]^{\frac{1}{2}} \tag{9-42}$$

Substituting $G = \tfrac{1}{2}E/(1 + \mu)$, $a = (EK_b/GK_t)^{\frac{1}{2}}$ and $K_b = (d^2/4)I_y$ (the approximate value for narrow I beams) into the foregoing, the following expression is obtained:

$$M_{cr} = \frac{\pi^2 EI_y}{L^2}\frac{d}{2}\left[1 + \frac{2}{1 + \mu}\left(\frac{L}{\pi d}\right)^2 \frac{K_t}{I_y}\right]^{\frac{1}{2}} \tag{9-43}$$

This can be further simplified for I-shaped beams by approximating:

$$I_y = 2\left(\frac{tb^3}{12}\right) = \frac{tb^3}{6} \tag{9-44}$$

$$K_t = 2\left(\frac{bt^3}{3}\right) \tag{9-45}$$

and

$$S_x = \frac{I_x}{\tfrac{1}{2}d} = bt\,d \tag{9-46}$$

Then using these approximations, the critical buckling stress F_{cr} is:

$$F_{cr} = \left(\frac{M_{cr}}{S_x}\right) = \left\{\left[\frac{\pi^2 E}{12(L/b)^2}\right]^2 + \left[\frac{\pi E}{\sqrt{18(1 + \mu)}}\frac{bt}{Ld}\right]^2\right\}^{\frac{1}{2}} \tag{9-47}$$

For deep girders with wide flanges the value of the second term in Eq. 9-47 is much smaller than the first term and may be neglected. In this case,

$$F'_{cr} = \frac{\pi^2 E}{12(L/b)^2} \tag{9-48}$$

Equation 9-48 is a simple Euler equation for the lateral buckling of the compression flange, neglecting the stiffening effect of the tenison flange and of the pure torsion of the section.

For relatively long, shallow, thick-walled beams, weak in torsion-bending, the value of the first term in Eq. 9-47 is much smaller than the second term and may be neglected. In this case:

$$F''_{cr} = \frac{\pi E}{\sqrt{18}} \frac{bt}{Ld} \sqrt{\frac{1}{1 + \mu}} \tag{9-49}$$

By using $E = 29.5$ times 10^6 psi and $\mu = 0.25$,

$$F_{cr} = \frac{19.5 \times 10^6}{Ld/bt} \tag{9-50}$$

Equation 9-50 takes into account the torsional stiffness of the beam but essentially neglects the effective stiffness of the compression flange acting as an independent column. Conventional proportions of rolled W$^{\rm F}$ beams are such that Eq. 9-50 applies in numerous cases.

The most significant of the limitations of Eq. 9-50 are the following.

(*a*) The beam is assumed to be subjected to pure bending and no allowance is made for varying conditions of loading and end support.

(*b*) Approximations of I_x, I_y, and K_t used in the derivation of Eq. 9-47 are limited to typical I shapes, but they are not valid for deep beams with small flanges.

(*c*) The critical stresses are derived assuming that all stresses are within the elastic limit and that stresses are proportional to strains. Thus computed values of f_{cr} exceeding elastic limit are not valid. In such instances, inelastic lateral buckling may occur after the beam has started yielding. This problem becomes important in structures designed on the basis of plastic strength and in such cases buckling stress may be determined by numerical methods.[2]

For *loading conditions other than pure flexure* the values of the critical maximum bending moment M_{cr} is somewhat larger than for pure flexure. For example, for a simply supported W$^{\rm F}$ beam subjected to a concentrated load W at midspan the critical bending moment $M_{cr} = WL/4$ is about 34% greater than that given by Eq. 9-42, provided the load is applied at the centroidal axis. For a simply supported W$^{\rm F}$ beam under a uniformly distributed load w, the critical bending moment $M_{cr} = wl^2/8$ is about 12% greater than that given by Eq. 9-42, provided the load is applied at the centroidal axis.

When the loads are not applied at the centroidal axes, the critical bending moment decreases when the load is applied on (or near) the compression flange and increases when the load is applied on (or near) the tension flange. The increased or decreased moment may be measured in terms of the critical value corresponding to centroidal axis loads. It is found that the ratio η defined as

$$\eta = \frac{\text{critical moment under given loading}}{\text{critical moment under centroidal loading}}$$

varies with the nature of loading and ratio L/a, where a is $(EK_b/GK_t)^{\frac{1}{2}}$. The typical values of η in Table 9-1 are based on the Timoshenko solution:

Table 9-1   Values of η for Simply Supported W Beams

Values of L/a		2	4	10	20
Concentrated Load	At top flange	0.63	0.71	0.85	0.92
	At bottom flange	1.57	1.39	1.17	1.08
Uniformly Distributed Load	At top flange	0.69	0.76	0.88	0.93
	At bottom flange	1.46	1.33	1.14	1.07

These relations apply to lateral buckling of simply supported beams. The critical value of maximum bending moment for a cantilever W beam of span L is about 40 to 65% greater, depending on the ratio L/a, than the critical moment for the beam under constant moment when simply supported. End fixity and intermediate restraints also increase the value of the critical maximum moment by an amount varying from 50 to 200%, depending on the nature of restraint and type of loading. A more detailed analysis of the lateral buckling of beams is beyond the scope of this text.

9-13 BUCKLING OF THIN PLATES

Steel structural shapes which effectively transmit compressive or bending loads have sections which consist of relatively thin elements. In the preceding discussion of buckling it has been assumed that no local instability would develop before the member buckles as a whole. Local buckling, if permitted to occur, may reduce significantly the load-carrying capacity of the member.

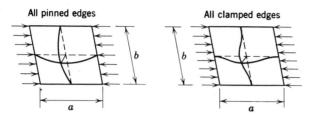

Fig. 9-29   Elastic buckling of thin rectangular plate.

For this reason it is necessary to consider performance of thin plate elements under various stresses.

If a rectangular flat plate is subjected to uniform compression it will buckle when a critical stress f_{cr} is reached (Fig. 9-29). The magnitude of the elastic buckling stress f_{cr} may be expressed as follows:

$$f_{cr} = \frac{C\pi^2 E_p}{(a/r)^2} = \frac{C\pi^2 E}{12(1 - \mu^2)}\left(\frac{t}{a}\right)^2 \tag{9-51}$$

where C is a coefficient which depends on edge restraints and aspect ratio a/b of the plate, E_p is equivalent elastic modulus of a plate in bending and is equal to $E/(1 - \mu^2)$, E is the usual elastic modulus, μ is the Poisson ratio, a is the length of the plate, r is the radius of gyration of the section about centroidal axis in the plane of the plate which is equal to $r = t/\sqrt{12}$, and t is plate thickness.

A long rectangular plate (Fig. 9-30) supported on all four edges will buckle into a number of waves so that the length of each wave approximately equals the width of the plate b, and the magnitude of the buckling stress is much more sensitive to changes in width b than to changes in over-all length a. Therefore it is convenient to define the critical buckling stress as

$$f_{cr} = \frac{C\pi^2 E}{12(1 - \mu^2)}\left(\frac{t}{a}\right)^2 = \frac{k\pi^2 E}{12(1 - \mu^2)}\left(\frac{t}{b}\right)^2 = k_c E \left(\frac{t}{b}\right)^2 \tag{9-52}$$

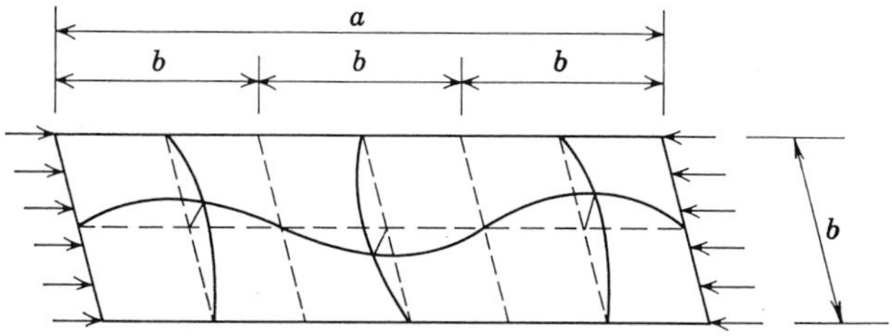

Fig. 9-30   Buckling of long rectangular plate.

where k_c depends on edge restraints, aspect ratio a/b, and Poisson ratio μ. When the aspect ratio $a/b \geq 3.0$, the value of k_c varies only slightly with a/b and it may be assumed a constant for given edge boundary conditions. Theoretical values of k_c for typical idealized edge conditions and for $\mu = 0.3$ are given in Table 9-2.

Table 9-2   Values of k_c for Long Rectangular Thin Plates Subjected to Uniform Compression

Unloaded Edge Conditions	k_c
Both edges simply supported	3.6
One edge simply supported, other fixed	4.9
Both edges fixed	6.3
One edge simply supported, other free	0.38
One edge fixed, other free	1.15

The load-carrying ability of a specific plate geometry may be increased considerably by simply fixing or stiffening the longitudinal sides of the plate.

In structural design it is advisable to have the dimensions of the individual plates of such a size that the buckling of the plate does not occur until the buckling stress is at the yield point stress of the material. This condition may be stated as follows:

$$f_{\text{cr}} = k_c E \left(\frac{t}{b}\right)^2 \geq F_y \qquad (9\text{-}53a)$$

or

$$\frac{b}{t} \leq \sqrt{\frac{k_c E}{F_y}} \qquad (9\text{-}53b)$$

The ratio of b/t will decrease with an increase in F_y, and therefore an element with a given width b using a steel with larger value of F_y will have to employ a larger thickness t to meet the requirement of Eq. 9-53b. Thus in this case possible economy of reducing steel area with the use of high yield strength steel would be impossible. However, if a stress limit f_m other than F_y should be set to prevent local buckling, it may be sufficient to limit b/t to

$$\frac{b}{t} \leq \sqrt{\frac{k_c E}{f_m}} \qquad (9\text{-}54)$$

The problem of setting limitations on width-thickness ratio is complicated by the inelastic behavior of steel plates, initial imperfections, and presence of residual stresses. The initial imperfections and residual stresses are greatly influenced by the manufacturing and fabrication of the structural elements.

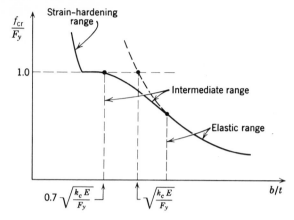

Fig. 9-31   Elastic and inelastic buckling of thin plate in compression.

Test results show[13] that buckling of steel plates under compression can occur in three ranges: elastic, strain-hardening, and intermediate ranges, as shown in Fig. 9-31. It is seen that due to imperfections and residual stresses the criterion of Eq. 9-52 would lead to an unconservative design, that is, one where buckling could occur at stresses significantly below yield.

Empirical evidence suggests that to achieve stable behavior of rectangular plates under compression up to yield strength F_y the width-thickness ratio b/t must not exceed $0.7\sqrt{k_c E/F_y}$. Also, the values of k_c must be taken to represent the actual edge restraint condition, which may fall somewhere between the theoretical values obtained for idealized conditions.

For typical shapes, shown in Fig. 9-32, the following values of k_c have been

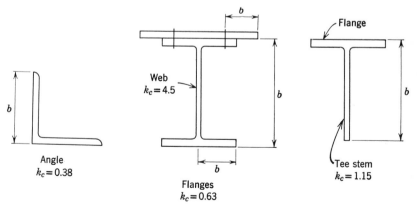

Fig. 9-32   Width of plate elements in typical structural shapes.

found to give results in reasonable agreement with experimental data.

Element	Angle	Flange	Stem of Tee	Web
k_c	0.38	0.63	1.15	4.5

These values were used to establish limits on b/t which are specified by the 1963 AISC Specification, Sec. 1.9 for elastic design.

Inelastic plate buckling under uniform compression may occur at stresses above proportional limit. An approximate magnitude of the critical stress in the inelastic range can be obtained from Eq. 9-53 by substituting effective modulus $\bar{E}$ for the elastic modulus E. This effective modulus can be taken as tangent modulus E_t or $\sqrt{EE_t}$, with the latter usually given as a better approximation.

Values of the effective modulus $\bar{E}$ depend on the stress-strain characteristics of the particular material, the intensity and state of stress, the distribution of stress, the shape of cross section, and the presence or absence of residual stresses. Effective modulus $\bar{E}$ may be expressed as a function of initial modulus E and the ratio of f_{cr}/F_y. The curve in Fig. 9-33 may be used to obtain an approximate value of $\bar{E}$ for structural steel.

In plastic design it is not sufficient that stresses in plate elements merely reach yield strength without local buckling. In order to ensure that the section develops adequate rotation capacity, large post-yield strains must be developed

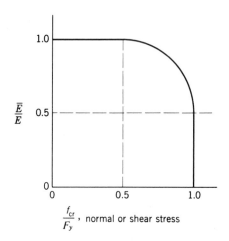

Fig. 9-33 Effective modulus $\bar{E}$ for Eq. 9-53. *Note:* Shear yield stress $= F_{vy}$.

$$F_{vy} = \frac{1}{\sqrt{3}} F_y.$$

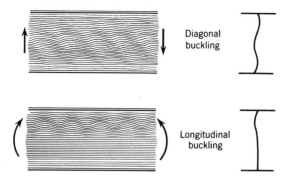

Fig. 9-34   Buckling of thin web.

without local buckling. Analytical and experimental studies show that such strains can be attained under static loads provided that $\sqrt{F_y/f_{\mathrm{cr}}}*$ does not exceed limit values ranging from 0.45 to 0.60 depending on the type of edge conditions. Also, in this relationship $f_{\mathrm{cr}}*$ is the buckling stress in the strain-hardening range and is a complex function of b/t, edge restraint, E, and μ. Therefore the relationships based on Eq. 9-53 would not be applicable and the provisions of the 1963 AISC Specification, Sec. 2.6 take the plastic behavior into account.

Thin web elements can buckle when subjected to bending stresses or to pure shear stresses (Fig. 9-34). The buckling stress under these loading conditions can be obtained from the following equations:

$$f_{v,\mathrm{cr}} = k_v E\left(\frac{t}{b}\right)^2 \tag{9-55}$$

and

$$f_{b,\mathrm{cr}} = k_b E\left(\frac{t}{b}\right)^2 \tag{9-56}$$

where $f_{v,\mathrm{cr}}$ and $f_{b,\mathrm{cr}}$ are shear and bending critical stresses, respectively; k_v and k_b are coefficients for shear and bending, Table 9-3, respectively, depending on edge conditions, aspect ratio, and Poisson ratio; b is the width of a long rectangular plate (short side); and t is plate thickness.

In standard rolled-steel shapes the depth-thickness ratio of the web is such that the plates will not buckle under the action of transverse loads. In built-up beams and plate girders thin webs may be subject to elastic buckling under bearing stresses and buckling of a plate loaded by compression stresses on one edge only must be considered (Fig. 9-35).

In an unstiffened web, as in Fig. 9-35*a*, provided the flange is braced against lateral buckling, two modes of web buckling are possible, depending on torsional rigidity of the flange or external restraint of its rotation. The critical compression stress is defined by:

$$f_{cr} = k_w E \left(\frac{t}{h}\right)^2 \tag{9-57}$$

where k_w is 1.8 for mode 1 and 5.0 for mode 2.

Table 9-3   Theoretical Values of k for Long Rectangular Thin Plates Subjected to Shearing or Bending Stresses

	k_v*	k_b
1. All edges simple supported	4.8	21.5
2. All edges clamped	8.1	35.7

* For plates with aspect ratio $(a/b)1.0$ the values of k_v can be approximated for condition 1 as $4.8 + 3.6(b/a)^2$, and for condition 2 as $8.1 + 5(b/a)^2$.

In a stiffened web (Fig. 9-35*b*), where the stiffeners are assumed nonbearing and are provided only to increase the stability of the web, the critical stress is difficult to determine. An approximate value for the critical stress in this case has been suggested:

$$f_{cr} = k_w E \left(\frac{t}{h}\right)^2 + 3.6E \left(\frac{t}{a}\right)^2 \tag{9-58}$$

where k_w has the same values as in Eq. 9-57.

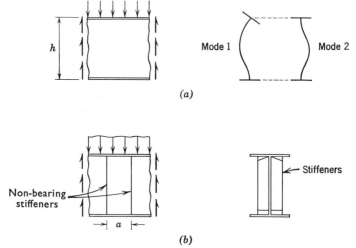

(a)

(b)

Fig. 9-35   Unstiffened and stiffened thin web. (*a*) Unstiffened web. (*b*) Stiffened web.

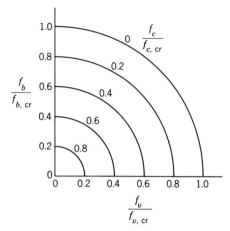

Fig. 9-36 Buckling under combined stress by Eq. 9-59.

Combinations of shearing, bending, and bearing compression stresses may occur at points adjacent to local concentrated loads. Buckling under action of combined stresses can be closely approximated by empirical interaction criteria. For flat-plate elements the following equation is suggested:

$$\frac{f_c}{f_{c,\mathrm{cr}}} + \left(\frac{f_b}{f_{b,\mathrm{cr}}}\right)^2 + \left(\frac{f_v}{f_{v,\mathrm{cr}}}\right)^2 \le 1.0 \qquad (9\text{-}59)$$

where subscripts c, b, and v, denote bearing, bending, and shear, respectively, and cr denotes critical buckling stress under the action of one type of loading only. The graphical solution of Eq. 9-59 is shown in Fig. 9-36.

9-14 POSTBUCKLING STRENGTH OF FLAT PLATES

Initial buckling of edge-supported plates does not result in their collapse. In this respect buckling of restrained plates (Fig. 9-37a) is different from that of slender rods or unrestrained plates. Initial buckling of a slender rod or unrestrained plate leads to its collapse very quickly because its lateral deflection is unrestrained. In an edge-supported plate, the lateral deflections due to buckling are partially restrained by transverse bending of the plate between the edge supports. In a buckled plate, the central strip A can deflect laterally without appreciable increase in load, but the supported edge strip at B remains straight and is capable of supporting additional load approximately up to yield stress of the material.

As the load gradually increases, the stress distribution over the width of plate with restrained edges varies (Fig. 9-37b). At low intensities the stress

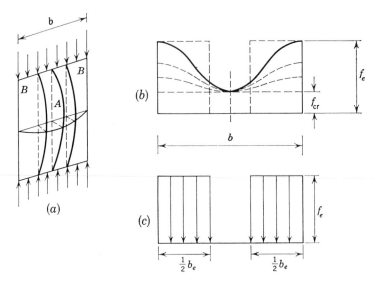

Fig. 9-37   Postbuckling strength of plates. (*a*) edge-supported plate, (*b*) stress distribution, and (*c*) effective width.

is distributed uniformly until it reaches the value of f_{cr}, the initial buckling stress of the plate. As the load increases, the central strip cannot carry additional stress, but the edge strips can. Therefore any increase in load beyond initial buckling will result in increased stresses in the edge strips but not in the central strip. Resulting stress distribution is shown in Fig. 9-37*b*. Load-carrying capacity, known as post-buckling strength, is reached when the stress f_e in the edge strips reaches some maximum value at which the entire plate deforms rapidly with little or no increase in load. For plate elements of a relatively short column, this maximum stress f_e may be compressive yield strength of the material and, for elements of long slender columns, this may be the critical column stress of the edge stiffener which may be considerably less than the yield value.

Exact theoretical determination of the compressive strength would have to account for large deflections, inelastic behavior of the material, variations introduced by eccentricities of loading, and slight irregularities in flatness of the plate. The complexity of these variables makes a purely mathematical analysis impractical. Therefore approximate solutions supported by test results[14,15] are used for practical design. The concept of "effective" width is based on replacing the plate width *b*, subjected to a varying stress distribution across the width (Fig. 9-37*c*), by an equivalent width of plate b_e, in which the total load *P* is carried by a uniformly distributed stress equal to the edge stress. The effective width b_e may be approximated as that width of the plate which just buckles when the compressive stress reaches the value f_e. Thus

$$f_e = k_c E \left(\frac{t}{b_e} \right)^2 \quad \text{or} \quad b_e = Ct \sqrt{\frac{E}{f_e}} \qquad (9\text{-}60a)$$

where $C = \sqrt{k_c}$ and postbuckling strength P_u is

$$P_u = f_e b_e t \qquad (9\text{-}60b)$$

For a long thin steel plate, pinned at all four edges, the theoretical value of C is 1.9. In actual structural members, the degree of restraint at edge supports is difficult to evaluate and in many instances the most reliable value of C must be obtained experimentally.

The use of post buckling strength is more applicable to thin plates or sheets than to the thicker plates usually required in heavy structures. However, the 1963 AISC Specification does recognize the phenomenon and makes provision for it as noted in Sec. 1.9.1, that is, "when a projecting element exceeds the width-to-thickness ratio prescribed, but would conform to same and would satisfy the stress requirements with a portion of its width considered as removed, the member will be acceptable." In the design criteria for stiffened thin web plate girders the postbuckling strength of plates in shear and in bending has been considered and is included in the AISC Specificatiors Sec. 1.10, as discussed in Chapter 11.

9-15 LOCAL BUCKLING OF TUBULAR MEMBERS

In using steel tubes or pipes for structural members two considerations may be of importance. First, local buckling should be prevented at stresses below yield strength, and second, a more severe restriction, is that the tendency to buckle locally should not reduce general buckling load of a tubular member.

Local buckling stress of a cylindrical shell with thin walls under uniform compression (Fig. 9-38) can be determined theoretically. Under ideal conditions this stress is

$$f_{cr} = kE \left(\frac{t}{R} \right) \qquad (9\text{-}61)$$

where R is mean radius, t is wall thickness, and k is 0.6. However, tests indicate that tubes can actually develop only a fraction of this stress because buckling of a cylindrical tube is highly sensitive to initial imperfections.[3] Also, imperfections resulting from fabrication indentations, joint seams, and similar disturbances can greatly reduce the buckling stress. Even for seamless round tubes, a more realistic estimate of local buckling stress is obtained by using $k = 0.12$ in Eq. 9-61.

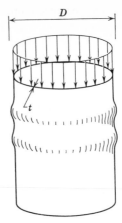

Fig. 9-38   Buckling of a cylindrical shell.

On this basis the ratio t/R should be greater than $F_y/0.12E = F_y/3,500,000$ in order to prevent local buckling at stresses below yield. This would limit the D/t ratio to no greater than $7,000,000/F_y$. Actually in most structural tubes the D/t ratio is limited to values considerably below this limit. Assuming $F_y = 33,000$ psi Wilson suggested a limit for D/t of 135, and most standard pipe sections limit D/t to $1.8\sqrt{E/F_y}$, which for $F_y = 33,000$ psi results in $D/t \leq 54$. American Iron and Steel Institute specifies a limit for D/t ratio of $3,300,000/F_y$, which is half that obtained from test results on carefully prepared specimens. One of the reasons for the substantially lower limits on D/t of practical structural shapes than would be required to prevent local buckling under ideal loading conditions is to obtain the radial rigidity necessary to minimize local damage in handling and in connections

PROBLEMS

1. Derive expressions for maximum stress in columns loaded as shown in Fig. P-1.

2. Plot a curve of critical buckling stress F_{cr} versus slenderness ratio L/r for a rolled-shape steel column. Due to residual stresses the stress–strain relationship in a steel column deviates from ideally linear elastic behavior as indicated by experimental results obtained from the concentric axial compression test, as follows:

Average stress, ksi	5	10	15	20	25	30
Average strain per 10^3	0.167	0.333	0.60	0.90	1.50	2.50

Use tangent-modulus modification of Euler's criterion.

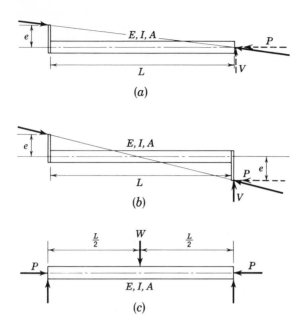

Fig. P-1

3. A series of 4 W 13 steel columns of different lengths were tested to determine axial compression strength. The plot of average failure stress P/A against the slenderness ratio L/r indicated that an empirical equation of the form P/A max $= A - B(L/r)^2$ would fit the data best. Typical test results are: $L_1 = 8'-3''$ and maximum $P_1 = 83.2$ kips; $L_2 = 3'-4''$ and maximum $P_2 = 111$ kips. Determine the empirical coefficients A and B in the foregoing equation.

4. Assuming that the empirical curve obtained in Prob. 6 is tangent to an Euler curve $C\pi^2 E/(L/r)^2$ at some value of $L/r = X$, determine this value of L/r and the end-fixity coefficient C characteristic of these column tests. Assume that $E = 29 \times 10^3$ ksi.

REFERENCES

1. Timoshenko, S., and J. Gere, *Theory of Elastic Stability*, McGraw-Hill Book Co., New York, 1961.
2. Bleich, F., *Buckling Strength of Metal Structures*, McGraw-Hill Book Co., New York, 1962.
3. Johnston, B. G., *Guide to Design Criteria for Metal Compression Members*, Column Research Council, 2nd ed. John Wiley and Sons, New York, 1966.
4. Huber, A. W., and L. S. Beedle, "Residual Stress and the Compressive Strength of Steel," *Welding J. Research Suppl.* (December 1954.)

5. Timoshenko, S., "Theory of Bending, Torsion, and Buckling of Thin-Walled Members of Open Cross-Sections," *J. Franklin Inst.* **239**, Nos. 3, 4, and 5 (1945).

6. Newmark, N. M., "Numerical Procedure for Computing Deflections, Moments, and Buckling Loads," *Trans. ASCE* **108** (1943).

7. Laurenson, R., and R. V. Bremer, "Critical Collapsing Load for Columns of Variable Cross-Section," *Civil Engineering* **18**, April 1948.

8. Tu, Shou Ngo, "Column with Equal Spaced Elastic Supports," *J. Aeronaut. Sci.* **11**, No. 1 (Jan. 1944).

9. Livesley, R. K., *Matrix Methods of Structural Analysis*, Pergamon Press, New York, 1964.

10. Davies, J. M., "Frame Instability and Strain Hardening in Plastic Theory," *J. Structural Div. Proc. ASCE* **92**, ST3 (June 1966).

11. Newmark, N. M., "A Simple Approximate Formula for Effective End Fixity of Columns," *J. Aeronaut. Sci.* **16**, No. 2 (Feb. 1949).

12. Galambos, T. V., "Inelastic Lateral Buckling of Beams," *J. Structural Div. Proc. ASCE* **89**, ST5 (Oct. 1963).

13. Ueda, Y., "Elastic, Elastic-Plastic, and Plastic Buckling of Plates With Residual Stresses," Ph.D. Dissertation, Lehigh University, 1962.

14. Winter, G., "Strength of Thin Steel Compression Flanges," *Trans. ASCE* **112**, 527 (1947).

15. Winter, G., and R. H. J. Pian, "Crushing Strength of Thin Steel Webs," *Cornell Univ. Eng. Expt. Sta. Bull.* **35**, Part I (1946).

10

Design of Compression Members

👾👾

10-1 INTRODUCTION

Chapter 9 discussed the fundamental theories of buckling which apply to compression members and beams as idealized members. As indicated in that chapter, column strength depends on the type of load, material properties, length of member, end restraint conditions, and shape of the member, as well as the imperfections in the material and shape. Usually the loading is idealized as either pure axial load or combined bending with axial load. Material properties are characterized by the modulus of elasticity and yield strength. Length L of the member is measured by the distance between supports, or by the effective length KL, which accounts for the end restraint conditions. There is no simple method of defining accurately all these variables under actual service conditions and empirical methods are used for the proportioning of columns.

This chapter will discuss the design of individual compression members in order to illustrate the detailed procedure involved in selecting appropriate proportions of such members.

10-2 COLUMNS SUBJECTED TO AXIAL LOADS

In proportioning columns subjected to axial load only the principal objective is to define an average stress $F_a = P_a/A$ which corresponds to the allowable load P_a. The use of an allowable average stress F_a is convenient in that, once it is known, the required cross-sectional area A for a given design load P is defined by

$$A = \frac{P}{F_a} \tag{10-1}$$

Allowable Stress Design. The determination of allowable stress F_a for columns subjected to axial load is based on the type of failure, such as: primary lateral buckling, elastic or inelastic, gradual yielding due to excessive local stress, local buckling, and torsional buckling. The shape and proportions of the commonly rolled sections are such as to prevent failure by local buckling or by torsional buckling. Thin sections which may fail by local buckling or torsion are discussed in Chapter 9. The two remaining types of failure are (*a*) lateral buckling, and (*b*) gradual yielding due to an excessive local stress.

Consider a column having an effective slenderness ratio (KL/r). Lateral buckling may occur when the load exceeds the critical load P_{cr} in either the elastic or the inelastic range. The average stress corresponding to this load is

$$F_{cr} = \frac{P_{cr}}{A} = \frac{\pi^2 \bar{E}}{(KL/r)^2} \tag{10-2}$$

where $\bar{E}$ is the effective modulus. In slender columns elastic buckling usually occurs when F_{cr} is a fraction of the yield strength F_y. For the A36 rolled steel shapes this fraction is usually less than 0.5, and therefore, for elastic buckling $F_{cr} \leq 0.5F_y$, $\bar{E}$ becomes E. The lowest value of slenderness ratio at which elastic buckling would occur is $(KL/r)_c = C_c$ and can be defined as follows:

$$C_c = \left(\frac{KL}{r}\right)_c = \left(\pi^2 \frac{E}{F_{cr}}\right)^{1/2} = \left(2\pi^2 \frac{E}{F_y}\right)^{1/2} \tag{10-3}$$

which is obtained from Eq. 10-2 by substituting $F_{cr} = 0.5F_y$ and solving for KL/r. The numerical values of C_c for steels of different yield strengths are given in Table 10-1.

This limiting value of slenderness ratio for elastic buckling has been adopted by the 1963 edition of the AISC Specification. For columns subjected to primary axial load which have effective slenderness ratios KL/r equal to or greater than C_c the AISC allowable stress is obtained from

Eq. 10-2 by dividing the critical stress by a factor of safety equal to 1.92. Thus for cases when elastic buckling controls, the allowable stress in psi is

$$F_e' = F_a = \frac{F_{\text{cr}}}{\text{F.S.}} = \frac{149 \times 10^6}{(KL/r)^2} \tag{10-4}$$

The numerical values of $F_a = F_e'$ defined by Eq. 10-4 are given in Table 10-1.

Inelastic buckling usually occurs at values of slenderness ratio KL/r below the C_c limiting value. The critical buckling stress may be determined by the procedure outlined in Sec. 9-3. However, the effective modulus or the tangent modulus for different steel sections is difficult to define and empirical equations are usually utilized to define the inelastic buckling stress. The critical stress can be generally approximated by the nth degree parabola (Eq. 10-5) having a horizontal tangent at $KL/r = 0$ with $F_{\text{cr}} = F_y$ and a common tangent with an elastic buckling stress curve at $KL/r = C_c$ with corresponding $F_{\text{cr}} = \lambda F_y$, where k, λ, and C_c values are to be determined.

The parabola can be written as follows:

$$F_{\text{cr}} = F_y - k\left(\frac{KL}{r}\right)^n \quad \text{for} \quad 0 < \left(\frac{KL}{r}\right) \le C_c \tag{10-5}$$

The elastic buckling stress is given by Eq. 10-2.

Differentiating Eqs. 10-5 and 10-2,

$$\frac{dF_{\text{cr}}}{d(KL/r)} = -nk\left(\frac{KL}{r}\right)^{n-1} \quad \text{for} \quad 0 < \left(\frac{KL}{r}\right) \le C_c \tag{10-6}$$

and

$$\frac{dF_{\text{cr}}}{d(KL/r)} = -\frac{2\pi^2 E}{(KL/r)^3} \quad \text{for} \quad \left(\frac{KL}{r}\right) \ge C_c \tag{10-7}$$

For $KL/r = C_c$:

$$F_{\text{cr}} = F_y - k(C_c)^n = \frac{\pi^2 E}{(C_c)^2}$$

or

$$k(C_c)^n + \pi^2 E(C_c)^{-2} - F_y = 0 \tag{10-8}$$

and

$$\frac{dF_{\text{cr}}}{d(KL/r)} = -nk(C_c)^{n-1} = -\frac{2\pi^2 E}{(C_c)^3}$$

or

$$nk(C_c)^{n-1} - 2\pi^2 E(C_c)^{-3} = 0 \tag{10-9}$$

For a value of n determined empirically from test results, values of k and C_c can be determined from Eqs. 10-8 and 10-9. Tests on steel columns indicated that a good agreement with experimental values can be obtained

Table 10-1   Allowable Stresses for Axially Loaded Structural Steel Columns

Slenderness Ratio, Kl/r	Minimum specified yield point*—psi													Main Members† F'_e or F_a ksi	Secondary Members‡ F_a ksi
	36,000	40,000	42,000	45,000	46,000	50,000	55,000	60,000	65,000	70,000*	80,000*	90,000*	100,000*		
Limiting Slenderness Ratio, C_c	126.1	119.6	116.7	112.8	111.6	107.0	102.0	97.68	93.84	90.43	84.59	79.75	75.66		
	Main Members Allowable Stress, ksi														
0	21.60	24.00	25.20	27.00	27.60	30.00	33.00	36.00	39.00	42.00	48.00	54.00	60.00	—	—
5	21.38	23.76	24.93	26.71	27.29	29.65	32.60	35.54	38.48	41.42	47.29	53.15	59.01	—	—
10	21.15	23.47	24.62	26.37	26.95	29.26	32.14	35.01	37.88	40.74	46.44	52.13	57.77	—	—
15	20.88	23.17	24.29	25.99	26.55	28.79	31.61	34.39	37.18	39.95	45.45	50.93	56.34	—	—
20	20.59	22.81	23.91	25.57	26.12	28.31	31.02	33.72	36.40	39.06	44.35	49.56	54.74	372.50	—
25	20.28	22.43	23.50	25.11	25.63	27.75	30.36	32.95	35.54	38.09	43.12	48.08	52.93	238.59	—
30	19.95	22.02	23.06	24.61	25.11	27.16	29.66	32.15	34.61	37.03	41.79	46.44	50.96	165.68	—
35	19.57	21.59	22.60	24.08	24.56	26.51	28.92	31.28	33.60	35.89	40.36	44.66	48.85	121.73	—
40	19.19	21.13	22.09	23.51	23.98	25.83	28.11	30.34	32.53	34.67	38.82	42.78	46.60	93.20	—
45	18.77	20.64	21.56	22.91	23.36	25.11	27.26	29.36	31.38	33.37	37.18	40.77	44.18	73.64	—
50	18.34	20.12	20.98	22.27	22.70	24.36	26.36	28.31	30.18	32.00	35.45	38.65	41.62	59.65	—
55	17.89	19.58	20.39	21.61	22.01	23.55	25.42	27.20	28.93	30.56	33.63	36.41	38.91	49.30	—
60	17.43	19.01	19.78	20.92	21.27	22.72	24.43	26.06	27.60	29.05	31.71	34.03	36.04	41.43	—
65	16.94	18.43	19.14	20.19	20.53	21.85	23.40	24.86	26.21	27.47	29.71	31.55	33.04	35.29	—
70	16.43	17.81	18.47	19.43	19.74	20.94	22.32	23.61	24.76	25.82	27.61	28.95	29.87	30.44	—

Kl/r														Kl/r > Cc (indep.)		
75	15.89	17.18	17.78	18.66	18.93	19.98	21.20	22.29	23.25	24.08	25.40	26.21	26.54	26.51		
80	15.36	16.52	17.05	17.84	18.08	19.02	20.04	20.93	21.68	22.28	23.09			23.28		
85	14.79	15.84	16.32	17.00	17.21	17.98	18.83	19.51	20.04	20.40				20.62		
90	14.19	15.12	15.56	16.13	16.31	16.93	17.57	18.04	18.32	18.42				18.40		
95	13.59	14.40	14.75	15.22	15.37	15.84	16.26	16.51						16.51		
100	12.98	13.64	13.93	14.29	14.40	14.70	14.90							16.51		
105	12.33	12.86	13.08	13.32	13.37	13.54								14.90		
110	11.68	12.06	12.19	12.32	12.33									13.51		
115	10.98	11.23	11.27											12.31		
120	10.28													11.27		
														10.35		

Kl/r > Cc

Allowable stresses are independent of the yield point.
See values at right.

Kl/r		
125	9.56	9.54	9.79
130		8.82	9.29
135		8.18	8.84
140		7.60	8.44
145		7.09	8.10
150		6.62	7.78
155		6.20	7.61
160		5.82	7.28
165		5.47	7.06
170		5.16	6.88
175		4.87	6.72
180		4.60	6.57
185		4.35	6.44
190		4.13	6.35
195		3.92	6.27
200		3.72	6.20

* Yield strength.

† Allowable stresses below single solid line are for main members with Kl/r ratios between 120 and 200. Stresses above double line are used only as F_e' values. Stresses below double line are used either as F_a or F_e' values.

‡ Allowable stresses are given for secondary members having Kl/r values of 120 or higher and are the same for all steels.

by taking $n = 2$. Then, from Eq. 10-9,

$$2k(C_c) - 2\pi^2 E(C_c)^{-3} = 0$$

or

$$k = \pi^2 E(C_c)^{-4} \tag{10-10}$$

Substituting this into Eq. 10-8,

$$2\pi^2 E(C_c)^{-2} = F_y$$

or

$$C_c = \left(\frac{2\pi^2 E}{F_y}\right)^{\frac{1}{2}} \tag{10-11}$$

Substituting these values of k and C_c into Eq. 10-5 with $KL/r = C_c$,

$$F_{\mathrm{cr}} = F_y - \frac{\pi^2 E}{(C_c)^2} = F_y - \frac{\pi^2 E F_y}{(2\pi^2 E)} = 0.5 F_y \tag{10-12}$$

It can be seen, therefore, that using a value of $n = 2$, that is, taking a quadratic parabola for the average of experimental values, requires that at $KL/r = C_c$ the critical stress $F_{\mathrm{cr}} = \lambda F_y = 0.5 F_y$, or that $\lambda = 0.5$. Based on the foregoing, for $0 < KL/r < C_c$

$$F_{\mathrm{cr}} = F_y - k\left(\frac{KL}{r}\right)^2 = F_y\left[1 - 0.5\left(\frac{KL}{rC_c}\right)^2\right] \tag{10-13}$$

The allowable stress corresponding to the foregoing value of critical stress is obtained as

$$F_a = \frac{F_{\mathrm{cr}}}{\mathrm{F.S.}} \tag{10-14}$$

For columns with slenderness ratios varying between 0 and C_c the factor of safety (F.S.) is varied between that of simple yielding, as in tension members, and that of elastic buckling, normally taken greater than that for tension members.

The 1963 AISC Specification provides for the following variation in F.S.:

$$\mathrm{F.S.} = \frac{5}{3} + \frac{3}{8}\left(\frac{KL}{rC_c}\right) - \frac{1}{8}\left(\frac{KL}{rC_c}\right)^3 \tag{10-15}$$

This expression approximates a quarter sine wave varying from F.S. $= 1.67$ at $KL/r = 0$ and F.S. $= 1.92$ when $KL/r = C_c$. The numerical values of the allowable stresses defined by Eq. 10-15, based on AISC Factor of Safety values for steels of different yield strengths are given in Table 10-1.

Members which carry no primary loads, but are used merely to brace other structural members or the structure as a whole, should be so attached at the ends that they may be considered essentially fixed against translation or rotation at the ends. In such instances, their capacity may be evaluated by using a value of $K < 1.0$, evaluating an appropriate KL/r, and then treating these members in the same manner as primary members, that is having the same allowable stresses as defined previously. The 1963 AISC Specification provides for an alternate method for secondary (bracing) members which have slenderness ratios $200 > L/r \geq 120$. In such instances, the values of $K = 1$ may be assumed and the allowable stresses increased by a factor

$$\frac{1}{1.6 - L/200r} \qquad (10\text{-}16)$$

These increased allowable stresses for secondary members based on the L/r values within the range $200 > L/r > 120$ are given in Table 10-1. Allowable stress variation with effective slenderness ratio KL/r is shown in Fig. 10-1.

The determination of AISC allowable loads for axial compression members may be summarized as follows:

(*a*) Allowable axial compression load $P_a = F_a \times A$, where A is the gross section of the member.

(*b*) Allowable stress F_a depends on the effective slenderness ratio KL/r and yield strength of steel F_y. In determining the allowable stress the largest

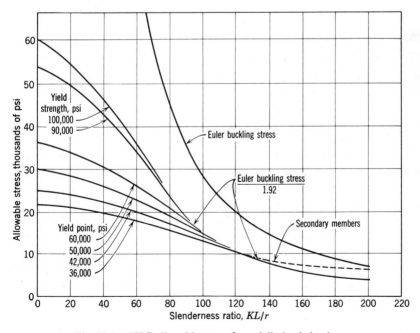

Fig. 10-1 AISC allowable stress for axially loaded columns.

KL/r of any unbraced segment must be used, where KL is the effective length and r is the corresponding radius of gyration.

(*c*) The actual unbraced length L is the distance center to center of adjacent members connecting to the column in a given plane, or the distance from such a member to the base of the column. The effective length factor K may be determined by rational methods of analysis. Approximate methods have been outlined in Secs. 9-10 and 9-11. A value of K may be less than unity if sidesway is prevented by appropriate diagonal bracing, shear walls, or attachment to a structural system having adequate lateral stability provided by shear walls or diagonal bracing. In a frame which depends upon its own bending stiffness for lateral stability, sidesway may occur and the effective length factor K, to be determined by suitable analysis, may be greater than unity, but in no instance shall it be taken less than unity.

(*d*) For compression members, the value of KL/r shall not exceed 200.

(*e*) For values of KL/r less than a limit value C_c defined by Eq. 10-3, the factor of safety varies with the slenderness ratio and therefore allowable stress F_a is defined most conveniently by tables or charts. Table 10-1 and Fig. 10-1 give these values for a variety of typical steels. More detailed tables for allowable stresses are given in the AISC Manual, in the Appendix to Sec. 5, pp. 5-58 to 5-93.

(*f*) For values of KL/r greater than a limit value C_c, the allowable stress is defined by Eq. 10-4, and the values corresponding to this formula are also shown in Fig. 10-1 and Table 10-1 as well as in the AISC Tables.

(*g*) For bracing and secondary members, when the unbraced length slenderness ratio L/r exceeds 120, allowable stresses may be obtained, using this unbraced length with a reduction in factor of safety (that is, increase in allowable stress values) to account for the conservative estimate of length. The stresses allowed in such cases are also shown in Fig. 10-1, Table 10-1, and in the AISC Tables.

In the design criteria just considered the load was assumed to be perfectly concentric with the axis of the member, which is considered perfectly uniform and straight. Actually, in real structures, the loading is eccentric with respect to the axis, although the magnitude of this eccentricity cannot be determined precisely. This eccentricity may be due partly to service load distribution and partly to initial crookedness and other imperfections of the column.

In a column subjected to load P with eccentricity e, producing a bending moment Pe, the maximum compressive stress due to combined effects of axial load and bending at a given section can be determined by considering the deflected shape of the member. Within the elastic range, the solution of this problem can be obtained analytically, and has been discussed in Sec. 8-13.

To account for small imperfections and initial crookedness of the member subjected essentially to axial loading, the "secant formula" is often used.

This formula is based on considering a pin-ended compression member with length L, cross section A, radius of gyration r, and equal eccentricities e of the load P at the ends. Initial yielding develops when the maximum stress f_{max} due to combined effects of axial load and bending corresponding to load P_y equals nP. In this case, n is an appropriate load factor applied to the maximum service load P. Using the "beam-column" analysis and the solutions given in Table 8-1, the maximum bending moment at midlength is

$$M_{max} = M_o \sec z = (nPe)\left(\sec \frac{L}{2r}\sqrt{\frac{nP}{AE}} \right) \qquad (10\text{-}17)$$

and the maximum stress is

$$f_{max} = \frac{nP}{A} + \frac{M_{max}}{(I/c)} = \frac{nP}{A}\left(1 + \frac{ec}{r^2}\sec \frac{L}{2r}\sqrt{\frac{nP}{AE}} \right) \qquad (10\text{-}18)$$

Letting $f_{max} = F_y$ and solving for P/A:

$$\frac{P}{A} = \frac{F_y/n}{1 + (ec/r^2)\sec [(L/2r)\sqrt{nP/AE}]} \qquad (10\text{-}19)$$

For a section with eccentricities about both principal axes 1–1 and 2–2, Eq. 10-19 can be generalized as follows:

$$\frac{P}{A} = \frac{F_y/n}{1 + (e_1 c_1/r_1{}^2)\sec [(L/2r_1)\sqrt{nP/AE}] + (e_2 c_2/r_2{}^2)\sec [(L/2r_2)\sqrt{nP/AE}]} \qquad (10\text{-}20)$$

The term P/A occurs in both rational and trigonometric form in Eqs. 10-19 and 10-20, and therefore no explicit solutions for P/A are possible. The equations may be solved graphically with appropriate charts, similar to Fig. 9-9, or numerically by trial and error.

For a special instance of opposite end eccentricities a formula similar to the secant formula may be derived from the fundamental beam column solution. This formula is

$$\frac{P}{A} = \frac{F_y/n}{1 + (ec/r^2)\operatorname{cosec} [(L/2r)\sqrt{nP/AE}]} \qquad (10\text{-}21)$$

It is convenient to assume that $e = e' + e''$, where e' is due to imperfections and e'' is the eccentricity due to known loading conditions. A reasonable approximation for e' has been proposed as follows: $e' = 0.1r^2/c + L/750$. This relationship can serve only as a first approximation. The actual magnitude of imperfection is subject to considerable variation. Another approximation of imperfection is given by $e' = 0.25r^2/c$. The two approximations just given are equivalent when $L/750 = 0.15r^2/c$ or $L = 112.5r^2/c$.

The secant and cosecant formulas (Eqs. 10-20 and 10-21) are based on the initial yielding criterion of an eccentrically loaded column. The reserve strength due to the plastic action may be considerable for some shapes, and therefore the actual factor of safety may be considerably in excess of the load factor n.

Because the secant formula does not give a direct solution for P/A, and because of the uncertainty regarding the values of e', similar empirical formulas have been widely used. The Gordon-Rankine formula, also known as the Rankine formula, is a semi-empirical formula, which with properly chosen coefficients can be made to follow rather closely the secant formula, or it can be made to fit a set of test data within certain ranges of L/r. The Rankine formula specifies an allowable average stress F in the following form:

$$F = \frac{P}{A} = \frac{f}{1 + k_1(L/r)^2} \tag{10-22}$$

The following Gordon-Rankine formula was recommended by the AISC Specification (1949) for design of building columns with values of $L/r > 120$:

$$\frac{P}{A} = \frac{18{,}000}{1 + \dfrac{1}{18{,}000}\left(\dfrac{L}{r}\right)^2} \tag{10-23}$$

Plastic Design.   Plastic design is used effectively only in continuous structures so that columns in such structures are generally subjected to appreciable bending. Therefore the plastic design of columns is usually governed by combined axial load and bending, which will be discussed in Sec. 10-3.

However, when bending moments are small, inelastic buckling may occur. For this reason, AISC Plastic Design Specification provides that the maximum axial load P at ultimate loading, which is taken as service load times a load factor of about 1.7, shall not exceed $0.6P_y$ where P_y is the yield load of the section. The slenderness ratio L/r of the columns which may develop a plastic hinge may not exceed 120. When columns do not develop a plastic hinge their slenderness ratio shall not exceed 200, and for $L/r > 120$, their capacity shall not exceed $[8700/(L/r)^2]P_y$.

Example 10-1.   Select an A36 steel W column with an effective length $KL = 15$ ft (with respect to both principal axes) carrying a load P of 1000 kips.

Solution

Estimate allowable stress $F_a = 19$ ksi

Required section area $A = P/F_a = 1000/19 = 52.6$ in.^{2}

From AISC Manual, pp. 1-13, try 14 W 176:

$$A = 51.73 \text{ in.}^2 \qquad r_y = 4.02 \text{ in.}$$

Therefore $KL/r = (15 \times 12/4.02) = 44.8$
From Table 10-1, p. 406: $F_a = 18.77$ ksi
Allowable load $P_a = A \times F_a = 51.73 \times 18.77 = 980$ kip < 1000 kip
   Section inadequate. Try 14 W^F 184:

$$A = 54.07 \text{ in.}^2 \qquad r_y = 4.04 \text{ in.}$$

Therefore $(KL/r) = (15 \times 12/4.04) = 44.6$
From Table 10-1, p. 406: $F_a = 18.8$ ksi
Allowable load $P_a = A \times F_a = 54.07 \times 18.8 = 1015$ kip > 1000 kip
   Section adequate. Use 14 W^F 184—A36 steel.

10-3 COLUMNS SUBJECTED TO AXIAL LOAD AND BENDING

Design of a member subjected to both axial compression and bending (beam-column) is generally performed by a trial procedure, whereby a section is assumed and checked for its adequacy by appropriate criteria. This checking for stresses and strength is generally a simple problem once specific criteria are established, as, for example, those of AISC. Selection of an optimum section, structurally and economically, requires more ingenuity. Some approximate methods of design will be presented here after the development of appropriate design criteria is discussed. Usually, these criteria take the form of interaction equations which were discussed in general terms in Sec. 8-14.

Allowable Stress Design. The basic form of the design criterion is a linear interaction equation

$$\frac{P}{P_a} + \frac{M_x}{M_a} \leq 1.0 \tag{10-24}$$

where P_a and M_a are the allowable values of axial load and bending moment when these act separately and P and M_x are the design values of axial load and bending moment acting simultaneously. The moment M_x is taken at a critical section and depends on both the primary bending moment due to transverse loads and end moments and the secondary bending moment due to axial load and the deflected shape of the member.

For design it is usually convenient to rewrite Eq. 10-24 in terms of stresses as follows:

$$\frac{f_a}{F_a} + \frac{f_{bx}}{F_b} \leq 1.0 \tag{10-25}$$

where F_a is the allowable stress for axial loading, F_b is the allowable stress for bending, and $f_a = P/A$ and $f_{bx} = M_x/S_x$.

Two modes of failure are considered: verification of local yielding at braced ends of the members, and verification of general yielding or instability.

To provide for the yielding limitation at a braced section where there is no contribution of secondary bending and where the axial load capacity is determined by the yield strength F_y of the material, the following allowable stresses are used in the AISC Specifications:

$$F_a = 0.6F_y \quad \text{and} \quad F_b = 0.66F_y \quad \text{or} \quad 0.60F_y$$

The bending stress f_{bx} is taken at the braced end as $f_{bx} = f_b = M_o/S$, where M_o is the primary bending moment at this braced end. The design criterion then becomes

$$\frac{f_a}{0.6F_y} + \frac{f_b}{F_b} \leq 1.0 \tag{10-26}$$

To provide for general yielding or instability between braced ends, the allowable stress F_a for axial load must be taken with due regard to effective slenderness ratio (KL/r), and yield strength F_y, as described in Sec. 10-2. The allowable stress F_b for bending is taken as an appropriate fraction (0.66 or 0.60) of yield strength. The definition of bending stress $f_{bx} = M_x/S_x$ is somewhat more complex because it depends on the evaluation of the moment M_x at the critical section. Theoretically, the value of M_x can be determined by an appropriate "beam-column" analysis discussed in Sec. 8-13. For development of design criteria approximate formulas have been developed for the more common cases.

In developing these approximate formulas the stress f_{bx} is represented as a multiple or a fraction of the primary compressive bending stress at the section under consideration, for example,

$$f_{bx} = \alpha f_b = \alpha \frac{M_{ox}}{S} \tag{10-27}$$

where M_{ox} is the primary moment at the section. In the absence of transverse loading between points of support M_{ox} is taken as the larger of the moments at the supports. When intermediate transverse loading is present, the maximum moment between points of support is used for M_{ox}.

The amplification factor α must account for the nonlinear increase in total bending moment and stress when all the loads, axial and transverse bending, are increased proportionately by a load factor n. If service axial load is P, the critical increase in primary moment will occur when the axial load is nP. Consider a pin-ended beam-column with equal end eccentricities e and service load P, so that primary moment $M_o = Pe$. The critical design condition will be represented by nM_x when the axial load is increased to nP. This moment occurs at midlength and can be closely approximated, similar to the deflection approximation given in Eq. 8-61, as

$$nM_x = \alpha n M_o = nM_o \frac{1}{(1 - nP/P_{cr})} \qquad (10\text{-}28)$$

where P_{cr} is the critical Euler load. The factor (nP/P_{cr}) is

$$\left(\frac{nP}{P_{cr}}\right) = \left(\frac{nP/A}{P_{cr}/A}\right) = \left(\frac{f_a}{F_e'}\right) \qquad (10\text{-}29)$$

where $f_a = P/A$ and $F_e' = (P_{cr}/n \cdot A)$ and is given by Eq. 10-4. The design moment M_x and corresponding bending stress $F_{bx} = M_x/S_x$ are

$$M_x = \alpha M_o = M_o \frac{1}{(1 - f_a/F_e')} \qquad (10\text{-}30)$$

and

$$f_{bx} = \frac{M_x}{S_x} = \frac{M_o}{(1 - f_a/F_e')S_x} = f_b \frac{1}{(1 - f_a/F_e')} \qquad (10\text{-}31)$$

In this particular case when the end moments are equal and there is no transverse load on the beam-column between the ends the amplification factor α is simply

$$\alpha = \frac{1}{(1 - f_a/F_e')} \qquad (10\text{-}32)$$

When the moments vary along the member the amplification factor can be written as

$$\alpha = \frac{C_m}{(1 - f_a/F_e')} \qquad (10\text{-}33)$$

where the coefficient C_m varies depending on the variation of bending moments along the member. Then, if the column is subjected to end moments M_1 and M_2 at the two ends, so that $M_2 = \beta M_1$ and $1 > \beta > -1$, and lateral translation of the ends is prevented, it can be designed for an equivalent primary moment $C_m M_1$ which is constant along the column. Various empirical expressions for C_m have been proposed[4]. One is

$$C_m = \sqrt{0.3 + 0.4\beta + 0.3\beta^2} \qquad (10\text{-}34)$$

The AISC, in order to simplify this expression, uses

$$C_m = 0.6 + 0.4\beta \geq 0.4 \qquad (10\text{-}35)$$

Both of these expressions are shown in Fig. 10-2. When lateral translation of the ends is not prevented, a different buckling mode becomes possible. For this reason, AISC places a more conservative limit on the value of C_m (Fig. 10-2),

$$C_m = 0.6 + 0.4\beta \geq 0.85 \tag{10-36}$$

The foregoing amplification factors considered beam-columns without transverse load applied between the ends. A case which is encountered in trusses when the compression chord is subjected to transverse loads between panel points follows. It is assumed that lateral translation is prevented at the ends. The loading under service conditions, the deflected shape and bending-moment diagrams, primary and total, are shown in Fig. 10-3. Let the critical condition correspond to load factor n, and the maximum moment nM_x will occur at a section corresponding to the maximum primary moment M_3. Then, when the loads are increased proportionately by factor n:

$$nM_x = nM_3 + nP \cdot ny \tag{10-37}$$

where $ny = ny_o\left(\dfrac{1}{1 - nP/P_{cr}}\right)$ from Eq. 8-61, where $P_{cr} = [\pi^2 EI/(KL)^2]$.
Then

$$M_x = \alpha M_3 = \left[1 + \frac{nPy_o}{M_3(1 - nP/P_{cr})}\right]M_3 = \frac{\left(1 - \dfrac{nP}{P_{cr}} + \dfrac{nPy_o}{M_3}\right)}{(1 - nP/P_{cr})} \cdot M_3 \tag{10-38}$$

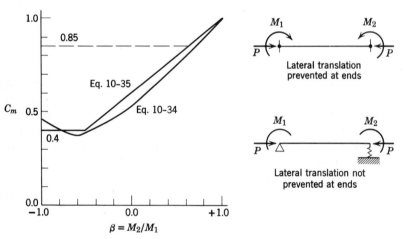

Fig. 10-2 Influence of Moment Gradient on C_M.

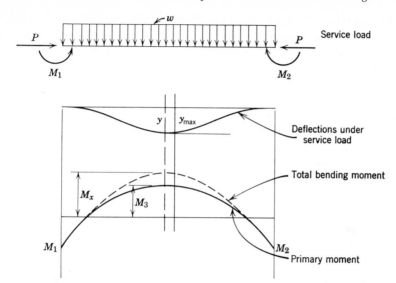

Fig. 10-3 Primary and total moment diagrams and deflected shape.

The amplification factor α can be expressed in terms of C_m, f_a, and F_e', as before:

$$\alpha = \left[1 + \frac{nPy_o}{M_3(1 - nP/P_{cr})} \right] = \left[\frac{1 - \dfrac{nP}{P_{cr}}\left(1 - \dfrac{P_{cr}y_o}{M_3}\right)}{(1 - nP/P_{cr})} \right] = \frac{C_m}{(1 - f_a/F_e')}$$

(10-39)

where

$$C_m = \left[1 - \frac{f_a}{F_e}\left(1 - \frac{P_{cr}y_o}{M_3}\right) \right]$$

(10-40)

The determination of AISC allowable loads for members subjected to combined compression and bending may be summarized as follows:

(*a*) The stresses due to axial load P and bending moment M at a section under consideration are computed in the conventional manner, so that $f_a = P/A$ and $f_b = M/S$. These stresses must satisfy an appropriate interaction formula.

(*b*) At points braced in the plane of bending the interaction formula is given by Eq. 10-26.

(*c*) In general, the stresses must satisfy an interaction equation of the form:

$$\frac{f_a}{F_a} + \alpha\frac{f_b}{F_b} \leq 1.0$$

(10-41)

In this equation F_a is determined as the allowable stress for the column if axial force P existed alone. Similarly, F_b is the allowable compressive bending stress if bending moment M existed alone. A discussion of the determination of amplification factor α follows.

(*d*) The general form of α is given by Eq. 10-33. However, when $f_a/F_a \leq 0.15$, the amplification factor α should be taken as unity. When $f_a/F_a > 0.15$, the amplification factor depends on the magnitude of axial stress f_a, the effective slenderness ratio (KL/r), the variation of bending moment along the member, and the presence or absence of sidesway. For special conditions values of the equivalent moment coefficient C_m and of the corresponding values of the amplification factor α can be obtained from Eqs. 10-33, 10-35, 10-36, and 10-40.

A preliminary estimate of section requirements based on AISC allowable stress criteria can be made by calculating an equivalent axial compression load P_{eq} corresponding to combined compression end bending. Then the required cross-sectional area can be estimated from Eq. 10-1, where P becomes P_{eq}, and F_a allowable stress is assumed for the preliminary design. The equivalent load can be defined from the interaction equation as follows:

$$\frac{f_a}{F_a} + \alpha \frac{f_b}{F_b} \leq 1.0 \tag{10-42}$$

or

$$\frac{Af_a}{AF_a} + \frac{\alpha f_b S}{F_b S} \leq 1.0 \tag{10-43}$$

or

$$Af_a + \frac{\alpha f_b \cdot S}{F_b} \times \frac{AF_a}{S} \leq AF_a \tag{10-44}$$

or

$$P + \frac{F_a}{F_b} \cdot B \cdot \alpha \cdot M \leq AF_a \tag{10-45}$$

where $B = A/S$ is a geometric characteristic of given section called the "bending factor," previously defined in Sec. 7-4. Also

$$\frac{P + \alpha(F_a/F_b) \cdot B \cdot M}{F_a} = A = \frac{P_{eq}}{F_a} \tag{10-46}$$

or

$$P_{eq} = P + \alpha \frac{F_a}{F_b} \cdot B \cdot M \tag{10-47}$$

When bending moments exist about both x and y axes of a column the equivalent load becomes

$$P_{eq} = P + \alpha_x \frac{F_{ax}}{F_{bx}} B_x M_x + \alpha_y \frac{F_{ay}}{F_{by}} B_y M_y \tag{10-48}$$

Then, assuming a value of F_a, the estimated required section becomes

$$A = \frac{P_{eq}}{F_a} \qquad (10\text{-}49)$$

The nearest economical section may be selected and the adequacy of this section can be verified for the combined P, M_x, and M_y loads. The values of $B_x = A/S_x$ and $B_y = A/S_y$ vary widely for different sections. For typical column sections the values of B_x and B_y are tabulated in the AISC Manual, pp. 3-13 to 3-56, and a graph of approximate values is shown in Fig. 7-10.

For W shapes the values of B_x vary between 0.18 and 0.25, and the values of B_y vary between 0.443 and 2.25, with the larger values corresponding to the smaller size shapes. Since these values have to be estimated for a preliminary design, in case the designer cannot make a judicious guess based on experience, values of $B_x = 0.25$ and $B_y = 0.75$ may be used.

The values of F_{ax}/F_{bx} and F_{ay}/F_{by} depend on the slenderness ratios which determine F_{ax} and F_{ay}. For the A36 steel, a value of 0.75 may be used as a first approximation. The amplification factors α_x and α_y depend on the end moments, axial stress f_a, and on slenderness ratios which determine F_{ex}' and F_{ey}'. As a first approximation values of α's can be taken as unity. Substituting these estimated values, Eq. 10-48 becomes

$$P_{eq} = P + (1.0)(0.75)(0.25)M_x + (1.0)(0.75)(0.75)M_y$$
$$= P + 0.2M_x + 0.6M_y \qquad (10\text{-}50)$$

The foregoing expression is intended only as a rough first approximation for W shapes that may be used only when the designer is unable to make a more judicious choice.

Example 10-2.   Select an A36 steel W section to carry the loads shown. The column, 12 ft long, is part of a rigid frame multistory building, braced against sidesway in the x-z plane but with possible sidesway in the y-z plane. Approximate analysis indicated that $(KL)_x = 1.25 \times 12 = 15$ ft, and that $(KL)_y = 12$ ft.

Solution
Using Eq. 10-50:

$$P_{eq} = P + 0.2M_x = 400 + 0.2 \times 2000 = 800 \text{ kip}$$

$$\text{Assume } F_a = 19 \text{ ksi}$$

$$\text{Estimated } A = \frac{P_{eq}}{F_a} = \frac{800}{19} = 42 \text{ in.}^2$$

Try 12 W 133 (see AISC, p. 1-15)

$$A = 39.11 \text{ in.}^2 \qquad r_x = 5.59 \text{ in.} \qquad r_y = 3.16 \text{ in.}$$
$$S_x = 182.5 \text{ in.}^3$$

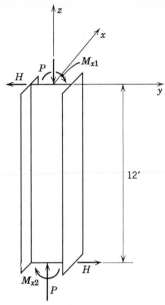

Fig. 1 $P = 400$ kip, $M_{x1} = 2000$ kip-in., $M_{x2} = 1800$ kip-in.

Compact section, adequately braced, $F_b = 24$ ksi.
Verify adequacy of 12 W 133 using Eq. 10-42:

$$f_a = \frac{P}{A} = \frac{400}{39.11} = 10.2 \text{ ksi}$$

$$f_b = \frac{M}{S} = \frac{2000}{182.5} = 11.0 \text{ ksi}$$

$$\left(\frac{KL}{r}\right)_x = \left(15 \times \frac{12}{5.59}\right) = 32.2$$

$$\left(\frac{KL}{r}\right)_y = \left(12 \times \frac{12}{3.16}\right) = 45.5$$

$$F_a\left[\text{for } \left(\frac{KL}{r}\right)_y = 45.5, \text{ see Table 10-1}\right] \quad F_a = 18.77 \text{ ksi}$$

$$F_e'\left[\text{for } \left(\frac{KL}{r}\right)_x = 32.2, \text{ see Table 10-1}\right] \quad F_e' = 144 \text{ ksi}$$

$$C_m(\text{see Eq. 10-36, } \beta = -\frac{1800}{2000} = -0.9) \quad C_m = 0.85$$

$$\alpha = (C_m)/(1 - f_a/F_e') = 0.85/(1 - 10.2/144) = 0.916$$

Then in Eq. 10-42:

$$\frac{f_a}{F_a} + \alpha \frac{f_b}{F_b} = \frac{10.2}{18.77} + 0.916 \frac{11.0}{24.0} = 0.544 + 0.42 = 0.964 < 1.0$$

Use section 12 W^F 133—*adequate.*

Plastic Design. Plastic design of columns in rigid frames may be effective when sidesway is prevented by diagonal bracing or by attachment to a structural system having ample lateral rigidity provided by shear walls or similar elements. When sidesway of the frame is permitted formation of plastic hinges leads to substantial reduction in the over-all lateral stability. The amount of lateral bracing required to prevent frame instability in high-rise building frames with plastic hinges has not yet been adequately defined. For this reason the 1963 AISC Specification limits the application of plastic design to continuous beams or to frames of only one or two stories in height.

Columns in a braced frame may be designed to satisfy an interaction equation of the general form:

$$\frac{M}{M_p} = H + K\left(\frac{P}{P_y}\right) + J\left(\frac{P}{P_y}\right)^2 \tag{10-51a}$$

In this equation, values of H, K, and J depend on the deflected shape and loading conditions as well as the slenderness ratio of the column. The values of these coefficients are given in tables in the Appendix to the AISC Specification, pp. 5-64–5-75. The values of M and P in Eq. 10-51a are those obtained from a plastic analysis of a frame subjected to ultimate loading, which is taken as service load times a load factor of 1.7, or other appropriate value. The values of M_p and P_y are the plastic moment and axial yield load capacities of the section, respectively.

For columns in one- or two-story rigid frames which are not laterally braced the compression load shall be such that

$$\frac{P}{P_y} \leq 0.5\left(1 - \frac{L}{70r}\right) \tag{10-51b}$$

Within these limits it is estimated that lateral frame instability will not develop.

10-4 DESIGN REQUIREMENTS

In the design of a compression member the maximum loads and the effective length of the member are usually known, whereas the shape and the dimensions of the cross section and the connections are to be determined. The required cross-sectional area A is determined by the load P or P_{eq} when compression is combined with bending and the allowable average stress F from the relation $A = P/F$; the allowable stress F depends on the slenderness ratio L/r and the end restraint conditions, all of which are not accurately known until the shape and size of the member have been determined. Hence the solution of this problem can only be accomplished by successive trials.

Slenderness Ratio.   In order to reduce the weight of the member, a high allowable average stress must be developed by the member. Because the allowable stress is decreased for high values of L/r, a minimum value of L/r for the given cross-sectional area is desirable, provided this does not conflict with other economic and stability considerations. With the value of L predetermined, the problem reduces to the selection of a practical section with the greatest value of r.

Usually if the compression loads are sufficiently high, an economical shape can be selected in a range of L/r values not in excess of 80. If compression loads are light, the areas calculated for adequate strength are small, and slender members may be justified in the range of L/r of 80 to 120. In using slender members, the effects of accidental loads, generally neglected in the analysis, and the effects of construction loads must be considered. Bracing members may often be used as a walkway for workmen during and after construction; they may be used as a support for some unforeseen auxiliary or temporary equipment; and there is also the possibility of vibration. For these reasons most specifications provide that the slenderness ratio L/r shall not exceed a specified maximum value, usually taken as 200.

Construction Loads.   Generally the loads imposed during construction are smaller than the full design loads, but the type of loading and the conditions of support are also different. During fabrication, transportation, and erection, the compression member may be subjected to a variety of loads, which may sometimes control the design of the member. Very often, only nominal connections are used during construction, and final bracing may be left out during initial phases of erection. Thus the adequacy of a compression member may have to be considered under an entirely different set of conditions from those existing in the final structure. Although it may seem superfluous to emphasize the need for investigating such conditions, many structural failures which have occurred during various construction phases could have been prevented by a careful check on the strength of the columns under construction loads.

Secondary Stresses.   Since most stress calculations are based on idealized conditions of loading and structural behavior, the actual stresses in members differ from the computed stresses. This difference arises from neglecting: certain known loads, the effects of joint rigidity, the general distortion of the

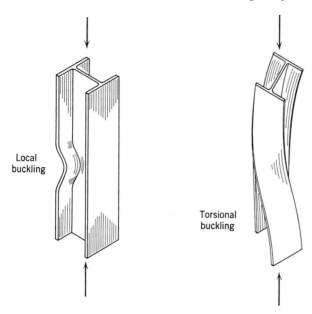

Fig. 10-4　Local and torsional buckling.

frame, and stresses induced by the fabrication and construction processes.
If the actual structure approximates closely the idealized structure, then the
computed stresses, known as primary stresses, closely approach the actual
stresses. If the assumptions made in neglecting various effects are not realized
in the structure, then the computed "primary" stresses may differ appreciably
from the actual stresses, because of "secondary" stresses existing in the
structure. In the design of conventional buildings secondary stresses are
generally neglected. In the design of bridges secondary stresses due to frame
distortion or due to joint rigidity are sometimes considered.

Local and Torsional Buckling. Compression member sections with relatively thin elements may fail by local buckling—often called "crippling"—or
by twisting (Fig. 10-4) if the sections are torsionally weak. The crippling
stress of a plate element having width b and thickness t depends on the
conditions of support along its edges, the modulus of elasticity E, and the
square of the ratio of thickness to width $(t/b)^2$. The problem of determining
the magnitude of the crippling stress is considered in Sec. 9-13.

To prevent premature buckling the width-thickness ratio is usually limited,
as shown in Eq. 9-53b. The AISC Specification provides for the several cases
of edge restraint, as shown in Fig. 10-5 and Table 10-2.

Torsional buckling is rarely critical for conventional shapes. "Box" or
tubular sections have a much greater torsional rigidity than "open" shapes

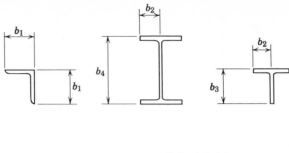

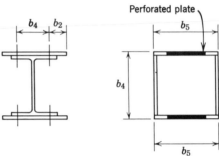

Fig. 10-5

such as *I*, *T*, and *Z* shapes and, for such tubular shapes, the torsional mode of buckling need rarely be considered. Sections with two axes of symmetry are torsionally more stable than sections with one or no axis of symmetry. Also, the torsional rigidity of open sections is approximately proportional to the cube of the thickness of the elements; therefore an increase in thickness will enhance the torsional rigidity of the section. When compression elements of column sections subjected to combined axial load and bending are in the

Table 10-2   Maximum Width-Thickness Ratios for Steel Plates in Compression

Type of Edge Restraint (Fig. 10-5)	b_1	b_2	b_3	b_4	b_5
Limit of $(b/t)\sqrt{F_y}$	2400	3000	4000	8000	10000
Limit of (b/t) for					
$F_y = 33$ ksi	13	16	22	44	55
36	13	16	21	42	33
42	12	15	20	39	49
46	11	14	19	37	47
50	11	13	18	36	45

plastic range, the effective torsional stiffness of the shape is reduced materially and the possibility of torsional buckling is enhanced. Although possibility of torsional buckling may be an important factor in plastic design procedures, it rarely affects the allowable stresses under service loads. The determination of the elastic torsional buckling load is discussed in Sec. 9-6.

Material Properties. The allowable stress F is usually determined by the criterion of safety against yielding with an adequate factor of safety. Therefore, using alloy steels with yield strengths higher than that of structural carbon steel would permit the use of higher allowable stresses, and a corresponding reduction in sectional area of the member. This increase in F_a is particularly significant for $L/r < 60$.

The reduction of area may result either in reduction of lateral dimensions for a constant thickness or in reduction of thickness for a constant lateral dimension. In the first instance, the section radius of gyration is reduced, resulting in an increase in slenderness ratio L/r; in the second instance the thickness-to-width ratio t/b is reduced. Because the modulus of elasticity of all steels is approximately constant, it follows that, with reductions in radius of gyration r or thickness t, the possibility of buckling is enhanced, either that of Euler buckling, or possibly local and torsional buckling. For $L/r > 100$ no benefit is obtained from using steels with higher yield strengths, insofar as compression elements are concerned.

10-5 CHOICE OF SECTIONS

Theoretically, an infinite number of compression members can be designed to carry a given load, but actually only a few will satisfy practical requirements in a given instance. It may be that there is no one section that will be "the most satisfactory." Two or three may be equally good and several others may be almost as good. The final choice may be determined perhaps by personal preference.

If a compression member is free to buckle in any direction, then evidently a pipe section (Fig. 10-6a) would be the most economical since it has the same value of r in all directions and has a high local buckling strength. A solid

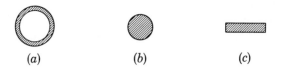

(a)   (b)   (c)

Fig. 10-6 Variation of r with shape (all areas equal): (a) pipe column, (b) solid-bar column, and (c) thin rectangular column.

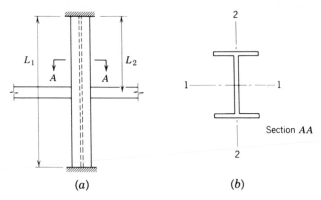

Fig. 10-7   Column with intermediate support.

round bar (Fig. 10-6*b*) has a much smaller *r* than a pipe of the same total cross-sectional area and hence is less economical; but it is better than a thin rectangular section, say a flat bar (Fig. 10-6*c*), which has a very small *r* in its narrow direction.

By furnishing intermediate supports to a compression member, the unsupported length L can be reduced, permitting use of a smaller section at a higher average stress. Sometimes it may be possible to furnish support only in one direction (Fig. 10-7*a*). Then the value of L will differ in two directions, and it might be economical to use sections with different *r*'s in the two directions in order to obtain approximately equal values of L/r (Fig. 10-7*b*). Also when a compression member carries different bending moments in two directions, it is desirable to choose a section with a greater value of r in the direction with greater bending moment.

In order to obtain large values of r it is desirable to distribute the material in a section as far away from the centroidal axes as possible; in doing so, however, the thickness of the material should not be reduced below a certain minimum.

Two types of sections are available: (*a*) standard rolled-steel sections and (*b*) built-up sections (as defined in Chapter 7 on tension members). Compression members made of formed light-gage steel can be used for some light and moderate loads, and will be discussed in Chapter 14.

Rolled sections cost less than built-up sections per unit weight and hence are used whenever feasible. Theoretically, rods and bars can withstand some compression, but for structural members longer than about 10 ft, their compressive strength is so small that it is not possible to use them.

Pipes are used for compression members carrying small and medium loads. These usually range from 3 in. diameter up to 12 in. diameter, although smaller pipes as well as larger ones are available. They come in standard, extra strong, and double extra strong classifications, with cross-sectional areas ranging from about 2 to 21 in.², as shown in the AISC Manual, p. 1-72, 1963. They are best adapted to welded construction, and a number of structures, such as towers, domes, and roof trusses, have been built utilizing pipe members. Also structural steel tubing, square or rectangular, can be used effectively to carry moderately light loads. Dimensions and section properties of such standard tubular sections are shown on pp. 1-73 to 1-75 of the AISC Manual, 1963 Edition.

Single-angle members are used for small trusses and bracing. Equal-leg angles are usually more desirable, but unequal-leg angles are also used. In either instance, attention must be paid to the least radius of gyration about one of the principal axes. Although angles with cross-sectional area up to about 17 in.² (8 × 8 × 1⅛ in.-angle) are available, the least radius of gyration for a given area is much smaller for angles than for other shapes, and hence they are not suitable for long members. For single-plane trusses, that is, where gusset plates in one plane only are used for connections, angles are commonly employed in order to simplify end connections.

Single channels, as well as single standard I sections, are seldom used as compression members because of the small value of r about the axis parallel to the web. But, if additional intermediate support is furnished in the weak direction, they might be economical sections. The channels and I sections are sometimes preferred because of the method of rolling at the mills the out-to-out dimension of these sections remains constant for a given depth, which is not so with the other rolled sections.

For steel buildings, the most common column section is undoubtedly the W shape. Their connection to beams is relatively easy and can be designed to be quite strong. The radii of gyration about the two axes are approximately equal to each other. A somewhat greater r in one plane is not entirely undesirable because columns are frequently subjected to heavier bending moments in one plane. W sections have a range of cross-sectional area from 3 to 125 in.², and least r from 1 to 4 in. The 14-in. W series alone furnishes areas from 9 to 125². It is commonly used in building design because it keeps the size of the columns within a reasonable limit and is preferred for architectural reasons. The clear depths (inside to inside of flanges) of the W sections are constant for a given size as a result of the practice of spreading the rolls; but the out-to-out depths vary for different weights of the same size, and hence W sections cannot be conveniently employed in a two-plane truss. Also, for splices in building columns of different weights filler plates are required to compensate for the change in depth.

The T section is suitable for small trusses and adapts itself well to welding. Other sections, such as the Z, are not generally used as compression members because of their low values of r and difficulties in the design of suitable connections.

The shapes and sizes of standard rolled sections are limited by considerations of economy and manufacturing process in the rolling mills. Therefore, whenever rolled sections cannot provide a suitable compression member, special built-up sections must be fabricated. Built-up sections may be used for one or more of the following reasons:

(*a*) To provide sufficiently large cross-sectional area, which cannot be obtained in any one rolled section.

(*b*) To provide a special shape and depth, which will facilitate connections between different members.

(*c*) To obtain a sufficiently large radius of gyration or a more desirable ratio of the radii of gyration in two different directions, which cannot be obtained in any rolled section.

It must be emphasized that, because of additional cost of fabrication for built-up sections and additional cost of material required for latticed compression members, rolled sections are more economical for the same area and slenderness ratio, if a normal connection can be designed for the member. However, in order to obtain in a monolithic shape the same area and slenderness ratio as that of a latticed column, the thickness of the material may have to be greatly reduced, and the local buckling strength may be so low as to make the monolithic shape impractical. The use of built-up sections is usually restricted to large structures where the compressive members are long and carry large compressive loads. Although the primary purpose of built-up sections is to provide a large radius of gyration and carry heavy loads, the arrangement of the section must permit ease of fabrication, connection, and maintenance, such as painting, as well as minimize lattice material.

The section used most frequently in roof trusses is a two-angle section placed back to back (Fig. 10-8*a*). This section is particularly economical when single gusset plates are used. The section is tied together by rivets spaced at certain intervals with fillers between the angles. In order to obtain radii of gyration more nearly equal about the two centroidal axes, unequal-leg angles are preferred for such a section.

Two-angle sections can also be used with outstanding legs spread (Fig. 10-8*b*). This section has greater stiffness in one plane. However, it has no advantage over the first regarding the least radius of gyration because this is largely controlled by the size of the outstanding leg, and it has the disadvantage of requiring additional lacing and riveting.

Starred angles (Fig. 10-8*c*) possess a definite advantage over the first two arrangements in having a relatively large radius of gyration. This type of section is suited to a single-plate truss and is simple to fabricate.

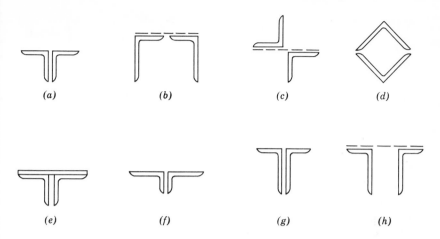

(a)   (b)   (c)   (d)

(e)   (f)   (g)   (h)

Fig. 10-8 Double-angle compression members (dotted lines show lacing).

An arrangement limited to welded sections is shown in Fig. 10-8*d*. Its primary advantage is adaptability to welded design, torsional rigidity, and high crippling strength. The radius of gyration of the box arrangement of two equal-leg angles is the same as that of the starred angles, whereas for unequal-leg angles the difference between the radii of gyration of sections shown in Figs. 10-8*c* and *d* is small.

Four-angle sections are often used in arrangements such as shown in Fig. 10-9. Shapes *a* and *b* may be used instead of W^F sections when it is desirable to maintain the over-all depth of section. Lacing is used, as in *a*, when the load on the member is small, whereas a web plate, as in *b*, may be used for larger loads. A web plate, although it decreases the least radius of gyration slightly, can be considered effective in carrying the load, whereas the lacing bars only serve as bracing. In order to obtain higher values of *r* in shapes *a* and *b*, the longer legs should be turned out, as indicated. Four angles arranged in a box, as shown in *c*, are sometimes used in order to

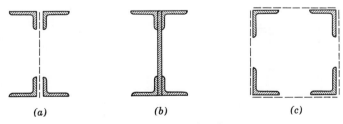

(a)   (b)   (c)

Fig. 10-9 Four-angle columns.

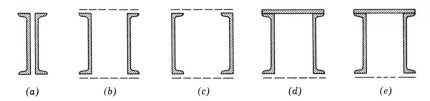

Fig. 10-10   Two-channel columns.

obtain a higher value of *r*. Such sections, however, require a large amount of lacing, and are economical only for long members carrying small loads.

Two channel sections back to back are occasionally used (Fig. 10-10), although they have a small value of *r* about the *y* axis and therefore are not considered advantageous in general. Two channels spaced apart constitute a good section. For compression members, the flanges of the channels are generally turned out (Fig. 10-10c) in order to obtain a larger value of *r* for a given spacing between the two channels, whereas, for most tension members as well as some compression members, which do not require a large value of *r*, the flanges are generally turned inward in order to minimize lacings and to obtain better connections at the joint (Fig. 10-10b, *d*, and *e*).

When a greater cross-sectional area is required or where stiffer bracing between the parts is necessary, the lacing bars on one or both sides are replaced by a solid or a perforated cover plate, as shown in Fig. 10-11. With a solid cover plate on one side only (Fig. 10-11d), the interior is still accessible for inspection and painting, although this is not so for solid cover plates on both sides. For this reason, members with solid cover plates are more commonly used in buildings, where maintenance painting is not required, whereas in bridge structures perforated cover plates are used. Plate and angle shapes are employed when the maximum-size channels do not provide sufficient area. Sections similar to those shown in Fig. 10-11 have a favorable value of radius of gyration, and can be arranged to obtain any desirable area. Furthermore, they can be easily connected and thus serve well as compression members of a bridge truss.

Occasionally I sections can also be built-up by welding (Fig. 10-12). This

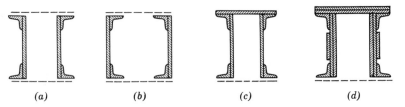

Fig. 10-11   Plate and angle columns.

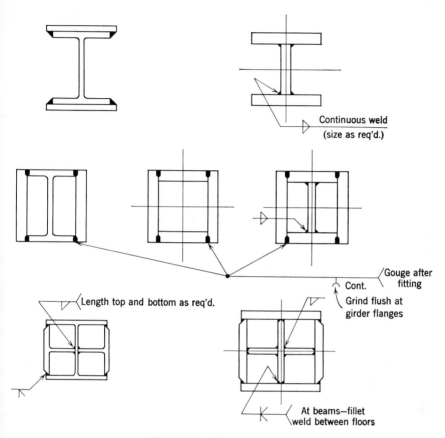

Fig. 10-12   Built-up columns.

is not often used because of the wide variety of I and W^F sections now available. But the sections shown in Fig. 10-12 are occasionally adopted. A built-up section for very heavy building columns is shown in Fig. 10-13.

10-6 LATTICED COLUMNS

Large compression members in bridges and other structures are sometimes built up using rolled shapes connected by an open lattice (Fig. 10-14). Three types of lattice are commonly used: lacing with or without battens (plates perpendicular to column axis), battens only, and perforated cover plates. The function of the lattice is to assure integral action of the solid longitudinal segments, usually called "main" segments. In designing latticed columns,

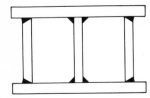

Fig. 10-13 A heavy built-up building column.

the following conditions must be considered: buckling of the column as a whole under axial load, buckling or yielding of individual segments of the column, strength of the lattice frame, and the distortion of the cross section (Fig. 10-15).

Buckling Strength—Axial Load. The buckling strength of latticed columns is smaller than that of solid columns having the same slenderness ratio and the same area, provided that the solid column does not buckle locally because of thinness of material. As indicated in Sec. 8-13 a column bent about axis xx is subjected to a shear force $V' = P \sin \theta$, which is the shearing component of load P (Fig. 8-34). This shearing force produces deformations in the lattice which tend to reduce the over-all stiffness of the column and therefore reduce the buckling strength of the column. Buckling strength for a pin-ended latticed column subjected to axial load can be expressed as

$$P_{\rm cr} = \frac{\pi^2 \bar{E}(I/k^2)}{L^2} = \frac{\pi^2 \bar{E} I}{(kL)^2} \qquad (10\text{-}52)$$

where I/k^2 is the reduced stiffness of the column, and $\bar{E}$ is the effective modulus of elasticity, which may be taken as the tangent modulus E_t. The

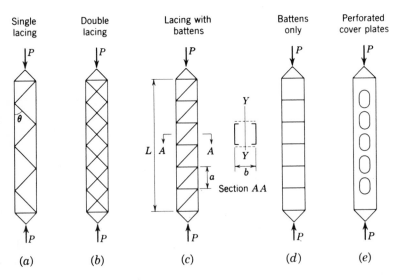

Fig. 10-14 Various latticed columns.

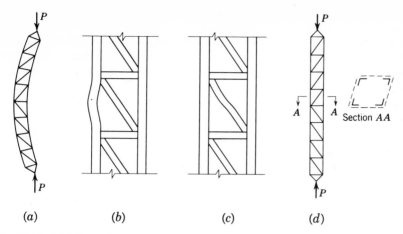

Fig. 10-15 (*a*) failure of column as a whole, (*b*) failure of a main segment, (*c*) failure of a lattice member, and (*d*) Distortion of cross section.

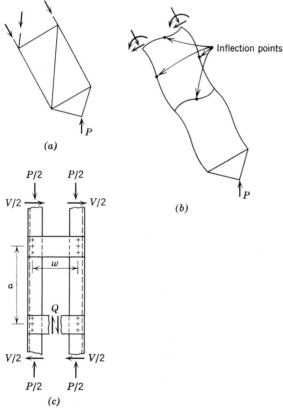

Fig. 10-16 Deflected shapes of latticed columns.

values of k can be derived theoretically[1] by approximating the shearing deformation of the lattice. In a column with lacing it is assumed that the lattice subjected to shear V acts as a simple pin-connected truss (Fig. 10-16a). In columns with battens only it is assumed that the shear V is carried by bending of the main segments and of the batten plates (Fig. 10-16b). Columns with perforated plates may be considered as columns with battens which are rigid, so that the shear is resisted essentially by the main segments (Fig. 10-16c). Perforated plates, unlike other types of lattice, contribute to the cross-sectional area and the flexural stiffness of the column, and these can be determined on the basis of net section properties of the perforated plates. Theoretical values of k for various latticed pin-ended columns follow.

Lacing without Battens

$$k = \sqrt{1 + \frac{\pi^2}{(L/r)^2} \frac{A}{A_d} \frac{1}{\cos \theta \sin^2 \theta}} \tag{10-53}$$

where L = distance between pinned ends of the column

r = radius of gyration of the over-all section of the column

A = cross-sectional area of the column

A_d = cross-sectional area of the diagonal lacing elements in a panel

θ = angle of inclination of lacing elements to the longitudinal axis of the members (Fig. 10-14a).

Table 10-3   Values of k for Single or Double Lacing

A/A_d \ L/r	$\theta = 60°$ Single Lacing				$\theta = 45°$ Double Lacing			
	30	50	70	90	30	50	70	90
5	1.07	1.03	1.01	1.01	1.08	1.03	1.02	1.01
20	1.26	1.10	1.05	1.03	1.28	1.11	1.06	1.04
50	1.57	1.24	1.12	1.08	1.60	1.25	1.14	1.08

Batten Plates Only

$$k = \sqrt{1 + \frac{\pi^2}{12(L/r)^2}\left(\frac{A}{A_b}\frac{ab}{r_b^2} + \frac{a^2}{r_z^2}\right)} \tag{10-54}$$

where A_b = cross-sectional area of battens

a = spacing of battens (Fig. 10-14)

b = distance between centroids of main segments

r_b = radius of gyration of batten

r_z = radius of gyration of main segment about its own centroidal axis

Table 10-4   Values of k for Batten Plates

$$\lambda = \frac{A}{A_b}\frac{ab}{r_b^2}$$

λ \ a/r_z	$L/r = 40$			$L/r = 60$			$L/r = 80$		
	10	20	30	10	20	30	10	20	30
0.0	1.03	1.10	1.21	1.01	1.04	1.10	1.00	1.03	1.06
10.0	1.03	1.10	1.21	1.01	1.04	1.10	1.00	1.03	1.06
100.0	1.06	1.12	1.23	1.02	1.12	1.12	1.01	1.04	1.07

When shear rigidity of the battens is small, the column stiffness may be reduced more than is indicated by the above values[1] of k.

Lacing with Battens

$$k = \sqrt{1 + \frac{\pi^2}{(L/r)^2}\left(\frac{A}{A_d}\frac{1}{\cos\theta \sin^2\theta} + \frac{A}{A_b}\tan\theta\right)} \qquad (10\text{-}55)$$

Perforated Plates

$$k = \sqrt{1 + \frac{\pi^2}{12(L/r)^2}\left(\frac{a}{r_z}\right)^2} \qquad (10\text{-}56)$$

where a = spacing of perforations, taken as 1.5 times the length of perforation. (For selected values of k, see Table 10-4, when $\lambda = 0$.)

Yielding or Buckling in Main Segments.   When a main segment of the column yields or buckles locally, a latticed column may fail before the critical buckling load P_{cr} is reached. The maximum stress in a main segment of a latticed column can be determined by considering two effects: (a) bending of the column as a whole, and (b) local bending of the segment acting as a separate column between lattice supports, as follows. For given loading conditions on the column, the stress at any point can be determined by simple theory, $f = (P/A) + (My/I)$, where the moment M is calculated using beam-column (Eq. 8-50). The cross-sectional area A and the moment of inertia I are those of the net section of the column. The segment that carries the greatest stress f may be considered as a separate column with unsupported length a. The stress f, as determined previously, together with effects of V, the shear component of axial load, and possible effects of crookedness, define the loading conditions of this column. Then the strength of this segment, whether governed by yielding or buckling, can be verified as described in the preceding sections of this chapter.

This procedure may require elaborate arithmetical calculations, although for simple loading conditions these are not mathematically complex. The advantage of this method is that the actual maximum stress in the column under design load is determined and can be compared to the yield strength of the material.

Shear Force on the Lattice. In order to assure integral action of main segments, the lattice must be capable of resisting the following loads: stresses due to external transverse loads on the column, stresses due to the deflected shape of the column, stresses due to shortening of main segments, and stresses due to "crushing" induced by the curvature of the column. Usually the latter two are negligible and are not considered in the design. Also the connections between the main segments and the lattice must be adequately designed. Shears V due to external loads are determined from a conventional shear diagram (Fig. 10-17). The shearing component V' caused by the axial load P depends on the local slope of the bent column and is

$$V' = P \sin \theta \simeq P\theta \tag{10-57}$$

For given loading conditions, the value of V' may be determined from the beam-column equation, or it may be estimated assuming the column shape to be a parabola, a sine curve, or the elastic curve of a uniformly loaded beam. The maximum end slopes θ corresponding to these assumptions are, respectively: $4\Delta/L$, $\pi\Delta/L$, and $3.2\Delta/L$, where Δ is the maximum deflection at midlength of the column.

The value of Δ is in part due to initial eccentricity and crookedness and in part because of secondary bending as a result of axial load. If the component of Δ due to initial eccentricity e is such that $ec/r^2 = 0.25$, and the component of Δ due to secondary bending is Δ', so that $\Delta' = m(f_b/E)(L^2/c)$ where m is a numerical coefficient depending on the type of loading producing the deflection, then

$$\Delta = e + \Delta' = 0.25\frac{r^2}{c} + m\frac{f_b}{E}\frac{L^2}{c} \tag{10-58}$$

Assuming $r/c = 0.8$, which represents an average r/c ratio for built-up columns, and assuming the deflected shape of the column to correspond to an elastic curve of a uniformly loaded beam, for which $m = \frac{5}{48}$ and $\theta_{max} = 3.2\Delta/L$, the following equation results from Eq. 10-57:

$$\frac{V'}{P} = \theta = 3.2\frac{\Delta}{L} = 3.2\left(0.25\frac{r^2}{c} + \frac{5}{48}\frac{f_b}{E}\frac{L^2}{c}\right)\frac{1}{L}$$

$$= 0.64\frac{r}{L} + 0.133\frac{f_b}{E}\frac{L}{r} \tag{10-59}$$

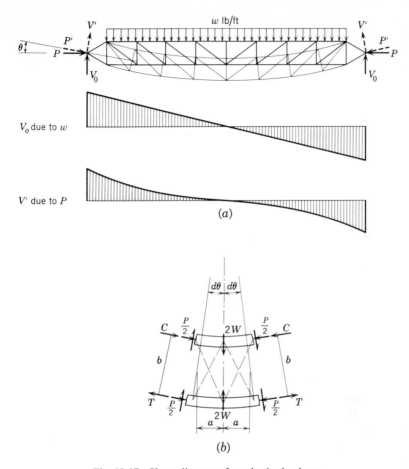

Fig. 10-17　Shear diagrams for a latticed column.

For a value of $f = 16,000$ psi at the design load P, and $E = 29 \times 10^6$ psi, the foregoing equation is simplified to the following:

$$\frac{V'}{P} = 0.64 \frac{r}{L} + \frac{1}{13,600} \frac{L}{r} \tag{10-60}$$

Equation (10-60) serves to indicate the general order of magnitude of the shear component due to axial load P. For example, if $L/r = 100$, $V' = 0.014P$. In practice, various codes define minimum values of V' to be considered in design as a specified percentage of axial load P; 2% is a common value.

In addition to shear loads, the lattice must resist secondary crushing loads.

Consider a free body of a panel of a latticed column shown in Fig. (10-17). The column is subjected to an axial load P and a bending moment $M = C_b = T_b$. Because of change in angle between sections, the compressive forces C have a resultant force $2W$ which is equal and opposite to the resultant $2W$ of the tension forces T. The resultant forces $2W$ between top and bottom segments exert a crushing force on the lattice.

Assuming that the moment M is constant over the length $2a$,

$$2\,d\theta = \frac{M}{EI}\,2a \tag{10-61}$$

and

$$2W = C2\,d\theta = \frac{Mc}{I}\frac{A}{2}\frac{M}{EI}\,2a = \frac{M^2}{EI}\frac{ac}{r^2} \tag{10-62}$$

where $c = b/2$, and A is the total cross-sectional area of the column. For conventional latticed columns W is only a fraction of 1% of axial load P and is usually negligible in the design of conventional structures. Only when the curvature M/EI becomes large must the effect of the crushing load be taken into account.

Distortion.   In addition to over-all buckling of the column and the strength of latticed webs, the possibilities of distortion in the plane of the cross section (Fig. 10-15d) must be considered. Usually such distortion is prevented by the use of transverse diaphragms or bracing.

Design Requirement for Column Lattice.   Type and arrangement of column lattice (Fig. 10-14) is determined by requirements of strength, fabrication, maintenance, and cost. With modern methods of automatic gas cutting and welding, perforated cover plates are often used in preference to other types of lacing.

When single lacing, double lacing, or lacing with battens is used (Fig. 10-14a, b, and c), the loads in the diagonals and battens caused by the total transverse shear $V = V_o + V'$ (Fig. 10-17) may be determined by assuming that the members of the lattice are hinged at the connections and resist axial forces only. These loads are easily computed and the diagonals and battens must be proportioned as tension or compression members to resist these loads safely. Also the connections between the lattice and the main segments must transmit these loads safely.

When a built-up column is tied with battens only the transverse force V induces secondary bending in the main segments and the battens. The bending moments at critical sections caused by V can be calculated, assuming location of the inflection points as shown in Fig. 10-16b, and the resulting axial and bending stresses in the main segments and in the battens should not

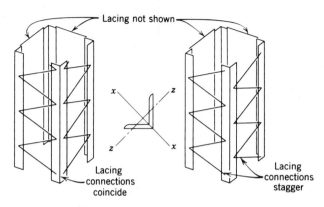

Fig. 10-18   Latticed columns with four angles.

exceed safe limits. Shearing stresses in the battens and in the main segments should also be checked. The total shearing force in the battens is $Q = Va/w$ (Fig. 10-16c); the shearing stress is $f_v = \frac{3}{2}Q/A_s$, where A_s is the total shearing area of the battens resisting force Q.

A special problem arises in a column built up with four angles at the corners and lacing members on all four sides (Fig. 10-18). If lacing connections in the two adjacent faces are made to coincide, then the angle is fully restrained at that point and would tend to buckle between the lacing connections in the weak direction, bending about axis zz. If lacing connections are staggered in the two adjacent faces, then the angle is not fully restrained at the connections and would tend to buckle about an axis other than its principal axis zz. Consider a column with several panels of staggered lacing bars, and assuming that the lacing bars do not deform, the critical buckling load for the angle with staggered supports can be defined as

$$P_{cr} = \frac{C\pi^2 E I_z}{L^2} \tag{10-63}$$

where the value of C depends on the proportions of the angle, particularly on the ratio of I_x/I_z. The proportions of an average angle shape are such that $I_x/I_z = 3.8$, which corresponds to $C = 1.75$. It is apparent from Eq. 10-63 that the effect of the staggered lattice is to increase the buckling strength of the angle, which indicates that such an arrangement is more efficient than the one using the unstaggered lattice.

When perforated cover plates are used (Fig. 10-14e), they should be proportioned to resist axial compressive stresses as well as shearing stresses.[2] Usually secondary bending stresses and stress concentrations around openings may be neglected. Similar to latticed columns, the allowable column stress should be determined from an appropriate specification or code. The length c (Fig. 10-19) of perforation should not exceed 20 times the least radius of gyration of a column segment in order to develop full value of the allowable stress. The allowable load on the column is equal to the product of allowable stress and the net section taken through the perforations.

The shape of the perforations should be ovaloid, elliptical, or circular. For the first two shapes, the long axis of the perforation should be in the direction of the column axis, and the spacing a of perforations shall not be less than $1.5c$. A member with staggered perforations should be treated as if the perforations were opposite to each other. Perforated and other cover plates must be detailed in such a way that axial compressive stresses do not cause failure by local buckling before the strength of column is reached. The criterion for design of these plate elements is obtained by equating the plate buckling stress (f_{cr}) plate and the column strength (f_{cr}) column. The plate buckling stress is defined as

$$(f_{cr})_{\text{plate}} = k\bar{E}\left(\frac{t}{w}\right)^2 \qquad (10\text{-}64)$$

where k is a coefficient determined by the type of loading, the conditions of edge support, and the proportions of the plate; $\bar{E}$ is the effective modulus of elasticity corresponding to the critical stress f_{cr}; and t/w is the ratio of thickness to width of the plate element. The strength of the column in general is

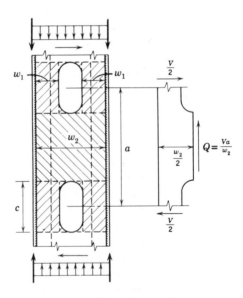

Fig. 10-19   Perforated cover plate.

defined as some function of modulus of elasticity $\bar{E}$, yield point stress F_y, and slenderness ratio L/r. Thus the proportions of plate element can be defined as a function of L/r, $\bar{E}$, and F_y. The general form of this function is complex, but it can be approximated with sufficient accuracy by two simple expressions[3] defining the following conservative design criteria for practically all types of structural steel:.

$$\frac{w}{t\sqrt{k}} = 19.2 \qquad \text{for} \quad \frac{L}{r} \text{ equal or less than } 60$$

$$\tag{10-65}$$

$$\frac{w}{t\sqrt{k}} = 0.32\,\frac{L}{r} \qquad \text{for} \quad \frac{L}{r} \text{ greater than } 60$$

The width-thickness ratio w/t may be thus determined when the values of k are known. For a perforated plate (Fig. 10-19), two regions are of interest. The element w_1, subjected to compression, is free along one edge and re-strained along the other, which is riveted or welded to the main segment. The restraint along the edge may vary between that equivalent to a longitudi-nal hinge, which permits the plate to rotate about the axis of the edge, and that of a continuous clamping connection, which prevents no rotation. The k values corresponding to these two conditions are $k = 0.38$, and $k = 1.15$, respectively; conditions realized in practice result in restraint that approaches the lower of the two values.

The element w_2 is restrained along both edges. These restraints may vary from a longitudinal hinge to a clamping connection, with corresponding k values varying from 3.6 to 6.3. Again practical conditions more often approach the lower value.

The values of w/t ratios for the various edge restraints are shown in Table 10-5.

These ratios agree reasonably well with the minimum thickness specified by the various codes for plate elements subjected to compression.

Another method for determining w/t ratio for a column plate element is to let the plate buckling stress f_{cr} equal a stress n times the allowable column

Table 10-5 Width-Thickness Ratios

Edge Restraint	k	$L/r \leq 60$	$L/r > 60$
One edge free, another hinged	0.38	12	$0.2L/r$
One edge free, another clamped	1.15	21	$0.35L/r$
Edges hinged	3.6	36	$0.6L/r$
Edges clamped	6.3	48	$0.8L/r$

stress f, and then solve for maximum w/t corresponding to this condition. Then,

$$\frac{w}{t} = \left(\frac{k\bar{E}}{f_{\mathrm{cr}}}\right)^{\frac{1}{2}} = \left(\frac{k\bar{E}}{nf}\right)^{\frac{1}{2}} \tag{10-66}$$

The principal difficulty in using this expression is the determination of appropriate values of $\bar{E}$ and k. For example, with $k = 3.6$, $f = 15$ ksi, $n = 1.8$, the value of $f_{\mathrm{cr}} = nf = 1.8 \times 15 = 27$ ksi. Assuming a value of $\bar{E} = 15 \times 10^3$ ksi (corresponding to $f_{\mathrm{cr}} = 27$ ksi), the width-thickness ratio is

$$\frac{w}{t} = \left(\frac{3.6 \times 15 \times 10^3}{18.15}\right)^{\frac{1}{2}} = 45 \tag{10-67}$$

This value is slightly less conservative than the limiting ratio indicated in Table 10-5.

10-7 SPLICES AND SPECIAL CONNECTIONS

Introduction. Since the strength of columns is greatly affected by the end conditions and the eccentricity of the load, the design of column splices, bases, caps, and brackets must be considered together with the design of column sections. The problem of beam-to-column connections, however, depends to a large extent on the design of beams, and therefore will be considered in a subsequent chapter, after beam design has been discussed.

Column Splices. Similar to a tension splice, the design load for a compression splice can be defined as (*a*) the computed force in the member, or (*b*) the maximum allowable load, which is usually greater than the computed force. If the actual load on the member is known accurately, it may be sufficient to design just for that force. On the other hand, designing the splice for a load less than the load-carrying capacity of the member reduces the overload capacity of the structure. Hence, it is often desirable to design the splices for the full strength of the member.

Whereas in a tension splice the entire load must be transmitted by the splice material, in a compression splice a part of the load may be transmitted by direct bearing between the column ends (Fig. 10-20). In order to assure direct bearing at a splice, the ends of columns should be milled where they join or bear. Since a large part of the total load in the column can be transmitted through the milled bearing area, the splice material may be designed to transmit only the remainder of the load. The exact proportion of the load transmitted by the splice plates cannot be determined accurately because perfect matching of the two column ends is difficult to achieve under normal construction conditions.

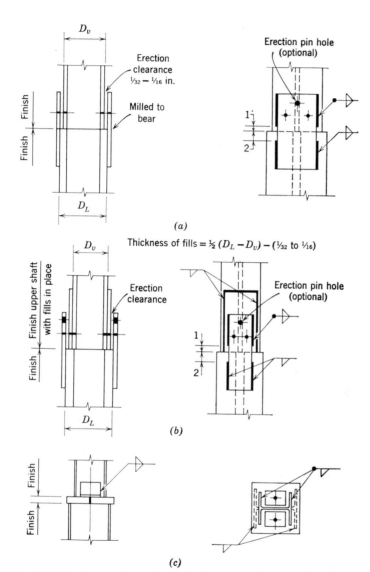

(a)

(b)

(c)

Fig. 10-20   Welded column splices. (*a*) depth of upper and lower sections nominally the same, (*b*) depth of upper section nominally 2 in. less than depth of lower section, and (*c*) butt plate splice. (From AISC Manual.)

Although no exact solution is possible, the general problem of load distribution between the column and the splice material may be illustrated by the following. Assuming perfect bearing contact between two ends of a W column with cross-sectional area A_c, subjected to a perfectly concentric load P, and spliced by two plates continuously welded to the outside flanges (Fig. 10-20), and having cross-section area A_s, the load distribution between the splices and the column is easily determined. Neglecting any bending effects at the splice section and assuming perfect bearing contact between column ends, the average stress f is

$$f = \frac{P}{A_s + A_c} \qquad (10\text{-}68)$$

the load transmitted through the column is

$$P_c = fA_c = \frac{P}{1 + A_s/A_c} \qquad (10\text{-}69)$$

and the load in the splice is

$$P_s = fA_s = \frac{P}{1 + A_c/A_s} \qquad (10\text{-}70)$$

It can be noted that, if A_s is very small, P_s is also very small, whereas if $A_s = A_c$, $P_s = P/2$, or the splice will resist half the total load. The assumptions underlying the foregoing calculations are all unconservative, since the load is not perfectly concentric, the two ends may not match perfectly, and bending effects at the splice section generally cannot be neglected. Depending on the discrepancies between the ideal case and the actual conditions in the structure, the splice material may be required to carry anywhere from 25 to 100 percent of the total load. Twenty-five percent of total load is considered to be a bare minimum design value necessary to assure a splice which can hold the column in place, whereas, if the column ends are not milled to bear, 100% of the load must be transmitted by the splice material.

Some members may be subjected to compression combined with bending sufficiently large to produce tension in some parts of the section. Then the part subjected to tension must be spliced so that it transmits all its tensile load; the part subjected to compression must also be spliced heavily. Hence in most instances it is advantageous to locate splices at sections of low bending. As a practical compromise, it may be desirable to make the splice plates equivalent to column flange material.

Requirements for splicing bridge compression members are given in AREA and in AASHO Specifications. AISC does not specifically provide for the design of column splices. For building columns the ends are often milled to bear, and, whenever practicable, splices are located near floor level; 18 in.

above floor level is commonly used for columns subjected to essentially axial loads. The portion of the load to be transmitted by the splice depends on the amount of bending in the column. Twenty-five percent of the load may be satisfactory if the amount of bending is very small at the splice; if the columns are designed to transmit lateral loads in buildings, large moments may be acting on the column, and the splices must be designed to transmit the combination of axial and bending loads, unless located near the points of inflection.

The allowable average stress for the column is usually taken as the allowable stress for the splice material. Thus the area of the splice material can be determined directly as a given proportion of the gross area of the column cross section. Although this procedure may not be entirely rational, it is justified primarily because of the uncertainty about actual stresses in the splice material.

Several typical column splices are shown in Fig. 10-21. The simplest type can be used when the outside dimensions of the columns are matched; then only two or four splice plates will be required. Because of the practice of spreading rolls, the inside flange dimensions of the columns are usually matched so that fillers are required only on the outside. When column sizes differ so greatly that a large part of the area cannot bear directly, some form of bearing butt plate will be required between the columns; the fillers could be quite thick, and extra lines of fasteners may be needed (Fig. 10-21*d*). Additional typical details for column splices are shown in AISC Manual of Steel Construction.

10-8 BRACKETS

Column brackets are required to support loads located eccentrically with respect to the center line of a column. For the lightest loads, up to 35 kip, an unstiffened seat angle, riveted or welded to the column, might serve the purpose (Fig. 10-22*a*). For heavier loads a stiffened angle seat or a split tee may be used (Fig. 10-22*b*, *c*, and *d*). When eccentricities are large, cantilever bracket designs, such as Figs. 10-22*e* and *f*, may be necessary.

A special problem occurs in the design of crane runways. If the crane girder loads are relatively small compared to the roof and floor loads on the column, it is economical to use a bracket seat connected to the flanges of the main column (Fig. 10-23). When girder loads are heavy, or when the center line of the crane runway is located far from the center of the column, the eccentric moments may become excessive, and it may be more economical either to step the column at the level of the girder (Figs. 10-24*a* and *b*) or to design a separate column to support the runway (Fig. 10-24*c*).

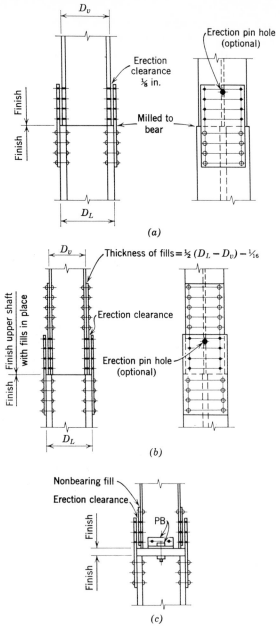

Fig. 10-21 Riveted and bolted column splices. (*a*) depth of upper and lower sections nominally the same, (*b*) depth of upper section nominally 2 in. less than depth of lower section, and (*c*) butt plate splice. (From AISC Manual.)

446

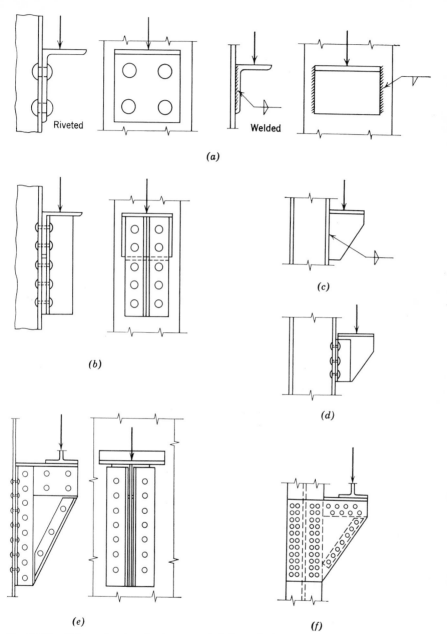

Fig. 10-22 Column brackets for beams and girders. (a) bracket without stiffeners (angle-seat type), (b) angle-seat bracket with stiffener, (c) welded angle and plate bracket, (d) riveted angle and plate bracket, (e) stiffened angle and plate bracket, and (f) plate bracket connected parallel to column flanges.

447

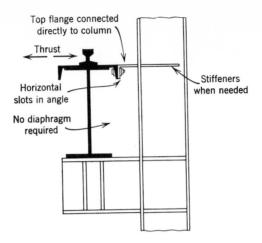

Fig. 10-23   Column bracket for crane girder.

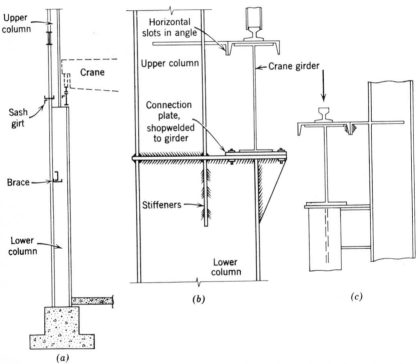

Fig. 10-24   Columns for crane girders. (*a*) a stepped-up column, (*b*) details of stepped-up column, and (*c*) two separate columns connected by ties.

10-9 COLUMN BASES AND CAPS

Column bases can be classified into two types; one transmitting direct load only; another carrying an appreciable bending moment in addition to direct load. When a column is subjected to direct load only, or when the shear load and bending moment at the base are negligible, the design of the base does not present special problems. Usually a steel plate is used to distribute the column load over a sufficient area to maintain the unit bearing stress within the allowable value for the concrete footing.

For small columns, the base plates can be shop-welded to the column (Fig. 10-25*a*). For large columns, the base plates are shipped loose, set to proper level, and grouted in place before the erection of columns. Anchor bolts, to secure the columns in position, are embedded in the concrete

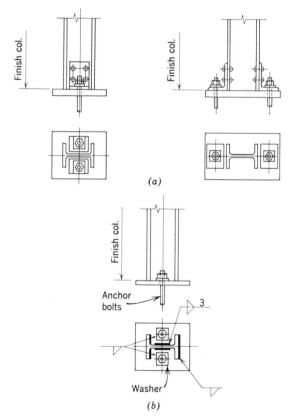

Fig. 10-25 Column bases for direct compressive loads. (From AISC Manual.)

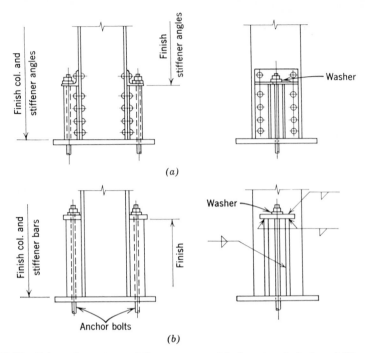

Fig. 10-26 Column bases transmitting moments. (*a*) riveted or bolted, and (*b*) welded. Base plate shop-welded to column shaft, or shipped as a separate piece. (From AISC Manual.)

footing and project above the base plate to engage lug angles, which are shop-welded to the column flanges or web (Fig. 10-25*b*).

If the base plate is shop-welded to the column care must be exercised to provide a uniform bearing under the plate in its final position. The top of the concrete can be prepared accurately to the proper level, and a thin lead plate can be inserted on top of the concrete to compensate for minor irregularities. Alternately the base plates can be set on wedged-shaped steel or hardwood shims and grouted after the entire first tier of columns is plumbed and leveled. Sometimes a thin steel plate equal in size to the steel base plate is grouted in place at the correct elevation and the base plate is set directly thereon.

For column bases transmitting moment, angles anchored to the footing may be used (Fig. 10-26*a*), provided the angles are designed for uplift forces. Stiffeners can be used to advantage for welded construction, as in Fig. 10-26*b*. The main problems in the design of the column bases are to determine the size of the plates and their thicknesses. The size is determined by the required

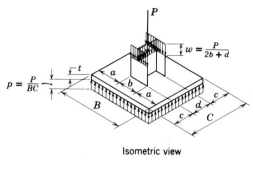

Isometric view

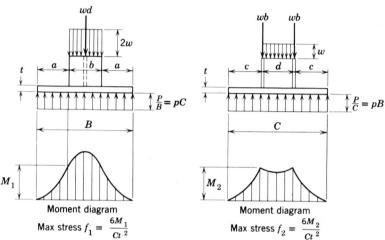

Moment diagram

Max stress $f_1 = \dfrac{6M_1}{Ct^2}$

Moment diagram

Max stress $f_2 = \dfrac{6M_2}{Ct^2}$

Fig. 10-27   Base-plate design—Method 1.

bearing area on the foundation; and the required thickness is determined so that bending stress in the plate does not exceed the allowable value.

The bending stress is determined as $f = Mc/I = 6M/bt^2$, where M is the bending moment in the plate. The actual value of M is highly indeterminate, primarily because the pressure distributions between the column and the plate and between the footing and the plate cannot be accurately determined analytically. Two approximate methods are used. One is based on the assumption that bearing pressures are uniformly distributed over all contact surfaces (Fig. 10-27). The other method is based on the assumption that the load between the plate and the column is uniformly distributed over some assumed equivalent area (Fig. 10-28). AISC recommends that the equivalent area should be defined by assuming $b = 0.80b$ and $d = 0.95d$, and that the bending moments be assumed uniformly distributed across the width of the base plates. For both methods the moments at the edges of the column are used for computing maximum fiber stresses. Although these assumptions do not agree with the elastic plate theory, they serve adequately as empirical procedures to determine base-plate thickness.

For roofs of some buildings and especially roofs of one-story structures, it has been found convenient to connect the roof beams to the top of the columns. This is accomplished by welding a cap plate to the column section and permitting the beam to be placed on it. The beam is held in place by two bolts through the flange and cap plate, as indicated in Figs. 10-29a and b.

When two beams are supported by the same column, as in Fig. 10-29c, the cap plate must be large enough to provide a sufficient bearing area for the beams. If the column section does not have sufficient depth, as in Fig. 10-29c, an alternate procedure is indicated in Fig. 10-29d. Although this figure

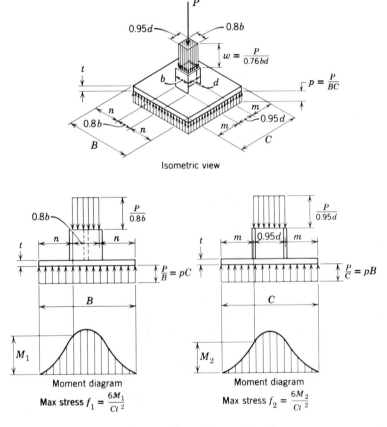

Isometric view

$$\text{Max stress } f_1 = \frac{6M_1}{Ct^2}$$

$$\text{Max stress } f_2 = \frac{6M_2}{Ct^2}$$

Fig. 10-28 Base-plate design—Method 2.

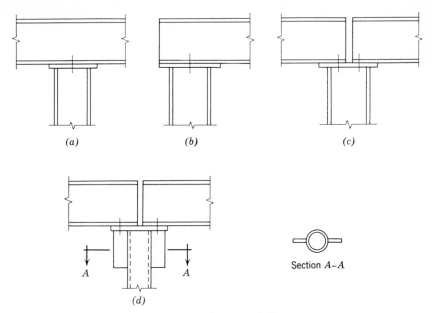

(a)   (b)   (c)

Section A–A

(d)

Fig. 10-29   Column cap designs.

illustrates two beams resting on a circular column, the method is easily adaptable to the inside flange column sections. The vertical stiffeners and welds must be designed to transmit their share of the beam reaction to the side of the column section.

Example 10-3.   A 14 W^F 78 steel column with an unsupported length of 16 ft rests on a concrete footing. Compute and choose the required thickness and size of the rolled base plate, in order to carry the full direct-load strength of the column, according to the AISC Specification. Derive the formula for t on p. 3-75 of the AISC Manual.

Solution

Maximum moment $M = \frac{1}{2}pm^2$ kip-in./in., assuming $m > n$, and p is bearing pressure.

$$f = \frac{M}{I}(c) = \frac{\frac{1}{2}pm^2}{\frac{1}{12}(1)(t)^3}(\tfrac{1}{2}t) = \frac{3pm^2}{t^2} = F_b$$

$$\therefore\ t^2 = F_b pm^2$$

$$t = \sqrt{\frac{3pm^2}{F_b}}$$

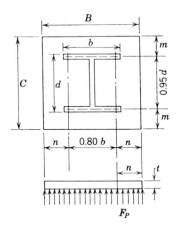

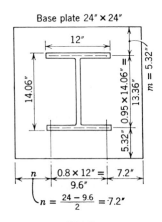

Fig. 1

Design Base-Plate. Properties of 14 W⁻ 78 from p. 1-12, AISC Manual:

$$A = 22.94 \text{ in.}^2, \qquad r = 3.00 \text{ in.}$$

$$L = 16 \text{ ft} = 16 \times 12 = 192 \text{ in.}$$

Hence

$$\frac{L}{r} = \frac{192}{3} = 64$$

For A36 steel, from Table 10-1:

$$F_a = 17 \text{ ksi}$$

Allowable concentric column load:

$$P = F_a \times A = 17 \times 22.94 = 390 \text{ kip}$$

Allowable bearing pressure $= F_p = 0.25f_c' = 675$ psi for 2700 psi concrete. Required area of base plate

$$A_p = \frac{P}{F_p} = \frac{390}{675} = 578 \text{ in.}^2$$

Using 24 in. × 24 in. base plate; $A_p = 576$ in.^{2}

$$p = \frac{390}{576} = 0.678 \text{ ksi}$$

From Fig. 1, $n > m$; $F_b = 27$ ksi;

$$\therefore \quad t = \sqrt{\frac{3 \times 678(7.2)^2}{27,000}} = 1.98 \text{ in.}$$

Use 2-in.-thick plate.

PROBLEMS

Note–Use A36 steel unless otherwise specified.

1. An interior building column is 20 ft long and is to support a 200-kip axial load. The column is supported laterally in all directions at the ends only. Using AISC Specification, select the most economical W section for this column.

2. A truss member 12 ft long is subjected to 55 kips compression. The member is to be built of two equal-leg angles back to back separated by $\frac{3}{8}$-in. gusset plate or filler. The member is supported laterally at the ends only. Using AISC Specification, choose the lightest angles for the section.

3. An alternative arrangement for the truss member of Prob. 2 is a "star" section as shown in Fig. P-3. Select the lightest angles permitted by AISC Specification.

Fig. P-3

4. A 10-ft-long main column is fixed at the bottom and free at the top. The column is to carry a 100-kip axial load. Select the lightest W section acceptable by AISC Specification.

5. The columns shown in Fig. P-5 are 10 ft long and simply supported at the ends, and they carry a 100-kip eccentric load P as shown. Select the lightest W sections (*a*) using conventional AISC Specification, and (*b*) using load factor 1.8 and AISC plastic design criteria.

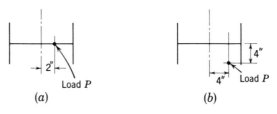

(*a*)   (*b*)

Fig. P-5

6. For the steel column of each floor shown in Fig. P-6, assuming no sidesway at each floor, select economical W sections: (*a*) neglecting effect of eccentricity, and (*b*) considering these effects. Use AISC Specification, and tabulate results of *a* and *b*.

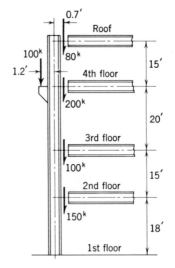

Fig. P-6

7. A five-story-building steel frame is shown in Fig. P-7. The design of the frame is based on the assumption that connections transmit no moment, and on the following gravity loads: Roof: dead load 30 psf, live load 20 psf; each floor: dead load 70 psf, live load 150 psf; exterior walls: dead load 30 psf. The column spacing a, b, may vary as follows: (1) $a = b = 15$ ft, (2) $a = b = 20$ ft, (3) $a = b$ $= 25$ ft, (4) $a = 15$ ft, $b = 20$ ft, (5) $a = 15$ ft, $b = 25$ ft, (6) $a = 20$ ft, $b = 25$

ft. Determine design loads for typical interior, exterior, and corner columns of the fifth floor and of the first floor. Consider permissible reduction in live loads as follows: The live load on any member supporting a loaded area of 150 ft² or more may be reduced at the rate of 0.08% per ft² of area supported by the member. The reduction R shall not exceed either 60% or the following value (in percent):

$$R = \frac{\text{dead load} + \text{live load}}{4.33 \text{ live load}} \, 100$$

Using AISC Specification, determine the size of W shapes required for the columns.

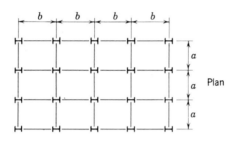

Plan

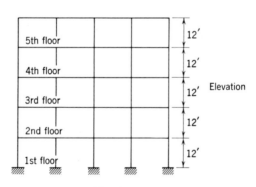

Elevation

Fig. P-7

8. The sections of a building column shown in Fig. P-8 are to be spliced. Determine the following: (*a*) thickness of filler plates, (*b*) load transmitted by filler plates, (*c*) number of ⅞-in. diameter A325 bolts or rivets required. Draw a sketch showing all dimensions of the filler plates and splice plates, and show rivet or bolt arrangement. (AISC Specification.)

9. The sections of a building column, as shown in Prob. 8, are to be spliced by welding instead of bolting. Select weld size for shop and field welds, and design

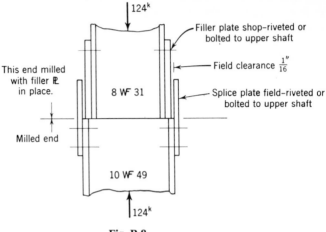

Fig. P-8

the connection. Draw a sketch showing all dimensions using weld arrangements, as shown in Fig. P-9. (AISC Specification.)

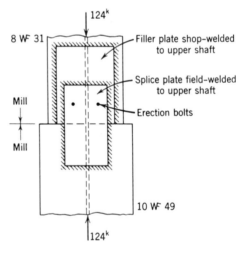

Fig. P-9

10. A top-chord member of a truss bridge shown in Fig. P-10 is subjected to an axial compressive load of 250 kip and a uniformly distributed transverse load of 2 kpf (including weight of member). According to the AASHO Specification, determine whether the section shown is adequate to carry the loads. Assume hinged ends for the truss member.

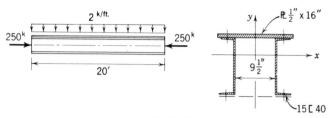

Fig. P-10

11. A section shown in Fig. P-11 consists of four $6 \times 6 \times \frac{7}{16}$-in. angles, two solid plates $24 \times \frac{1}{2}$-in., and two perforated plates 26 in. wide. The length of the member is 100 ft, and it is to carry an axial compression load of 500 kip. Determine the thickness of the perforated plates and the permissible size and spacing of ovaloid perforations. Use AASHO Specification.

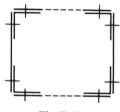

Fig. P-11

REFERENCES

1. Timoshenko, S., and J. Gere, *Theory of Elastic Stability*, McGraw-Hill Book Co., New York, 1961.
2. Johnston, B. G., "Guide to Design Criteria for Metal Compression Members," Column Research Council, 2nd ed., John Wiley and Sons, New York, 1966.
3. White, M. W., and B. Thurlimann, "Study of Columns with Perforated Plates," *AREA Bull.* **531** 1956.
4. Massonnet, C. E., and M. A. Save, *Plastic Analysis and Design*, Vol. 1, Blaisdell, New York, 1965.

11

Design of Beams and Girders

ᗡᗡᗡ

11-1 INTRODUCTION

One of the principal load-carrying members ın steel frames is a beam or girder. The theoretical considerations of beam and girder behavior were presented in Chapter 8. This chapter will discuss the practical aspects of designing beams and girders and thus will develop other factors pertinent to design, such as allowable stresses in bending, shear and bearing, deflection considerations, stiffening of thin girder webs, connections, and local details.

11-2 ROLLED BEAMS—DESIGN CONSIDERATIONS

The principal considerations in the design of rolled beams are:

(*a*) Proportioning for strength in bending with due regard to the stability of the compression flange, and also to the adequacy of the selected shape to develop required strength in shear and in local bearing.

(*b*) Proportioning for stiffness, with due regard to the deflections of the member and prevention of excessive deformation under service conditions.

(*c*) Proportioning for economy, particularly the selection of a beam size and grade of steel which would lead to an economic design.

The usual situation confronting a designer is to select a beam size for a given span and load which will satisfy these three requirements.

Rods, bars, angles, and T sections are rarely used as beams because of their inherent weakness in resisting bending.   Channels can be used to carry light loads, except that they are weak in the lateral direction and are generally braced or supported laterally in some manner.   Usually the most efficient shapes are I shapes, either rolled or built up.

Two types of beams are currently rolled: American Standard (Fig. 11-1a) and Wide Flange shapes (Fig. 11-1b).   American Standard beams, the first beam sections rolled in the United States, are rolled in sizes varying from 3 to 24 in. in depth.   For each given depth, there are two to five sections of varying weight, depending on web thickness. Their variety is not nearly as great as the W^F shapes, which vary in depth from 4 to 36 in., with 2 to 39 weights for each depth.

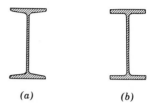

Whereas the standard beams of a given depth maintain a constant distance out-to-out of flanges, the W^F shapes keep a constant distance between the insides of flanges.   The weight of the standard beams is increased by spreading only the web rolls, thus increasing the web thickness without appreciable increase in the flange thickness.   The thickness of the W^F flanges, on

Fig. 11-1 Rolled beam sections. (*a*) American Standard section. (*b*) Wide Flange section.

the other hand, can be substantially increased since both the web and flange rolls can be spread.

In addition to furnishing a wider choice of sections the W^F shapes possess several other desirable features in comparison with the standard beams. For the same section modulus, W^F shapes are lighter than standard beams because more material is concentrated in the flanges where it can resist bending moment more efficiently than in the web.   As indicated by the name "wide flange," the flanges of the W^F shapes are wider, thus resulting in greater lateral stability and easier connection of the flanges to other members. These flanges maintain a uniform thickness along the width, whereas those of the standard beams are tapered.   The uniform thickness of the flanges facilitates riveting or welding of connections.

In addition to the regular W^F shapes there are Miscellaneous Beams which are lighter than the regular sections and can be economically employed for light construction.   Another type of a light beam, an expanded or "castellated" beam, is obtained by splitting the web of a standard section in a predetermined pattern and rejoining the segments in such a way as to produce a regular pattern of holes in the web (Fig. 11-2).   It is obvious that the section modulus of the expanded section is increased appreciably over that of the original shape without increasing the amount of material or weight used.   The cost of fabrication may offset the saving in material cost for a given section modulus, particularly when the beam is produced by simple

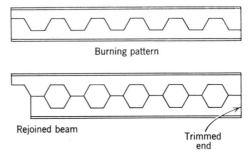

Burning pattern

Rejoined beam

Trimmed
end

Fig. 11-2   Castellated beam.

hand operations. Also, a beam with web holes develops significant second-ary stresses in the web and flange elements, and an increase in depth may impose limitations in allowable stress due to lateral torsional buckling. However, recent progress[1] in developing efficient production methods and refinement in methods of stress analysis and design criteria for castellated beams may lead to increased use of these beams.

11-3 ROLLED BEAMS—BENDING

The choice of a proper rolled section to be used for a beam is often based on its ability to carry the maximum bending moment M without exceeding the allowable unit fiber stress.

The required section modulus $S = I/c$ is given by $S = M/F_b$, where M is the maximum bending moment and F_b is the allowable bending stress. If the maximum distance c from the neutral axis to the extreme fiber is the same for both the tension and the compression flanges, the flange with the smaller allowable stress will control the design. If the allowable stress is the same for both flanges, then the flange with the greater c will control the design. If both the allowable stress and the value of c differ, it is often necessary to investigate both tension and compression flanges, using their respective values of F_b and c.

The basic allowable bending stress F_b, tension or compression, is taken as a fraction of the yield strength F_y. For buildings designed according to AISC Specification F_b is taken as $0.60F_y$, and for highway bridges (AASHO Specification) F_b is $0.55F_y$. These values of allowable stress are based on the condition that the member is adequately braced against lateral buckling and its flexural capacity is at least equal to the initial yield moment M_y, defined in Sec. 4-2. The basic allowable stresses may be decreased when special conditions, such as buckling or fatigue, control the design or they may be increased when dynamic loading or plastification controls.

In building design it is often possible to proportion flexural members in such a way that they develop full plastic moment M_p, defined by Eq. 4-1, without local or lateral buckling. For such members the increase in bending capacity is defined by the shape factor $k = M_p/M_y$, and therefore the allowable moment computed on the basis of elastic section modulus $S = I/c$, may be increased by this factor. This is accomplished by increasing the basic allowable stress to $F_b = k(0.60F_y)$. The shape factor k for W and I beams loaded in the plane of the web is generally in excess of 1.12, and for design purposes the allowable stress is taken as $F_b = 1.1(0.60F_y) = 0.66F_y$.

Use of this increased allowable stress involves the following conditions: (*a*) the compression elements of the profile must be so proportioned that local buckling will not occur before full plastification of the section, (*b*) the member must be adequately braced against lateral buckling, and (*c*) the loads must not cause failure before full plastification is attained. The first requirement is met by the so-called "compact" sections which are defined below; the second requirement is defined by placing an upper limit *l* on the distance between points of lateral bracing support of the compression flange; the third requirement is met by the condition that the plane of loading coincide with an axis of symmetry.

The "compact" section is defined by placing limits on the width-thickness ratios of compression elements so that large plastic strains can develop before inelastic buckling. These limits are clearly lower than those required to prevent elastic buckling. For the width/thickness ratios of various plate elements (Fig. 11-3), the following limits Table 11-1 are proposed by the AISC Specification. For standard rolled shapes the dimensions are listed in the AISC Manual, pp. 1–6 to 1–25, where the noncompact sections are indicated by appropriate annotation.

The condition of adequate bracing of the compression flange in W and I-shaped sections is satisfied in accordance with AISC Specification when the distance *l* between points of lateral bracing does not exceed either $2400b/\sqrt{F_y}$

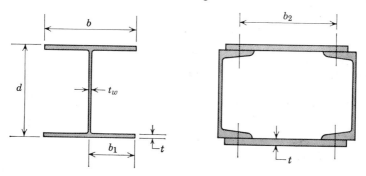

Fig. 11-3 Plate elements in compact shapes.

Table 11-1   Compact Section Proportions, AISC

Yield Strength, ksi		36	40	42	46	50
b_1/t	$1600/\sqrt{F_y}$	8.5	8	8	7.5	7
d/t_w	$13{,}300/\sqrt{F_y}$	70	66	64	62	60
b_2/t	$6000/\sqrt{F_y}$	32	30	30	28	27

or $20 \times 10^6 A_f/dF_y$, where b is the full flange width, A_f, is the area of compression flange, and d is the depth of the section. For different values of yield strength F_y, the lateral bracing limits are shown in Table 11-2.

The foregoing restrictions do not apply to box-sections which have a large torsional rigidity, and where lateral buckling is not critical. The reduction in allowable bending stress due to possible lateral buckling of W and I-shaped beams is examined in the following.

Although the basic allowable stress $F_b = 0.60F_y$ is based upon the assumption of sufficient lateral support for the compression flange to develop yielding, this support can not be provided in all cases. When the support is inadequate, lateral buckling occurs before yielding and the allowable bending stress must be reduced accordingly.

The general case of buckling for the combined action of bending and twisting is very complex. For certain shapes subjected to pure flexure approximate expressions for the critical bending moment were given in Sec. 9-12. For the particular case of W or I-shaped beams this critical moment was defined by Eq. 9-43, and with further approximations (see Eqs. 9-44, 9-45, and 9-46) the critical buckling stress was defined by Eq. 9-47, which follows:

$$F_{cr} = (M_{cr}/S_x) = \left\{ \left[\frac{\pi^2 E}{12(l/b)^2} \right]^2 + \left[\frac{\pi E}{\sqrt{18(1+\mu)}} \frac{bt}{ld} \right]^2 \right\}^{1/2} \tag{11-1}$$

or it can be written as

$$F_{cr} = [(F_{cr}')^2 + (F_{cr}'')^2]^{1/2} \tag{11-2}$$

where F_{cr}' and F_{cr}'' correspond to the first and second terms in Eq. 11-1, and are also defined by Eqs. 9-48 and 9-49, respectively.

Table 11-2   Lateral Bracing for Increased Allowable Stress, AISC

Yield Strength, ksi	36	40	42	46	50
$l/b = 2400/\sqrt{F_y}$	13	12	12	11.5	11
$ld/A_f = 20 \times 10^6/F_y$	545	500	480	436	400

In many practical cases one of the two values, F_{cr}' or F_{cr}'', predominates. F_{cr}' represents essentially the torsion-bending resistance of the compression flange or its independent strength as a column; it is the dominant term in deep girders with wide flanges. F_{cr}'' represents the pure torsional resistance of the section and is the dominant term for long, shallow, thick-walled W^F beams. The actual critical stress is greater than either F_{cr}' or F_{cr}'', although the larger of these two often approximates the critical stress with sufficient accuracy. Therefore for practical design either F_{cr}' or F_{cr}'' is used to develop allowable stress values. The value of F_{cr}'' defined by Eq. 9-50 and reduced by a factor of 0.6 results in one of the AISC criteria:

$$F_b = \frac{12 \times 10^6}{ld/bt} \tag{11-3}$$

The value of F_{cr}' defined by Eq. 9-48 is strictly applicable for the elastic case. For the inelastic region a parabolic expression is adopted, similar to the approximation of column capacity in the inelastic range given by Eq. 10-13. This equation is

$$F_{cr}' = F_y[1 - 0.5(Kl/rC_c)^2] \tag{11-4}$$

where Kl is the effective length of the column, r is the radius of gyration, and C_c is the limit slenderness ratio beyond which elastic buckling develops and is equal to $\sqrt{2\pi^2 E/F_y}$. Using this form and a factor of 0.6 for the allowable value, AISC defines the second limit for the lateral buckling stress as

$$F_b = 0.6F_y\left[1 - \frac{(l/r)^2}{2C_c^{\,2}C_b}\right] \tag{11-5}$$

which is similar to Eq. 11-4, except that in place of effective length factor K, a moment gradient factor C_b is used. In Eq. 11-5 the radius of gyration r is that for the section of an effective compression flange. The specification prescribes that the effective section is a T formed by the flange plate plus $\frac{1}{6}$ of the area of the web, and r is taken with respect to the axis in the plane of the web.

The C_b moment gradient factor is introduced because the basic equations are developed for beams which are simply supported and subjected to a uniform moment, whereas actual conditions of loading and restraint usually result in a variable moment which influences the critical stress F_{cr}. When no transverse loads are carried between lateral braces (Fig. 11-4), C_b is given conservatively as

$$C_b = 1.75 - 1.05\left(\frac{M_1}{M_2}\right) + 0.3\left(\frac{M_1}{M_2}\right)^2 \quad \text{but not more than 2.3} \tag{11-6}$$

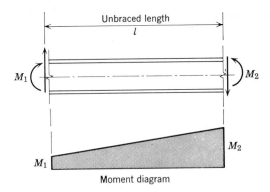

Moment diagram

Fig. 11-4   Unbraced segment of beam.

where M_1 = smaller end moment at end of unbraced length

M_2 = larger end moment at end of unbraced length

M_1 and M_2 are taken about the strong axis of the member.

Ratio of end moments M_1 and M_2 is positive for single curvature and negative for reverse curvature.

When the bending moment at any point within the unbraced length is larger than the moment at each end, the ratio M_1/M_2 is taken as unity. It is also conservative to take the value of C_b equal to unity, thus avoiding the detailed calculation.

Equation 11-5 can be simplified without loss of accuracy for I girders and W shapes by taking $r = 0.26b$, so that it becomes

$$F_b = A - \frac{B}{C_b}\left(\frac{l}{b}\right)^2 \tag{11-7}$$

where A and B are coefficients depending on F_y as in Table 11-3.

The AASHO Specification for highway bridges provides limits for allowable bending stresses in the compression flange which are similar to Eq. 11-7, except that C_b is taken as unity. The appropriate values of coefficients A and B, and the limits for maximum values of l/b are given in Table 11-3.

The stresses defined by Eqs. 11-3 and 11-5 represent the separate contributions to stability of the compression flange lateral stiffness and of the torsional stiffness of the section, and therefore each value separately underestimates the permissible limit stress based on lateral buckling. However, as a conservative limit stress for design of beams where lateral buckling may occur, the larger of the values given either by Eq. 11-3 or Eq. 11-5 can be used

To illustrate the determination of AISC allowable bending stresses the values for two W shapes and a small welded girder will be considered

Table 11-3   Coefficients *A* and *B* in Eq. 11-7

	Yield Strength, ksi	36	40	42	46	50
AISC	*A*	22	24	25	27.5	30
	B	0.0105	0.0121	0.0135	0.0169	0.0193
AASHO	*A*	20	—	23	25	27
	B	0.0075	—	0.0102	0.0122	0.0144
	Maximum (*l/b*) value	36	—	34	32	30

(Fig. 11-5).   The rolled shapes are 16 W^F 96 and 16 W^F 36 beams, and the welded girder is 30 in. with web thickness $\frac{5}{16}$ in., and flanges 7 × $\frac{1}{2}$ in. All three shapes are fabricated using A36 steel, and meet the requirements for compact sections shown in Table 11-1.

For different laterally unbraced lengths *l* the allowable stresses are calculated from Eqs. 11-3 and 11-5 (Fig. 11-5).

For 16 W^F 96 the allowable stress is defined by OABDK envelope (Fig. 11-5). Thus for *l/b* less than 13 (see Table 11-2) the allowable stress is $F_b = 0.66F_y$.

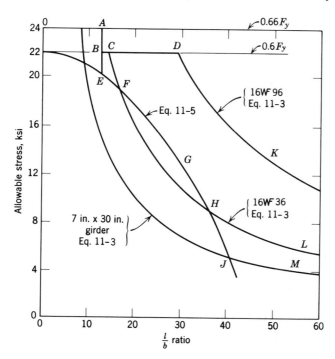

Fig. 11-5   AISC allowable stresses for selected shapes.

For l/b larger than 13 but less than 29 the allowable stress is $F_b = 0.6F_y$. For l/b values greater than 29 the allowable stress is governed by Eq. 11-3.

For 16 W 36 the allowable stress is defined by OABCFGHL envelope (Fig. 11-5). As mentioned previously, for l/b less than 13 the allowable stress is $F_b = 0.66F_y$; when $13 < l/b < 14.5$, $F_b = 0.60F_y$. For $14.5 < l/b < 17$ the allowable stress is governed by Eq. 11-3; when $17 < l/b < 35F_b$ is governed by Eq. 11-5; for $l/b > 35$ the allowable stress is again governed by Eq. 11-3.

For the 7 × 30 girder the allowable stress is defined by OAEGJM envelope (Fig. 11-5). For $l/b < 13$ the allowable stress is $F_b = 0.66F_y$; for $13 < l/b < 41$, the allowable stress is governed by Eq. 11-5, and when $l/b > 41$, F_b is governed by Eq. 11-3.

It should be noted that these envelopes are discontinuous because the limit allowable values are given by empirical equations which represent a conservative estimate of this limit and are obtained by simplifying actual behavior of the beam.

Section moduli of rolled sections are given in the AISC Manual, pp. 1–6 to 1–57. In order to facilitate the choice of the lightest section, the AISC Manual (pp. 2–4 to 2–5) gives a table of beam sections arranged in the order of decreasing values of section modulus if their depth is greater or if more material is concentrated in the flanges. For convenience these values are also given in Appendix C, Table C-1. In this table the values of flange width b, and of ratio d/bt are tabulated together with the values of section modulus S.

The lightest section in any series, for example, 18 W 50, 12 W 40, is usually the one most readily obtainable. These sections are identified in the table by bold face type. In the field it is difficult to tell the difference between two similar sections of different weights, such as between 12 W 40 and 12 W 45. By calling for the lightest section in the series, the engineer is assured that the furnished section is not weaker than the one specified.

Example 11-1.   A steel floor beam in a building has a span of 20 ft. simply supported and carries a uniformly distributed load of 1.2 kip/ft, including its weight. The compression flange is restrained against lateral buckling. Using AISC Specification select an economical section. Consider bending moment only. Assume A36 steel.

Solution

Maximum bending moment:

$$M_{max} = \frac{wL^2}{8} = \frac{1.2 \times \overline{20}^2}{8} = 60 \text{ kip-ft} = 720 \text{ kip-in.}$$

Allowable bending stress:

$$F_b = 0.66F_y = 24 \text{ ksi (AISC Sec. 1.5.1.4)}$$

Required section modulus:

$$S = \frac{M}{F_b} = \frac{720}{24} = 30 \text{ in.}^3$$

From Table C-1, Appendix C, or AISC Manual, p. 2-5:

Section modulus, in.³	34.9	38.1	34.1	30.8
Section	14 B 26	16 B 26	12 W⁻ 21	10 W⁻ 29

Use the lightest section—14 B 26 or 16 B 26 depending on depth desired which may be governed by deflection limitations or architectural considerations.

Example 11-2. Select a steel shape for the floor beam of Example 11-1 if lateral supports for the compression flange are provided only at the ends. All other conditions of Example 11-1 apply.

Solution
Using 14 B 26 with lateral supports at ends, verify allowable stresses.
Section 14 B 26 compact. Unsupported length of compression flange $l = 20$ ft $= 240$ in.
Flange width $b = 5.03$ in.

$$\frac{l}{b} = \frac{240}{5.03} = 47.8 > 13 \qquad \text{(AISC Sec. 1.5.1.4.4)}$$

Therefore Max $F_b = 0.6F_y = 22$ ksi—but allowable stress based on lateral buckling must be verified (Eqs. 11-3 and 11-7).
From Eq. 11-3:

$$F_b = \frac{12 \times 10^6}{ld/bt} = \frac{12 \times 10^6}{240 \times 6.61} = 7560 \text{ psi}$$

From Eq. 11-7 (also see Table 11-3, and with $C_b = 1.0$):

$$F_b = A - \frac{B}{C_b}\left(\frac{l}{b}\right)^2 = 22 - 0.0105\left(\frac{240}{5.03}\right)^2 < 0 \quad \text{(not applicable)}$$

F_b, governed by Eq. 11-3, is 7.56 ksi. $\therefore$ 14B 26 inadequate to carry load. Redesign based on lower allowable stress, try $F_b = 18$ ksi.
Required section modulus:

$$S = \frac{M}{F_b} = \frac{720}{18} = 40 \text{ in.}^3$$

From Table C-1, Appendix C:

Try 12 W⁻ 40, $S = 51.9$ in.³, $d/bt = 2.89$, $b = 8.0$.

From Eq. 11-3:

$$F_b = \frac{12 \times 10^6}{ld/bt} = \frac{12 \times 10^6}{240 \times 2.89} = 17,300 \text{ psi}$$

From Eq. 11-7:

$$F_b = 22 - 0.0105\left(\frac{l}{b}\right)^2 = 22 - 0.0105\left(\frac{240}{8}\right)^2 = 12.6 \text{ ksi}$$

F_b, governed by Eq. 11-3 is 17.3 ksi.

Allowable moment $M = F_b S = 17.3 \times 51.9 = 900$ kip-in. > 720 kip-in., therefore 12 W⁻ 40 is satisfactory.

11-4 ROLLED BEAMS—SHEAR AND BEARING

The web of beams and girders must be investigated for shearing stress, bearing stress, and effect of combined shear and bending.

The maximum shearing stress occurs at the neutral axis and is defined by $f_v = VQ/It$. For sections having an I shape, such as W⁻ rolled beams or built-up girders, this value is only slightly greater than the average shearing stress V/A_w. For purposes of design the average stress is often used; it is assumed that the allowable value is specified on the basis of the average rather than the maximum shearing stress value.

In calculating the average shearing stress, the effective web area is considered as $A_w = h_e t$, where h_e is effective depth of beam taken as the distance between flange centroids. For a rolled beam the outside dimension of the beam is often used.

The allowable shearing stress in the webs of rolled shapes is taken as a fraction of yield strength. AISC Specification for buildings prescribes the allowable shear stress $F_v = 0.40F_y$, and the AASHO Specification for highway bridges prescribes $F_v = 0.33F_y$.

Thus for various steels the allowable shear stresses are

Table 11-4 Allowable Shear Stresses

Yield Strength F_y, ksi	36	40	42	46	50
AISC F_v, ksi	14.5	16	17	18.5	20
AASHO F_v, ksi	12	13	14	15	17

Where concentrated loads are applied, bearing or "crippling" stresses must be considered. These are defined as $f_c = P/A_c$, where A_c is the effective area at the toe of the fillet included within 45° lines drawn from the edges of the

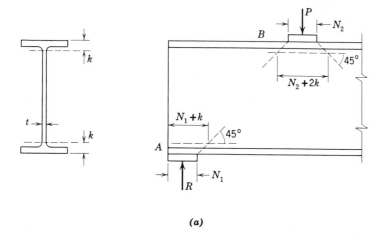

(a)

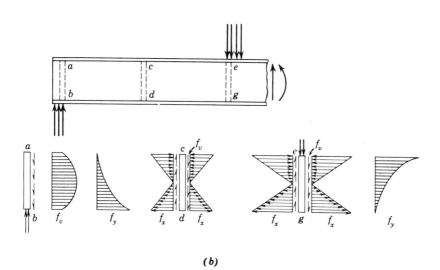

(b)

Fig. 11-6 Concentrated loads on beams. (*a*) effective bearing area for beams and (*b*) combinations of stresses in beams and girders.

bearing. For rolled beams, the distance to the toe of the fillet is tabulated in the AISC Manual so that the effective area can be easily calculated as (Fig. 11-6*a*)

$$A_c = t(N_1 + k) \qquad \text{or} \qquad A_c = t(N_2 + 2k) \tag{11-8}$$

AISC Specification for buildings provides that this bearing stress f_c shall not exceed the allowable value $F_a = 0.75F_y$. If local bearing stress exceeds the allowable value, bearing stiffeners should be provided. Design of these stiffeners is discussed in Sec. 11-21.

AASHO Specification for highway bridges provides that suitable stiffeners shall be provided for webs of rolled beams at bearings when the shear stress in the web adjacent to the bearing exceeds 0.75 of the allowable shear stress.

Combinations of shearing, bending, and bearing stress may occur at points adjacent to concentrated loads, as shown in Fig. 11-6b. A factor of safety n against local yielding of a plate element subjected to normal and shearing stresses (Fig. 11-9a) may be defined with reasonable accuracy by the Hencky–Von Mises criterion:

$$(f_x^2 + f_y^2 - f_x f_y + 3f_v^2)^{1/2} = F_y/n \tag{11-9a}$$

where F_y is the tensile yield strength of the plate. The bending, bearing, and shearing stresses in the web element vary within the beam or girder and depend on the type of loading. The magnitude of the normal and shearing stresses cannot be accurately determined in all instances because of the unknown effects of load concentration and the post-yielding or post-buckling stress redistribution. Therefore only approximate values of these stresses can be used for design purposes with a suitable factor of safety. In many instances a factor of safety of 1.5 may be sufficient.

The minimum tensile yield strength of A36 steel is 36 ksi and, using Eq. 11-1, the pure shear yield stress F_{vy} is $F_y/\sqrt{3} = 20.7$ ksi. Thus allowable shear stress of 14.5 ksi provides a factor of safety against yielding of 1.43. Using a factor of safety of 1.5, the criterion for local yielding due to combined stresses f_b and f_v may be defined using Eq. 11-9a, which leads to

$$(1.5f_b)^2 + 3(1.5f_v)^2 = F_y^2 \tag{11-9b}$$

11-5 DEFLECTIONS

Permissible deflections of beams are limited by various functional requirements. Excessive deflections are undesirable because:

(a) they may produce cracks in ceilings, floors, or partitions.

(b) they may cause discomfort to the human users of the structure.

(c) they are indicators of the lack of rigidity which might result in vibration and overstress under dynamic loads.

(d) they may produce distortion in connections and lead to high secondary stresses.

(e) they may result in poor drainage of the roof, increasing loads due to "ponding."

With the development of higher strength steels and the emphasis on large span beams in structures, control of deflections assumes particular significance. The designer can reduce deflections by increasing the depth of the member, reducing the span, providing greater end restraints, or by other means.

In general, beam deflection is a function of the loading, span, modulus of elasticity, and the geometry of cross section. The majority of steel beams have a constant cross section and the deflections of such beams under the service loads can be calculated in a relatively simple manner. For many typical cases the AISC Manual lists expressions for the deflections (pp. 2-120 to 2-135).

A relatively simple method for calculating deflections can be developed by modifying the basic formula for simply supported beams under uniformly distributed loads. The deflection at midspan of such a beam is

$$\Delta = \frac{5WL^3}{384EI} = \frac{5ML^2}{48EI} = \frac{5fL^2}{48Ec} \qquad (11\text{-}10a)$$

where W = total load, M = maximum moment = $WL/8$, f = maximum fiber stress = Mc/I, L = span length, E = modulus of elasticity of the material, I = moment of inertia of the beam, generally using the gross rather than the net section, and c = distance from neutral axis to the extreme fiber whose stress is f. From Eq. 11-10a, it can be observed that for a given stress f, the maximum deflection in a beam of constant section is independent of I, except insofar as I may affect the values of the maximum fiber stress f.

Substituting the value of E for steel as 30,000 ksi into Eq. 11-10a leads to:

$$\Delta = K_w \frac{f}{48} \frac{L^2}{6000c} \qquad (11\text{-}10b)$$

where K_w is the coefficient which accounts for the actual load distribution, and f is the maximum bending stress (ksi) assuming a simply supported beam. The values of coefficients K_w for simple span beams and constant section are given in Table 11-5.

For symmetrical beams in which $d = 2c$, and with $L = 12l$, where l is span in feet and L is span in inches, Eq. 11-10b can be simplified as follows:

$$\Delta = K_w f \frac{l^2}{1000d} \qquad (11\text{-}10c)$$

Deflections in restrained beams, once the end moments at the supports are known, can be determined using Eq. 11-10c and the principle of superposition.[2] Deflection Δ of a restrained beam can be represented as that due to transverse load on a simply supported span, Δ_{sim}, reduced by the effects

Table 11-5   Loading Coefficients $K_w^{(2)}$

Load	Beam	K_w
(1) Uniform load		1.00
(2) Uniform moment		1.20
(3) Load at midspan		0.80
(4) Loads at $\dfrac{L}{3}$		1.02
(5) Loads at $\dfrac{L}{4}$		1.10
(6) Moment at end		0.60

of the end restraint moments M_L and M_R at left and right supports, respectively. This can be written as follows:

$$\Delta = \Delta_{\text{sim}} - \Delta_L - \Delta_R \qquad (11\text{-}11)$$

where Δ_L and Δ_R are deflections due to moments M_L and M_R, respectively. If the maximum stresses due to M_L and M_R are f_L and f_R, respectively, and the maximum stress due to given load on a simply supported span is f_{sim}, then

$$\Delta = \Delta_{\text{sim}} - \Delta_L - \Delta_R = [K_w f_{\text{sim}} - K_L f_L - K_R f_R]\frac{l^2}{1000d} \qquad (11\text{-}12)$$

From Table 11-5, $K_L = K_R = 0.6$, so that:

$$\Delta = [K_w f_{\text{sim}} - 0.6(f_L + f_R)]\frac{l^2}{1000d} \qquad (11\text{-}13)$$

Example 11-3.   Find the deflection at midspan for the beams selected in Example 11-1.

Solution
From Eq. 11-10c:

$$\Delta = K_w f \frac{l^2}{1000d}$$

For 14 B 26: $K_w = 1.0$, $l = 20$ ft, $d = 13.89$ in., $S = 34.9$ in.[3]

$$f = \frac{M}{S} = \frac{720}{34.9} = 20.6 \text{ ksi}$$

$$\Delta = 20.6 \frac{(20)^2}{1000 \times 13.89} = 0.595 \text{ in.}$$

For 16 B 26: $K_w = 1.0$, $l = 20$ ft, $d = 15.65$ in., $S = 38.1$ in.[3]

$$f = \frac{M}{S} = \frac{720}{38.1} = 18.9 \text{ ksi}$$

$$\Delta = 18.9 \times \frac{(20)^2}{1000 \times 15.65} = 0.483 \text{ in.}$$

Example 11-4. Find the deflection at midspan of a 16 WF 58, A36 steel, 30 ft long restrained beam, loaded uniformly with $w = 2.5$ kip/ft, when $M_L = 2000$ kip-in and $M_R = 1880$ kip-in.

Solution

The simple span moment

$$M_{\text{sim}} = \frac{wl^2}{8} = \frac{2.5 \times 30 \times 30 \times 12}{8} = 3375 \text{ kip-in.}$$

and the stress

$$f_{\text{sim}} = \frac{M_{\text{sim}}}{S} = \frac{3375}{94.1} = 35.8 \text{ ksi}$$

The stresses f_L and f_R are:

$$f_L = \frac{M_L}{S} = \frac{2000}{94.1} = 21.2 \text{ ksi}$$

and

$$f_R = \frac{M_R}{S} = \frac{1880}{94.1} = 20 \text{ ksi}$$

The value of $K_w = 1.0$, $l = 30$, $d = 16$.
Then

$$\Delta = [35.8 - 0.6(21.2 + 20)] \frac{30 \times 30}{1000 \times 16} = 0.62 \text{ in.}$$

In most building codes, the live load deflection of beams which carry the plastered ceiling is limited to $1/360$ of the span length. Only the live-load deflection need be limited to the foregoing value since deflection due to dead load takes place before plastering. Besides, dead-load deflections are often counterbalanced by providing camber in beams. Camber is secured by "gagging" beams cold. Extremely small cambers may not be permanent, and maximum camber is also limited in order to avoid serious overstressing during the cambering operations (AISC Manual, p. 1–95 to 1–101).

Specifying a maximum deflection/span ratio, or specifying a minimum depth/span ratio below which the allowable bending stress f is to be proportionately reduced, are essentially equivalent methods, as can be seen from

Eq. 11-14 which is derived from Eq. 11-10a:

$$\frac{\Delta}{L} = \frac{5fL}{48Ec} = \frac{5}{24}\frac{f}{E}\frac{L}{d} \tag{11-14}$$

For example, AISC Specification provides that L/d of beams and girders supporting flat roofs shall be not greater than $600,000/F_b$, whether designed as simple or continuous spans. Substituting this limit in Eq. 11-14 the equivalent deflection limitation for simply supported uniformly loaded beams is 1/240 of the span. This limitation is presumably aimed at a minimum requirement of rigidity to prevent failures due to ponding.[3,4]

The AISC Commentary on the Specification further recommends that L/d of fully stressed beams and girders in floors should be, if practicable, not greater than $800,000/F_y$. If members of less depth are used, the allowable unit stress should be decreased in the same ratio as the depth is decreased from that recommended. For fully stressed roof purlins the L/d, if practicable, should not be greater than $1,000,000/F_y$, except when roofs have a slope greater than 3 in 12. Generally for continuous beams, the same span/depth limitations are recommended as for simple spans.

11-6 BIAXIAL BENDING

If the plane of the resultant load does not coincide with a principal plane of the beam section (Fig. 11-7) bending about both principal axes will result. For symmetrical sections the stress is easily calculated by the principle of superposition, so that, if M_1, $M_2 =$ bending moments about major and minor axes 1 and 2, respectively, and S_1, $S_2 =$ corresponding section moduli, then the maximum fiber stress is given by

$$f = f_1 + f_2 = \frac{M_1}{S_1} + \frac{M_2}{S_2} \tag{11-15}$$

For unsymmetrical sections the method described in Sec. 8-3 should be used.

For design purposes Eq. 11-15 can be written in the following form:

$$f = \frac{M_1}{S_1}\left[1 + \frac{M_2}{M_1}\left(\frac{S_1}{S_2}\right)\right] = \left(\frac{M_1}{S_1}\right)(1 + \beta) \tag{11-16}$$

where $\beta = M_2 S_1/M_1 S_2$. The ratio S_1/S_2 varies between somewhat narrow limits for given types of beams—for American Standard beams and for narrow W^F shapes it is of the order of 10; for very wide W^F, it can be as small as 2.5. The ratio M_2/M_1 is known from the loading conditions. The optimum choice of a section is such that β is relatively small, say about $\frac{1}{4}$.

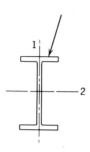

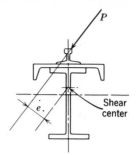

Fig. 11-7 Biaxial bending.

Fig. 11-8 Section reinforced
with channel on top flange.

If M_2/M_1 is a small quantity—1/50 or smaller—a narrow W^F beam can be used, β is of the order of 0.2, and bending about the minor axis is not excessive. If M_2/M_1 is of the order of 1/10, a wide W^F beam should be used in order to minimize the magnitude of the resultant stress.

An alternative solution is to reinforce one flange of the beam by a channel (Fig. 11-8). This solution is especially advisable when the lateral loads are not applied at the centroid of the beam, that is, when torsion might be produced in a symmetrical beam section. As an example, for girders supporting overhead cranes (Fig. 11-8), transverse loads are applied at the level of the top of the rail. A built-up section with a channel on top will have a shear center above the centroid of the W^F beam and therefore will reduce the resultant torque on the cross section.

Calculation of stresses in a section subjected to torsion has been discussed in Secs. 8-6 and 8-7. In practice such calculations are frequently omitted, but instead it is assumed that bending due to lateral force is taken entirely by the upper flange.

11-7 REINFORCED, TAPERED, AND BUILT-UP BEAMS

When available rolled-beam sections do not have sufficient strength to resist the external bending moment they may be reinforced along the entire length or part of it.

For flange reinforcement, plates and channels are most conveniently used (Fig. 11-9). Such reinforcement also helps to increase the resistance against lateral buckling.

Existing beams can often be conveniently strengthened by welding plates to the flanges. In such instances, the design should be made so as to avoid overhead welding as much as possible. For example, if the welding is to be

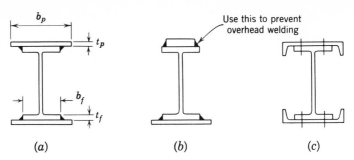

Fig. 11-9   Reinforced rolled beams.

done with the beam in its final position, the design shown in Fig. 11-9*b* is preferred to that in Fig. 11-9*a*.

One economical application of flange reinforcement is to add plates to portions of a beam where high bending moments exist. By reinforcing cantilever or continuous spans in the regions of high bending moment, smaller rolled sections can be used. The size and length of the plates can be tailored to suit the moment diagram (Fig. 11-10). When the allowable stress differs for the tension or compression flanges the amount of top and bottom reinforcement may also differ.

The additional plates can be connected to the rolled section by riveting, welding, or bolting. The required strength of the connection is determined by the amount of shear between the connecting elements, as obtained by the formula $q = VQ/I$, where q is the shear load per inch of span to be resisted by the connectors.

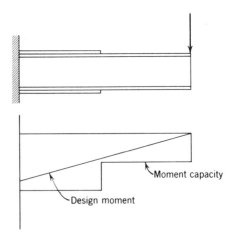

Fig. 11-10   Beam with cover plates.

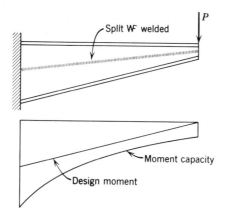

Fig. 11-11 A tapered beam.

Beams with variable depth may be desired for architectural or economic reasons (Fig. 11-11). Such beams are made from plates by welding them together, or from rolled shapes. Tapered beams lend themselves well to roof structures. For a simply supported tapered beam under a uniformly distributed load the controlling section may not be at the midspan but somewhere near the third-points depending on the taper.

11-8 CONTINUOUS BEAMS—ELASTIC DESIGN

Continuous beams are often more economical than simple beams of the same span. This becomes apparent when it is noted that the maximum moment in a continuous beam is much less than that in a corresponding simple beam. For example, comparing the interior span of a continuous beam with a simple beam under a uniformly distributed load (Fig. 11-12),

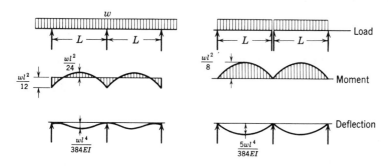

Fig. 11-12 Comparison between simple and continuous beams.

it is apparent that the maximum moment of the former is about two-thirds of the latter. When variable-depth sections are used, the amount of material required for the beam is approximately proportional to the total area of the moment diagram. On this basis material required for a continuous beam is about 40% of that required for a simple span beam. Continuous beams are also more rigid. For the same beam section (Fig. 11-12), the maximum deflection of the continuous beam is only 20% that of the simple beam. These advantages may not be as overwhelming when partial loadings are considered, but they do indicate that continuous beams are often more rigid and economical.

In the design of continuous beams based on the elastic theory the moments and shears in the beams may be obtained by various methods, usually described in standard textbooks on structural analysis and it is assumed that the reader is acquainted with one or more of these methods. After determination of the moments and shears, the beam is proportioned so that no section is overstressed under the design loads. If a beam is to be of uniform section, it is evidently controlled by the sections of maximum moment and shear. If it is to be of variable section, economy can be obtained by choosing a rolled shape sufficiently strong for the larger part of the beam and reinforcing it at points of high moment or shear.

The AISC Specification recognizes the greater ultimate capacity of continuous beams compared with simple beams and permits a $\frac{1}{10}$ reduction of the negative moment but provides for a larger positive moment as follows:

"Beams and girders which meet the requirements of a compact section and are continuous over supports or are rigidly framed to columns by means of rivets, high strength bolts or welds, may be proportioned for $\frac{9}{10}$ of the negative moments produced by gravity loading which are maximum at points of support, provided that, for such members, the maximum positive moment shall be increased by $\frac{1}{10}$ of the average negative moments. This reduction shall not apply to moments produced by loading on cantilevers."

Such a provision seems to be logical for beams of constant section subjected to distributed loads because of their plastic reserve strength. However, where concentrated or repeated loads are encountered, the possibility of fatigue failure must be investigated. When high residual stresses are "locked up" in the structure, brittle fractures may result from repetitions of high stress, particularly at low temperatures. For such structures, stress rather than strength may govern the design. This is one reason why plastic design has not yet been accepted in the bridge design specifications.

Example 11-5.   A continuous beam in a building floor with two spans of 28 ft each is to be designed for a uniform load of 3.8 kip/ft, including its own weight. Consider only full loading on both spans, neglecting the effect of partial loadings. AISC Specification. A36 steel.

Solution

Assume a uniform moment of inertia for the entire length in calculating moments. Using the elastic theory the maximum moments are (AISC Manual, p. 2–123):

$$+M_{max} = \tfrac{9}{128}wL^2 = \tfrac{9}{128} \times 3.8 \times \overline{28}^2 = 210 \text{ kip-ft} = 2520 \text{ kip-in.}$$

$$-M_{max} = -\tfrac{1}{8}wL^2 = -\tfrac{1}{8} \times 3.8 \times 28^2 = -372 \text{ kip-ft} = -4460 \text{ kip-in.}$$

Moments adjusted for continuity, (AISC Sec. 1.5.1.4.1):

$$-M = \frac{9}{10}(-4460) = -4014 \text{ kip-in.}$$

$$+M = 2520 + \left(\frac{4460 + 0}{2}\right)\frac{1}{10} = 2743 \text{ kip-in.}$$

The moment diagram is plotted in Fig. 1.

Choosing a section for the maximum positive moment and reinforcing

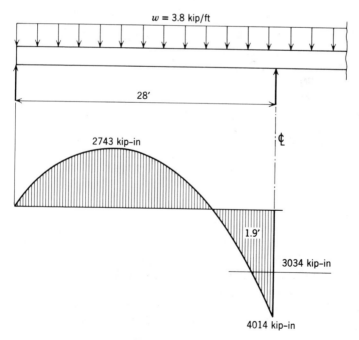

Ex. 11-5   Fig. 1

with cover plates for the maximum negative moment (note beam is restrained against lateral buckling top and bottom):

Section modulus required:

$$S = \frac{M}{F_b} = \frac{2743}{24} = 114 \text{ in.}^3$$

Span/depth ratio $= 800,000/F_y$

$$d_{min} = \frac{LF_y}{800,000} = 15.1 \text{ in.}$$

Use 16-in. compact section or deeper.

From AISC, pp. 2–4, the following sections may be used:

$$21 \text{ W}^F 62 \qquad S = 126.4 \text{ in.}^3$$
$$18 \text{ W}^F 64 \qquad S = 117.0 \text{ in.}^3$$
$$16 \text{ W}^F 71 \qquad S = 115.9 \text{ in.}^3$$

Use 21 W^F 62 (assume no limitation of depth),

$$\therefore \quad M_0 = 24 \times 126.4 = 3034 \text{ kip-in.}$$

Cover plates to provide additional moment ΔM at interior support.

$$\Delta M = M - M_0 = 4014 - 3034 = 980 \text{ kip-in.}$$

The lever arm of cover plates is approximately $h = 21.5$ in.

Cover plate area A_c required:

$$A_c = \frac{\Delta M}{hF_b} = \frac{980}{21.5 \times 24} = 1.87 \text{ in.}^2$$

To facilitate welding, use a cover plate about 1 in. narrower than the flange width. Use a plate 7 in. wide, $w_c = 7$ in.

Required thickness:

$$t_c = \frac{A_c}{w_c} = \frac{1.87}{7} = 0.27 \text{ in.}$$

Use a $\frac{3}{8}$ in. plate.

Required length of cover plate:

Negative moment capacity of unreinforced section $= 3034$ kip-in. From the bending-moment diagram, this moment occurs at the section 1.9 ft from support. Extend the cover plate at least 1 ft beyond the point of theoretical cutoff, making the total required length of cover plates $2(1.9 + 1) = 5.8$ ft. Use cover plates 6 ft long.

The weld required for cover plates is governed by the maximum shear V_{max} at the support. From elastic analysis (AISC Manual, p. 2–123):

$$V_{max} = 0.625wL = 0.625 \times 3.8 \times 28 = 66.5 \text{ kip}$$

Required weld strength per inch $q = V_{max}Q/I$, where Q of the cover plate is

$$A_c \frac{h}{2} = 2.46\left(\frac{21.37}{2}\right) = 26.27 \text{ in.}^3$$

I of reinforced section:

$$I = I_0 + 2A_c\left(\frac{h}{2}\right)^2 = 1326.8 + 2 \times 2.46\left(\frac{21.37}{2}\right)^2 = 1888 \text{ in.}^4$$

$$q = \frac{V_{max}Q}{I} = \frac{66.5 \times 26.27}{1888} = 0.92 \text{ kip/in.} \quad \text{(per cover plate)}$$

On each side of the cover plate only 0.46 kip/in. is required. Using continuous weld of minimum size $\frac{3}{16}$ in. provides shear strength of 1.8 kip/in., which is more than required.

Note: Maximum stress at the support based on the computed moment of inertia I of the reinforced section should be checked, as follows:

$$f = \frac{M}{I/c} = \frac{4014}{1888/10.8} = 23.04 \text{ ksi} < 24.0 \text{ ksi}$$

11-9 CONTINUOUS BEAMS—PLASTIC DESIGN

Another approach to the design of continuous beams is to base it on strength rather than stress. It is known that a continuous beam will not collapse even when it is overstressed at certain sections (Sec. 4-8). Methods have been developed for determining the strength of continuous beams which take into account their load carrying capacity beyond the initial yielding. These methods are known by various names: ultimate design, limit design, plastic design, collapse method of design, etc. In general terms, the methods are based on the fact that a continuous beam will collapse only when the number of plastic hinges developed is such that the structure is transformed into a hinge-link mechanism. These plastic methods of design are described in detail in several publications[5,6,7] and rules for such plastic design are incorporated in the 1963 AISC Specification.

Design of continuous beams of uniform section on the basis of plastic strength, which permits redistribution of moments after initial yielding, shows appreciable economy of material relative to conventional elastic design. A continuous beam of variable section "tailored" to follow the moment diagram may develop several plastic hinges at the same time, thus precluding any appreciable redistribution of moments after initial yielding. Therefore little increase in economy of material is obtained by using plastic design for beams of variable cross sections.

Example 11-6.   Determine the required section for a two-span continuous beam of 28-ft spans supporting a design uniform load of 3.8 kip/ft. Use A36 steel and AISC Specification, Part 2 (Plastic Design).

Solution

Load factor $= 1.70$ (AISC Sec. 2.1).

Ultimate load $= 1.70 \, (3.8) = 6.45$ kip/ft.

Assume mechanism as shown in Fig. 1 and consider left half of the structure. Plastic hinges formed a distance x from the ends and at the interior support. Using virtual work method based on virtual rotation θ:

External work $= \sum$ Force $\times$ displacement

$$= 2 \left(6.45 \times \frac{28}{2} \right) \left(\theta \frac{x}{2} \right) \times 12 = 1080 \theta x \text{ kip-in.}$$

Internal work $= \sum$ Moment $\times$ rotation

$$= M_p \left(\frac{x}{28 - x} \right) \theta + M_p \left(\frac{28}{28 - x} \right) \theta$$

$$= \left(\frac{28 + x}{28 - x} \right) M_p \theta$$

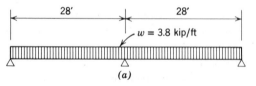

(a)

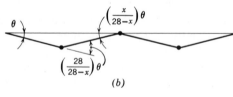

(b)

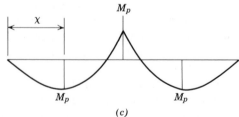

(c)

Ex. 11-6   Fig. 1

Equating Ext. work and Int. work:

$$1080\theta x = \left(\frac{28 + x}{28 - x}\right) M_p\theta$$

or

$$M_p = 1080x\left(\frac{28 - x}{28 + x}\right)$$

For maximum value of M_p:

$$\frac{dM_p}{dx} = 0$$

Reduces to:

$$x^2 + 56x - 784 = 0$$

and $x = 11.59$ ft.

Consequently

$$M_p = 1080 \times 11.59\left(\frac{16.41}{39.59}\right) = 5200 \text{ kip-in.}$$

Requires "plastic" section modulus $Z = M_p/F_y = 5200/36 = 144$ in.[3]
From Table, p. 2–7 of AISC Manual:

Use 18 W 70

Note: For the same continuous beam designed by the allowable stress AISC procedure, Example 11-5, the section used was 21 W 62 with two cover plates $72 \times \frac{3}{8}$ in.

11-10 ECONOMICS OF BEAM SELECTION

The task of selecting a rolled shape has become one of economics of material as well as load-carrying capability. With several grades of steel available in shapes, the selection of the most economical one may require a brief study.

Such a study is presented in Table 11-6. The table indicates that for a span of 20 ft and a distributed load of 38 kips, the rolled beam size required to satisfy the strength requirements varies from a 16 W 36 for A36 steel to a 14 B 17.2 for the A514 grade steel. The corresponding deflection ranges from 0.510 to 1.548 in., and in some instances exceeds the allowable building deflection of 0.667 in. However, assuming that deflection is not a major consideration, and that the choice of steel is based principally on the price of the material, the order of economy is designated by the relative price. It is seen that the A572–50 steel is the most economical, with A441 as a close second choice. If deflection is the limiting criterion, then A572–42 steel

Table 11-6   Relative Economy of Various Steels in Bending

$W = 38$ kips

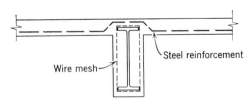

	A572					A514	
Steel	A36	42	50	A242	A441	Grade B	Grade F
Yield point, psi	36,000	42,000	50,000	50,000	50,000	100,000*	100,000*
Allowable stress, psi	22,000	25,000	30,000	30,000	30,000	60,000	60,000
Section required	16 WF 36	14 WF 34	14 WF 30	14 WF 30	14 WF 30	14 B 17.2	14 B 17.2
Relative weight	1.000	0.944	0.834	0.834	0.834	0.478	0.478
Deflection, in.†	0.510	0.671	0.788	0.788	0.788	1.548	1.548
Relative deflection	1.000	1.316	1.545	1.545	1.545	3.035	3.035
Relative price††	1.000	0.997	0.925	1.161	0.958	1.170	1.347

* Yield strength.   †† Based on 1967 prices
† 1/360 of span = 0.667 in.
Note: Local buckling must be investigated.

could be used as well as A36 because the difference in price is slight. The table is based on selecting specific rolled shapes from each grade of steel. However, it may be economical to fabricate a beam of a higher strength steel to be approximately the depth of another beam and thereby reduce the deflection and become competitive.

Similar studies can be carried out to evaluate influence of other factors on economics of design, including effect of continuity, of variation in layout, and use of built-up and composite beams.

11-11   ENCASED BEAMS

Construction which consists of steel beams totally surrounded by concrete is referred to as "encased beam" construction, (Fig. 11-13). The natural bond existing between the concrete and steel becomes the medium of inter-action of the two materials. This interaction was realized when it was observed that the deflection of bridge decks was much less than the calculated values using the steel beams only as the load-carrying member. Tests were

Fig. 11-13   Encased beam.

conducted in the United States in 1921 and in Canada and Europe in 1922. The first provision for the encased beam was introduced in the 1952 Edition of the AISC Specification.

The 1963 Edition of the AISC Specification defines an encased beam as:

"Beams totally encased 2 inches or more on their sides and soffit in concrete poured integrally with the slab may be assumed to be inter-connected to the concrete by natural bond, without additional anchorage, provided the top of the beam is at least $1\frac{1}{2}$ inches below the top and 2 inches above the bottom of the slab, and provided that the encasement has adequate mesh or other reinforcing steel throughout the whole depth and across the soffit of the beam."

The composite encased beam is considered to be homogeneous by transforming the concrete compression areas into an equivalent steel area. This is accomplished by dividing the concrete area by the ratio of the moduli of elasticity, $n = E_s/E_c$.

The analysis and design of encased beams are based upon the elastic theories of flexure, and the usual assumptions of beam behavior prevail. The steel beam is proportioned to support all the dead loads applied before the concrete hardens, and the encased beam, steel and concrete acting together support all the dead and live loads applied after the concrete has hardened. The bending stresses for loads acting on the encased beam are computed on the basis of the moment of inertia of the composite section. Concrete in tension is neglected. The allowable tension stress in the steel due to all loads may be taken as $0.66F_y$, where F_y is the yield point of the steel.

If dead loads are supported temporarily on shoring, the final computed stresses should include these loads together with dead and live loads added after the concrete has hardened. As an alternate, the AISC Specification permits the steel beam alone to be proportioned to resist all loads at a bending stress of $0.76F_y$ without shoring.

11-12 COMPOSITE BEAMS

The term composite beam construction for buildings and bridges defines a system in which interaction of a concrete slab with a steel beam is accomplished by means of a mechanical device called a shear connector. The concrete slab becomes the compression flange of the composite beam and the steel section resists the tension stresses. The tension portion of the beam is usually not encased. The shear connectors may be in the form of channels, studs, or spirals serving to transfer the longitudinal shear from the concrete to the steel and also serving to hold the concrete from uplifting.

The use of composite beams results in cost savings because it makes a more efficient structure by taking advantage of weight reduction and multi-purpose service of the slab. The slab is used not only to transfer the load to the beam, but also is called upon to assist in carrying the load in conjunction with the steel beam.

The AASHO Specification for bridges includes a section on composite design of girders and bases its analysis on the elastic theory. The proportions of the steel beam and concrete slab are established on the basis of the moment of inertia of the composite section. The size of the shear connectors is selected and spaced according to the principles of elastic behavior and longitudinal shear distribution. Thus the spacing of connectors will be a function of the shear transfer according to the usual expression VQ/I. The connectors in bridge girders have a higher factor of safety against yielding than those in buildings and do not permit a slip of the connector to occur under design loads. Fatigue considerations should be taken into account when designing welded connectors and the allowable shearing stresses of the American Welding Society should be used.

The AISC Specification for buildings bases its composite design criteria on the ultimate load-carrying capacity of the assembly. Investigations on shear connectors and their behavior in composite beam action have established the fact that slippage of the connectors does take place but can be included in design calculations.

The basic assumptions for the analysis and design of a composite beam are:

(*a*) The concrete slab is continuously connected throughout the entire length of the steel beam.

(*b*) The slip of the shear connector is directly proportional to the load on the connector.

(*c*) A linear distribution of strain exists over the depth of the member.

(*d*) The slab and beam do not become separated vertically at any point along the beam.

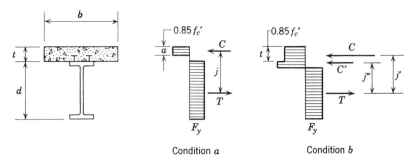

Condition *a*   Condition *b*

Fig. 11-14   Stress distribution in composite beam.

The ultimate design concept is based on the fact that the concrete slab fails by crushing as a result of the compressive stresses. At the instant of failure, a uniform stress distribution may be assumed to be acting on the concrete and steel. The assumed stress distribution is indicated in Fig. 11-14.

Two conditions of stress distribution may occur: (*a*) when the neutral axis lies within the thickness of the concrete and (*b*) when the neutral axis lies below the slab within the steel. Condition *a* indicates that the concrete is sufficient to resist all the compression and *b* indicates that the slab is insufficient to resist the compressive forces alone and must share the load with the steel beam.

For condition *a*, the equations of equilibrium are

$$T = A_s F_y$$
$$C = 0.85 f_c' b a \qquad \qquad (11\text{-}17)$$
$$C = T$$
$$\therefore \quad a = \frac{A_s F_y}{0.85 f_c' b}$$

$$M_{\text{ult}} = Tj = T\left(\frac{d}{2} + t - \frac{a}{2}\right) \qquad (11\text{-}18)$$

For condition *b*, the equations of equilibrium are

$$C = 0.85 f_c' A_c$$
$$T = A_s F_y - C'$$
$$C' = A_{sc} F_y$$
$$C' = T - C$$
$$M_{\text{ult}} = Cj' + C'j''$$

where A_{sc} = area of steel in compression.   (11-19)

The symbols are noted in Figure 11-14 and their values are determined from the geometry of the cross section and the ultimate stress distribution.

The foregoing discussion was presented for determining the moment capacity of the composite beam when subjected to a positive bending moment. In general, for negative moments, the steel beam acting alone plastically is considered to resist the moment. In some instances, the tension reinforcement in the slab is considered to be acting with the tension flange provided the studs are continued in the negative moment region.

Tests have indicated that the ultimate moment of a composite beam is not affected by the manner of construction. That is, the steel beam need not be shored while the concrete is hardening. Therefore the composite beam cross section may be designed on the basis of the total applied dead and live loads. However, as a safeguard, the AISC Specification limits the section modulus of the composite section with respect to the tension flange to not more than the following:

$$S_{tr} = \left(1.35 + 0.35\frac{M_L}{M_D}\right)S_s \tag{11-20}$$

where S_{tr} = section modulus of the transformed section

S_s = section modulus of steel beam alone

M_L = live-load moment

M_D = dead-load moment

provided that the steel beam acting alone can support the loads before the concrete has hardened at the usual allowable stresses prescribed for the steel.

The total end shear on the composite beam is carried by the web and end connections of the steel beam alone.

Although tests have indicated that the shear connectors fail by tension, for design purposes the allowable connector stresses are expressed in terms of an equivalent shear load.

The allowable horizontal shear loads, designated q, for a concrete with stone aggregate are listed in the AISC Specification and Table 11-7.

As a protection against fire effects, it is recommended that all shear connectors have at least one inch of concrete cover in all directions.

Table 11-7   Allowable Loads on Shear Connectors

Connector	Allowable Horizontal Shear Load (q) kips (Applicable only to concrete made with ASTM C33 aggregates)		
	$f_c' = 3000$	$f_c' = 3500$	$f_c' = 4000$
$\frac{1}{2}''$ diam. × 2″ hooked or headed stud	5.1	5.5	5.9
$\frac{5}{8}''$ diam. × $2\frac{1}{2}''$ hooked or headed stud	8.0	8.6	9.2
$\frac{3}{4}''$ diam. × 3″ hooked or headed stud	11.5	12.5	13.3
$\frac{7}{8}''$ diam. × $3\frac{1}{2}''$ hooked or headed stud	15.6	16.8	18.0
3″ channel, 4.1 lb	4.3w	4.7w	5.0w
4″ channel, 5.4 lb	4.6w	5.0w	5.3w
5″ channel, 6.7 lb	4.9w	5.3w	5.6w
$\frac{1}{2}''$ diam. spiral bar	11.9	12.4	12.8
$\frac{5}{8}''$ diam. spiral bar	14.8	15.4	15.9
$\frac{3}{4}''$ diam. spiral bar	17.8	18.5	19.1

w = length of channel in inches.

The required number of connectors is based on the beam behavior at ultimate moment. As the load is increased on the beam, some of the shear connectors will begin yielding and a redistribution of load between connectors takes place before failure occurs at the ultimate moment. The ultimate horizontal shear force to be transferred from the concrete to the steel or vice versa is the ultimate compressive load in the concrete or the ultimate tension load in the steel, whichever is the smaller. This ultimate shearing load must be resisted by the connectors in the distance between the end of the beam or point of zero moment to the point of maximum moment. The connectors may be equally spaced in the distance noted.

The required number of connectors n is determined by dividing the nominal shearing load V_h by the allowable shear load q of the connector to be used and expressed as $n = V_h/q$. V_h is equal to C or T (Fig 11-15).

The nominal shearing load to be used for the determination of the type and number of connectors is the smaller of the normal loads on the concrete or steel. If the concrete fails first, its load is computed by the equation

$$V_h = \frac{0.85 f_c' A_c}{2} \tag{11-21}$$

where V_h = shearing load on connectors

    f_c' = specified compressive strength of the concrete

    A_c = area of effective concrete flange

The factor of 2 reduces the ultimate load to an allowable load for the usual elastic design procedures.

If the steel is the critical material, the design shearing load is determined as follows:

$$V_h = \frac{A_s F_y}{2} \tag{11-22}$$

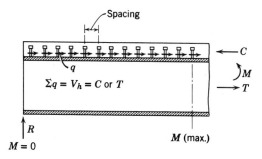

Fig. 11-15

where A_s = total area of the steel beam

F_y = yield point of the steel

The composite beam with the concrete compression flange behaves as a T-beam. As such the theoretical effective width of the flange is a function of the span of the beam, Poisson's ratio, and the shape of the moment diagram. The various specifications which permit composite beam design prescribe the manner by which the effective width shall be determined. The AISC Specification outlines the requirements for the effective width of the concrete slab to be the smallest of the following when the slab is on both sides:

1. Not more than $\frac{1}{4}$ of the beam span
2. Projection beyond steel flange to be not more than $\frac{1}{2}$ the clear distance to the adjacent beam
3. Not more than 8 times the slab thickness

For beams on only one side, the projection beyond the beam shall be:

1. Not more than $\frac{1}{12}$ of the beam span
2. Not more than $\frac{1}{2}$ the clear distance to the adjacent beam
3. Not more than 6 times the slab thickness

The AASHO Specification has somewhat similar requirements for bridge beams.

Example 11-7. Design an interior beam for composite action for the 30 ft × 32 ft bay with the following data, assuming the beam to be simply supported.

Design Data:

4 in. concrete slab
Live load = 100 psf
Partitions = 20 psf

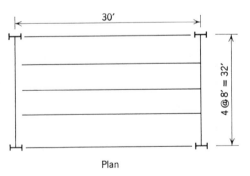

Plan

Fig. 1

A36 steel beam
F_b is noted in solution
$f_c' = 3000$ psi, $n = 10$
$E = 29,000$ ksi

Solution

Beam load:

$$
\begin{aligned}
\text{Live load} &= 100 \text{ psf} \\
\text{Partitions} &= 20 \\
\text{4-in. slab} &= 50 \\
\text{Total loading} &= \overline{170} \text{ psf}
\end{aligned}
$$

$$
\begin{aligned}
\text{Load to beam} = 170 \times 8 \text{ ft} &= 1360 \text{ plf} \\
\text{Estimate steel beam weight} &= 40 \\
\text{Total dead} + \text{live loading} &= \overline{1400} \text{ plf} \\[4pt]
\text{Total load on 30-ft beam} &= 42.0 \text{ kips}
\end{aligned}
$$

Noncomposite Beam Design

$$ M = \frac{WL}{8} = 157.5 \text{ kip-ft} $$

Allowable unit stress for compact sections, adequately braced (AISC Sec. 1.5.1.5.1.)

$$ F_b = 24.0 \text{ ksi} $$

$$ \text{Req'd } S. = \frac{(157.5)(12)}{24} = 78.8 \text{ in.}^3 $$

A 16 W̅ 50 and an 18 W̅ 45 will provide sufficient strength to satisfy the load requirement. Both beam sections also satisfy the requirements of a compact section, namely, that projecting elements have a width-to-thickness ratio no greater than $8\frac{1}{2}$ (with due consideration for tolerances) and a web depth-to-thickness ratio less than 70 (AISC Sec. 2.6).

The 16-in. beam was selected for minimum depth and the 18-inch beam was selected for minimum weight. The difference in the exact beam weight and assumed beam weight does not alter the design.

Live-load deflection:

The partition load will be considered with the live load in the calculation of the live-load deflection.

Beam Section	Deflection
16 W̅ 50	0.92 in.
18 W̅ 45	0.85 in.

Allowable live-load deflection (AISC Sec. 1.13) $1/360$ span $= 1.00$ in.

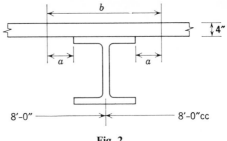

Fig. 2

Composite Design: (AISC Sec. 1.11).   On the basis of the noncomposite design a 16 W^F 40 beam section is assumed. Design is based on construction without shoring.

Effective width of concrete (AISC Sec. 1.11.1) (Fig. 2)

$$b \leq \tfrac{1}{4} \text{ beam span length} = \tfrac{1}{4} \times 30 = 12 \quad = 90.0 \text{ in.}$$

$$a \leq \tfrac{1}{2} \text{ clear beam spacing} = \tfrac{1}{2}(8 \times 12 - 7) = 44.5 \text{ in.}$$

$$a \leq 8 \text{ times slab thickness} = 8 \times 4 \qquad = 32.0 \text{ in.}$$

Minimum effective width $b = 2 \times 32 + 7 = 71$ in.

Properties of Transformed Section: (AISC Sec. 1.11.2.2).   Concrete width must be transformed to an equivalent width of steel, (Fig. 3).

A modular ratio, $n = 10$, will be used.

$$\therefore \quad b/n = 71/10 = 7.1 \text{ in.}$$

$$y_t = \frac{(11.77)(8.00) + (28.4)(18.00)}{11.77 + 28.4} = 15.07 \text{ in.}$$

Moment of Inertia—Composite Section

$$I_{tr} = 515.5 + (11.77)(7.07)^2 + \tfrac{1}{12}(7.1)(4)^3 + (28.4)(2.93)^2$$

$$I_{tr} = 1385.5 \text{ in.}^4$$

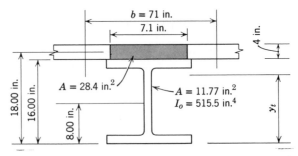

Fig. 3   Position of center of gravity—composite section.

Section Modulus—Tension Flange

$$S_{tr} = 1385.5/15.07 = 91.9 \text{ in.}^3$$

Maximum Moments:

Live load moment $M_L = (\tfrac{1}{8})(0.960)(30)^2 = 108.0$ kip-ft

Dead load + weight of beam = 0.440 klf

Dead load moment = $\tfrac{1}{8}(0.440)(30)^2 = M_D = 49.5$ kip-ft

Maximum Section Modulus of Transformed Section: (AISC Sec. 1.11.22)

$$S_{tr} = (1.35 + 0.35 M_L/M_D)S_s$$

$$S_{tr} = \left(1.35 + 0.35\frac{108.0}{49.5}\right)64.4 = 136.1 \text{ in.}^3$$

$$S_{tr} = 136.1 > 91.9 \quad \therefore \text{ O.K.}$$

Stress Calculations: Dead Load on Steel Beam only: (AISC Sec. 1.11.2.2)

$$F_{t,c} = (49.5)(12)/64.4 = 9{,}200 \text{ psi, Tension and Compression} < 24{,}000 \text{ psi}$$

AISC Design Procedure: (Sec. 1.11.2.2). The AISC Specification states that "When Shear Connectors are used the composite section shall be proportioned to support all of the loads without exceeding the allowable stresses as prescribed."

Steel, tension flange, $f_t = \dfrac{157.5 \times 12{,}000}{91.9} = 20{,}600$ psi

Steel, compression flange, $f_c = \dfrac{(157.5)(12{,}000)(0.93)}{1385.5} = 1270$ psi

Concrete, compression $f_c = \dfrac{(157.5)(12{,}000)(4.93)}{(1385.5)(10)} = 672$ psi

Live Load Deflection—Composite Design (AISC Sec. 1.13)

$$\text{Deflection} = \frac{(5)(960)(30)^4(12)^3}{(384)(29 \times 10^6)(1385.5)} = 0.44 \text{ in.}$$

Shear Connectors: (AISC Sec. 1.11.4). The total horizontal shear to be resisted between the point of maximum positive moment and each end of the steel beam may be taken as the *smaller* value of Eq. 11-21 or 11-22.

Eq. 11-21:
(Concrete critical)
$$V_h = \frac{0.85f_c'A_c}{2} = \frac{(0.85)(3000)(71 \times 4)}{2} = 362 \text{ kip}$$

Eq. 11-22:
(Steel critical)
$$V_h = \frac{A_sF_y}{2} = \frac{(11.77)(36,000)}{2} = 212 \text{ kip}$$

Number of Connectors Each Side of Maximum Moment: (AISC Sec. 1.11.4)

$$n = \frac{V_h}{q}$$

q = Allowable shear load for one connector, or one pitch of a spiral bar obtained from Table 11.7.

A 3-in. maximum height of connector is recommended in order to provide $1''$ of cover in the 4-in concrete slab.

Stud: Use $\frac{5}{8}$-in. × $2\frac{1}{2}$-in. $q = 8.0$ kip (stone concrete-ASTMC33)
$n = 212/8 = 26.5$, use 27
Spacing for single row $= (15 \times 12)/27 = 6.67$-in.
Use $6\frac{1}{2}$ in. spacing

Channels: Use 3[4.1 $q = 4.3w$ kip
Assume w (length of channel in inches) $= 3$ in.
$q = (4.3)(3) = 12.9$ kip
$n = 212/12.9 = 16.4$ Use 17
Spacing $= 180/17 = 10.6$ in. Use $10\frac{1}{2}$ in.

Spirals: $\frac{1}{2}$-in. diameter bar $q = 11.9$ kip
n (contact points) $= 212/11.9 = 17.8$, Use 18
Spacing $= 180/18 = 10$-in.

Recommended maximum spacing of approved shear connectors not to exceed 24 in.

Beam Connection to Column: (AISC Sec. 1.11.3).   The connection of the beam to the column must be designed to carry the total dead and live load on the beam. Maximum shear: $V = 21$ kip. Web shear stress, $f_v = 21/(16)$ $(0.307) = 4275$ psi which is less than the F_v of 14,500 psi permitted for A36 steel.
A check of the capacity of the standard beam connections should be made.

Comparison of Composite and Noncomposite Designs:

Item	Composite	Noncomposite	
Steel Beam Size	16 Wᶠ 40	16 Wᶠ 50*	18 Wᶠ 45†
Live Load Deflection	0.44 in.	0.92 in.	0.85 in.

* Increase in weight 25%.
† Increase in weight 12%.

11-13 PLATE GIRDERS

Plate girders are beams built up of steel plates and shapes, which may be riveted, welded, or bolted together (Figs. 11-16, 11-17). Whereas rolled beams

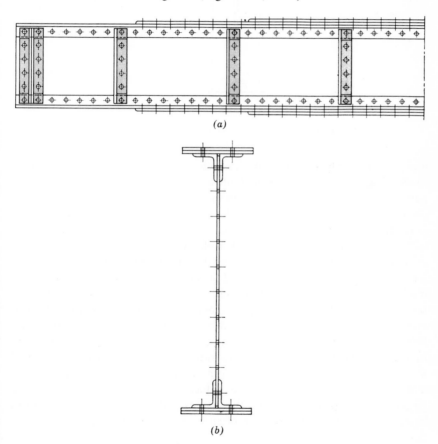

(a)

(b)

Fig. 11-16 Riveted or bolted plate girder. (*a*) elevation, (*b*) section.

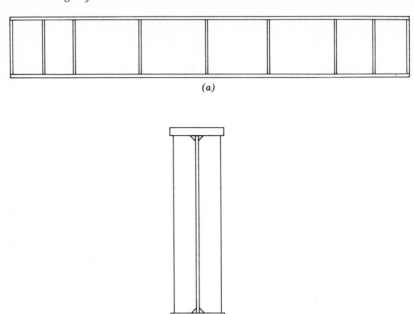

(a)

(b)

Fig. 11-17   Welded plate girder. (*a*) elevation, (*b*) section.

are available only in standard sizes which may not always be the most economical, plate girders can be built to any desired proportion to suit the particular requirements. The saving in material due to this better proportioning, however, may be offset by the increase in cost of fabrication. Thus, in general, for smaller beams where the saving in material is minor compared with the increase in fabrication cost, rolled beams are cheaper. For heavier construction, where the available rolled beams are not sufficient to carry the load, plate girders have to be used. For intermediate cases, say with section moduli between 500 and 1100 in.³, either rolled beams or plate girders, or rolled beams reinforced with plates may be the most economical. For ordinary loading, rolled beams would be more economical for simple spans below 30 ft, and plate girders for spans above 70 ft. For spans between these limits, the two types may compare rather closely—heavier loads would tend to favor plate girders. For modern highway bridges, where continuous and cantilever layouts are used to reduce the maximum moments, plate girders may not be required until the span exceeds about 80 ft because rolled beams with reinforcing plates can often be economically employed for such designs.

Before the development of welded construction, plate girders were limited to spans not in excess of 150 ft. Since 1945, however, numerous plate girders spanning 200 to 300 ft have been built, and the girder bridge over the Save river in Belgrade has a span of 780 ft.

Generally, welded plate girders weigh less than riveted girders. The amount of saving may vary from 5 to 15%, depending on the governing specifications. Saving in material, however, does not necessarily mean saving in total cost, if the fabrication cost is high. With the advent of automatic welding and development of ASTM A36 steel which permits use of normal procedures for welding of thick plates, the unit cost of fabrication has been reduced considerably. Besides the typical sections for plate girders shown in Figs. 11-16, and 11-17 other sections, such as shown in Fig. 11-18, may be used.

Riveted plate girders can be built up of rolled beams and channels (Fig. 11-18a). The use of rolled beams saves some riveting work, whereas the use of channels increases the lateral rigidity which is desirable if the girder is not laterally supported, or if bending exists also in the horizontal plane. Figure 11-18b illustrates a delta-girder section and Fig. 11-18c shows a box girder which may be used where heavy loads are encountered and the depth of girder is limited.

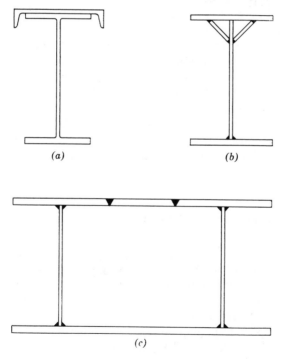

(a)   (b)

(c)

Fig. 11-18   Miscellaneous girder sections.

In riveted and bolted girders, holes must be provided for insertion of fasteners. Elimination of holes for welded girders is an advantage, as it eliminates a costly fabrication operation and avoids the need of determining the effect of holes on stresses in the girders.

For riveted girders, some specifications still call for the use of net area, thus requiring more material. In other specifications, use of gross area for riveted design, provided the hole areas do not exceed 15% of flange areas, indirectly increases the design stress in riveted girders by as much as 15%. Whether such an increase is warranted depends on a number of conditions, such as type of loading, and frequency and magnitude of overloads.

In welded girders, generally only plates are used to build up the flanges, whereas in riveted girders, angles are almost always required. The elimination of angles from the flanges increases the effective depth of the girder since more steel is now concentrated near the extreme fiber. This may also result in thicker and/or wider flanges, thus increasing the lateral buckling strength of the compression flange. The omission of flange angles increases the clear depth of the web. If the design of the web is controlled by the criteria for buckling, either a thicker web must be used for the welded girders or more stiffeners will be necessary.

A riveted girder requires angle stiffeners, with one leg of the angle riveted to the web. In a welded girder, only plates are required for stiffening, thus saving one leg of the angle. The stiffeners in welded girders can be sheared plates, requiring no milling, crimping (Fig. 11-19), or filler plates. For riveted girders, filler plates are needed under the stiffening angles when the practice of crimping is not permitted by specifications, or when fabricating shops do not have machines for crimping.

In riveted girders splicing of flanges and web plates requires additional material for the splice plates. In welded girders use of butt welds for splicing flanges and web plates results in saving of material and usually in reduction of cost.

11-14 PROPORTIONING OF PLATE GIRDERS

When designing a plate girder to resist a given bending moment M, it is desirable to maximize the lever arm of the internal forces, so that the material required for the flanges is minimized. The area of the web required to resist a given shear V depends on the allowable shear stress F_v. If this stress were independent of the depth and thickness of the web, then the lightest girder would be one with a deep thin web. In this case the depth may be determined by available headroom, deflection limitations, available plate sizes, aesthetic proportions, site transportation, or construction conditions. In reality deep

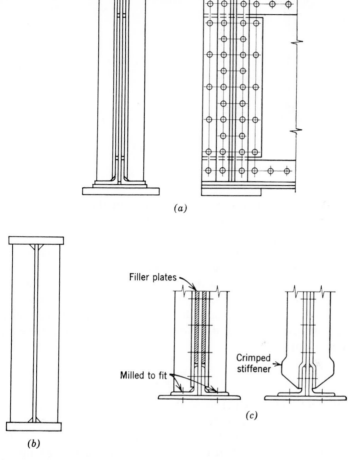

Fig. 11-19 (*a*) angle stiffeners with filler plates for riveted girders, (*b*) plate stiffener for welded girders, and (*c*) Milled stiffener with filler plates and crimped stiffener.

thin plates buckle at low stresses, and thus the allowable shear stress is not independent of web proportions.

Assuming allowable shear stress F_v and the effective depth of girder h_0 are known, the thickness t of the web is given by the following:

$$t = \frac{V}{F_v h_o} \tag{11-23}$$

Determination of actual allowable stress F_v is discussed in Sec. 11-18. If the

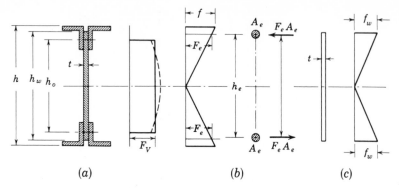

Fig. 11-20   Preliminary proportioning of girder web and flanges.

computed value of t is less than the minimum thickness determined by practical considerations, or is so small as to require a large number of stiffeners, a thicker plate may have to be adopted.

Having chosen the size of web plate, the effective flange area to resist the moment can be computed from the relation $M = F_e A_e h_e$ (Fig. 11-20), where F_e is the allowable fiber stress at the level of the flange centroid (about 90 or 95% of the allowable stress at the extreme fiber), h_e is the effective depth for the flanges (which is approximated by the distance between the centroids of the flanges, about 90 to 95% of the over-all depth), and A_e is the equivalent flange area. Thus

$$A_e = \frac{M}{F_e h_e} \qquad (11\text{-}24)$$

This equivalent flange area A_e is made up of the actual area of one flange, reduced for rivet holes if required, plus a fraction of the web area. Most of the flange area is concentrated near the extreme fiber where it can efficiently resist bending moment, whereas the web area is distributed through the entire depth and is subjected to a triangular variation of bending stress. The moment resistance M_w of the web can be defined as follows (Fig. 11-20)

$$M_w = \frac{f_w}{2} \frac{A_w}{2} \frac{2}{3} h_w = \tfrac{1}{6} f_w h_w A_w = k' f_e h_e A_w \qquad (11\text{-}25)$$

where A_w is the area of the entire web and f_w is extreme fiber stress for the web.

From Eq. 11-25 it can be seen that one-sixth of the total web area can be considered as effective in resisting moment M_w with lever arm h_w and stress f_w. If rivet holes are to be deducted, and reduced stress F_e and effective depth h_e are to be considered, the value of k' varies approximately from one-eighth to one-sixth. Assuming $k' = $ one-sixth, the area required *for each* flange A_f will be

$$A_f = A_e - \frac{A_w}{6} \tag{11-26}$$

This gives a preliminary design for the maximum-moment section. Flanges for other sections will usually be smaller, but additional material will be required for web stiffeners and splices. The total weight of the girder of constant depth can be approximated by assuming the maximum-moment cross section along the entire span. This weight may be reduced if the section can be varied along the span by reducing flange area or web thickness. However, if stiffeners are required they may add 5 to 10% of girder weight, and splices and other details may add another 5 to 10%.

11-15 OPTIMUM GIRDER DEPTH

Although the depth of a girder is often limited by considerations other than the economy of the girder material, sometimes it may be determined by the condition of minimum weight. In any case, it is desirable to know the optimum depth for economy so that, even if it cannot be adopted, it will serve as a guide to be considered together with other requirements.

The optimum depth of a girder will depend on such factors as the type of girder—whether riveted or welded, whether single web or box-shaped—the ratio of shear to moment, and others, which affect the proportion of flange, web, and stiffener materials. Some expressions may be derived to obtain the optimum depth of girders based on required moment resistance M and essentially without regard to other related elements in the structure. In order to derive these expressions, it is necessary first to set up an expression giving the weight of the girder in terms of the depth. Let

(*a*) $\bar{A}_{fg} = $ average gross flange area for entire span, including flange splices $= c_1 \times$ net flange area at midspan $= c_1 A_f$

(*b*) $\bar{A}_w = $ average gross web area for entire span, including web splices and stiffeners $= c_2 \times$ gross web area $= c_2 ht$, where h is the depth of the web (taken as the depth of girder) and t the thickness of the web assumed to be constant for entire span

(*c*) $A_n = $ net web area at point of maximum moment $= c_3 \times$ gross web area $= c_3 ht$

(*d*) $M = $ maximum design moment, in.-lb

(*e*) $c_4 h = $ distance between centroids of the flanges

(*f*) $c_4 F = $ allowable stress at the centroids of flanges where F is the allowable fiber stress at the extreme fiber.

Values of coefficients c_1 through c_4 are discussed later in this section. The net flange area A_n required for M:

$$A_n = A_e - \frac{A_w}{6} = \frac{M}{c_4{}^2 hF} - c_3 \frac{ht}{6} \qquad (11\text{-}27)$$

where $c_3(ht/6)$ represents the approximate equivalent flange area furnished by the web. The average gross cross-sectional area for the entire girder is

$$A_g = 2c_1 A_n + c_2 ht = 2c_1\left(\frac{M}{c_4{}^2 hF} - c_3 \frac{ht}{6}\right) + c_2 ht \qquad (11\text{-}28)$$

In order to determine the value of h for minimum A_g, it is necessary to differentiate A_g with respect to h and to equate this derivative to zero. Since t is sometimes a variable, Eq. 11-28 should be considered for the following three cases:

Case 1.   When shear is heavy and controls the design of the web, the total web area required is constant for any depth h; then $ht = $ a constant $ = k_1$, and

$$A_g = 2c_1\left(\frac{M}{c_4{}^2 hF} - c_3 \frac{k_1}{6}\right) + c_2 k_1 \qquad (11\text{-}29)$$

which evidently indicates that A_g is a minimum when h has the greatest possible value. This is natural because, if the web material ht is constant, the deeper the girder, the smaller will be the flange weight or the total weight of the girder. Actually, when depth h exceeds certain limits additional stiffeners and thicker web plates will be required to prevent web buckling. Then c_2 and k_1 are no longer constant and this increasing depth h no longer results in weight reduction.

Case 2.   When t is a constant, such as a required minimum thickness, then differentiating Eq. 11-28 with respect to h and equating it to zero leads to

$$h = \left[\frac{6c_1 M}{c_4{}^2 tF(3c_2 - c_1 c_3)}\right]^{\frac{1}{2}} \qquad (11\text{-}30)$$

Case 3.   When web plate is relatively thin, then the thickness might have to increase with h in order to meet minimum buckling requirements. When thickness t is a constant fraction of depth, $t = h/k$, Eq. 11-28 is transformed into the following:

$$A_g = 2c_1\left(\frac{M}{c_4{}^2 hF} - c_3 \frac{h^2}{6k}\right) + c_2 \frac{h^2}{k} \qquad (11\text{-}31)$$

and, from $dA_g/dh = 0$,

$$h = \left[\frac{3c_1 kM}{c_4{}^2 F(3c_2 - c_1 c_3)}\right]^{\frac{1}{3}} \qquad (11\text{-}32)$$

If t does not vary directly as h, a similar method may be used to obtain the equation for optimum h, but it will not be derived here.

Equations (11-30) and (11-32) can be applied when the constants c_1, c_2, c_3, c_4, and k are known.

The value of c_1 will depend on whether full-length cover plates are required, and whether gross or net area of the flanges is considered as effective in resisting the moment. Gross area is used for welded girders and also for some riveted girders. It further varies with the magnitude and the shape of the bending moment diagram. Approximate values of c_1 vary from 0.75 to 1.0.

The value of c_2 also varies with several factors: the specification for stiffener spacing, whether crimped stiffeners or fillers are used, whether girders are riveted or welded. Some approximate values are given as follows:

Riveted girders without stiffeners	$c_2 = 1.0$
Riveted girders with crimped stiffeners	$c_2 = 1.4$
Riveted girders with straight stiffeners with fillers	$c_2 = 1.6$
Welded girders	$c_2 = 1.3$

The value of c_3 is 1.0 if no rivet holes are to be deducted from the web; if 20% of the web is to be deducted for holes, then $c_3 = 0.80$.

The average value of c_4 for riveted girders is about 0.9; for welded girders c_4 is about 0.95.

The value of k varies with the ratio of angle leg to web depth. In general, for riveted and welded girder, k varies from 120 to 180.

For a typical girder, if $c_1 = 0.95$, $c_2 = 1.5$, $c_3 = 0.85$, $c_4 = 0.90$, $k = 150$, Eqs. 11-30 and 11-32 reduce to the following:

$$h = \left[\frac{6 \times 0.95M}{0.90^2 tF(3 \times 1.5 - 0.95 \times 0.85)} \right]^{\frac{1}{2}} = 1.35 \left(\frac{M}{Ft} \right)^{\frac{1}{2}} \quad (11\text{-}33)$$

$$h = \left[\frac{3 \times 0.95 \times 150M}{0.90^2 F(2 \times 1.5 - 0.95 \times 0.85)} \right]^{\frac{1}{3}} = 5.35 \left(\frac{M}{F} \right)^{\frac{1}{3}} \quad (11\text{-}34)$$

The two expressions for depth h as functions of (M/F) or (M/Ft) are plotted in Fig. 11-21.

It should be noted that, although these equations are interesting and often useful, slight variation from the obtained optimum values will result in very little additional weight. Hence no refinements should be attempted when applying these equations. Rather, they should serve as a guide, to be studied together with other requirements of girder depth. Also, there is only a slight difference between the riveted and the welded girders, the average optimum depth for the welded girders is a few percent less than the riveted ones.

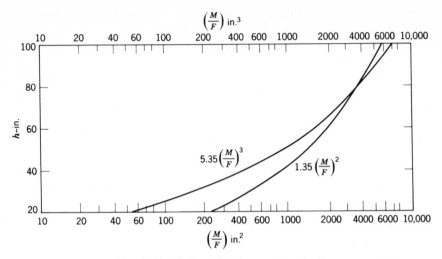

Fig. 11-21 Estimated optimum girder depth.

For plate girders of the usual proportions, the requirements of depth/span ratio and deflection are met with no difficulty. Only if a girder is limited to a very shallow depth will it have to be checked for such requirements.

11-16 PLATE GIRDER WEB BUCKLING

Unstiffened Web. Economical proportions of plate girder webs may be obtained with relatively thin webs, which have a possibility of buckling before yielding. Therefore, buckling must be considered in girder web design. Three principal types of buckling have to be considered: shear or "diagonal" buckling, bending or "longitudinal" buckling, and bearing or "vertical" buckling. In addition buckling under the action of combined stresses should theoretically be considered.

The basic considerations in determining the critical buckling stresses are briefly presented in Sec. 9-13, and theoretical values of elastic buckling stresses are defined in Eqs. 9-55, 9-56, and 9-57. An empirical expression for buckling under combined stresses is given in Eq. 9-59.

Girders with thin webs, if unstiffened, have low buckling strength. Assuming web plates ideally flat and simply supported at all edges and using a factor of safety n against buckling permits determination of critical design shear stress F_v as a function of depth/thickness ratio h/t. For example using Eq. 9-55, with $k_v = 4.8$ and $n = 1.5$:

$$F_v = \frac{f_{vcr}}{n} = \frac{4.8E}{1.5}\left(\frac{t}{h}\right)^2 = 3.2E\left(\frac{t}{h}\right)^2 \qquad (11\text{-}35)$$

To provide this factor of safety against initial shear buckling, from Eq. 11-35, the ratio h/t theoretically must not exceed the following:

$$\left(\frac{h}{t}\right) \leq \left(\frac{3.2E}{F_v}\right)^{\frac{1}{2}} = \frac{9700}{\sqrt{F_v}} \tag{11-36}$$

Similarly, for bending stress, using Eq. 9-56, with $k_b = 21.5$ and $n = 1.5$:

$$F_b = \frac{f_{bcr}}{n} = \frac{21.5E}{1.5}\left(\frac{t}{h}\right)^{\frac{1}{2}} = 14.3E\left(\frac{t}{h}\right)^{\frac{1}{2}} \tag{11-37}$$

To provide this factor of safety against initial flexural buckling, from Eq. 13-37, the ratio h/t theoretically must not exceed the following:

$$\frac{h}{t} \leq \left(\frac{14.3E}{F_b}\right)^{\frac{1}{2}} = \frac{20,600}{\sqrt{F_b}} \tag{11-38}$$

Under distributed transverse loads buckling of a web plate can occur as shown in Fig. 9-35a. For the modes of buckling shown the buckling stress is defined by Eq. 9-57. Using this equation with $k_w = 2.0$ and $n = 1.2$:

$$F_c = \frac{f_{cr}}{n} = \frac{2.0E}{1.2}\left(\frac{t}{h}\right)^2 = 1.67E\left(\frac{t}{h}\right)^2 \tag{11-39}$$

Therefore to provide this factor of safety against vertical buckling, from Eq. 11-39, the ratio h/t theoretically must not exceed the following:

$$\left(\frac{h}{t}\right) \leq \left(\frac{1.67E}{F_c}\right)^{\frac{1}{2}} = \frac{7000}{\sqrt{F_c}} \tag{11-40}$$

The presence of bending, shearing and bearing stresses tends to reduce the magnitude of critical buckling stresses. The effect of combined stresses can be evaluated approximately using Eq. 9-59. Using the boundary condition coefficients k_v, k_b, and k_w selected for Eqs. 11-35, 11-37, and 11-39, the interaction equation can be written in the following form:

$$\frac{f_c}{2E(t/h)^2} + \left(\frac{f_b}{21.5E(t/h)^2}\right)^2 + \left(\frac{f_v}{4.8E(t/h)^2}\right)^2 \leq 1.0 \tag{11-41}$$

or

$$\left(\frac{h}{t}\right) \leq \sqrt{\frac{4.8E}{F}} = \frac{11,900}{\sqrt{F}} \tag{11-42}$$

where effective stress

$$F = \left\{f_v^2 + \left(\frac{f_b}{4.5}\right)^2 + f_c\left[11.4E\left(\frac{t}{h}\right)^2\right]\right\}^{\frac{1}{2}}$$

Stiffened Web. The load-carrying capacity of unstiffened thin webs is only slightly greater than the initial buckling strength. However, when the thin webs are stiffened their load-carrying capacity is greatly increased. The stiffeners serve a dual purpose: they increase the initial buckling load and also enable the web to carry shear loads in excess of initial buckling due to the so-called "tension field" effect. This post-buckling behavior of stiffened web plates is discussed in Sec. 11-17.

Theoretical values of initial buckling strength of stiffened plates can be calculated using appropriate values of k in Eqs. 9-55 and 9-56. Use of these expressions is based on the assumptions of ideally flat elastic plates so stiffened that each panel behaves as an independent plate with idealized edge support conditions. In order to approximate this behavior the stiffeners must be sufficiently rigid to prevent any significant lateral bending.

The effect of various arrangements of stiffeners on allowable stresses is shown in Figs. 11-22 and 11-23.

The shear stresses plotted in Fig. 11-22 are calculated using a factor of safety of 1.5 against buckling (Eq. 9-55) or yielding (Eq. 11-9a). Five conditions are considered: *A*—unstiffened web, *B*—web stiffened so that spacing of the stiffeners is equal to depth of the web, *C*—spacing of the stiffeners equal to half the web depth, *D*—web stiffened with one longitudinal stiffener at 0.2 the depth of the web, and *E*—corresponding to combined stiffeners of conditions *B* and *D*.

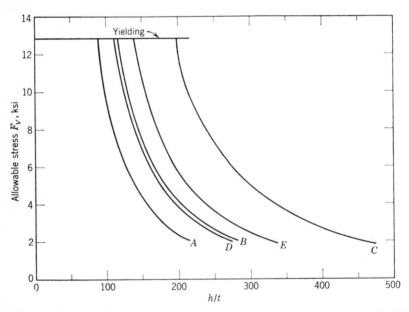

Fig. 11-22 Effect of intermediate stiffeners on allowable shear stress. *A.* No stiffeners. *B.* Transversè stiffeners, spacing $= h$. *C.* Transverse stiffeners, spacing $= 0.5h$. *D.* Longitudinal stiffener at $0.2h$. *E.* Stiffeners as in *B* and *D* combined.

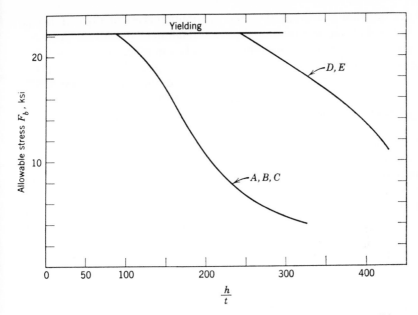

Fig. 11-23 Effect of stiffeners on allowable bending stress. Stiffener conditions *A-E* as in Fig. 11-22.

It can be seen that the most effective method of stiffening is to provide closely spaced transverse stiffeners as in *C*. Furthermore, a single longitudinal stiffener as in *D* is slightly less effective than an equivalent length of stiffeners placed transversely, as in *B*.

The bending stresses plotted in Fig. 11-23 are calculated using the same factor of safety of 1.5 and the same five stiffening conditions *A* through *E* as in the case of shear. In this instance it can be seen that transverse stiffeners have little influence on buckling stress due to bending, but that a longitudinal stiffener effectively increases buckling stress.

When both shearing and bending stresses are developed, it is apparent that transverse stiffening would be essential when shearing stresses predominate and that longitudinal stiffeners can be effective when significant bending stresses are developed.

11-17 POSTBUCKLING STRENGTH OF GIRDER WEBS

As the shearing stress in a stiffened girder web increases from initial zero value to its ultimate capacity, the behavior of the web is characterized by three stages. When the shearing stress is less than the buckling stress f_{vcr}

the web does not buckle and the stress distribution is essentially that defined by the simple beam theory. For precise evaluation of the critical value f_{vcr}, the bending stresses in the web should be taken into account, but as a first approximation these bending stresses may be neglected, and the buckling stress defined by Eq. 9-55 can be used. Thus during this stage the web is assumed to resist pure shear stress which is characterized by equal tension and compression principal stresses acting along diagonals at 45 degrees with the beam axis.

If the shearing stress is increased beyond the value of the buckling stress the web does not collapse and can resist additional load. The reserve strength of an initially buckled stiffened web is due to the "tension field" action of the web, which can be represented as the action of flange-stiffener truss frame with the web behavior equivalent to tension diagonals (Fig. 11-24a). In the second stage, when shear stress exceeds the initial buckling value, part of the load is resisted by the conventional beam action and part by the "tension field" action. This stage is often referred to as "partial tension field."[8,9,10,11]

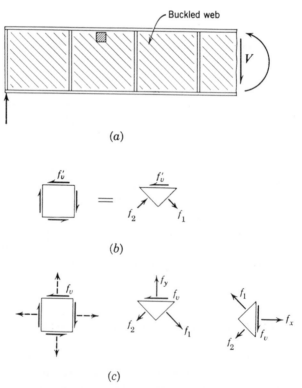

(a)

(b)

(c)

Fig. 11-24   Tension field web stresses.

The third stage develops as the girder approaches its ultimate capacity. In this stage portions of the web may be yielding, and also the secondary stresses and deformation of girder flanges and stiffeners have considerable influence on the actual stress distribution in the girder and on the mechanism of failure.

To clarify some of the essential mechanisms contributing to the post-buckling strength of plate girder webs the idealized "partial tension field" behavior is considered below.

Before buckling an element within the web is subjected to principal stresses f_1 and f_2 (Fig. 11-24b) such that:

$$f_1 = f_2 = f_v' = V_1/ht \tag{11-43}$$

where f_1 is tensile and f_2 is compressive principal stress. Buckling occurs when $f_v = f_{vcr}$ at a total shear force $V_{cr} = f_{vcr}ht$.

After buckling, as the shear force V increases, the compressive principal stress $f_2 = f_{vcr}$ cannot increase, but the tensile principal stress f_1 can increase. This increase in tensile principal stress tends to produce horizontal and vertical tensile stresses in the web, producing the "tension field" effect. The stresses induced in the web are shown in Fig. 11-24c. If α is the angle of principal stresses and $f_2 = f_{vcr}$ defined by Eq. 9-55, then using equilibrium conditions[8] (Fig. 11-25),

$$f_v = (f_1 + f_2) \sin \alpha \cos \alpha \tag{a}$$

$$f_1 = \frac{f_v}{\sin \alpha \cos \alpha} - f_2 \tag{b}$$

$$f_2 = f_{vcr} \tag{c} \qquad (11\text{-}44)$$

$$f_x = f_v \cot \alpha - f_{vcr} \tag{d}$$

$$f_y = f_v \tan \alpha - f_{vcr} \tag{e}$$

The "tension field" stresses f_x and f_y tend to pull the flanges and the end stiffeners together (Fig. 11-26).

The flanges are kept apart by the stiffeners, and the end stiffeners are kept apart by the flanges. As a result of this action, the flanges, in addition to primary flange forces, resist secondary bending and axial compression. The intermediate stiffeners, which are not stressed before buckling, are subjected to compressive loads due to diagonal tension developed after buckling. The loads on the web-to-flange and stiffener-to-flange attachments are also increased, as a result of diagonal tension.

Theoretical determination of the stress distribution in and strength of a buckled web is complicated. An approximate method for determining shear

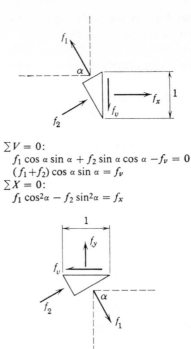

$$\Sigma V = 0:$$
$$f_1 \cos \alpha \sin \alpha + f_2 \sin \alpha \cos \alpha - f_v = 0$$
$$(f_1 + f_2) \cos \alpha \sin \alpha = f_v$$
$$\Sigma X = 0:$$
$$f_1 \cos^2\alpha - f_2 \sin^2\alpha = f_x$$

$$\Sigma Y = 0:$$
$$f_1 \sin^2\alpha - f_2 \cos^2\alpha = f_y$$
$$\Sigma X = 0:$$
$$(f_1 + f_2) \cos \alpha \sin \alpha = f_v$$

Fig. 11-25 State of stress in a buckled web.

strength of a thin-webbed plate girder will be outlined. Two types of failure are considered: rupture by shear and rupture by diagonal tension.

In a riveted girder, rupture by shear may take place at the web-to-flange connection, or web-to-stiffener connection, where the rivet holes reduce the effective cross section of the web, and introduce stress concentrations. Neglecting the effect of bending stress on the web plate, the shearing rupture strength V_{su} of a riveted girder is defined by

$$V_{su} = f_{vu}ht = k_r\left(1 - \frac{d_r}{p_r}\right)f_{su}\, ht \qquad (11\text{-}45)$$

where f_{vu} is the effective stress at rupture, f_{su} is the ultimate shearing strength of web material, h and t are the web depth and thickness, respectively, d_r and p_r are rivet diameter and pitch, respectively, and k_r is the rivet stress concentration factor, varying from 0.85 to 0.95. Diagonal tension rupture occurs when maximum tension stress f_{max} equals tensile strength of the material f_t. Because of stress concentration at the corners of the web panel, particularly at the flange-to-web connections, the maximum tensile stress $f_{max} = k f_1$ where k is a stress concentration factor. When $k f_1 = f_t, f_1 = f_t/k$, and, from Eq. 11-45, diagonal tension rupture occurs at a load V_{tu} defined by

$$V_{tu} = f_v ht = \left(\frac{f_{tu}}{k} + f_{vcr}\right) ht \sin \alpha \cos \alpha \qquad (11\text{-}46)$$

The stress concentration factor k varies depending on type of web-to-flange attachment—welded or riveted—and on deformations of web, flanges, and stiffeners.

The angle of principal stress depends on the deformations of flanges, stiffeners, and web, and on the ratio of shear buckling stress f_{vcr} to the applied shear stress f_v, $r = f_{vcr}/f_v$. The angle may be defined in terms of r and a numerical coefficient η characteristic of the girder dimensions,

$$(\eta + 1) \tan^4 \alpha - r(\eta + 1) \tan^3 \alpha + r \tan \alpha - 1.0 = 0 \qquad (11\text{-}47)$$

where

$$\eta = \frac{bt}{A_s}\left[1 + \left(\frac{1}{I_{fc}} + \frac{1}{I_{ft}}\right)\frac{b^3 A_s}{720h}\right] \qquad (11\text{-}48)$$

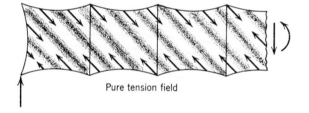

Pure tension field

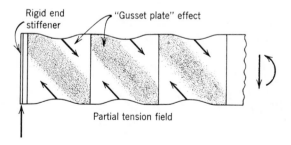

Rigid end stiffener   "Gusset plate" effect

Partial tension field

Fig. 11-26 Effect of tension field on flanges and stiffeners.

and A_s is the effective area of stiffener, I_{fc} and I_{ft} are the moments of inertia of compression and tension flanges, respectively, b is the stiffener spacing, and h is the web depth. When $I_{fc} = I_{ft} = I_f$, the expression for η is simplified, as follows:

$$\eta = \frac{bt}{A_s}\left(1 + \frac{b^3 A_s}{360 I_f h}\right) \tag{11-49}$$

For given proportions of the girder, η can be readily computed, and, if the value of $r = f_{vcr}/f_v$ is known or estimated, the angle of principal stress α can be determined from Eq. 11-47. Solution of this transcendental equation is greatly simplified by the use of Fig. 11-27. The ultimate strength of the girder is then determined from Eq. 11-46.

Another approximate solution proposed for steel plate girders is based on the assumption that the ultimate shear capacity V_u is

$$V_u = V_{cr} + V_t \tag{11-50}$$

where V_t is the pure tension field capacity corresponding to $V_{cr} = 0$. Then from Eq. 11-46, with $k = 1.0, f_{vcr} = 0$:

$$V_t = (f_t \sin \alpha \cos \alpha)ht = f_{vt}ht \tag{11-51}$$

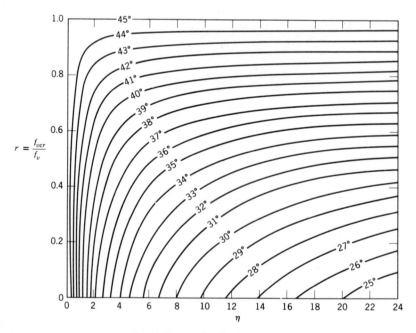

Fig. 11-27 Angle of principal stress.

where f_t is the diagonal tension field stress component defined by an approximate yield criterion[10] for the girder web plate as follows:

$$\frac{f_t}{F_y} + \frac{f_{v\text{cr}}}{F_{vy}} = 1.0 \tag{11-52}$$

Using notation $(f_{v\text{cr}}/F_{vy}) = C_v$, the diagonal tension field stress f_t becomes

$$f_t = (1 - C_v)F_y \tag{11-53}$$

Also in this solution the angle of principal stresses α is approximated as follows:

$$\sin 2\alpha = \frac{h}{\sqrt{a^2 + h^2}} = \frac{1}{\sqrt{1 + R^2}} \tag{11-54}$$

where a is the stiffener spacing and R is the aspect ratio a/h. Then the tension field contribution to the shear capacity is

$$V_t = f_{vt}ht = [(1 - C_v)F_y ht]\frac{1}{2\sqrt{1 + R^2}} \tag{11-55}$$

The ultimate shear capacity is then taken as the sum of buckling shear V_{cr}, and the tension field shear V_t (Eq. 11-55).

In girders with unbuckled webs, the deformation due to shear is usually small. As a result of buckling, the shear load is resisted partly by diagonal tension, and causes web distortions and additional strains in flanges and stiffeners. These deformations contribute to the increase in girder deflection, and the increase in load-carrying capacity of the girder due to diagonal "tension field" is accompanied by a decrease in its stiffness. Deflection of a buckled plate girder, accounting for the diagonal "tension-field" effects on the web, flanges, and stiffeners, can be obtained by the method of strain energy. The total strain energy U in the girder is equal to the sum of strain energies in the flanges, web, and stiffeners; that is, the work due to a single load P applied at the point where the deflection is desired is $W_p = U = U_f + U_w + U_{st} = \frac{1}{2}P\Delta$. The deflection Δ is then $\Delta = 2U/P$. For other loading conditions where $W \neq \frac{1}{2}P\Delta$, a least-work method or a virtual-work procedure can be used to evaluate deflection in a similar manner. Expression for U_f, U_w, and U_{st} are based on classical principles of strain energy and will not be derived here. In girders with very thin webs, the actual deflection may be three to four times that computed by conventional flexure theory neglecting shear deformation, or two to four times that computed by conventional flexure theory accounting for shear deformation.

11-18 DESIGN SPECIFICATIONS FOR PLATE GIRDERS

Some specifications, such as AASHO, are based on the assumption that a reasonable factor of safety against elastic buckling must be provided. Others, such as AISC, are based on the assumption that a suitable factor of safety against inelastic postbuckling failure must be provided.

Experimental studies carried out on thin-web steel girders[10,11,12,13] indicate that postbuckling strength can be insured by proper design of web, flanges, and stiffeners. However, these studies have been carried on under essentially static loading conditions without significant repeated loadings. Therefore the second method, accepted for design of buildings (AISC Specification), has not been accepted for bridge structures.

AISC Specification. Bending resistance of plate girders is based on moment of inertia of the gross section, provided that the holes in the flanges do not exceed 15% of the flange area. When this limit is exceeded the excess only must be deducted from the gross section area.

The *minimum web thickness* is established by the condition that the web should not buckle under the vertical in-plane compressive stresses resulting from the curvature of the beam in bending (Fig. 11-28). The web must resist these compressive stresses by plate action and the critical buckling stress f_{cr} may be obtained from Eq. 9-52 using $a = h$ and $C = 1.0$, so that

$$f_{cr} = \frac{\pi^2 E}{12(1 - \mu^2)}\left(\frac{t}{h}\right)^2 = 26{,}200{,}000 \left(\frac{t}{h}\right)^2 \tag{11-56}$$

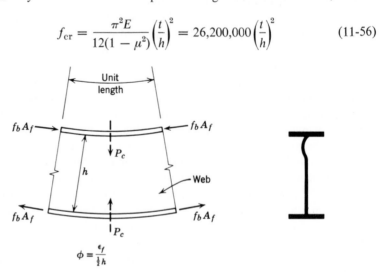

Fig. 11-28 Crushing loads on the web.

The vertical compression stress due to flange curvature is

$$f_c = \frac{P_c}{t} = \frac{f_b A_f \phi}{t} = \frac{f_b A_f 2\epsilon_f}{ht} = 2f_b \epsilon_f \frac{A_f}{A_w} \qquad (11\text{-}57)$$

where P_c is the crushing force per unit length of the girder, f_b is the bending stress in the flange, A_f is the flange area, ϵ_f is the strain in the flange, A_w is the web area. For stability of the web, with f_{cr} defined by Eq. 11-56:

$$f_c \leq f_{cr} \qquad (11\text{-}58)$$

or

$$\frac{h}{t} = \sqrt{\frac{26,200,000 A_w}{f_b \epsilon_f A_f}} \qquad (11\text{-}59)$$

Evaluation of this limiting depth/thickness ratio requires definition of the quantities under the radical in Eq. 11-59. The minimum value of A_w/A_f is

Table 11-8   Maximum Depth-Thickness Ratio

F_y, ksi	33	36	42	46	50
h/t	345	320	282	260	243

assumed to be 0.5, and the maximum value of f_b is taken as F_y. The magnitude of the maximum strain ϵ_f in the flanges depends on the residual stresses due to fabrication. For the purposes of evaluating the limiting h/t ratio ϵ_f is taken as

$$\epsilon_f = \frac{(F_y + f_r)}{E} = \frac{F_y + 16,500}{E} \qquad (11\text{-}60)$$

where residual stress f_r is taken as 16,500 psi. Then, substituting the foregoing values into Eq. 11-59,

$$\frac{h}{t} \leq \frac{14,000,000}{\sqrt{F_y(F_y + 16,500)}} \qquad (11\text{-}61)$$

With depth h of the girder established, Eq. 11-61 defines the minimum thickness that can be used. In practice, depending on required shear resistance, over-all depth limitation, and economy of stiffening the web, a thickness greater than the minimum value may lead to an optimum design.

The *allowable shearing stress* in plate girder web is limited by one of the following criteria:

(a) Factor of safety of 1.45 against yielding in pure shear.

(*b*) Factor of safety of 1.67 against elastic or inelastic buckling in pure shear.

(*c*) Factor of safety of 1.67 against web yielding in postbuckling range when shear is carried partly by developing "tension field" stresses.

Unstiffened Webs. Yielding in pure shear occurs at a stress F_{vy} which is closely approximated by $F_y/\sqrt{3}$. Using a factor of safety of 1.45, the allowable shearing stress F_v based on criterion (*a*) is

$$F_v = \frac{F_{vy}}{1.45} = 0.4F_y \tag{11-62}$$

For different steel grades these limit stresses are listed in Table 11-4.

Buckling in pure shear controls the allowable stress of relatively thin unstiffened webs, or webs where the transverse spacing of stiffeners is such that a/h exceeds 3. Elastic buckling stress is calculated from Eq. 9-55 using a value of $k_v = 4.8$:

$$f_{vcr} = 4.8E\left(\frac{t}{h}\right)^2 = C_v F_{vy} \tag{11-63}$$

where C_v is the ratio of shear buckling and shear yield stresses. With a factor of safety of 1.67 the allowable stress for unstiffened web is

$$F_v = \frac{f_{vcr}}{1.67} = 85 \times 10^6 \left(\frac{t}{h}\right)^2 = \frac{C_v F_y}{2.89} \tag{11-64}$$

where

$$C_v = \frac{245 \times 10^6}{F_y(h/t)^2} \tag{11-65}$$

When the elastic buckling stress given by Eq. 11-63 exceeds the proportional limit stress in shear taken as $0.8F_{vy}$, the elastic behavior no longer represents actual conditions and the inelastic buckling stress f'_{vcr} can be obtained approximately from the following

$$f'_{vcr} = \sqrt{0.8F_{vy}f_{vcr}} = C'_v F_{vy} \tag{11-66}$$

The allowable shear stress for unstiffened web is then:

$$F_v = \frac{f'_{vcr}}{1.67} = 4800\left(\frac{t}{h}\right)\sqrt{F_y} = \frac{C'_v F_y}{2.89} \tag{11-67}$$

where

$$C'_v = \frac{13,900}{\sqrt{F_y}(h/t)} \tag{11-68}$$

Stiffened Webs. These webs can fail by any of the three mechanisms: yielding, buckling, or postbuckling tension field behavior. The allowable stresses depend on h/t—web slenderness, a/h—stiffener spacing ratio, and F_y—yield strength of the material.

The total allowable shear stress based on postbuckling behavior is based on the approximate solution given by Eq. 11-50, so that it can be expressed by two terms representing allowable stress component for prebuckling behavior and allowable stress component for tension field behavior. Then

$$F_v = F_{vo} + F_{vt} \qquad (11\text{-}69)$$

The upper limit for allowable stresses is, as before, $0.4F_y$. Buckling of a stiffened web in the elastic range is given by Eq. 9-55, where the value of k_v is variable and depends on the stiffener spacing ratio a/h. Assuming simply supported edges for welded girders the k_v values* are

$$\text{for } a/h > 1: \quad k_v = 4.8 + 3.6(h/a)^2 \qquad (a)$$
$$\text{for } a/h < 1: \quad k_v = 3.6 + 4.8(h/a)^2 \qquad (b)$$
$$(11\text{-}70)$$

With these values elastic buckling stress is

$$f_{vcr} = k_v E \left(\frac{t}{h}\right)^2 = C_v F_{vy} \qquad (11\text{-}71)$$

and corresponding prebuckling allowable based on factor of safety of 1.67 is

$$F_{vo} = \frac{f_{vcr}}{1.67} = 17.5 \times 10^6 k_v \left(\frac{t}{h}\right)^2 = \frac{C_v F_y}{2.87} \qquad (11\text{-}72)$$

where

$$C_v = \frac{50.6 \times 10^6 k_v}{F_y (h/t)^2} \qquad (11\text{-}73)$$

when the elastic buckling stress exceeds the proportional limit stress in shear taken as $0.8F_{vy}$, inelastic criteria should be used, and the value of inelastic shear buckling stress can be obtained from Eq. 11-66. The prebuckling allowable shear stress is then

$$F_{vo}' = \frac{f_{vcr}'}{1.67} = 2200 \left(\frac{t}{h}\right) \sqrt{k_v F_y} = \frac{C_v' F_y}{2.89} \qquad (11\text{-}74)$$

where

$$C_v' = \frac{6350 \sqrt{k_v}}{\sqrt{F_y}(h/t)} \qquad (11\text{-}75)$$

---

* These coefficients are evaluated as $k_v = k \dfrac{\pi^2}{12(1 - \mu^2)}$, where k values are often used in general theoretical derivations and in AISC Manual.

The tension field stress f_{vt} is obtained from Eqs. 11-55, as follows

$$f_{vt} = \frac{(1 - C_v)F_y}{2\sqrt{1 + R^2}} \tag{11-76}$$

where R is stiffener spacing ratio a/h and C_v is the ratio of shear buckling stress to shear yield stress. Using factor of safety of 1.67, the tension field component of allowable stress becomes

$$F_{vt} = \frac{f_{vt}}{1.67} = \frac{(1 - C_v)F_y}{3.34\sqrt{1 + R^2}} \tag{11-77}$$

It should be noted that when value of C_v defined by Eq. 11-73 exceeds 0.8, it should be replaced by C_v' defined by Eq. 11-75. In Eq. 11-77 the value of C_v' shall not exceed unity.

The total allowable stress F_v is the sum of prebuckling and tension field components and is

$$F_v = \frac{F_y}{2.89}\left[C_v + \frac{1 - C_v}{1.15\sqrt{1 + R^2}} \right] \tag{11-78}$$

where C_v' is to replace C_v when the latter exceeds 0.8.

For A36 steel the allowable stress in shear is plotted in Fig. 11-29 for both stiffened and unstiffened webs.

The contribution of the web to the bending strength of the girder will depend upon the h/t ratio because very thin webs may buckle at loads well below full plastification or yielding. The critical buckling stress for a plate in bending is given in Eq. 9-56 where the values of k_b vary between 21.5 and 35.7, depending on the edge support conditions. The limiting ratio h/t which permits development of yielding without buckling is such that $f_{bcr} = F_y$, and therefore, from Eq. 9-56, h/t is

$$\frac{h}{t} = \sqrt{\frac{k_b E}{f_{bcr}}} = \sqrt{\frac{k_b E}{F_y}} \tag{11-79}$$

The edge support conditions for the web vary, but when the compression flange is prevented from local and lateral buckling, the value of k_b may be assumed equal to 32. Using this value and assuming bending stress $F_b = 0.6F_y$, Eq. 11-79 becomes

$$\frac{h}{t} = \frac{24,000}{\sqrt{F_b}} = \beta_0 \tag{11-80}$$

The limiting values of $\beta_0 = h/t$ for various grades of steel and based on full allowable bending stress F_b, are indicated in Table 11-9.

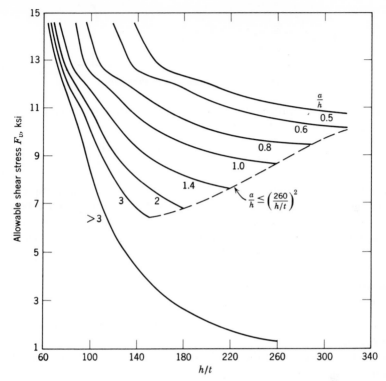

Fig. 11-29   AISC allowable shear stress for A36 steel thin web girders.

A thin web, with $h/t > \beta_0$, subjected to pure bending deflects laterally from its original position (Fig. 11-30), and therefore does not support its full share of the load unless stiffened by longitudinal stiffeners. In order to evaluate properly the bending moment capacity of a girder without a longitudinal stiffener either reduced section should be used with full allowable bending stress, or reduced allowable stress may be used with full section

Table 11-9

Yield Point, psi	β_0
33,000	170
36,000	162
42,000	152
46,000	145
50,000	138

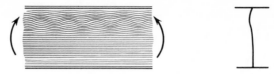

Fig. 11-30 Lateral deflection of thin web.

modulus. The latter procedure is convenient and when $h/t > \beta_0$ a reduced allowable bending stress F_b' should be used:

$$F_b' \leq F_b\left[1.0 - 0.0005\,\frac{A_w}{A_f}\left(\frac{h}{t} - \frac{24{,}000}{\sqrt{F_b}}\right)\right] \qquad (11\text{-}81)$$

The influence of combined shearing and bending stresses in the compression zone of the girder web may cause some premature buckling, but generally this is considered to have negligible effect on the ultimate capacity of thin web girders. However, consideration must be given to the combined effect of shearing and bending stresses in the tension zone where premature yielding may limit the capacity.

Analytical and experimental studies have shown that interaction of shear and bending can be neglected if

(a) the shearing stress in the web is not greater than 0.6 the allowable value F_v; or

(b) the bending stress in the web is not greater than 0.75 of the maximum allowable value.

Because in many cases high shear does not occur at the section of high moment, the interaction may not be critical. When a section is subjected to high shear and high moment a linear interaction between conditions (a) and (b) is recommended, as shown in Fig. 11-31. This effect may also be expressed by the following design conditions:

$$f_b \leq \left(1.375 - 0.625\,\frac{f_v}{F_v}\right)F_b \qquad (11\text{-}82)$$

or

$$f_v \leq \left(2.2 - 1.6\,\frac{f_b}{F_b}\right)F_v \qquad (11\text{-}83)$$

In Eq. 11-82 F_b is the allowable bending stress taking into account lateral instability of the compression flange. The flange of the girders must be proportioned to prevent local buckling before yield stress is reached. For flat plate flanges this can be achieved when width-thickness ratio does not exceed $3000/\sqrt{F_y}$. Appropriate limits for b/t are shown in Table 10-2.

In selecting minimum web thickness limits for the web panel proportions have been selected to prevent excessive distortion in fabrication, erection, and handling. These limits are

$$\frac{a}{h} \le \left(\frac{260}{h/t}\right)^2 \quad \text{and} \quad \frac{a}{h} \le 3.0 \qquad (11\text{-}84)$$

For a girder of given depth and subjected to given loading conditions appropriate web thickness and stiffener spacing may be determined from allowable stress charts such as shown in Fig. 11-29.

The influence of tension field is to induce diagonal tensile stresses in the web. In the intermediate and end panels it is assumed that "gusset plate" action precludes bending in the flanges (Fig. 11-26). In order to prevent bending in the end stiffeners or end posts due to tension field effects it is recommended that the web in the end panels (or in panels containing holes) be so proportioned that the plate will not buckle. Because the effective edge support condition of the plate is uncertain, the web panel proportions required to prevent buckling cannot be defined with precision.

The usual limits for h/t or a/t to prevent buckling are given as $11{,}000/\sqrt{f_v}$, although sometimes a more conservative limit of $9000/\sqrt{f_v}$ is recommended.

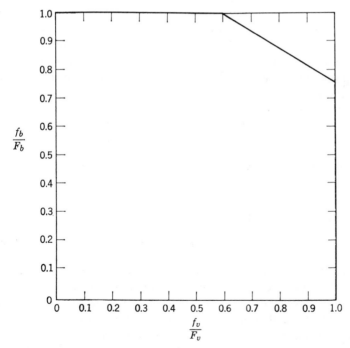

Fig. 11-31 AISC shear and bending allowable stress interaction in thin web girders.

The vertical component of the tensile stresses in the web due to tension field produces compression in the intermediate stiffeners. If the shear component of the tension field is $f_{vt} = V_t/ht$, the vertical component of the tensile stress in the web is $f_x = f_{vt} \times \tan \alpha$, and the compression force in the stiffener is $P_{st} = f_x \times at$. Using Eqs. 11-54 and 11-76 it follows that

$$P_{st} = (1 - C_v)F_y(\sqrt{1 + R^2} - R) \, at \, \frac{1}{2\sqrt{1 + R^2}} \qquad (11\text{-}85)$$

or

$$P_{st} = \frac{(1 - C_v)F_y ht}{2}\left(R - \frac{R^2}{\sqrt{1 + R^2}}\right) \qquad (11\text{-}86)$$

where $R = a/h$, and F_y = yield strength of web steel.

For stiffeners which are not symmetrical with respect to the web, eccentricities are introduced which produce bending of the stiffener. The combined stress due to direct compression P_{st} and moment $P_{st} \times e$, where e is the eccentricity of the load, may be expressed as

$$f_{st} = \frac{P_{st}}{A_{st}}D \cdot \qquad (11\text{-}87)$$

where D is the stress amplification factor which depends on the asymmetry (eccentricity) of the stiffener and on its shape. For pairs of stiffeners symmetrical with respect to the plane of the web $D = 1.0$, for single angle stiffeners on one side of the web $D = 1.8$, and for single plate stiffeners on one side of the web $D = 2.4$.

The required cross-section area for the stiffener is calculated from the condition that $f_{st} \leq F_y{'}$, where $F_y{'}$ is the yield strength of the stiffener steel and may be different from the F_y value for the web steel. Then

$$A_{st} = \frac{P_{st}D}{f_{st}} \geq \frac{P_{st}D}{F_y{'}} = \frac{(1 - C_v)F_y ht D}{2F_y{'}}\left(R - \frac{R^2}{\sqrt{1 + R^2}}\right) \qquad (11\text{-}88)$$

If $(F_y/F_y{'}) = Y$, and $(R - R^2/\sqrt{1 + R^2}) = K_r$ then

$$A_{st} = \tfrac{1}{2}(1 - C_v)ht D Y K_r \qquad (11\text{-}89)$$

Values of required stiffener area A_{st} assuming $D = Y = 1.0$ are shown in AISC Manual tables in Appendix 5 for different grades of steel.

In addition to carrying compression loads the web stiffeners must provide the required stiffness so that the assumed edge conditions for the web panels are realized. For this reason a minimum value of moment of inertia I_s of a single stiffener or a pair of stiffeners with reference to an axis in the plane of the web is usually specified. The AISC Specification recommends that

$$I_s = \left(\frac{h}{50}\right)^4 \qquad (11\text{-}90)$$

The attachment of the intermediate stiffener to the web must be sufficient to resist the tension field forces which produce the maximum stresses in the stiffener.

The shear transfer is not uniform along the length of the stiffener, and it is usually required that sufficient connectors be provided to build up the force P_{st} over a third of the depth h. Then the shear transfer q_u per inch of web depth is

$$q_u = \frac{P_{st}}{h/3} \qquad (11\text{-}91)$$

which is reduced by a factor of safety of 1.67 for usual allowable stress design. For such cases the design shear transfer is

$$q = \frac{3P_{st}}{1.67h} = \frac{1.8A_{st}F_y}{h} \qquad (11\text{-}92)$$

This is usually approximated by

$$q = h\sqrt{\left(\frac{F_y}{3400}\right)^3} \qquad (11\text{-}93)$$

Web crippling may occur in the girder when the applied loads, concentrated or uniformly distributed on the flange, are not supported directly by bearing stiffeners. Concentrated loads which are light enough to fall within limits given in Sec. 11-4 may be treated as distributed over full panel length a within which these loads act provided $a \leq h$. These loads together with any distributed loads on the flange produce a compressive stress on the edge of the web which should not exceed safe limits based on consideration of elastic buckling. An approximate expression for the critical stress is given in Eq. 9-58, and the AISC design recommendation are based on a factor of safety of 2.67 against such buckling.

Example 11.8. Design a plate girder for a building with a limited depth of 72 in., A36 steel, with the lateral support at points of concentrated loads. The loads and dimensions are indicated in Fig. 1. AISC Specification.

Solution

$$R = 3(25) + 80 = 155 \text{ kip}$$

$$M_1 = \frac{101 + 155}{2} \times 18 = 2304 \text{ kip-ft}$$

$$M_{max} = 2304 + \tfrac{21}{2}(7) = 2377.5 \text{ kip-ft}$$

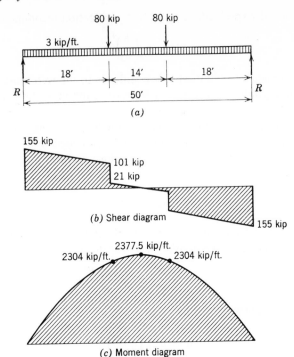

Ex. 11-8   Fig. 1.

Preliminary Web Design.   Assume web depth, $h = 70$ in.

For no reduction in flange stress $h/t = 24,000/\sqrt{22,000} = 162$ (AISC Sec. 1.10.6). Corresponding thickness of web $= \frac{70}{162} = 0.43$ in. Min. thickness of web $= \frac{70}{230} = 0.22$ in. (AISC Sec. 1.10.2).

Try web plate $70 \times \frac{3}{8}$ in.;

$$A_w = 70 \times 0.375 = 26.25 \text{ in.}^2; \quad h/t = \frac{70}{0.375} = 187$$

Preliminary Flange Design.   Assume flanges $\frac{7}{8}$ in. thick and allowable $F_b = 22$ ksi

$$\text{Required } A_f \simeq \frac{2377.5(12)}{70.88(22)} - \frac{1}{6}(26.25) \simeq 14.0 \text{ in.}^2$$

Try plate $17 \times \frac{7}{8} = 14.9$ in.$^2 > 14.0$ in.^{2}
Check local buckling:

$$\frac{b}{t} = \frac{8.5}{0.875} = 9.7 < 16.0 \qquad \text{(AISC Sec. 1.9)}$$

Trial Girder Section

Section	A	$\bar{y}$	$A\bar{y}^2$	I_0	I_{gr}	Weight
1-Web PL 70 × $\frac{3}{8}$	26.25	–	–	10,719	10,719	89.3
2-Flg PLs 17 × $\frac{7}{8}$	29.80	35.44	37,501	–	37,501	101.2
				$I_{\text{effective}}$ =	48,220	190.5

$$S_{\text{furn}} = \frac{48,220}{35.88} = 1345 \text{ in.}^3$$

$$S_{\text{req'd}} = \frac{2377.5(12)}{22} = 1295 \text{ in.}^3$$

Check lateral buckling:

$$\text{at midspan max } f_b = \frac{2377.5(12)}{1345} = 21.2 \text{ ksi}$$

Moment of inertia of flange plus $\frac{1}{6}$ web about Y-Y axis

$$I_{0y} = \frac{7}{8} \times \frac{17^3}{12} = 358 \text{ in.}^4$$

$$A_f + \tfrac{1}{6}A_w = 14.9 + \tfrac{1}{6}(26.25) = 19.3 \text{ in.}^2$$

$$r = \sqrt{\frac{I_{0y}}{A_f + \tfrac{1}{2}A_w}} = \sqrt{\frac{358}{19.3}} = 4.29 \text{ in.}$$

Check bending stress in 14 ft long panel:

$$M_{\text{max}} > M_1 \text{ and } M_2 \quad \therefore \quad C_b = 1 \qquad \text{(See Eq. 11-6)}$$

$$\frac{l}{r} = \frac{14 \times 12}{4.29} = 39.2 < 40$$

Stress reduction according to Eq. 11-5 can be neglected

$$\therefore \quad \text{allowable } F_b = 22.0 \text{ ksi} > 21.2$$

Allowable flange stress in 14 ft panel (Eq. 11-81):

$$F_b' = F_b\left[1.0 - 0.0005 \frac{A_w}{A_f}\left(\frac{h}{t} - \frac{24,000}{\sqrt{F_b}}\right)\right]$$

$$F_b' = 22.0\left[1.0 - 0.0005\left(\frac{26.25}{14.9}\right)(187 - 162)\right]$$

$$F_b' = 22.0(0.978) = 21.5 \text{ ksi} > 21.2$$

Check allowable f_b in 18-ft long panels (Eq. 11-6):

$$C_b = 1.75 - 1.05\frac{M_1}{M_2} + 0.3\left(\frac{M_1}{M_2}\right)^2$$

$$M_1 = 0; \qquad \frac{M_1}{M_2} = 0; \qquad \therefore \ C_b = 1.75$$

$$f_b = \frac{2304(12)}{1345} = 20.6 \text{ ksi}$$

$$\frac{l}{r} = \frac{18 \times 12}{4.29} = 50.3 > 40.0$$

Allowable stress in 18-ft panel:

$$F_b = 22,000 - \frac{0.679}{1.75}(50.3)^2 = 21,020 \text{ psi} = 21.02 \text{ ksi} \quad \text{(From Eq. 11-5)}$$

$$F_b = \frac{12,000 \times 14.9}{(18 \times 12)(71.75)} = 11.60 \text{ ksi} < 21.02 \quad \text{(From Eq. 11-3)}$$

Eq. 11-5 governs   21.02 ksi > 20.6 ksi
Check flange stress reduction in 18-ft panels (Eq. 11-81):

$$F_b' = 21.02(0.978) = 20.55 \text{ ksi}$$

20.55 ksi < 20.6 ksi (negligible understrength)

Use: Web   1 PL 70 × $\frac{3}{8}$
           Flange   2 PLs 17 × $\frac{7}{8}$

Stiffener Requirements.   End panel spacing:

$$f_v = \frac{155}{26.25} = 5.90 \text{ ksi}$$

for end panels,

$$a = \frac{11,000t}{\sqrt{f_v}} = \frac{11,000(\frac{3}{8})}{\sqrt{5900}}$$

$$a = 53.8 \text{ in.}$$

Space stiffeners 53 in. from each end of girder.
Provide bearing stiffeners at concentrated loads:
Clear distance between end stiffeners and concentrated loads =

$$(18 \times 12) - 53 = 163 \text{ in.}$$

Check intermediate stiffener requirements:

$$\frac{h}{t} = 187 < 260$$

Check allowable F_v in 163 in. panel:

$$\frac{a}{h} = \frac{163}{70} = 2.33 > 1.0$$

$$k = 5.34 + \frac{4.00}{(2.33)^2} = 5.34 + 0.74 = 6.08$$

$$C_v = \frac{45,000,000 \times 6.08}{36,000(187)^2} = 0.218 < 0.80$$

Allowable shear stress,

$$F_v = \frac{36}{2.89}(0.218) = 2.71 \text{ ksi}$$

Vertical shear 53 in. from end of girder:

$$V = 155 - 3(\tfrac{53}{12}) = 141.8 \text{ kip}$$

then

$$f_v = \frac{141.8}{26.25} = 5.4 \text{ ksi} > 2.71 \quad \text{(does not satisfy)}$$

Space intermediate stiffeners at $(\tfrac{163}{2}) = 81.5$ in. on centers;
Maximum spacing between intermediate stiffeners:

$$\frac{a}{h} = \left(\frac{260}{187}\right)^2 = 1.93$$

$$a = 1.93(70) = 135 \text{ in.} > 81.5 \text{ in.} \quad \text{satisfies}$$

for $\quad \dfrac{a}{h} = \dfrac{81.5}{70} = 1.16 \quad$ and $\quad \dfrac{h}{t} = 187: \quad F_v = 8.7 \text{ ksi} > 5.4 \quad$ ksi

Check interaction at concentrated loads:

$$f_v = \frac{101}{26.25} = 3.84 \text{ psi}$$

Maximum allowable bending tensile stress $= \left[0.825 - \left(0.375 \times \dfrac{3.84}{8.7}\right)\right] \times 36$

$$= 0.66(36) = 24.0 \text{ ksi} > 22.0 \text{ ksi}$$

Use: End spacing = 53 in.
Intermediate spacing = $81\tfrac{1}{2}$ in.

Check 14-ft panel:

$$\frac{h}{t} = 187$$

$$V = 21 \text{ kip}$$

$$f_v = \frac{21}{26.25} = 0.8 \text{ ksi}$$

$$a = 14 \times 12 = 168 \text{ in.}$$

$$\frac{a}{h} = \frac{168}{70} = 2.40 > 1.0$$

$$k = 5.34 + \frac{4.00}{(2.40)^2} = 6.03$$

$$C_v = \frac{45,000,000(6.03)}{36,000(187)^2} = 0.215 < 0.80$$

Allowable shear stress

$$F_v = \frac{36}{2.89}\left[0.215 + \frac{1 - 0.215}{1.15\sqrt{1 + (2.40)^2}}\right] = \frac{36}{2.89}(0.477) = 5.95 \text{ ksi} > 0.8 \text{ ksi}$$

Check web crippling assuming flange rotation prevented:

$$\text{Compressive stress} = \frac{3.00}{12 \times 0.375} = 0.67 \text{ ksi}$$

$$\text{Allowable compressive stress} = \left[5.5 + \frac{4}{(2.40)^2}\right] \times \frac{10,000}{(187)^2} = 1.77 \text{ ksi} > 0.67$$

$\therefore$ No stiffener required at midspan.
Space stiffeners as shown in Fig 2.

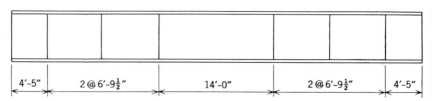

4'-5" | 2 @ 6'-9½" | 14'-0" | 2 @ 6'-9½" | 4'-5"

Ex. 11-8   Fig. 2.

Stiffener Size

Intermediate stiffeners:

$$\frac{a}{h} = 1.16, \qquad \frac{h}{t} = 187$$

$A_{st} = 0.098(26.25) = 2.57$ in.^{2} required

from page 529: $F_v = 8.7$ ksi, $f_v = 5.4$ ksi

Actual area required $= \dfrac{5.4}{8.7} \times 2.57 = 1.6$ in.^{2}

Try 2 bars, $3\frac{1}{2} \times \frac{1}{4} = 1.75$ in.$^2 > 1.6$ in.^{2}

Check width/thickness ratio:

$$\frac{b}{t} = \frac{3.5}{0.25} = 14 < 16$$

Check moment of inertia:

$$I_{req} = (\tfrac{70}{50})^4 = 3.84 \text{ in.}^4$$

$$I_{furn} = \tfrac{1}{12}(0.25)(7.38)^3 = 8.38 \text{ in.}^4 > 3.84 \text{ in.}^2$$

minimum length required $= 70 - (4 \times \frac{3}{8}) = 68\frac{1}{2}$ in.

For intermediate stiffeners: Use 2 bars $3\frac{1}{2} \times \frac{1}{4}$

Bearing stiffeners: at end reaction
Try 2 plates $7\frac{1}{2} \times \frac{1}{2}$ (Fig. 3)
Check width/thickness ratio:

$$\frac{b}{t} = \frac{7.5}{0.5} = 15 < 16$$

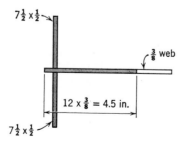

Ex. 11-8   Fig. 3.

Check compressive stress (end bearing):

$$I = \tfrac{1}{12}(\tfrac{1}{2})(15.38)^3 = 151 \text{ in.}^4$$

$$A_{\text{eff}} = (2 \times 7\tfrac{1}{2} \times \tfrac{1}{2}) + (4.5 \times \tfrac{3}{8}) = 9.19 \text{ in.}^2$$

$$r = \sqrt{\frac{151}{9.19}} = 4.05 \text{ in.}$$

$$l = \tfrac{3}{4}(70) = 52.5 \text{ in.;} \qquad \frac{l}{r} = \frac{52.5}{4.05} = 12.95$$

allowable compressive stress:

$$F_a = 21.05 \text{ in.}^2$$

actual compressive stress

$$f_a = \frac{155}{9.19} = 16.9 \text{ ksi} < 21.05 \text{ ksi}$$

Use 2-PL$7\tfrac{1}{2} \times \tfrac{1}{2}$ for bearing stiffeners at ends and at concentrated loads.

Welds for Bearing Stiffeners. Assume $\tfrac{3}{16}$-in. weld: allowable load = 1.8 kip/in. × 2 sides = 3.6 kip/in.

at reaction: $\dfrac{155}{2} = 77.5$ kip; length of weld = 77.5/3.6 = 21.5 in.

maximum spacing of intermittent welds = $16(\tfrac{1}{2}) = 8$ in.

Use 11 welds: $(\tfrac{3}{16}) \times 2$ in. at $6\tfrac{1}{2}$ in. spacing, each side, (Fig. 4)

at concentrated loads: $\dfrac{80}{2} = 40$ kip; length of weld = 40/3.6 = 11.1 in.

Use 11 welds: $(\tfrac{3}{16}) \times 1\tfrac{1}{2}$ in. at $6\tfrac{1}{2}$ in. spacing, each side, (Fig. 4)

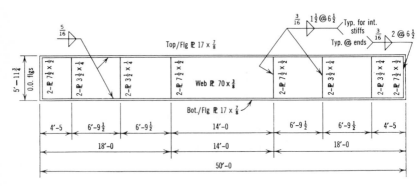

Ex. 11-8   Fig. 4.

17 x $\frac{7}{8}$

70 in.

17 x $\frac{7}{8}$

Ex. 11-8 Fig. 5.

Welds for Intermediate Stiffeners:

minimum weld size $= \frac{3}{16}$ in.

length $= 1\frac{1}{2}$ in. (intermittent)

Use $(\frac{3}{16}) \times 1\frac{1}{2}$ in. at $6\frac{1}{2}$ in. spacing, (Fig. 4).

Weld for connecting flanges to web: (Fig. 5).

$$\text{Shear force } q = \frac{VQ}{I}$$

$$V = 155 \text{ kip}$$

$$Q = 17(\tfrac{7}{8})(35.44) = 525 \text{ in.}^3$$

$$I = 48,220 \text{ in.}^4$$

$$q = \frac{155(525)}{48,220} = 1.69 \text{ kip/in. for 2 welds}$$

$$= 0.85 \text{ kip/in per weld}$$

Use $\frac{5}{16}$-in. weld, allowable is 3.00 kip/in. per weld, (Fig. 4).

Details of locations of bearing stiffeners under the concentrated loads prohibit maximum intermediate stiffener spacing. A $\frac{5}{16}$-in. web with stiffeners should be investigated as the next trial section.

11-19 AASHO DESIGN SPECIFICATION FOR PLATE GIRDERS

The AASHO method is based on the condition that the web is sufficiently thick or adequately stiffened so that it will not buckle at loads below the service load. The critical stress for a thin plate is increased by stiffeners which divide the plate into panels and thus increase the buckling stress.

The ratio of the theoretical buckling load to the service load, referred to as safety factor, may vary from 1.2 to 1.5.

The AISC method is based on the postbuckling strength of a thin web girder. In this case the inelastic postbuckling strength of the stiffened web is taken as the basis for design, on the condition that the ratio of the ultimate capacity to the service load is taken to be not less than 1.67. It is possible that in some cases the girders designed by this second method will not buckle at service loads, but in other cases buckling can develop at loads equal to or smaller than service loads. In the latter instance, however, such buckling is not related to any actual failure.

A summary of AASHO Specifications for design of plate girders is presented in the following without derivation.

Bending stresses in girders shall be determined using the formula $f_b = Mc/I$. The tensile stresses shall be computed using I_{net}, and the compressive stresses using I_{gross}. Both moments of inertia shall be calculated with respect to the gravity axis of the gross section.

For welded girders each flange shall consist of a single plate, and the flange area may be varied as necessary by varying width and thickness of plates joined by full penetration butt welds. The full width/thickness ratio of compression flange plate shall not exceed $3250/\sqrt{f_b}$, and in no case be greater than 24.

For riveted girders flange angles shall form as large a part of the area of the flange as practicable. For flange angles in compression the leg width/thickness ratio shall not exceed $1625/\sqrt{f_b}$ and in no case be greater than 12.

The gross area of the compression flange, except for composite design, shall be not less than the gross area of the tension flange. Cover plates shall be of equal thickness or shall decrease in thickness from the flange angles outward. Cover plate thickness shall not exceed that of the flange angles.

At least one cover plate of the top flange shall extend the full length of the girder except when the flange is covered with concrete.

Legs of angles 6 in. or greater in width connected to web plates shall have two lines of rivets. Cover plates 14 in. or wider shall have four lines of rivets.

Web plate thickness of girders without longitudinal stiffeners shall not be less than $h\sqrt{f_b}/23{,}000$, and in no case less than $h/170$. Web plate thickness of girders with a longitudinal stiffener shall not be less than $h\sqrt{f_b}/46{,}000$, and in no case less than $h/340$.

The longitudinal stiffener, if used, shall be located at $h/5$ from the compression flange. The thickness of this stiffener shall not be less than $b'\sqrt{f_b}/2250$, where b' is width of the stiffener. Also, the stiffener will be so proportioned that its moment of inertia I_s taken about the web shall not be less than

$ht^3[2.4(a^2/h^2) - 0.13]$, where a is the spacing of transverse stiffeners and t is web thickness. Longitudinal stiffeners are usually placed on one side of the web plate. If transverse stiffeners are placed on the same side of the web plate as the longitudinal stiffener, the latter need not be continuous and may be cut at the intersection with transverse stiffener. However, it may be advantageous to place transverse stiffeners on one side of the web plate, and the longitudinal stiffener on the other side.

Web plate thickness of girders without transverse stiffeners shall not be less than $h\sqrt{f_v}/7500$, and in no case less than $h/150$. Thus when shearing stress f_v exceeds $[7500(t/h)]^2$ transverse stiffeners must be provided.

The spacing a of transverse stiffeners shall be not greater than $11,000t/\sqrt{f_v}$, and in no case greater than clear unsupported depth of girder web between the flanges. Thus the shearing stress f_v shall not be greater than $[11,000(t/a)]^2$. The spacing of the first two stiffeners at the ends of simply supported girders shall be one-half of the foregoing limit.

Intermediate transverse stiffeners shall be made of plates for welded girders and shall be made of angles for riveted plate girders. When stiffeners are used in pairs, one fastened on each side of the web plate, they should have a tight fit at the compression flange. A single stiffener fastened on one side of the web plate shall be fastened to the compression flange. Some clearance between stiffeners and the tension flange is usually provided.

Moment of intertia of transverse stiffener shall not be less than λat^3, where t is web thickness, and $\lambda = [2.4(h/a)^2 - 1.83]$ but not less than 0.46. When transverse stiffeners are used on both sides of the web, the moment of inertia shall be taken about centerline of the web. When single stiffener is used the moment of inertia is taken about the edge in contact with the web plate.

The width of a plate stiffener or of the outstanding leg of an angle stiffener shall not be less than 2 in. plus $\frac{1}{30}$ of the depth of the girder, and preferably it shall be not less than $\frac{1}{4}$ of the flange width. The thickness of the stiffener shall be not less than $\frac{1}{16}$ of the width, and preferably not less than web thickness.

Bearing stiffeners must be provided at end and intermediate bearing supports and at points of concentrated loads producing compression in the web. These stiffeners shall be placed on both sides of the web and extend to the outer edges of the flange plates. Bearing stiffeners shall be designed as columns and their connection to the web shall be designed to transmit the entire end reaction from the bearings to the web. The stiffeners shall be ground to fit against the flange through which they receive the load or welded* to this flange. The bearing stress shall not exceed the allowable limit

---

* Effect of this welding on fatigue performance of the girders should be considered, see 11-21.

of $0.55F_y$, and the compression stress on the effective column section shall not exceed the allowable value corresponding to the slenderness ratio of the bearing stiffener including effective web area.

11-20 GIRDER FLANGES AND FLANGE CONNECTIONS

For a riveted plate girder, the conventional flange consists of two angles and one or more cover plates (Fig. 11-32a and b). The angles preferably should have unequal legs, with the shorter legs placed along the web. This tends to increase the effective depth and also the total flange width. For girders with heavy shears, a large number of rivets may be required to connect the angles to the web and it may be necessary to use angles of equal legs for connection to the web or even to place the longer legs along the web.

Occasionally, small girders are designed with flanges consisting of angles only. However, most specifications call for at least one cover plate for each flange, serving to tie together the outstanding legs of the angles; some other specifications only require a top plate to extend the whole length. A top flange without cover plate may allow water to collect between the flange angles and the web, resulting in corrosion even if paint is liberally used.

When heavy flanges are required, it is advantageous to put as much area in the plates as possible, both for increasing the effective depth and for cutting the plates to conform to the moment diagram. But the area of cover plates is sometimes limited by code to 70% of the total flange area.

It is considered good practice that all cover plates in a riveted girder should have the same thickness and should not be thicker than the angles. If unequal thickness is used, the outer ones should not be thicker than the inner ones. The width of plates should also be the same. For outdoor girders exposed to weathering, plate width should not decrease toward the top in order to avoid stagnant water and resultant corrosion.

A welded girder flange usually consists of one plate, however, the area of the flange may be varied by varying the width and the thickness of the flange plates (Fig. 11-32c and d). Plates up to 2.5 in. thick are used in welded girders, although for riveted girders the usual limit is about 1 in. as determined by the available power punch. The use of several plates for flanges in welded girders is to be avoided because longitudinal welds required for connecting these plates mean a considerable increase in labor.

When girder flanges become extremely heavy, special designs may be necessary. The compression flange of girders should be designed with lateral supports at intervals in order to avoid buckling. Sometimes it is not possible to furnish such supports—as for crane girders. Then it may be desirable to use channels or built-up angles in order to increase the lateral rigidity, as shown in Fig. 11-18.

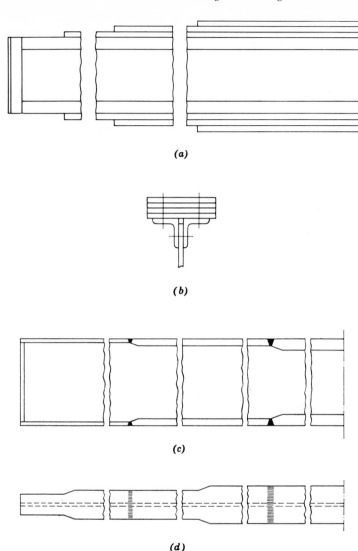

Fig. 11-32 Variable flange plates in girders. (*a*) elevation-riveted or bolted girder, (*b*) flange with cover plates, (*c*) elevation-welded girder, (*d*) variable width flange plate.

Once the depth of the girder has been chosen and a preliminary design made, the extreme fiber stresses must be verified in accordance with applicable specifications.

If the compressive flange is not laterally supported, the allowable stress based on the gross section should be checked against the allowable value, as discussed in Sec. 11-3.

If bending moments vary considerably along the length of the girder, it is desirable either to vary the depth of the girder providing relatively constant flange areas, or to vary the flange area using constant girder depth.

A change in flange area can be provided by varying flange plate proportions or by use of cover plates (Fig. 11-32). Determination of locations where flange proportions could be safely changed must be based on the variation in the bending moment. An algebraic solution can be obtained if the design moment diagram can be expressed by a single algebraic expression. This can be done for simple loading conditions; when variable loading must be considered, an envelope of critical moments can be constructed, but it may not be defined by a continuous function $M(x)$. In such cases a graphical solution is most advantageous.

In any event, for selected flange A_f and web proportions, the resisting moment M_r can be calculated, and the point where this flange becomes adequate can be found from the moment diagram either algebraically or graphically. For algebraic solution it is necessary to solve the equation $M(x) = M_r$ for x. For a graphical solution it is necessary to find intersection of $M(x)$ and M_r (Fig. 11-33). The intersection of the M_r lines with moment diagram defines the points of theoretical change in A_f or "cutoff" points.

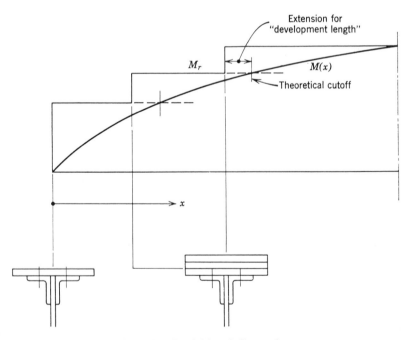

Fig. 11-33 Partial length flange plates.

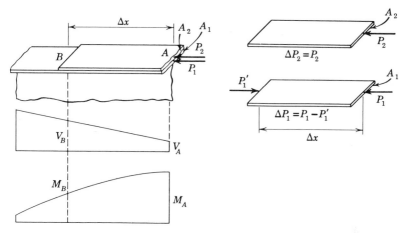

Fig. 11-34 Flange-web connection adjacent to a cover plate cutoff.

Most specifications call for some minimum extension of the larger flange area beyond these "theoretical cutoff" points. This is recommended to provide so-called "development length" for gradual redistribution of stress which would be caused by a discontinuity in flange area. It is particularly important to provide sufficient fasteners or welds to transmit the stresses carried by the "cutoff" area.

Let point A be the point of theoretical cutoff for a cover plate A_2 with a girder having a continuous flange A_1, where A_1 and A_2 represent the cross-sectional areas (Fig. 11-34). Let Δx be the transition length of the cover plate A_2 extending from point A to point B.

Assuming that the stresses at sections A and B can be defined by $f_A = (M_A/I_A)y_A$ and $f_B = (M_B/I_B)y_B$, the connection between the terminal cover plate and the flange must be designed for the load ΔP_2:

$$\Delta P_2 = (f_A - f_B)A_2 = \left(\frac{M_A}{I_A}\,y_A - \frac{M_B}{I_B}\,y_B\right)A_2 \qquad (11\text{-}94a)$$

This can be approximated conservatively by

$$\Delta P_2 = \frac{M_A - M_B}{I_B}\,y_B A_2 = \frac{V\,\Delta x}{I_B}\,y_B A_2 = \frac{VQ_2}{I_B}\,\Delta x = \frac{V}{h_e}\left(\frac{A_2}{A_1 + A_2}\right)\Delta x$$
$$(11\text{-}94b)$$

The connection between the web and the flange must transmit the shear force due to the variation of the flange load along the span, as shown in

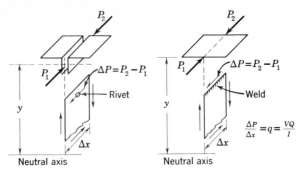

Fig. 11-35 Load transmission between flange and web.

Fig. 11-35, in which

$$\Delta P = \frac{\Delta M}{I}\,\bar{y}A_f = \frac{(V\,\Delta x)\bar{y}A_f}{I} = \frac{VQ_f}{I}\Delta x \simeq \frac{V}{h_e}\Delta x \qquad (11\text{-}95)$$

where h_e is effective depth approximately equal to I/Q_f. If $\Delta x = p$, where p is the rivet pitch in a single row of rivets, then the load per rivet is $\Delta P = R$. Conversely, for a given size of rivet with rivet value R, the required spacing p in a single row of rivets is $p = R/(V/h_e)$. For welded connection, the required strength of a continuous weld per inch is $q = \Delta P/\Delta x = V/h_e$,

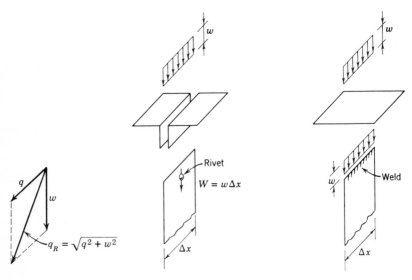

Fig. 11-36 Local transverse load on flange of girder.

which determines the size of the weld. Often the weld between the flange and the web is made by an automatic process and is a continuous fillet weld.

It has been assumed that the transverse loads bearing directly on the flanges are negligible. When concentrated loads exist, they are transmitted to the web by bearing stiffeners so that no special provision for strengthening web-to-flange connection is necessary. Distributed or minor concentrated loads may be transmitted by the web-to-flange connection as shown in Fig. 11-48. These loads W are superimposed on the normal shear which must be transmitted by the connections, so that, if the transverse load per inch of flange is w, the resultant load per inch on the rivet or weld (Fig. 11-36) is q_R:

$$q_R = (q^2 + w^2)^{1/2} \qquad (11\text{-}96)$$

When the transverse load w is due to moving wheel loads on rails attached to the flange, arbitrary lengths effective in supporting the wheel load are assumed. For cranes, effective lengths of 18 to 30 in. per wheel are often adopted.

Girder flanges carry normal stresses due to bending moment and therefore whenever possible, for economy of material, flange splices should be located at sections other than those of maximum moment. Various specifications differ as to the minimum area of flanges to be spliced. AISC permits the splice to be made for the actual stress in the flange but not less than 50% of its allowable capacity. AASHO provides for splice strength to be not less than 75%, and AREA to be not less than 100% of the spliced part. Thus the area of the flange splice must be adequate for transmitting the actual load in the cut part, but its minimum area varies between 50 and 100% of the splice part, depending on the specification. Moreover, AASHO and AREA require that no more than one part shall be spliced at the same cross section and that flange angle splices shall consist of two angles, one on each side of the girder.

In riveted or bolted girders two types of flange splice are commonly used (Fig. 11-37). One is a direct splice, in which one or both flange angles are spliced by inserting a so-called "bosom angle" (Fig. 11-37*b* and *c*). The other is an indirect splice when only one flange angle is spliced and the load in each of its legs is shared by splice material on both sides (Fig. 11-37*d*).

The loads transmitted by fasteners in flange splices are somewhat indeterminate (Fig. 11-37*a*). Consider a moment diagram in the region of a flange splice. In region I, the splice material must pick up the stresses from the main flange due to moment M_m and also must pick up the moment-increment stresses due to shear V. In region II, the main flange must pick up the stresses from the splice material due to moment M_0 and also must pick up the moment-increment stresses. These moment-increment stresses due to shear give rise to the indeterminacy of the fastener loads.

The following method may be used for determining the fastener loads.

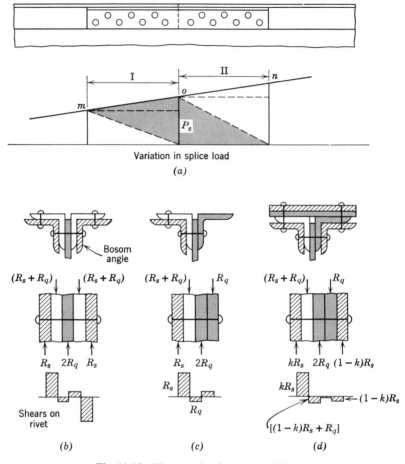

Fig. 11-37 Flange splice for riveted girders.

The total load P_s to be transmitted by the splice is fA_1, where $f = M_0 y/I$ and A_1 is the portion of the flange area being spliced. Assuming n fasteners in the splice, the load per fastener is $R_s = P_s/n = fA_1/n$. In a direct splice the load is transmitted to the splicing element by single shear in the fastener. In an indirect splice the portion of the load transmitted to each of the splicing elements may be obtained by considering the fastener as a simple beam, so that the shears are kR_s and $(1 - k)R_s$, Fig. 11-37.

These shears must be combined with moment-increment loads on fasteners giving due regard to sign and shear planes on which they act. The flange load increment in distance p is $(V/h)p$ and the shear in the flange to web fastener is $R_q = \frac{1}{2}(V/h)p$.

The maximum shear R on the fastener can be obtained by combining values of splice and moment-increment shears on the corresponding planes in the fastener, and it is

$$R = kR_s = \frac{kfA_1}{n} \tag{11-97a}$$

or

$$R = (1 - k)R_s + R_q = (1 - k)\frac{fA_1}{n} + \frac{1}{2}\frac{V}{h}p \tag{11-97b}$$

whichever is greater. If the allowable load on the fastener is R_v, then equating $R = R_v$, the required number of fasteners n is

$$n = \frac{kfA_1}{R_v} \quad \text{or} \quad n = \frac{(1 - k)fA_1}{R_v - \frac{1}{2}(V/h)p} \tag{11-98}$$

whichever is greater. Equation 11-98 defines the number of fasteners required in the splice without an allowance for an increase required by most specifications for indirect splices. The provision of two extra fasteners for each intervening plate required by AASHO and AREA would usually result in a more conservative design than that indicated by Eq. 11-98. Flange splices are not usually located at sections of maximum moment, although they may be located at sections near cover-plate cutoffs where the unit stress in flange angles may be near the maximum allowable. For this reason it is important to check the net section of the flange at the splice.

Various cover plate splices are shown in Fig. 11-38(a–f). Dotted lines show the path of load transfer. When one cover plate only is used and it has to be spliced, a simple splice—or so-called "patch plate"—is used as in a. When two cover plates are used and one must be spliced, it can be done conveniently near the cutoff point of the second cover plate, simply by extending the second cover a short distance as in b. When two cover plates are used and must be spliced, it is preferable not to cut them at the same point as in c but to stagger the splices as in d. This requires a longer splice plate but eliminates the second splice plate and results in an over-all economy of material and fabrication cost. When three cover plates are used, the splices may be staggered either as shown in e or as in f. The first is commonly used for shop splices and the second for field splices.

In welded girders, flange splices are generally made by butt welding (Fig. 11-39). For field splices the web-to-flange connection adjacent to the splice is left out until after the butt welding is completed.

When welded flange splices and web splices are located close together, the sequence of welding is of particular importance. Consider the splice shown in Fig. 11-40. In both the web and the flanges, welding should proceed symmetrically in order to minimize distortions. This can best be accomplished

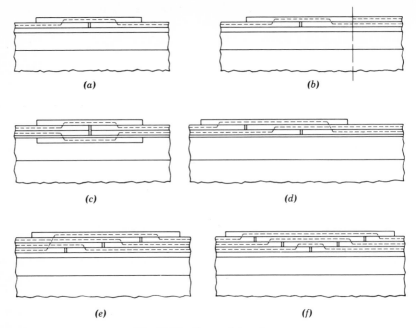

Fig. 11-38   Cover plate splices.

by two welders working on the opposite elements of the girder. Furthermore it is preferable to make the larger welds first, so that the welds can shrink upon cooling without great restraint. Usually the flanges have the heavy welds; therefore they are welded first, at least partially. Then webs are welded together and if necessary the flange welds are completed. All details of welding procedure should be arranged to minimize distortion and residual stresses.

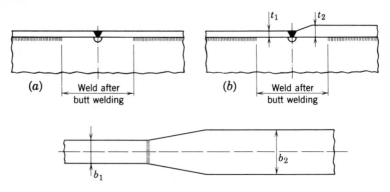

Fig. 11-39   Welded flange splices.

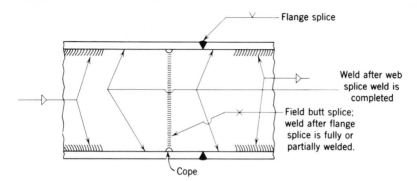

Fig. 11-40 A welded splice.

11-21 WEB STIFFENERS

Stiffeners may be of two types: (*a*) bearing stiffeners at the supports or other points of load concentration, serving to distribute the reaction or load to the whole depth of the web, and (*b*) transverse or longitudinal stability stiffeners to prevent web buckling or to increase postbuckling strength.

Bearing stiffeners should have a close fit against the loaded flanges and should extend as closely as possible to the edge of the flange plates or angles. Only the portion of the stiffener outside the flange angle fillet or the flange-to-web welds (Fig. 11-19) will be considered effective in bearing. Bearing stiffeners are somewhat similar to columns. But because of the variation of load along their height, and because of their interaction with the web, exact analysis is difficult. The AISC Specification defines the effective area of the column as consisting of the stiffeners plus a strip of the web having a width equal to 25 times its thickness (or 12 times its thickness for end stiffeners). The effective length of the column shall not exceed three-fourths of the length of the stiffeners. The AASHO Specification defines the effective column section as the two stiffeners with a strip of the web having a width equal to 18 times its thickness. For stiffeners where two or more pairs of plates or angles are used, the effective column area shall include the strip enclosed by the stiffeners on each side plus a width of not more than 18 times the web plate thickness. The effective length of the column is not specified and use of full stiffener height is recommended. In addition it is required that the bearing stiffener plates shall not be thinner than $b'/\sqrt{F_y}/21{,}800$.

Bearing stiffeners at points of reaction or concentrated loads must usually be placed symmetrically about the web. Load-bearing stiffeners should be designed to transmit the total load to the web. Uniform distribution of the shear in the web may be assumed for design of the connection.

Because of the presence of flange angles in riveted girders, the angle stiffeners must be crimped, or else fillers will be required (Fig. 11-19). Crimping the angle stiffeners is usually permitted for stability stiffeners but not for bearing stiffeners.

In riveted girders, the presence of fillers, if any, must be considered. If loose fillers are used, that is, fillers having no additional rivets attaching them to the web, the length of rivets may be excessive, resulting in bending of the rivets. In such cases the number of rivets may have to be increased to reduce the load per rivet. With tight fillers, that is, fillers having additional rivets attaching them to the web, the rivets attaching the filler only must be checked. The load to be carried by the filler rivets may be assumed equal to $Pt_f/(t_f + t_w)$, where P is the total load on the stiffener, and t_f and t_w are thickness of fillers and web, respectively.

Stability stiffeners may be used in pairs or as single stiffeners employed on one side of the web. Usually the function of stability stiffeners is to provide lateral stiffness. Increased effectiveness of stiffeners may be obtained by using torsionally rigid shapes which assure greater restraint against rotation of the panel edges.

To assure that the stiffeners provide adequate rigidity so that each panel may be considered acting independently from another empirical rules have been proposed for the magnitude of their moment of inertia I_s. The stiffener requirements based on AISC and AASHO Specifications are given in Secs. 11-18 and 11-19 of this chapter.

As noted in Sec. 11-16 the transverse stiffeners effectively increase the buckling resistance of the web to shear, and longitudinal stiffeners effectively increase buckling resistance of the web to bending. This has been demonstrated in several analytical and experimental studies[14,15,16,17] and consequently AASHO Specification provides for the use of longitudinal stiffeners. The contribution of longitudinal stiffeners to postbuckling strength and to the ultimate capacity of plate girders subjected to combined shear and bending has not been clearly established, and therefore use of longitudinal stiffeners is

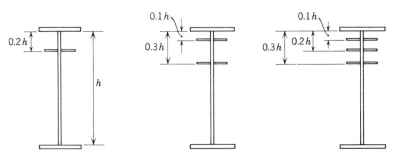

Fig. 11-41   Longitudinal web stiffeners in the compression zone.

not contemplated by the AISC Specification. Longitudinal stiffeners to prevent web buckling are most effective in deep girders with thin webs. Although usually only one stiffener placed at 0.2h is used, other arrangements (Fig. 11-41) may be effective in girders with depths greater than 10 ft.

Welded stiffeners are usually made of plates, whereas riveted stiffeners are of angles to facilitate connection. There is no standard practice for the amount of rivets or welds connecting stability stiffeners to the web. Usually the smallest rivets suitable for the stiffener angle can be used with a spacing up to 8 diameters. Welds can be discontinuous minimum welds, with weld length about one-third of the total. In girders subjected to repeated loads, to minimize fatigue effects, stiffeners should not be welded to the tension flange, and all load bearing welds must be continuous.

11-22 WEB SPLICES

Splices for girder webs should be avoided whenever possible. But there are conditions when splicing of girders is unavoidable. One is the available length of plates and shapes; another is the length limit imposed by the transportation facilities from the fabricating shop to the site of the structure. Occasionally, the capacity of the erecting crane may set the maximum weight of one piece to be handled.

Transportation facilities vary greatly with local conditions. Where good highways lead from the fabricating shop to the site, special arrangement can be made to transport long and heavy pieces. Where direct railroad transportation is used, the length of the pieces is governed by tunnel and bridge clearances, especially on curves. Sometimes it is a matter of balancing the extra cost of the splice against the additional cost of transporting heavier and longer pieces.

When flange-to-web connections are made in the field, the location of web splices should be staggered with that for flange splices. However, length limitations may require the location of both splices rather close together; for example, if the maximum length is limited to 40 ft, an 80-ft girder might have to be spliced at the center for both the web and the flanges. Alternatively, one 40-ft and two 20-ft sections may be used, requiring two splices instead of one. Where heavier transportation and erection facilities are available, flange-to-web connections are usually made in the shop. Then web-and-flange splices may have to be located close together in order to simplify fitting in the field.

Girder web transmits primarily shearing stresses, and web splices are most efficiently located at points of small shear, although practical requirements often dictate otherwise. In general, the shear force to be spliced in the web is much smaller than the shear capacity of the web. Then the question often arises as to whether the web should be spliced for the design shear at the section or for its shear-carrying capacity. Some engineers believe that it is desirable to make the splice at least as strong as the basic components; others argue that such additional strength would only be wasted since failure of the structure will start at some other weaker link rather than at the spliced section. In a statically determinate structure, where the maximum shear at the spliced section can be accurately predicted, designing for that shear may be sufficient, since such a design will yield a splice with the usual factor of safety, whereas any overdesign will not necessarily increase the over-all carrying capacity of the structure. On the other hand, in a statically indeterminate structure where forces may be redistributed because of plastic action,

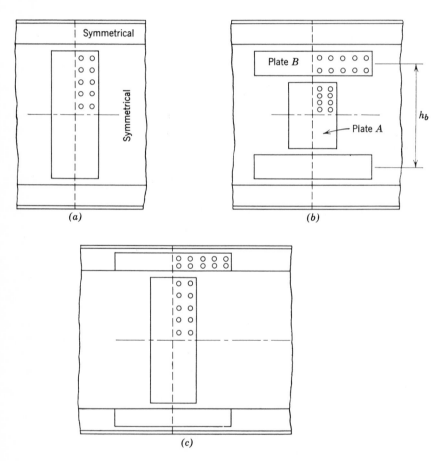

Fig. 11-42   Types of riveted web splices.

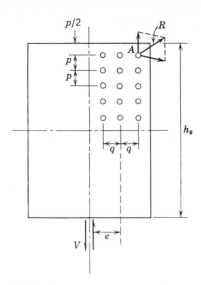

Fig. 11-43 Moment on rivet group at web splice.

increasing the strength of the splice may increase the ultimate strength of the entire structure. In such an instance, some additional safety might be justified if the additional expense involved is relatively small.

Most riveted or bolted web splices, except those for very heavy girders, are controlled by minimum dimension requirements rather than stress computations. For example, two splice plates are usually employed one on each side of the web; the splice plates must have not less than the minimum thickness and must extend the entire depth of the girder from flange to flange. In all instances the net section through the splice plates must provide the required area to resist the shear and the required section modulus to resist the bending moment safely.

In order to prevent damage to the web due to driving the splice rivets, a pair of stiffener angles is normally provided at the center line of the splice. Sometimes only one angle may be used.

Three types of riveted web splices may be employed (Fig. 11-42). Sometimes a web splice may be so located as to transmit shear only, although usually it must transmit a combination of shear and moment.

When a web splice is to transmit a pure shear V (with moment equal to zero at the section), the rivets should be designed to resist a force V applied at the centroid of the rivet group (Fig. 11-43). This means that the rivets should be designed to transmit load V, with an eccentricity e. When the depth h of the web is much greater than the eccentricity e, the design is often made for a direct shear V, neglecting the eccentricity. In this instance, for a given rivet diameter the rivet value R is known and the required number of rivets is simply V/R. Bolts may be substituted for the rivets.

If in addition to shear V there is a moment M at the splice section, then the portion of the total moment carried by the web must be transmitted by the web splice. This moment M_w (Fig. 11-43) may ·be approximated as follows:

$$M_w = M \frac{A_w/6}{A_f + A_w/6} \qquad (11\text{-}99)$$

where A_f and A_w are areas of each flange and the web, respectively. Consider a group of rivets in a splice plate resisting shear V and moment $M_s = Ve + M_w$, which consist of m rows spaced at q in. with n rivets in each row spaced at p in. The most highly loaded rivet A (Fig. 11-43) carries a load R which may be approximated as follows (see Chapter 5, Eqs. 5-25 and 5-36):

$$R = \left\{ \left[\frac{V}{mn} + \frac{6M_s q}{n(n^2 p^2 + m^2 q^2)} \right]^2 + \left[\frac{6M_s p}{m(n^2 p^2 + m^2 q^2)} \right]^2 \right\}^{1/2} \qquad (11\text{-}100)$$

For a given rivet diameter, the rivet value R and the minimum spacing p and q are known and, if we note that np is approximately equal to the depth of splice plate h_s, the required value of m can be determined from Eq. 11-100. In many instances, the vertical component of rivet load due to bending is small and may be neglected, so that Eq. 11-100 may be rewritten as

$$R = \left\{ \left(\frac{V}{mn} \right)^2 + \left[\frac{6M_s p}{m(n^2 p^2 + m^2 q^2)} \right]^2 \right\}^{1/2} \qquad (11\text{-}101)$$

When $mq \ll np$, it may be neglected in the denominator of the second term, so that

$$m = \frac{1}{R} \left[\left(\frac{V}{n} \right)^2 + \left(\frac{6M_s}{nh_s} \right)^2 \right]^{1/2} \qquad (11\text{-}102)$$

Since the depth h_s of the splice plate may be substantially less than the depth h of the web, the load on the extreme rivet A is probably greater than indicated by Eq. 11-99, in which it is assumed that the distance from the extreme fiber to rivet A is $\frac{1}{2}p$. This overloading can be easily seen by examining the load transfer from the web plate to the splice rivets shown in Fig. 11-44. A better design of a web splice for moment and shear is obtained by increasing the number of rivets near the edges of the splice plates, as shown in Fig. 11-42b. This design is used in deep heavy girders. In this case it is assumed that splice plate A transmits all the shear V and no moment, and that splice plates B transmit all the web moment M_w and no shear. The load P_b on each of the splice plates B can be determined as force in a couple M_w

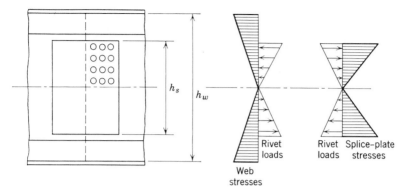

Fig. 11-44 Load transfer from web plate to splice rivets.

having a lever arm h_b, so that $P_b = M_w/h_b$. If the depth h_b is known, the loads P_b are determined and the corresponding number of rivets is easily found, thus defining the length of the plates B. Because the load in the plate A is the total shear, the number of rivets is simply determined and the width of the plate can be chosen. In some instances, it is desirable to splice the outer portions of the web plate directly by placing splice plates over the flange angles, as in Fig. 11-42c. In this case each splice plate is designed to resist the stresses that would exist in the uncut web under the splice plates, and the method of design outlined for splice in Fig. 11-42a would apply.

In welded girders the best web splices are obtained by butt welding in the shop. Particularly in large girders, webs may be formed of plates of various widths or thicknesses which are butt-welded together along both transverse and longitudinal seams. When field splices are required, the best results can be obtained by riveted or high-strength bolted splices designed in accordance with the methods just described. In field-welded splices, the flange-to-web-connections on each side of the splice must be left out and made after the butt welds, to minimize residual stresses, and copes must be provided for each of the connections, as shown in Fig. 11-40.

11-23 PLATE GIRDERS WITH MULTIPLE STEELS

Availability of several strength grades of steel makes it possible to design a plate girder for the most efficient use of all the steels. The highway bridge designers have capitalized on this principle by incorporating a high-strength steel for the web and flanges in a region of maximum moment in order to maintain a constant depth girder. By this method it has been possible to eliminate the haunches at the piers, thus facilitating fabrication, reducing

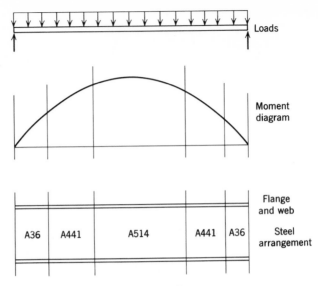

Fig. 11-45

weight, improving appearance, and generally reducing the cost of the total structure. Greater economies are realized in the longer spans.

The selection of the grade of steel is dependent upon the stress requirements and the price of the steel. Generally, when thick plates are required for a low-strength steel, it is advisable to investigate the possible use of a thinner plate of the next-higher-strength steel. The use of a thinner plate will reduce weight and shipping costs, and fabrication of thinner plates is more conveniently accomplished.

In building applications, the large transfer-girders over lobbies may be more efficiently designed using several grades of steel to fit the bending-moment diagram. Large girders supporting boilers may also benefit from a consideration of the use of several grades of steel.

To illustrate the manner of using several grades of steel in a plate girder, a simple beam supporting a uniform load is illustrated in Fig. 11-45. The bending-moment diagram is the usual parabola with the maximum moment at the center. In the region of maximum moments and stresses, the high-strength constructional alloy steel A514 is used for the web and flanges. As the moments and stresses become smaller, a high-strength low-alloy steel (A441) or equivalent is used. Toward the ends of the beam as the stresses become the least, A36 steel is used to advantage.

For continuous beams, the moment diagram would dictate the selection of steels in the same manner.

A "hybrid" beam or girder is one that is fabricated by using a stronger steel in the flanges than in the web. Studies have indicated that hybrid beams may be more economical than beams fabricated with one strength grade of steel throughout.

There are two approaches to the design of a hybrid beam. The first is the application of conventional design methods to proportion the web and flanges within the allowable stresses prescribed by a specification. This is accomplished by limiting the maximum bending allowable stress of the web and then using a higher-strength steel for the flanges. However, this is not the most economical use of the steel in the flanges or web and therefore does not develop the full capabilities of such a beam. This beam is considered to have elastic behavior.

The second approach to the design of a beam for maximum load-carrying ability is to permit the web to yield while the flange continues to remain elastic for a time and finally yields into the plastic bending condition. Investigations[18,19] have been conducted to verify the behavior of these beams and to study their ultimate strength. Static and dynamic tests have been conducted on a limited number of specimens to determine the behavior for

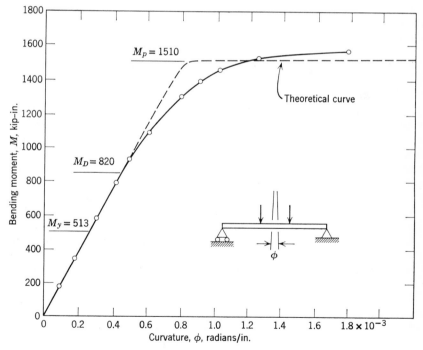

Fig. 11-46 Moment-curvature relationship for a hybrid beam.

all types of load applications. These preliminary tests indicate that hybrid beams develop the moment of the section based on the yield strength of the flange steel and that the web yields without rupture to accommodate this condition. Slight reduction in the moment capacity is observed when a concentrated load on the flange is sufficiently large to cause crippling in the web.

A typical moment-curvature plot for a hybrid beam having A7 steel (33,000 psi) in the web and A514 grade steel (100,000 psi) in the flanges is indicated in Fig. 11-46. The plot indicates that the hybrid beam develops its full plastic moment without premature rupture.

Studies on the characteristics of local buckling, lateral buckling, web crippling, column action, composite action with a concrete slab, flanges of different grades of steel, and combined actions of bending and compression are continuing.

PROBLEMS

Note–Use A36 steel unless otherwise specified.

1. A 10 W^F 25 beam, 30 ft long, simply supported at the ends, is to carry a uniformly distributed load. The compression flange is restrained laterally throughout the full length of the beam. Using AISC Specification, determine the allowable live load.

2. An 18 I 54.7 15-ft cantilever is loaded with a concentrated load P at the end. The beam is prevented from lateral buckling. Using AISC Specification, determine the permissible load P as controlled by flexure and by shear.

3. A W^F steel beam spans 30 ft between columns and overhangs 8 ft at one end. It supports a concentrated moving load of 40 kips, which can be placed anywhere on the beam. Assume that the weight of the beam and the additional dead load will produce a moment equal to 5% of live-load moment at the section. The beam is laterally supported along the entire length. Using AISC Specification, select an economical section; check for flexure and shear, and compute the length of bearing required at each support.

4. For the conditions of Prob. 3 select a section using AASHO Specification. Also check for flexure and shear, and determine the required length of bearing.

5. A W^F beam is rigidly fixed at one support and simply supported at the other, spanning 16 ft. The beam supports a uniformly distributed load of 4 kpf and is prevented from lateral buckling. Using AISC Specification, determine the required size of section.

6. What section is required for the beam of Prob. 5, using AISC plastic design criteria and assuming a load factor of 1.70, consider A36 and A441 steels.

7. The following simply supported floor beams support a 5-in. concrete slab and a live load of W_L psf. Determine (*a*) the allowable live load W_L psf, assuming that a 22-ksi bending stress in the beam is permissible, (*b*) the maximum total deflection Δ_T corresponding to the total load $W = W_D + W_L$, (*c*) the maximum deflection Δ_L due to live load W_L, (*d*) the ratio of live-load deflection Δ_L to span length L, and (*e*) the allowable live load W_L' permissible by AISC Specification governing the depth-span ratio.

Floor-beam span, ft	30	25	20
Floor-beam spacing, ft	12	10	8
Sections	12 W 79	10 W 77	8 W 48
	12 W 65	10 W 60	8 W 31

8. Select an economical W section for each of the beams indicated in Fig. P-8. The load in all cases is 3 kip/ft uniformly distributed. Use AISC Specification. Use A36 and A441 steels.

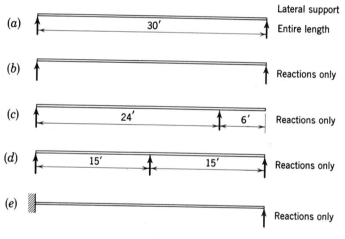

Fig. P-8

9. Select an economical W section for beams *d* and *e* in Fig. P-8, using AISC plastic design criteria, assuming a load factor of 1.70. If any additional lateral bracing is required for plastic design, determine the location of such bracing.

10. A 15 I 42.9 beam is simply supported at the ends and has a span of 25 ft. It is supported laterally at the ends only. Using AISC Specification, determine the allowable bending moment that the beam can resist.

11. Select an economical W section for the beam loaded as shown in Fig. P-11. Lateral restraint is provided at the supports only. Neglect the weight of the beam. Check the beam for shear and length of bearing. (AISC Specification.) Use A441 steel.

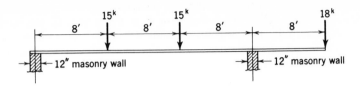

Fig. P-11

12. Compute the allowable load per foot for a 10 I 25.4 beam 23 ft long, fixed at the ends with lateral support at the ends only. (AISC Specification.)

13. A gable bent is used as a frame for an industrial building. The loads and reactions on the frame are shown in Fig. P-13, Draw the bending-moment diagram for the frame and select an economical W^F required for the frame girders. Use AISC Specification. Lateral support is provided at the eaves and at the ridge only.

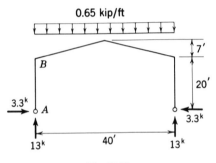

Fig. P-13

14. Three alternate roof framing plans are proposed, as shown in Fig. P-14. Using AISC Specification, select the most economical W^F sections for each system if the beams are fully supported laterally and carry uniformly distributed loads as follows: dead load = 0.3 kip/ft, live load = 0.8 kip/ft.

15. Select the most economical W^F section for the system shown in Fig. P-14*b* using AISC plastic design criteria, assuming load factor of 1.70 for dead and live loads, respectively.

16. A 21 W^F 62 beam reinforced by cover plates and laterally supported throughout on a 20-ft simple span carries an 80-kip concentrated load at a point 5 ft from one of the supports. Neglecting the weight of the beam and using AISC Specification: (*a*) select the size of cover plates, (*b*) specify the cutoff points for the cover plates, (*c*) specify the welded connections between the cover plates and the beam, and (*d*) specify an alternative bolted connection between the plates and the beam, using ⅞-in. bolts.

17. An 18 W^F 70 beam reinforced by cover plates and laterally supported throughout on a 30-ft simple span carries two concentrated loads 40 kips each at the third-points of the beam. Neglecting the weight of the beam and using AISC

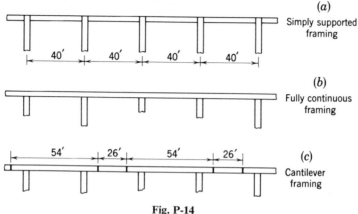

(a)
Simply supported
framing

40′   40′   40′   40′

(b)
Fully continuous
framing

54′   26′   54′   26′

(c)
Cantilever
framing

Fig. P-14

Specification: (*a*) select the size of cover plates, (*b*) specify the cutoff points for the cover plates, (*c*) specify the bolted connections between the plates and the beam, and (*d*) specify an alternative welded connection between the plates and the beam.

18. A 36 W^F 230 beam is reinforced by two cover plates top and bottom, with 18-in × $\frac{3}{4}$-in. inner plates and 20-in. × $\frac{3}{4}$-in. outer plates. The maximum shear on the section is 700 kips. Using AISC Specification, detail (*a*) welded connections between the outer and inner cover plates, and the inner cover plates and the beam, and (*b*) an alternate connection using 1-in. A325 bolts.

19. Repeat Probs. 16, 17, and 18 using AASHO Specification.

20. For the beams loaded as shown in Fig. P-20 and laterally supported throughout, determine the following: (*a*) economical W^F size without cover plates (AISC Specification), (*b*) economical W^F size with cover plates (AISC Specification), showing size of cover plates, their length, and connection to W^F beam, and (*c*) economical size of W^F without cover plates (AISC plastic design criteria), using load factor of 1.70.

21. A traveling crane in an industrial building is shown in Fig. P-21. The lifting capacity of the crane is 5 tons and the weight of the trolley and hoist is 0.5 ton. The trolley rides on the bridge which is formed by two W^F beams braced at the top. Using AISC Specification, select economical W^F shapes for the bridge beams, under the following conditions: (*a*) neglecting lateral and longitudinal forces, (*b*) considering the effects of lateral and longitudinal forces but neglecting the effect of twisting, and (*c*) considering the effects of lateral and longitudinal forces and of twisting.

22. An eave strut consists of 8 [11.5 and a 6-in. × 4-in × $\frac{3}{8}$-in. angle riveted together, as shown in Fig. P-22. The strut spans 18 ft between the columns and carries uniformly distributed loads as follows: $W_x = 40$ lbs/ft and $W_y = 150$ lbs/ft. Determine the maximum stress in the strut and show the position of the neutral axis. Neglect the effect of twisting. Assume that the strut is simply supported at the columns.

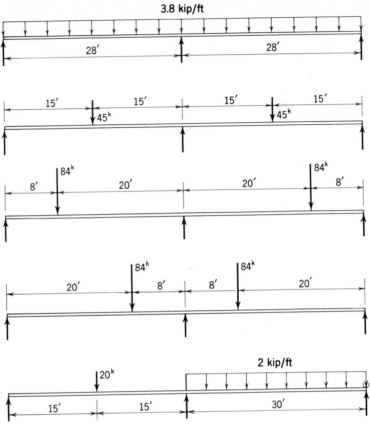

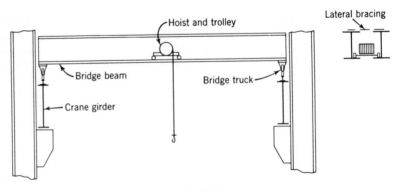

Fig. P-20

Fig. P-21

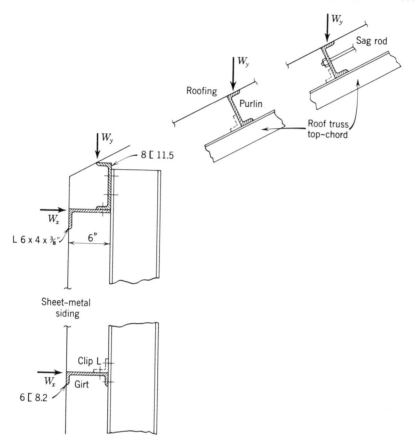

Fig. P-22

23. A 6 [8.2 is used as a girt to support sheet metal siding and spans 18 ft between columns, as shown in Fig. P-22. The girt carries a uniformly distributed horizontal load W_x of 80 lbs/ft. Determine the maximum bending stress in the girt: (*a*) neglecting the weight of the siding, and (*b*) considering the vertical load of 6 lbs/ft due to the weight of the siding. Neglect the effect of twisting and assume that the girt is simply supported at the columns.

24. Assuming that the girt of Prob. 23 is fully fixed against any rotation and twisting at the columns, determine the maximum stresses in the girt: (*a*) neglecting the effect of twisting, and (*b*) considering the effect of twisting.

25. Roof purlins shown in Fig. P-22 are spaced 7.5 ft on centers and span 18 ft between trusses. Assume the purlins to be simply supported. The loading on the purlins is as follows: roofing 5 psf, live load 15 psf. Estimate the purlin weight as 8 lbs/ft and select an economical channel size. Consider the channel to be laterally supported throughout its length and neglect the effect of twisting. AISC Specification.

26. Select an economical channel size for a purlin similar to that of Prob. 25 except that a sag rod midway between trusses prevents lateral displacement of the purlin at that point. Neglect the elongation of the sag rod. AISC Specification.

27. Select economical channel sizes for the conditions of Probs. 25 and 26 if the purlin is continuous over two bays, that is, 36 ft long.

28. For the following girders submit calculations and design sketches showing all dimensions of the girder section, stiffeners, and connections of the various elements: (*a*) welded girder 45-ft span, uniformly distributed live load of 10.5 kip/ft, laterally supported compression flange (AISC Specification), (*b*) riveted girder as above, (*c*) welded girder as above except made up of 36 W⁻ 300 core with cover plates, and (*d*) girder as in *c* above except bolted (A325 bolts).

29. Design a welded building girder supporting loads as shown in Fig. P-29. Submit calculations and sketches showing all the necessary information regarding girder web, flanges, stiffeners, and connections including splices. For the purpose of this design assume that maximum length of available bars and plates is 30 ft. Vary flange and web section along the span for optimum design; keep depth constant. Use AISC Specification. Assume the compression flange to be supported continuously.

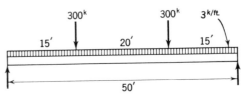

Fig. P-29

30. Design a riveted or bolted girder for the condition of Prob. 29.

31. Design a 55-ft simply supported riveted or bolted plate girder supporting 500 kips at midspan and a 5 kip/ft uniformly distributed load (which includes girder weight). The girder compression flange is supported laterally. Limlt the over-all height to 66-in., and use 1-in.-diameter rivets, or A325 bolts. Submit calculations and a sketch of the girder. (AISC Specification.)

REFERENCES

1. Boyer, J. P., "Castellated Beams," *AISC Engineering Journal* **1,** No. 3 (July 1964).
2. Hooper, I., "Rapid Deflection Calculations," *Contemporary Steel Design* **2,** No. 1, AISI, New York.
3. Chinn, J., "Failure of Simply Supported Flat Roofs by Ponding," *AISC Engineering Journal* **2,** No. 2, New York (1965).
4. Marino, F. J., "Ponding of Two-Way Roof Systems," *AISC Engineering Journal* **3,** No. 3, New York (1966).

5. Commentary on Plastic Design in Steel, ASCE Manual of Engineering Practice No. 41, 1961.
6. Massonnet, C. E., and M. A. Save, *Plastic Analysis and Design*, Vol. 1, *Beams and Frames*, Blaisdell, New York, 1965.
7. Beedle, L. S., *Plastic Design of Steel Frames*, John Wiley and Sons, 1958.
8. Langhaar, H., "Theoretical and Experimental Investigations of Thin Webbed Plate Girder Beams," *J. Appl. Mechanics* **10** (1943).
9. Denke, P. H., "Analysis and Design of Stiffened Shear Webs," *J. Aero. Sci.* **17**, 217 (1950).
10. Basler, K., "Strength of Plate Girders in Shear," *Journ. Str. Div., ASCE Proc.* **87**, No. ST7 (October 1961).
11. Basler, K., and B. Thurlimann, "Strength of Plate Girders in Bending," *J. Str. Div., ASCE Proc.* **87**, No. ST6 (August 1961).
12. Basler, K., "Strength of Plate Girders in Combined Bending and Shear," *J. Str. Div., ASCE Proc.* **87**, ST7 (October 1961).
13. Cooper, P. B., M. S. Law, and B. T. Yen, "Welded Constructional Alloy Steel Plate Girders," *J. Str. Div., ASCE Proc.* **90**, No. ST1 (February 1964).
14. Kerensky, O. A., A. R. Flint, and W. C. Brown, "The Basis for Design of Beams and Plate Girders in the Revised British Standard 153," *Proc. Inst. Civil Engrs., London* **5**, part III, August 1956.
15. Massonnet, C., "Stability Considerations in the Design of Steel Plate Girders," *J. Str. Div., ASCE* **86**, No. ST1 (January 1960).
16. Rockey, K. C., "Web Buckling and the Design of Webplates", *The Structural Engineer* (February and September 1958). Also: K. C. Rockey and I. T. Cook, "The Buckling Under the Bending of a Plate Girder Reinforced by Multiple Longitudinal Stiffeners," *Inst. J. Solids and Struct.* **1** (1965).
17. Cooper, P. B., "Strength of Longitudinally Stiffened Plate Girders," *J. Str. Div. ASCE Proc.* **93**, No. ST2 (April 1967).
18. Frost, R. W., and C. G. Schilling, "The Behavior of Hybrid Beams Subjected to Static Loads," Applied Res. Lab. U.S. Steel Corp. Report, 1963.
19. Schilling, C. G., "Web Crippling Tests on Hybrid Beams," *J. Str. Div. ASCE Proc.* **93**, No. ST1 (February 1967).

12

Design of Buildings

ΦΦΦΦΦΦΦΦΦΦΦΦΦΦΦΦΦΦΦΦΦΦΦΦΦΦΦΦΦΦΦΦΦΦΦΦΦΦΦΦΦΦΦΦΦΦΦΦΦΦ

12-1 INTRODUCTION

Design of steel buildings is examined in this chapter from the standpoint of structural design. Various problems related to mechanical, electrical, and architectural requirements which are encountered in practice in connection with the design of steel buildings are not included in this book. Nevertheless, the designer of steel buildings must familiarize himself with such matters as foundations, types of floors, wall and roof coverings, exterior and interior finish materials, types of door and window framings, provisions for staircases, elevators, pipe shafts, natural or artificial lighting, heating, air conditioning, and various other problems encountered in current building design.

The principal types of loads considered in building design are: (*a*) dead loads, which include the weight of the steel frame and of the floors, walls, roof covering, mechanical and electrical installations and partitions; (*b*) live loads on floors, which include all temporary loads; (*c*) snow, wind, and earthquake loads; (*d*) miscellaneous loads, such as overhead crane loads, effects of possible foundation movement, temperature variation, and other dimensional changes in the various elements of the structure. They are discussed in the following sections.

12-2 DEAD LOADS

In order to compute the dead loads in a building, a preliminary sketch of the building must be made, showing the structural and architectural layouts. Then the dead load can be estimated as follows:

(*a*) The areas of walls and partitions multiplied by their weight per square foot.

(*b*) The area of floors multiplied by their weight per square foot.

(*c*) The weight of framework, including beams, columns, and roof trusses, estimated at 2 to 3 lb pcf of building enclosure.

(*d*) Allowance for various items such as staircases and elevator shafts.

This method is close enough when estimating the weight of the building as a whole, as when the total weight is needed to compute earthquake forces or the loads on the foundations. When designing individual members of a building, the dead load tributary to that member may consist of various items which must be carefully considered. The weights of various building materials in pounds per square or cubic foot are given in most engineering handbooks; for example, AISC Manual. Some values are listed in Table 12-1.

Different formulas have been empirically devised for estimating the weight of buildings, but they are seldom satisfactory because they cannot take into consideration all the factors that affect the loads. Hence it is desirable to estimate the thickness and size of the different constituents and to compute their weight. Another guide for estimation is obtained by comparison with similar structures.

To estimate the weight of steel trusses, the following two formulas are often used:

$$w_s = C\left(\frac{wL}{f}\right)^{1/2} \tag{12-1}$$

$$w_s = C'\left(\frac{wL}{f}\right)^{2/3} \tag{12-2}$$

where w_s = weight of steel for trusses and bracing, pounds per square foot of projected roof area

w = total dead and live load on the truss, including its own weight, pounds per square foot of projected roof area

L = truss span in feet between center lines of supports

f = allowable stress in tension, kips per square inch

C = numerical constant, which is of the order of $\frac{1}{4}$ to $\frac{1}{3}$

C' = numerical constant, which is of the order of $\frac{1}{8}$ to $\frac{1}{6}$.

Equation 12-1 is valid for small spans and loads, say for wL/f less than 150; beyond this value, Eq. 12-2 should be used.

Many buildings require fireproofing, which might materially increase their weight. When concrete is used for this purpose, a minimum cover of 2 in. is often specified for structural steel; materials, such as vermiculite, may give adequate fireproofing with much smaller weight than concrete.

Table 12-1   Weights of Materials
Pounds per cubic foot

Masonry		*Water, Snow*	
Marble, granite	140–165	Water	62.5
Brick masonry	100–150	Snow, fresh	5
Concrete, normal	150	Snow, packed	10 and up
Concrete, lightweight	90–120	Snow, wet	40–50
Metals		*Miscellaneous*	
Steel	480	Sand	100–120
Aluminum	165	Glass	160
Brass	530	Asphalt	80–100
		Mortar	100
	Timber		
Redwood	26	Pine	35–40
Douglas fir	32	Oak	54

12-3   LIVE LOADS

Structural framing for roofs is generally designed for uniformly distributed loads, produced by wind or snow. Wind load will be discussed in the next section. The amount of snow load depends on the location of the structure, its orientation with respect to the direction of sunshine and wind as well as the slope of the roof. Although snow loads are specified in the local building codes, Table 12-2, recommended by the Housing and Home Finance Agency,[1] gives a fairly good picture of specifications across the United States. These recommendations were partly based on a study conducted by the U.S. Weather Bureau and reported in their pamphlet *Snow Load Studies*.[2]

Most building codes state that floors and their framing should be designed for the live loads they may be expected to carry, and the codes usually specify the minimum live loads to be used for certain types of buildings. Since it is difficult to foretell what the floor of a building may be expected to carry in the future, the minimum specified live loads are often used in design, except for certain warehouses and industrial buildings where the live loads are known to be heavier than the specified minimum.

In general, it can be said that for apartments, hospitals, and dwellings, the live loads vary from 40 to 70 psf; for hotels and office buildings, they vary from 50 to 100 psf; for public buildings, from 75 to 125 psf. Assembly halls and corridors of public buildings are designed for 100 to 150 psf. Factories are designed for 100 to 200 psf. Warehouses vary most widely, depending on the material to be stored; light warehouses are sometimes specified to have a minimum of 125 to 150 psf, whereas heavy ones may be designed to carry 400 psf or more.

The National Bureau of Standards of the U.S. Department of Commerce has issued a pamphlet entitled American Standard Building Code Requirements of Minimum Design Loads in Buildings and Other Structures,[3] which recommends certain sets of values. The same organization published another pamphlet, Live Loads on Floors in Buildings,[4] which presents and discusses some of the data obtained by actual surveys. The results of the

Table 12-2   Minimum Uniformly Distributed Vertical Live Loads on Roofs

Roof Slope*	Loads, psf of Horizontal Projection			
	3 in 12	6 in 12	9 in 12	12 in 12 or more
Southern States	20	15	12	10
Central States	25	20	15	10
Northern States	30	25	17	10
Great Lakes, New England†	40	30	20	10
Mountain areas†	40	30	20	10

* For flat roofs used for sun decks or promenades use 60 psf minimum as determined by load due to a crowd of people rather than by snow load.

† Great Lakes and New England and Mountain areas include the northern portions of Minnesota, Wisconsin, Michigan, New York, and Massachusetts; the states of Vermont, New Hampshire, and Maine; and also the Appalachians above 2000 ft elevation, the Pacific Coast range above 1000 ft, and the Rocky Mountains above 4000 ft.

surveys seemed to indicate that the specified live loads for building floors are generally on the conservative side, especially when large areas are considered. Over small areas actual loads may often exceed the minimum values specified in the code. It is unusual for all floors of a multistory building to be simultaneously loaded to full design values over a large area of the floor. Therefore a reduction in the total live load is usually permissible for the design of columns and foundations of multistory structures.

The reduction of live loads over a large tributary area is permitted in most building codes. Actual surveys showed that, while localized loads might exceed the specified values, the average loads over a large area were always much smaller. Also, although there may be many repetitions of loads over a small area, the total load over a large area varies within narrow limits and with less frequency. For large tributary areas, the dead load becomes a relatively greater portion of the total load, and the live load a relatively small portion. Hence any increase in live load will not significantly affect the stresses. The National Bureau of Standards[3,4] recommends the following rule for live-load reduction:

"For live loads of 100 pounds or less per square foot, the design live load on any member supporting 150 sq. ft. or more may be reduced at the rate of 0.08% per sq. ft. of area supported by the member, except that no reduction shall exceed neither R as determined by the following formula nor 60%:

$$R = 100 \times \frac{D + L}{4.33L} = 23.1\left(1 + \frac{D}{L}\right) \qquad (12\text{-}3)$$

in which R = reduction in percent

D = dead load per square foot of area supported by the member

L = basic design live load per square foot of area supported by the member.

For live loads exceeding 100 psf no reduction shall be made, except that the design load on columns may be reduced 20%."

The percentage of load reduction varies for different members of a building, such as between beams and columns. Very often, a straight percentage reduction, say 20 or 40%, is applicable to columns carrying more than one story, although it would be more reasonable to apply an increasing rate of reduction as the number of stories increases. For small loaded areas, the specified minimum uniform load is often insufficient, as heavy concentrated loads, such as safes or mechanical equipment, may exist in any commercial building. For design purposes, such concentrated loads must be placed at points where they produce the greatest stresses. Concentrated loads of 2000 lb for dwelling hotels and apartment houses, and of 5000 lb for offices, public buildings, and assembly halls appear to be reasonable. Such concentrations are considered to be spread over an area of several square feet.

For industrial buildings with heavy machinery, the actual weight of the equipment, together with its vibration effect, must be considered. Warehouses served by fork-lift trucks should be designed for their heavy wheel loads. The effect of such concentration may be serious on short spans, especially if repeated a sufficient number of times. Floors designed for uniform loads as high as 200 psf have been damaged as a result of moving concentrated loads.

12-4 WIND LOADS

The wind pressure on a surface depends on the wind velocity, the slope of the surface, the shape of the surface, the protection from the wind offered by other structures, and the density of the air, which decreases with the altitude and temperature. All other factors remaining unchanged, the pressure due to wind is proportional to the square of the velocity and the density of the air

$$p = C_d q = C_d \tfrac{1}{2} V^2 d \tag{12-4}$$

where p is the pressure on a surface in pounds per square foot, V is the velocity of wind in feet per second, d is the density of air in slugs per cubic foot, C_d is a numerical coefficient (called drag coefficient), and q is the dynamic pressure equal to $\tfrac{1}{2} V^2 d$. At sea level, d is 0.002378 slugs per cu ft, and for wind velocity V expressed in miles per hour, $q = 0.00256 V^2$ psf.

Wind velocities may reach values up to or greater than 150 mph which corresponds to dynamic pressure q of about 60 psf. Pressure as high as this is exceptional, and, in general, values of q of 20 to 30 psf are commonly used for wind loads on buildings. Values of C_d depend on the size and shape of the structures, and for some common shapes are given below in Table 12-3.

Equation 12-4 defines the average pressure normal to the wind. The distribution of this pressure on the surface varies significantly, particularly on surfaces inclined to the wind. Usually the wind is assumed to be horizontal, and the inclination of the surface is measured with respect to a horizontal plane. Two equations, Newton's and Duchemin's, were commonly used to account for inclination of the surface, but these were valid for special cases of flat plates and do not apply to prismatic structures. Wind-tunnel tests on structures, initiated by Eiffel in Paris, and extended by various investigators, have shown that the pressure distribution on a building is characterized by the negative pressures (suction) which exist on the major portion of the roof and the leeward side of the building. These negative pressures actually are gage pressures below the atmospheric pressure. It is apparent that the positive pressures, which are approximately equal to dynamic pressure q at points where the surface is normal to the wind, are rapidly decreasing with the slope and reach zero pressure for a slope about 45°. The negative pressures reach a maximum in the neighborhood of the ridge, generally forward of it, and are practically constant behind the ridge. In addition to the surface

Table 12-3   Drag Coefficients C_d

Shape	C_d	Structure
Square plate (normal to wind)	1.1	Signboards
Long narrow plate (normal to wind)	2.0	Girders
Long cylinders (vertical axis)	0.8	Chimneys, stacks
Short cylinders (vertical axis)	0.7	Tanks
Square long prism (vertical axis)	1.3	Tall buildings
Sphere	0.2	Tanks

pressures, there may exist positive or negative pressures inside a building, depending on the location and orientation of the openings.

Most building codes specify a basic value of wind force in pounds per square foot but allow variations for different heights of buildings as well as different shapes. For example, the Uniform Building Code of the International Conference of Building Officials specifies 15 psf for buildings below 60 ft high and 20 psf for buildings above 60 ft, 30 psf for tanks and signs since they are likely to be more exposed than ordinary buildings, but permits a 40% reduction for circular tanks. On the other hand, wind loads on open frames should be computed on 1.5 times the area.

A comprehensive study was made by an ASCE committee on wind forces, and a final report was published in the ASCE *Transactions*.[5] In the report it was recommended that, for tall buildings, the wind load should be 20 psf up to 300 ft height; for the portion above that limit, an increase of 2.5 psf should be made for every 100 ft increase in height. It further recommends that roofs and walls of buildings be designed for varying pressures, positive and negative, depending on their slope. If α is the slope of the roof to the horizontal, in degrees, the recommended wind forces on the windward slope are as shown in Fig. 12-1. On the leeward slope, a suction of 9 psf is

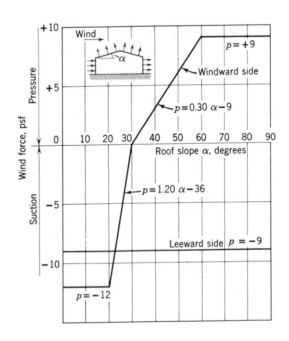

Fig. 12-1 Wind force on buildings. (ASCE committee on wind forces.)

recommended for all values of alpha in excess of zero. These recommendations are based on an assumed wind velocity of about 78 mph with due allowances for drag and suction effects.

Although this static approach to wind load design generally yields sufficient strength for structures, it may not be satisfactory for tall buildings, especially with respect to the comfort of occupants and the permissible horizontal movements or drifts which might result in the cracking of partitions and glasses. A frequently applied empirical rule limits the drift in each story to $0.001H$ where H is the story height. In tall buildings, this limitation seems to be somewhat too strict to comply with. Comfort of the occupants is related to the frequency and amplitude of the vibrations,[6] which in turn depend on the natural frequencies of the building and the gust fluctuations of the wind rather than the steady wind pressure.[7] However, the static wind load as specified by building codes may be conveniently used for a preliminary design, except for very unusual buildings.

12-5 EARTHQUAKE LOADS

There are two basic objectives in seismic design. One is to protect the public from loss of life and serious injury and to prevent buildings from collapse and dangerous damage under a maximum intensity earthquake; the other is to insure buildings against any but very minor damages under moderate to heavy earthquakes. Seismic loads are specified so that these two objectives can be attained within reason and without excessive cost. Generally speaking, these design loads represent the effects of moderate to heavy earthquakes expected for the region but may be exceeded during an extraordinary earthquake, in which case it is necessary to rely upon the reserve capacity in the structure.

Earthquake resistance calls for energy absorption or ductility rather than strength only. If a building is able to deflect horizontally several times the amount under the seismic design load and still maintain its vertical-load-carrying capacity, it will be able to absorb earthquakes considerably heavier than the design earthquake. The deflection or ductility may be provided in the plastic range of the materials and components of the building. If such ductility is present, collapse of the building can be prevented even if the building may be seriously damaged. Earthquake resistant design must utilize structural materials, details, and types of construction which will minimize damage under a strong-motion earthquake. Thus, in addition to seismic load design, the ductility and plasticity of a building should be given due consideration.

The seismic loads on the structure during an earthquake are internal inertia effects which result from accelerations to which the mass of the system is subjected. Actual loads depend on the following factors:

(*a*) The intensity and character of the ground motion as determined at the source and its transmission to the building

(*b*) The dynamic properties of the building, such as its mode shapes and periods of vibration and its damping characteristics

(*c*) The mass of the building as a whole or of its components.

Great progress in earthquake engineering[8,9,10] throws considerable light on the earthquake effects on buildings and is reflected in the seismic design codes. However, numerous uncertainties still exist. Among these are the probable intensity and character of the maximum design earthquake, the damping characteristics of actual buildings, the effects of soil-structure interaction, and the effects of inelastic deformations. Discussion of the fundamentals of earthquake engineering and of their relationship to practical design are beyond the scope of this book.

For convenience in design, an earthquake is translated into an equivalent static load acting horizontally on the building. Although it is not possible to predict the maximum earthquakes at a location, history and experience together with geological observations have shown that the maximum probable earthquakes do vary with different areas and different seismic design loads can be specified. Thus the United States has been divided into four zones of approximately equal seismic probability. Zone 3 is the heavy-earthquake zone, Zone 2 the moderate-earthquake zone, Zone 1 the light-earthquake zone, and Zone 0 with practically no earthquakes expected.

These concepts are incorporated in the 1967 Uniform Building Code[11] to which the reader is referred for details. Briefly summarized, the Code specifies a minimum lateral seismic force for building design as follows:

$$V = ZKCW \qquad (12\text{-}5)$$

where $Z = 1$ for Zone 3, $\frac{1}{2}$ for Zone 2, and $\frac{1}{4}$ for Zone 1. K has a basic value of unity, but may vary from 0.67 for ductile buildings to 3.00 for elevated tanks. $C = 0.10$ for one-story and two-story buildings and can be determined from the following formula for other buildings:

$$C = \frac{0.05}{\sqrt[3]{T}} \qquad (12\text{-}6a)$$

where T is the fundamental period of vibration of the structure in seconds in the direction considered and frequently may be computed from the following empirical formula:

$$T = \frac{0.05h_n}{\sqrt{D}} \qquad (12\text{-}6b)$$

where D is the dimension of the building in feet in a direction parallel to the applied forces and h_n is the height in feet above the base to the uppermost level. In buildings with complete moment resisting frames, $T = 0.10N$, where N = total number of stories above exterior grade.

In order to take into account the dynamic nature of earthquake forces, up to 15% of the total force V as computed from Eq. 12-5 is considered as concentrated at the top story, with the remaining distributed throughout the height of the building in proportion to the product of the weight at each story times its height above the base. The force F_t concentrated at the top is

$$F_t = 0.004V\left(\frac{h_n}{D_s}\right)^2 \leq 0.15V \qquad (12\text{-}7)$$

where D_s is the plan dimension in feet of the lateral force resisting system.

For parts or portions of buildings, the seismic force is given by a slightly different formula:

$$F_p = ZC_pW_p \qquad (12\text{-}8)$$

where W_p is the weight of the part or portion of a structure, and C_p varies from 0.10 for roofs and floors acting as diaphragms to 1.00 for cantilever parapets, ornamentations, and appendages, and up to 2.00 for attached exterior elements.

Although these provisions are largely empirical, they do represent an acceptable method for computing seismic loads. Corresponding to these loads, the normal allowable stresses can be increased by one-third, and under the action of these seismic loads the building structure still may not reach yielding. Should the seismic force greatly exceed these design loads, the plastic behavior and ductility of the structure will be called into action, thus providing reserve energy absorption in preventing collapse or catastrophic damage.

12-6 TYPES OF BUILDINGS

Steel buildings may be subdivided into three categories: one-story buildings, multistory commercial buildings, and special buildings characterized by very large spans. Some industrial buildings, such as power plants and chemical or processing plants, are of an intermediate type, since they require a multistory structure on a portion of the building area and a single-story structure on the rest.

For many years one-story industrial buildings were of the so-called "mill-building" type. They were characterized by the use of trusses to support the roof, large spans between columns, few if any interior walls or partitions, and the existence of industrial equipment within the building, such as cranes.

Currently many industrial buildings are built as the rigid-frame type (Fig. 12-2). The main advantage of the conventional mill building is the economy of the roof since the trusses can be built at relatively low cost. Mill buildings have numerous disadvantages, such as unfavorable lighting conditions, need for extensive bracing, and an appearance that is not generally pleasing. Many of these disadvantages may be eliminated by using rigid-frame designs.

The exterior walls of industrial buildings may be precast or cast-in-place concrete, concrete block, brick masonry, or metal sheathing. If metal sheathing is used it is supported by horizontal girts attached to the columns. The walls are usually nonbearing but must be strong enough to resist the lateral forces due to wind or earthquake. In ordinary one-story buildings this requirement is easily met. If girts are used, they must be designed to transmit the lateral loads to the columns or to other supporting members. Interior walls are often made of wood- or metal-framed partitions with gypsum or other types of paneling. Whatever the "skin" material selected, it should be weather-tight, corrosion-resistant, fireproof, and economical.

Typical multistory buildings, usually called tier buildings, are characterized by a regular floor layout with rectangular bays (Fig. 1-3). They are representative of apartment buildings, office buildings, hotels, public buildings, and

Fig. 12-2 Rigid frame transit shed, span 200 ft. (Courtesy Harbor Department, Port of Long Beach, Calif.)

Fig. 12-3 Alexander Memorial Coliseum, Georgia Institute of Technology, Atlanta. (Courtesy AISC.)

other similar types. The flooring or floor slabs are supported on beams (or joists) and girders at each tier. The girders and some of the beams are supported by columns, which are generally continuous over the tiers. Sometimes a large area in the lower tiers must be free of columns, as in theaters, auditoriums, and some other public buildings. In such cases it is necessary to discontinue some of the columns of the upper tiers and to support these columns by large girders or trusses.

Special buildings such as gymnasia, auditoria, transportation terminals, and hangars often require large areas unobstructed by columns (Fig. 12-3). These structures can be framed using arch bents, long-span trusses, rigid frames, girders, or suspension systems.

12-7 ONE-STORY INDUSTRIAL BUILDINGS

To facilitate handling of materials and "flow" and supervision of the work in a one-story industrial building, usually few walls and partitions are provided. Floor slabs are laid directly on the ground and whenever special mechanical equipment is required, it is placed directly on special foundations in order to eliminate possible vibration of the structure due to operating machinery. When the desirable layout of the building is established, the spacing of columns and the framing of the roof system is selected. Economical design is obtained when bay lengths are much smaller than the spans of the trusses; generally bay lengths vary from 15 to 30 ft, whereas truss spans may vary from 50 up to 100 ft or more. Various types of trusses are used, the choice depending on the spans between the rows of columns, the allowed clearances, the nature of the roofing material, and the type of lighting and ventilation. Various types of roofing may be selected, such as corrugated-metal deck laid on purlins, wooden sheathing with tar-and-gravel, concrete slab, or various composition roof materials.

Designs of one-story buildings for lateral forces due to wind or earthquake are not usually critical. However, provision of a proper bracing system to resist these lateral forces is extremely important. Lateral forces in both the longitudinal and the transverse directions must be considered.

Roof covering and exterior walls must be designed for wind forces—pressure or suction. The spacing of supporting elements, such as roof purlins or wall girts, is determined by the shearing or bending strength of the covering, and purlins or girts are designed so that they can adequately transmit the lateral forces to the main frame of the building. In some building layouts, the type of roof or wall covering may be selected so that purlins or girts can be eliminated altogether.

Crane Girders. Craneways in industrial buildings present special design problems related to crane girders and crane columns. For a crane of given capacity, which includes hoist, trolley, and crane bridge on a roller truck, the type of crane girder and column arrangement may be selected from the following considerations: magnitude of vertical, lateral, and longitudinal forces for which the structure is to be designed; girder span between columns;

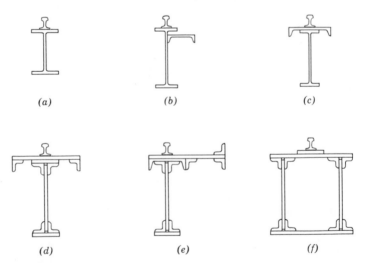

(a)   *(b)*   *(c)*

(d)   *(e)*   *(f)*

Fig. 12-4   Sections for crane girders.

clear height required above the floor; and clearance below roof truss or girder. For the clearance between crane trucks and building column, it is recommended that at least 18 in. be provided in order to prevent injuring a workman in between.

In selecting a trial section for a crane girder, the following must be considered: maximum permissible stress, maximum permissible deflections, and lateral stability and torsional rigidity. The sections commonly used are shown in Fig. 12-4 (*a–f*).

The section shown in Fig. 12-4*a* is a standard W which is economical whenever the magnitude of the loads and the spans permit it. To increase its capacity to resist lateral loads and to increase the torsional rigidity, reinforcing channels may be used as in *b* and *c*. Note that section *c* has a plane of symmetry; lack of symmetry in *b* should be taken into account in calculating bending stresses. Section *d* shows a riveted girder with a symmetrical cross section and section *e* shows a similar unsymmetrical girder with reinforcement of the top flange to increase lateral and torsional rigidity and load capacity. When lateral loads are high, and lateral stability and torsional rigidity are important, box sections, such as *f*, are highly advantageous.

In addition to stress, deflection, and stability limitations, the following special problems must be considered in the design of crane girders.[12]

(*a*) Because of moving loads, with accompanying dynamic effects and possible stress reversals, the connections should be designed with considerable conservatism. Rivets connecting flange angles to the web must be designed not only for the shear flow but also for a portion of the direct wheel load. Welds should be designed conservatively in view of the possibility of a large number of cycles of stress variation and of impact.

(*b*) End bearing must be adequately provided for, particularly the manner in which end reaction is carried through the stiffeners into the supporting column.

(*c*) Girder ends rotate a varying amount, depending on the position of the load on the girder. Any tendency to restrain this rotation would impose large moments on the end connections. The large moments and the accompanying fatigue effects may cause serious trouble. The best design practice is to avoid restraint of the girder ends against rotation.

(*d*) Occasionally the crane girder is used for longitudinal bracing of the building as a whole. This is accomplished by providing knee braces between the columns and the girder in the longitudinal vertical plane. Variation in girder deformation with changing position of the load and vertical and longitudinal dynamic effects may cause fatigue failures in such bracing connections. For this reason the girder should not be used for bracing the building unless the connections are properly designed.

(*e*) Wearing plates between the rail and girder flange are advisable to minimize the possibility of reduction in flange area due to wear.

Crane Columns. Typical details of columns supporting crane girders are shown in Figs. 10-23, 10-24. The design of crane columns is governed by the condition of maximum crane load combined with maximum roof load and lateral wind load. The crane load includes gravity and vertical and lateral dynamic effects.

Loads other than crane loads may occur simultaneously at all bays and sidesway in each bent should be considered because little or no restraint to sidesway is provided by adjacent bays. On the other hand, maximum crane loads occur only at one bay at a time and, if the building is fully braced in the plane of the roof-truss bottom chords, the loaded bent is largely prevented from sidesway by the adjacent bays. Actually, the problem of determining load distribution is statically indeterminate, but an approximate solution can be obtained assuming the following:

(*a*) Stiffness of roof truss (with knee brace) can be assumed infinite relative to stiffness of the column. Therefore the column may be assumed fixed against rotation at the top; that is, at the level of the roof truss bottom chord.

(*b*) With wind loads on the walls and roof, in addition to dead and live loads on the roof, the column can be considered as free to sidesway at the top.

(*c*) With local crane loads acting, the column is largely prevented from sidesway by the resistance of adjacent bays, which can be utilized effectively if the bents are braced.

These assumptions simplify calculation of moments in the column without considering the highly indeterminate space frame action of a mill building. The crane column can be isolated from the space frame and analyzed as a column subjected to axial load and bending. When the axial loads and bending moments for the crane column have been determined, its adequacy may be appraised, using methods described in Chapter 10. In this method it is assumed that all maximum loads will occur simultaneously. Although the frequency of such coincidence may be slight, the column should be checked for such a condition. However, some overstressing or increase in allowable stresses may be permitted.

Bracing. The simplest design for transverse force is to brace each bent independently of the others. This design has the advantage that, during erection, each bent is stable transversely as soon as completed. The structural behavior of a bent subjected to lateral loads depends on the fixity of the columns at the base, provision of a rigid connection between column and truss (such as a knee brace), and the nature of the soil under the foundations.

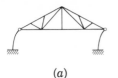 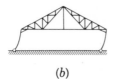 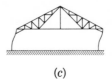

(a) (b) (c)

Fig. 12-5 Alternatives for bracing a building bent.

Three alternatives for bracing a bent are shown in Fig. 12-5. The choice of a particular method depends on the height of the building and the truss span. As the height of building increases, the overturning moment due to lateral forces increases. Fixing the column base (Fig. 12-5a) may be economical when the vertical column load and the size of foundation are large and the overturning moment is relatively small. The vertical column load and the foundation size increase with increase in truss span; hence fixed-base columns may be economical for relatively low and wide buildings. Resisting the overturning moment by a rigid bent with knee braces (Fig. 12-5b) imposes additional loads on the entire roof truss and requires additional material for the truss. Thus the use of knee braces may be economical when the truss span is relatively small. For a tall building it may be economical to provide "fixity" at both ends of the column (Fig. 12-5c).

This assumption that each bent acts independently as a frame is not really correct, since the building acts as a space frame. Because of various combinations of vertical and lateral loads on different bents, the amount of sidesway may differ, resulting in a "racking" of the building roof and walls. Experience shows that a full bracing frame in the plane of the roof-truss bottom chord is highly desirable. Furthermore, if the end walls are sufficiently rigid, the lateral forces can be carried by this horizontal bracing frame to the end walls, with only a small proportion of total lateral load carried by the bents. If the end walls are largely open, then each bent must be designed to take its full share of the lateral load.

Design for the longitudinal wind forces usually requires bracing of the sloping roof panels, in addition to the horizontal bracing in the plane of the roof-truss bottom chord. The horizontal bracing transmits lateral forces to the side walls. These side walls may be designed as shear panels, such as reinforced masonry walls, or cross-bracing must be provided in the wall between bents. Conventional practice is to brace end panels of the side walls. This type of bracing restrains length changes in the walls due to temperature variation, and may introduce significant stresses in the frame. To avoid this effect, bracing can be provided in the bays near the center of the building, thus permitting relatively free change in length between the center and the ends of the building.

12-8 ROOF TRUSSES AND PURLINS

The type of roof truss chosen for a building is largely governed by the required pitch and the lighting conditions. Some common types are shown in Fig. 12-6. The Fink truss is suited to a large pitch, the Howe or Pratt truss to a medium pitch, and the Warren truss to a small pitch. Skylights can be mounted on top of most trusses when desired. For unsymmetrical layouts to receive natural lighting, the sawtooth arrangement is often employed.

The spacing between trusses is determined by the required column spacing and by considerations of minimum cost for the structure as a whole. The usual economical spacing ranges between 15 and 30 ft, with the lower limit for short truss spans and the higher limit for long spans of about 100 ft or over.

The panel layout of a truss will depend on the spacing of the purlins, since it is economical to place the purlins at panel points so as to avoid bending in the top chords. Thus for trusses of medium and long spans, subdivision of the main panels may be desirable, as in Fig. 12-6a. For large trusses, however, it is sometimes economical to have two purlins for each panel, since the large top-chord section may be able to carry the bending

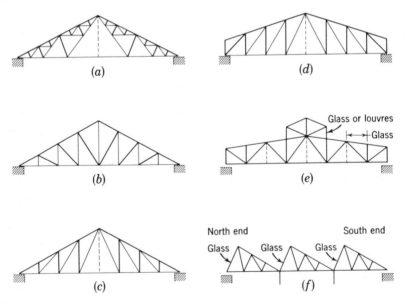

Fig. 12-6 Roof trusses. (*a*) Fink truss, (*b*) Howe truss, (*c*) and (*d*) Pratt truss, (*e*) Warren with skylight, and (*f*) Sawtooth.

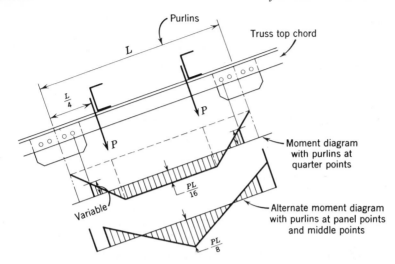

Fig. 12-7   Purlin loads on roof trusses.

moment efficiently. Figure 12-7 shows an arrangement where two purlins are placed at the quarter points of a panel. This will lead to reduced bending moments compared to the alternative case when a purlin is located at the middle point.

Two angles back to back or a structural tee form the most common section for the members of a roof truss. When the load is light and the span short, a single angle section will often suffice and may be used in spite of its lack of symmetry. This is especially true for web members that carry only nominal stresses. For heavy roof trusses, some of the built-up sections shown in Chapter 7, Figs. 7-4 and 7-5, may be required.

For convenience in fabrication, it is often economical to have the same section continuous over several panels of a truss chord, even though the computed stresses may differ appreciably for the several panels. For small trusses which are fabricated in the shop and shipped as a unit to the site of erection, there are often no chord splices except at the ridge. Medium-sized and large trusses are shipped in several parts; then it is often convenient to change the section of the chord at the field splices. Fink trusses are often shipped in four parts (Fig. 12-8b): two major parts (1) and (2), the center portion (3) of the bottom chord, and the tie (4) in the center of the truss. Thus the top chord is made up of only one uniform section and the bottom chord of two different sections.

In general, small and medium trusses of symmetrical design are lifted at the ridge during erection (Fig. 12-8a). In order to prevent buckling of the bottom chord, it is necessary to proportion it to carry the compressive stresses developed during hoisting. An empirical relation is given by $b/L = \frac{1}{125}$, where b is the width of the bottom chord at its center and L the span length.

Stresses in truss members are determined for various loading conditions. If there are no other loads than dead, wind, and snow loads, the following combinations are investigated:

(*a*) Dead load and snow load.
(*b*) Dead load, wind load on one slope.
(*c*) Dead load, wind load, and one-half snow load.

In regions with little or no snow, combination (*a*) would consist of dead load plus specified minimum live load. All dead, live, and snow loads are considered uniformly distributed (with the exception of known concentrations such as machines and equipment on top of the roof), and the analysis is quite simple. Combinations (*b*) and (*c*) can be obtained by adding the stresses due to wind on one slope to a proportion of the stresses in (*a*). The effect of suction due to wind is seldom considered in conventional designs of trusses, although, it may control the design of roofing and its attachment to the frame.

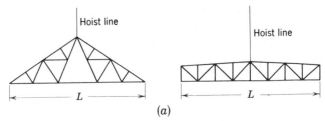

(*a*)

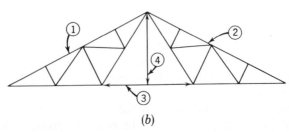

(*b*)

Fig. 12-8   (*a*) Erection of truss as one unit, and (*b*) shipping pieces for Fink truss.

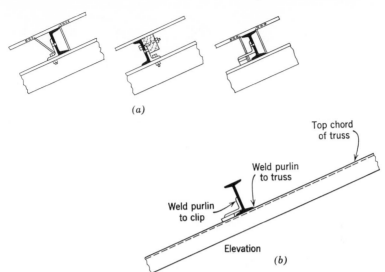

(a)

Top chord
of truss

Weld purlin
to truss

Weld purlin
to clip

Elevation

(b)

Fig. 12-9 Purlin details. (*a*) Channel purlins, and (*b*) I-beam purlins.

If knee braces are installed between the truss and the columns, the equilibrium of the truss is modified, and the stresses in the members near the supports are greatly increased. These additional stresses are especially large for Fink and other similar trusses whose depth is small near the supports, since at the supports the moment introduced through the knee braces is a maximum. Strictly speaking, with the presence of the knee braces, the truss is far from being one simply supported on columns and should be analyzed as a frame under the action of horizontal as well as vertical loads. However the stiffness of the columns is relatively small, so that the effect of continuity under symmetrical loads is usually negligible.

Purlins are secondary beams spanning between the trusses and transmitting the loads from the roof covering to the trusses. They are spaced at 2 to 5 ft or more, depending on the covering material. Purlins are generally made of I-beams or channels (Fig. 12-9), although angle and 2-bar purlins are occasionally employed.

Purlins are designed as simple beams, continuous beams, or cantilever beams. The simple beam design yields the greatest moments and deflections, and continuity cannot be easily ensured through the purlin splices; thus, the cantilever design is often preferred. When continuous or cantilever design is used, it is sometimes desirable to decrease the width of the exterior bay so as to reduce the maximum moments in the end span of the purlins. Figure 12-10 indicates that a reduction of the end span to $0.82L$ brings the moments to the same values as those in the interior spans.

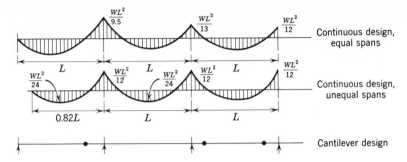

Fig. 12-10   Continuous and cantilever purlins.

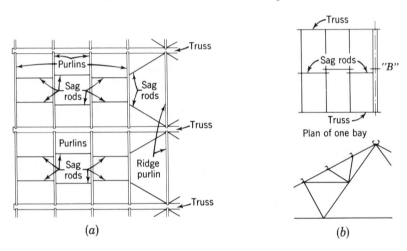

Fig. 12-11   Sag-rods for roof purlins.   (a) oblique sag rods, (b) straight sag rods, and (c) details at ridge purlin.

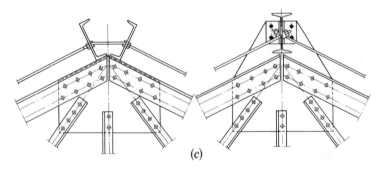

When the slope of the roof is appreciable, the component of the vertical load acting in the weak plane of the purlin must be considered. Since the rigidity of an I-beam or channel is quite small about its weak axis, sag rods are often installed in the plane of the slope. Generally, there are two sag rods in each bay, which are connected to the ridge purlins (Fig. 12-11b). The ridge purlins are made up of an I-beam or two channels (Fig. 12-11c), and must be stronger than the other purlins in order to resist the loads transmitted by the sag rods. When corrugated steel covering is used, it may be assumed that the sheet transmits these components, and sag rods may be dispensed with.

The load from the sag rods or from the corrugated sheets being known, its vertical components acting on the ridge purlin can be computed, considering maximum symmetrical loading on the two slopes. The horizontal component is computed assuming load on one slope only, producing an unbalanced pull. It should be noted that these loads from the sag rods are merely estimates because the sag rods are tightened in place by means of nuts on the threaded ends and therefore the actual load distribution between the various purlins is not well known.

Even though the sag rods considerably reduce the span length of the purlins in the weak direction, stresses in that direction may not be negligible; they are in the order of 10 to 20% of the stresses in the major direction. When designing the purlin sections, it is usual to choose a section modulus somewhat greater than that required for bending only in the stronger plane and check it for the combined effect of bending in two directions. Two or three trials should result in a satisfactory section.

12-9 RIGID FRAMES

A rigid frame resists external loads essentially by virtue of bending moments developed in the ends of members. Thus the connections in a rigid frame must transmit moment as well as thrust and shear. In general, rigid frames may be classified as single-story or multistory frames, and single-span or multispan frames, as shown in Fig. 12-12. In a narrow sense "rigid frame" usually refers to a single-story single-span frame, and the following remarks will be largely applicable to such frames.

Rigid frames may be made of rolled shapes or built-up members, with riveted, bolted, or welded connections. With careful design, attractive and economical structures may be obtained for spans varying from 30 to 200 ft. In some instances rigid-frame construction may require slightly greater amounts of steel than a truss-column frame, but the simplicity and speed of erection and possible reduction of wall height usually result in appreciable

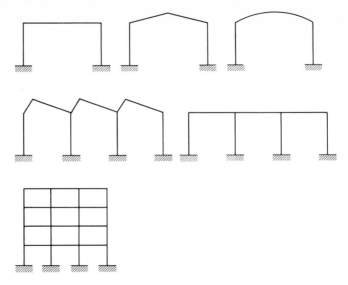

Fig. 12-12   Types of rigid frames.

savings. Also, the use of welding and "plastic" method of design may achieve further economies, so that the use of rigid frames becomes economically advantageous.

In designing rigid frames the following procedure may be followed:

(*a*) Determine the shape and general proportions of the frame: that is, column height, span, roof shape.

(*b*) Select general form of construction: rolled shapes, built-up members, riveted, bolted, or welded.

(*c*) Select the type of roof construction: joist, purlins, metal deck, or wood.

(*d*) Determine the spacing of the frames.

(*e*) Select the type of base for frame columns: free or restrained rotation.

(*f*) Determine loads on the frame and distribution of force and moments in the frame.

(*g*) Proportion members and connections, including design of frame knees.

(*h*) Design bracing and splices.

(*i*) Consider secondary stresses due to temperature variations, settlement of supports, and shear and direct stress.

(*j*) Design bases and foundations for the frames.

The spacing of rigid frames depends on the type of building, roof loads, and to some extent the span of the frame. For ordinary roof loads, and frame spans of 30 to 200 ft, the spacing between frames may vary from 15 to about 40 ft.

Before proceeding with the analysis of forces and moments in the frame, the restraints at column bases must be established. The forces that may be developed at a column base are the shear, thrust, and moment reactions (Fig. 12-13). When a base is designed to resist moment, it is called a "fixed" base and, when no moment can be transmitted, it is called a "hinged" base. Although the column may be rigidly attached to the footing, some rotation may occur on account of the movement of the footing and deformation of surrounding soil. Thus unless the frame is to be constructed on rock or other extremely rigid foundation, the use of a fixed base may not be warranted. A condition of "hinged" base may be obtained simply by providing anchor bolts in the plane of the neutral axis only or, in heavy frames, by actually using a steel pin.

For frames with large span-to-height ratios, and for spans in excess of 60 to 80 ft, the horizontal thrust at the base may be sufficiently large to cause lateral movement of the foundation. For these frames, provision of ties between column bases or, more commonly, between column footings, is desirable.

For a given rigid frame, with known loads and support conditions,

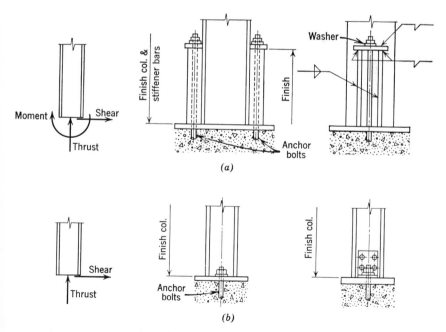

Fig. 12-13 Column bases for rigid frame. (*a*) fixed base, and (*b*) hinged bases.

determination of reactions and internal forces and moments is a statically indeterminate problem. Solution of this problem requires consideration of the load–deformation relations of the frame components. If the load–deformation relations are linear, that is, the frame material is elastic and the stresses do not exceed elastic-limit values, then the internal forces and moments can be determined by using methods based on the theory of elasticity. If plastic deformations take place, that is, local deformations increase without increase in local stress, then forces and moments can be determined by using methods based on the theory of plasticity.

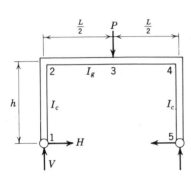

Elastic solutions are based on the conditions of continuity and equilibrium, and the assumption that $M_{max} < M_{yield}$. Plastic solutions are based on the conditions of local plastification (formation of plastic hinges when $M_{max} = M_p$), formation of collapse mechanism (proper number and location of plastic hinges to cause collapse at the smallest possible load), and conditions of equilibrium. Details of elastic and plastic methods of analysis are outside the scope of this book; they are treated in numerous other textbooks and publications.

Fig. 12-14 A rigid frame.

Some of the basic principles of plastic behavior and general remarks about plastic design are included in Chapter 4. It should be noted that plastic behavior of multistory frames, particularly those subjected to lateral forces, has not been fully evaluated, and the 1963 AISC criteria are not applicable to such structures.

The following example illustrates the differences in the distribution of internal forces and moments obtained by the elastic and plastic methods. Consider rigid frame bent with a single concentrated load at midspan of the girder (Fig. 12-14). Analysis based on elastic behavior results in the following reactions and internal forces and moments:

$$M_1 = M_5 = 0, \qquad V_1 = V_5 = \frac{P}{2}, \qquad H_1 = H_5 = \frac{M_2}{h}$$

$$M_2 = M_4 = \frac{3}{8}\left[\frac{PL}{2(h/L)(I_g/I_c) + 3}\right] \qquad (12\text{-}9)$$

$$M_3 = \frac{PL}{4} - M_2$$

where P is the prescribed "working" load, so that the nominal stresses due to forces and moments do not exceed "allowable" stresses.

The analysis based on plastic behavior may be carried out as follows. At collapse, "plastic hinges" form at points 2, 3, and 4. If $I_c < I_g$, a plastic hinge forms in the column top at $M_2 = M_p = 1.1S_cf_y$, where S_c is the section modulus and 1.1 is taken as the shape factor for the steel wide-flange shape. Another plastic hinge forms at midspan of the girder at $M_3 = M_p = 1.1S_gf_y$, where S_g is the girder section modulus. From equilibrium condition at collapse load P_u:

$$M_2 + M_3 = \frac{P_u L}{4} \qquad (12\text{-}10)$$

and

$$M_2 = \frac{1}{4}\left(\frac{P_u L}{S_g/S_c + 1}\right), \qquad M_3 = \frac{P_u L}{4} - M_2 \qquad (12\text{-}11)$$

and

$$M_1 = M_5 = 0, \qquad V_1 = V_5 = \frac{P_u}{2}, \qquad H_1 = H_5 = \frac{M_2}{h} \qquad (12\text{-}12)$$

P_u, the collapse load, may be taken as kP, where k is a load factor to be selected depending on the degree of safety required (see Chapter 4).

Note that, for a given span L, for an elastic solution, the moments in the members depend on the ratios h/L and I_g/I_c, whereas, for a plastic solution, the values of the moments are independent of h/L but depend on the ratio S_g/S_c only.

In a special case when $L = 2h$, $I_g = 2I_c$, $S_g = 1.5S_c$,

	M_2	M_3	H
Elastic case	$0.075PL$	$0.175PL$	$0.15P$
Plastic case	$0.1kPL$	$0.15kPL$	$0.2kP$

Having determined the reactions and moments at the joints, a free-body diagram for each member may be drawn (Fig. 12-15). The proportioning of these members must conform to the requirements discussed in Chapters 4, 8, 9, 10, and 11 dealing with elastic and plastic behavior and design of beams, girders, and columns. If working loads and elastic methods of analysis have been used, then proportions must be such as to meet allowable stress provisions of appropriate codes. If ultimate loads and plastic methods of analysis have been used, then the strength of the member, defined by yielding, or general or local buckling, must be used as a criterion.

To develop fully the advantages of rigid-frame construction and the

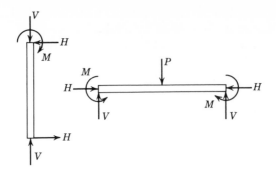

Fig. 12-15   Free-body diagrams of beam and column of Fig. 12-15.

strength of its elements, they should be properly braced to prevent lateral buckling. In particular, segments heavily stressed in compression should be adequately braced.

In some instances, purlins supporting deep metal deck roofing may have sufficient rigidity to provide adequate bracing for the frames. The effectiveness of this type of bracing is usually limited to rigid frames having a span $L = 40$ ft or less.

Rigid-frame structures having a span L greater than 40 ft should have bracing struts or frames in order to prevent lateral buckling of compression elements. Such bracing should be provided at the knee and at appropriate locations along girder and column members.[13]

The number and location of splices in a rigid frame is determined by the available sizes of shapes and plates, and by transportation and erection requirements. Design of the splices for columns and girders is discussed in Chapters 10 and 11.

Usually the internal forces used in the design of rigid frames are determined for known conditions of external loading. Such conditions as foundation movements and temperature variations may produce significant stresses in the structure. It is advisable to check the effect of these secondary conditions on forces and moments in the frame, particularly for large structures. Generally speaking, these secondary effects will influence the elastic behavior more than the ultimate strength.

12-10   RIGID-FRAME KNEES

A special problem in the design of rigid frames is the design of the "knee" transmitting shear, thrust, and moment. Two basic types of knees are used: one having straight flanges and the other having curved flanges (Fig. 12-16).

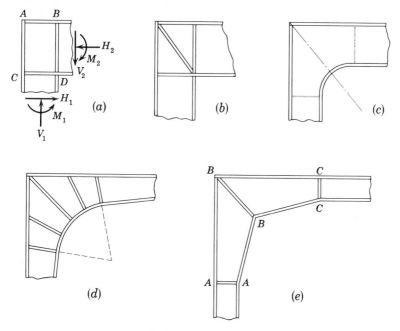

Fig. 12-16   Rigid-frame knees.

Both types may be stiffened or unstiffened, although a knee with curved flanges, which is commonly used with large frames, normally requires radial stiffeners.

The stresses in the knee can be obtained with sufficient accuracy by calculating conventional normal and shearing stresses around the boundary $ABCD$ resulting from flange forces $F = (P/A + MC/I)A_f$, and from web stresses as shown in Fig. 12-17.[14,15] The combined effect of shearing and normal stresses in the column and in the girder web outside of the knee should not exceed allowable values.

In order to prevent high local stresses at D, the concentrated flange loads F_2 and F_3 must be transmitted gradually to the web of the knee, and thus stiffeners should be provided along lines CD and BD.

A simple, and slightly conservative, approximation of shearing stress in the knee web is obtained by assuming that the bending moment and the thrust in the members are resisted by flanges

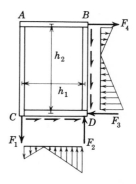

Fig. 12-17   Stresses in a rectangular rigid-frame knee.

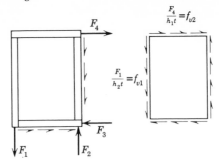

Fig. 12-18 Approximate analysis of a rectangular rigid-frame knee.

only, whereas the web resists only the shears. Thus, from Fig. 12-18,

$$F_1 = \frac{M_1}{h_1} - \frac{V_1}{2}, \qquad F_2 = \frac{M_1}{h_1} + \frac{V_1}{2}$$

$$F_3 = \frac{M_2}{h_2} + \frac{H_2}{2}, \qquad F_4 = \frac{M_2}{h_2} - \frac{H_2}{2}$$

(12-13)

and, consequently,

$$f_{v1} = \frac{F_1}{h_2 t} \quad \text{and} \quad f_{v2} = \frac{F_4}{h_1 t}$$

(12-14)

When shear stress computed by Eq. 12-14 exceeds the allowable value F_v, either reinforcing doubler plates or diagonal stiffeners are used. In the latter case, the stress distribution may be considered as a combination of two systems: one in pure shear and the other in pure diagonal compression. The deformed shape of the knee web (Fig. 12-19) must be such that shortening of the web along AD due to shear must be the same as shortening of the stiffener carrying load P_s.

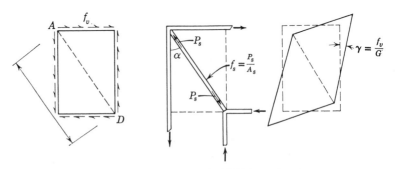

Fig. 12-19 Diagonal stiffener for rigid-frame knee.

Using this condition,

$$f_v = \frac{F_4}{h_1 t + (E/G)A_s \sin^2 \alpha \cos \alpha} \tag{12-15}$$

and

$$f_s = \frac{P_s}{A_s} = \frac{F_4}{(G/E)h_1 t \sin \alpha \cos \alpha + A_s \sin \alpha} \tag{12-16}$$

where A_s is the cross-sectional area of the diagonal stiffener.

Stresses in haunched connections such as shown in Fig. 12-16 are difficult to determine. An approximate but reasonably accurate estimate of the stresses may be made by considering curved beams with nonparallel flanges.[21,22] Stress distribution in curved beams is considered in Chapter 8 where it is shown that the stress f_n normal to a radial section is

$$f_n = a + \frac{b}{R + y} = \frac{P}{A} - \frac{M}{A - RZ}\left(\frac{Z}{A} - \frac{1}{R + y}\right) \tag{12-17}$$

where R is radius of curvature of the centroidal axis and Z is characteristic property of the cross section defined by Eq. 8-42. For a segment with non-parallel flanges (Fig. 12-20), the radial section is replaced by the so-called cylindrical section AB. This section is defined by a circular arc perpendicular to the boundary tangents at A and B. In segments with included angle 2α between tangents not greater than $45°$, the normal stress f_n' is approximately

$$f_n' = \frac{1}{\cos \alpha}f_n \quad \text{and} \quad f_n'' = \frac{1}{\cos \alpha'}f_n \tag{12-18}$$

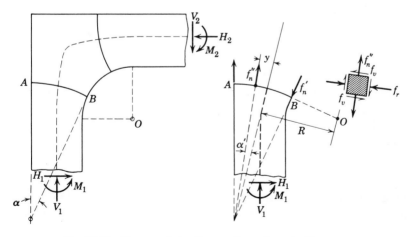

Fig. 12-20 Stress analysis of a curvilinear haunched knee.

In addition to the tangential normal stress f_n', large radial stresses f_r may be induced if the radius of curvature is small. These radial stresses may be calculated using Eq. 8-45, or by replacing f_n with f_n'. To prevent buckling in the knee web due to these stresses, radial stiffeners (Fig. 12-16d) should be used.

Shear stress f_v on section AB may be approximated by neglecting the curvature of the flanges and considering only the effect of the taper of the flanges defined by angle α. Here also the method outlined in Chapter 8 may be used.

Stresses in trapezoidal haunched knees such as shown in Fig. 12-16e, may be determined by the methods outlined in Chapter 8 for tapered-beam sections. An important element in the design of such knees is to provide transverse stiffeners at all sections where flange directions are discontinuous, as along sections AA, BB, CC (Fig. 12-16e). Calculated nominal stresses in all haunched knee connections should not exceed allowable stresses.

The use of curvilinear haunched knees may be desirable in very large rigid frames. Usually, however, rectangular or trapezoidal knees may be just as effective and often more economical.

In order to assure that failure of a rigid frame does not occur in the knee connection, the latter should be proportioned to develop full plastic moment of jointed members. If the girder and the column have different cross sections, the strength of only the weaker member needs to be developed.

The web of an unstiffened knee (Fig. 12-18) must be sufficiently thick to develop shear yield stress $f_{vy} = f_y/\sqrt{3}$; that is, it must not buckle due to shear at a stress $f_{v,cr}$ less than f_{vy}. Assuming this condition to be satisfied, then the web must develop in shear the full plastic moment M_p. This condition will be met if the web thickness t_r is

$$t_r = \frac{\sqrt{3}\,S}{h_1 h_2} \tag{12-19}$$

where S is the section modulus of the smaller of the two members connected by the knee.

If the actual web thickness t is less than the required thickness t_r, then a reinforcing double plate must be used,[15] such that its thickness t_d is

$$t_d = t_r - t \tag{12-20}$$

or a diagonal stiffener should be used, as in Fig. 12-19. For a stiffener having a total flange width b, the required thickness t_s is approximately

$$t_s = t_d \frac{h_1 h_2}{b h_3}$$

Of course, ratio b/t_s should be such as to prevent the possibility of local buckling at a stress below yield point.

12-11   DESIGN OF A RIGID FRAME*

Example 12-1.   A two-story building for a fire station is designed with timber framing on the upper story supported by steel rigid frames having a span of 58 ft 0 in. in the lower story (Fig. 1). The frames are spaced 13 ft 6 in. on centers and have a layout as shown (Fig. 2). Design of the frame, first using the elastic theory and then using the plastic theory is described. The Uniform Building Code, International Conference of Building Officials, is followed for the elastic design, and AISC Specification, Part 2, is followed for plastic design.

(*a*) *Design Loads.*   Assume both the roof and the floor loads to be uniformly distributed on the frame.

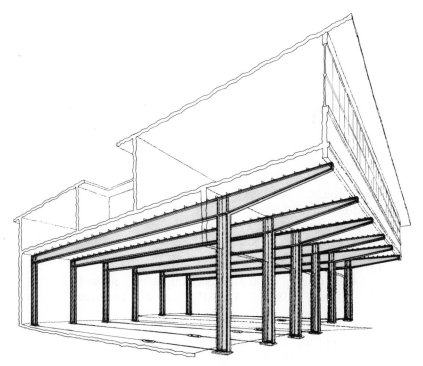

Fig. 1   Rigid-frame structure.

* This design was originally supplied through the courtesy of Milton G. Leong, consulting structural engineer, Berkeley, Calif., but has been revised for A36 steel and 1963 AISC specification.

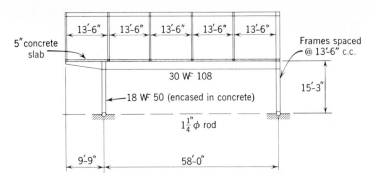

Fig. 2   Frame dimensions.

Roof Loads
  Dead Load

Roofing	4 psf
1-in. diagonal sheathing	2.5
2-in. rafters	4.5
Acoustic tile	1
Total	12 psf

  Live Load (UBC Sec. 2305)
    Basic loading of 20 psf reduced to 12 psf for tributary
      area over 600 sq ft (58′ × 13.5′ = 783 sq ft)   12 psf

Total dead load + live load	24 psf
24 × 13.5′	324 plf

First Floor Loads:
  Dead Load

5-in. concrete slab	62 psf
Partitions	10 psf

  Live Load (UBC Sec. 2306)
    Basic loading of 40 psf reduced by 0.08% per sq ft,
      0.08% × 58 × 13.5 = 63%, but not to exceed
      60% or

$$R = 23.1\left(1 + \frac{DL = 72}{LL = 40}\right) = 65\%$$

    40 × 40%   16 psf

Total dead load + live load	88 psf
88 × 13.5′	1185 plf

Total Loads on Frame:

Roof, assuming uniform distribution	324 plf
Floor	1185
Frame, assumed	91
Total	1600 plf

(b) *Elastic Design.* Composite action exists between the 5-in. concrete slab and the steel frame, and also between the concrete encasement and the columns but may be neglected in the calculations, being on the side of safety. All lateral loads are transmitted to shear walls through the concrete slab acting as a horizontal diaphragm and need not be considered in design of the steel frame.

Assuming sections for the frame members, the relative stiffnesses are computed:

Girder: 30 WF 108, $I = 4461$ in.[4]

$$\frac{I}{L} = \frac{4461}{58 \times 12} = 6.4 \text{ in.}^3$$

Columns: 18 WF 50, $I = 800.6$ in.[4]

$$\frac{I}{L} = \frac{800.6}{15.25 \times 12} = 4.4 \text{ in.}^3$$

Modified for hinged ends of columns,

$$4.4 \times \tfrac{3}{4} = 3.3 \text{ in.}^3$$

It will be close enough to assume that the stiffness ratio is 2 to 1, for girder to column (Fig. 3a).

Fixed-end moments are:

For 58-ft span,

$$\frac{wL^2}{12} = 1.6 \times \frac{58^2}{12} = 450 \text{ kip-ft}$$

For 9.75-ft cantilever,

$$\frac{wL^2}{2} = 1.6 \times \frac{9.75^2}{2} = 76 \text{ kip-ft}$$

The moment distribution for the frame is carried out as in Fig. 3b. The end negative moments for the girder are 212 and 270 kip-ft. The maximum positive moment is approximated by the moment at midspan, Fig. 3c.

$$\frac{1.6 \times 58^2}{8} - \frac{212 + 270}{2} = 675 - 241 = 434 \text{ kip-ft}$$

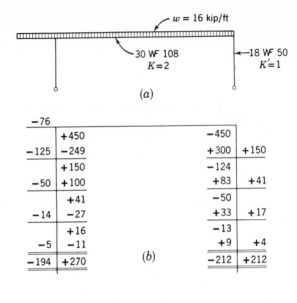

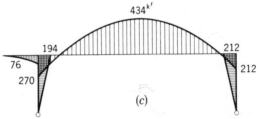

Fig. 3   (*a*) frame and loads, (*b*) moment distribution, and (*c*) elastic moment diagram.

The section of the girder is determined by the maximum moment of 434 kip-ft, which at an allowable stress of 24,000 psi for a compact section would require a section modulus of

$$\frac{434 \times 12,000}{24,000} = 217 \text{ in.}^3$$

A 24 WF 94 has a section modulus of 220.9 in.³ and is therefore satisfactory.

The columns are subjected to both direct load and bending moment. One column has a moment of 212 kip-ft and would require a section modulus of

$$\frac{212 \times 12,000}{24,000} = 106 \text{ in.}^3$$

As an approximation, assume that the effect of direct load would result in an additional section modulus requirement of 30%, that is, $S = 1.3 \times 106 = 138$ in.[3] An 18 W^F 50 with a section modulus of 89 in.[3] is not sufficient. It is proposed to try a 21 W^F 68 having the following properties:

$$S = 139.9 \text{ in.}^2, \quad A = 20.02 \text{ in.}^2, \quad r = 1.74 \text{ in.}, \quad KL/r = 183/1.74 = 1.06$$

and the allowable compressive stress $F_a = 12.20$ ksi. For bending about the strong axis, $KL = 2 \times 183/8.59 = 42.6$, and the effective critical stress $F_e' = 82.1$ ksi. The column direct loads, with distributed load on the girder of 1.6 kip/ft are:

$$P_1 = 1.6 \times 67.8^2/2 \times 58 = 63.4 \text{ kip}$$
$$P_2 = 1.6 \times 67.8 - 63.4 = 45.1 \text{ kip}$$

Then, checking one column, and using $C_m = 0.85$:

$$\frac{f_a}{F_a} + \frac{C_m f_b}{\left(1 - \dfrac{f_a}{F_e'}\right) F_b} = \frac{45.1/20.02}{12.20} + \frac{0.85 \times 212 \times 12}{139.9[1 - (2.2/82.1)]24} = 0.83 < 1.0$$

and at end:

$$\frac{f_a}{0.6F_y} + \frac{f_b}{F_b} = \frac{2.25}{21.6} + \frac{18.18}{24.0} = 0.85 < 1.0$$

checking the other column:

$$\frac{f_a}{F_a} + \frac{C_m f_b}{\left(1 - \dfrac{f_a}{F_e'}\right) F_b} = \frac{63.4/20.02}{12.20} + \frac{0.85 \times 184 \times 12}{139.9[1 - (3.16/82.1)]24} = 0.85 < 1.0$$

and at end:

$$\frac{f_a}{0.6F_y} + \frac{f_b}{F_b} = \frac{3.16}{21.6} + \frac{16.64}{24.0} = 0.83 < 1.0$$

Therefore the section is satisfactory for both columns. Adoption of the 21 W^F 68 for the columns and the 24 W^F 94 for the beam instead of the originally assumed 18 W^F 50 and 30 W^F 108 would change the moments in both the girder and the columns. Revised calculations will not be presented here. It suffices to mention that, after recalculation and also trials with other sections, it was found that the lightest practical sections were 21 W^F 68 for the columns and 24 W^F 94 for the girder as required by the elastic design procedure.

The maximum horizontal reaction, using the originally computed moments, is

$$H = \frac{212}{15.3} = 13.9 \text{ kips}$$

This can be carried by heavy anchor bolts and unsymmetrical footings. It is preferred to carry them by tie rods embedded in the ground-floor slab. The net cross-sectional area required for the rod is

$$A_{net} = \frac{13.9}{22.0} = 0.63 \text{ in.}^2$$

Use a $1\frac{1}{8}$-in. rod with net area of 0.693 in.^{2} and gross area of 0.994 in.^{2} The rod is prestressed to counteract the dead-load horizontal reactions.

(*c*) *Plastic Design.* The equilibrium method of plastic analysis is used for this design. In order to avoid excessive deflections, the 24 W 94 section is retained for the girder, and the column section will be modified as permitted. The design is based on rules for plastic design. It is usual to apply a load factor of 1.70 for such a steel-frame design. Thus the ultimate load for plastic design is set at

$$w_{ult} = 1.6 \text{ kip/ft} \times 1.70 = 2.72 \text{ kip/ft}$$

which will yield a cantilever moment of

$$M_{uc} = \frac{2.72 \times 9.75^2}{2} = 129 \text{ kip-ft}$$

and a simple beam maximum moment of

$$M_{us} = \frac{wL^2}{8} = \frac{2.72 \times 58^2}{8} = 1144 \text{ kip-ft}$$

The plastic moment capacity of 24 W 94 girder is given by a form factor 1.15:

$$M_{pg} = 1.15 \times F_y \times S$$
$$= 1.15 \times 36 \times 220.9$$
$$= 9145 \text{ kip-in.}$$
$$= 762 \text{ kip-ft.}$$

To supply sufficient strength at the midspan point of the girder, the required plastic moment capacity of the column M_{pc} is given by the relation (Fig. 4):

$$\frac{129 + M_{pc} + M_{pc}}{2} = 1144 - 762$$

$$M_{pc} = 317.5 \text{ kip-ft}$$

The required section modulus of the columns is

$$S = \frac{M_{pc}}{1.15 F_y} = \frac{317.5 \times 12}{1.15 \times 36} = 92.02 \text{ in.}^3$$

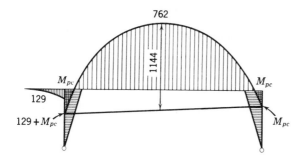

Fig. 4 Plastic moment diagram.

Check the effect of direct axial load on the column, assuming 18 W^F 55, with $S = 98.2$ in.^{3} and $A_s = 16.19$ in.^{2}

Maximum $P_{axial} = 2.72 \times 67.8^2/(2 \times 58) = 107.78$ kip,

$$P_y \text{ of short column} = AF_y = 16.19 \times 36 = 583 \text{ kip}$$

$$\frac{P}{P_y} = \frac{107.78}{583} = 0.18$$

which is greater than the allowable limit of 0.15. Hence the section modulus required should be increased by the ratio

$$\frac{1}{1.18(1 - P/P_y)} = 1.033$$

resulting in

$$1.033 \times 92.02 = 95.06 \text{ in.}^3$$

18 W^F 55 is sufficient.

The slenderness ratio of 18 W^F 55 in weak direction is

$$\frac{183}{1.61} = 114$$

The allowable ratio is given by the following formula, since P/P_y is less than 0.6:

$$\text{Allowable } \frac{L}{r} = \sqrt{\frac{8700}{P/P_y}} = \sqrt{\frac{8700}{0.18}} = 219.85$$

The plastic-moment capacity of the column is not affected by the shear, since the length of 15.25 ft is greater than $4d = 4 \times 1.5 = 6.0$ ft.

It is also necessary to check for shear in the girder:

$$V_{ult} = 0.00055 \, F_y t_w d$$
$$V_{ult} = 0.00055 \times 36{,}000 \times 0.516 \times 24.29 = 248 \text{ kips}$$

which is greater than the ultimate shear load of

$$V_{\text{load}} = 2.72 \times 29 = 78.88 \text{ kip}$$

The proportions of the cross sections are checked for local buckling: 24 W 94 girder:

$$\frac{b}{t} = \frac{9.061}{0.872} = 10.4 < 17$$

$$\frac{d}{t_w} = \frac{24.29}{0.516} = 47.07 < 70$$

Columns are encased in concrete, and no check for local buckling is necessary.

(*d*) *Comparison of Two Designs.*   Depending on the choice of proper load factors for the plastic design and the allowable stresses for the elastic design, divergent results can be obtained. The values chosen for this example being the usual ones, the plastic design method yielded slightly more economical results. The saving in this particular case is

Elastic design, total: 8170 lb of steel per frame

Plastic design, total: 7826 lb

$$\text{Saving} = \frac{8170 - 7826}{8170} = 4.2\%$$

(*e*) *Detailed Design.*   Based on the plastic design, the frame is drawn up using welded connections, as shown in Fig. 5. The details for the column section, its base and footings are shown in Fig. 6.

(*f*) *Lateral Support for Compression Flanges.*   The top flange of the middle portion of the girder is supported laterally by the concrete slab. The end portions of the girder carry a relatively small negative moment. Hence the girder flange is not in danger of buckling.

12-12   MULTISTORY BUILDINGS

A typical multistory building is shown in Fig. 1-3. Framing in a multistory building consists of columns, girders, and beams, which support floor and roof loads. The columns are generally continuous from story to story, and the girders and beams are connected to the columns. Exterior and interior walls may be made of masonry or concrete, and these often are surfaced with decorative finish such as marble, metal, or fine wood paneling. The modern trend in multistory building is to fill large areas of exterior walls with glass. Although this reduces the weight of the walls and may have a pleasing appearance, it may present problems in the transmission of lateral forces

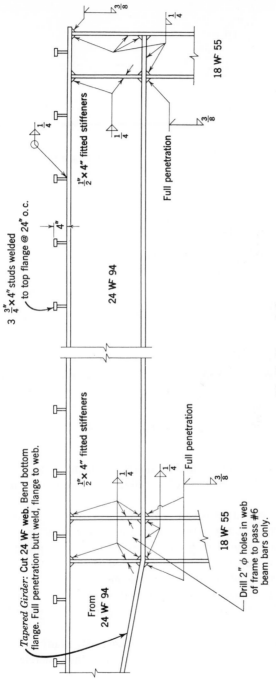

3 $\frac{3''}{4} \times 4''$ studs welded to top flange @ 24" o.c.

$\frac{1}{4}$

$\frac{4''}{}$

24 WF 94

$\frac{3}{8}$

$\frac{1}{4}$

$\frac{1''}{2} \times 4''$ fitted stiffeners

$\frac{1}{4}$

18 WF 55

Full penetration

$\frac{3}{8}$

Tapered Girder: Cut **24 WF web.** Bend bottom flange. Full penetration butt weld, flange to web.

$\frac{1''}{2} \times 4''$ fitted stiffeners

$\frac{1}{4}$

$\frac{1}{4}$

Full penetration

$\frac{3}{8}$

18 WF 55

From 24 WF 94

Drill 2" φ holes in web of frame to pass #6 beam bars only.

Fig. 5 Frame design.

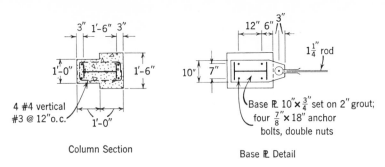

4 #4 vertical
#3 @ 12"o.c.

Column Section

Base ℞ 10"× $\frac{3}{4}$" set on 2" grout;
four $\frac{7}{8}$"× 18" anchor
bolts, double nuts

Base ℞ Detail

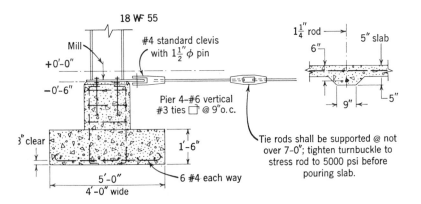

Fig. 6   Details for column.

and in thermal insulation, and often increases the maintenance cost appreciably because of the high cost of washing glass walls or windows in tall buildings.

In buildings with a relatively small number of stories, say under ten, bearing walls may be used, in which case some beams or girders may be omitted. Usually, nonbearing walls are used and are supported on beams, which are called wall beams. When located at the edges of a floor, these beams are called spandrels or spandrel beams.

Partitions are sometimes made of hollow masonry, or various panels using metal, wood, gypsum, or plaster, and other specially developed products.

Floors are usually made of concrete slabs, and many special materials, such as metal decks, foam concrete, or lightweight concrete, may be used economically.

In order to provide adequate fire protection, the steel frame members must be covered with a fire-resistant material. It is common to encase all columns and beams in concrete, although gypsum or other fireproofing materials may also be used.

Beams and girders used in multistory buildings depend to a large extent on the type of floor system used. Steel beams (sometimes called joists) may be economically replaced by concrete joists poured monolithically with the slab, or by prestressed precast concrete floor units.

Steel columns used in multistory buildings generally have a small slenderness ratio because the distances between floors are relatively small and the loads are usually high. Typical built-up sections are shown in Fig. 10-12. The gravity loads on the columns in each tier are proportioned accordingly. When the structure is designed to resist the lateral forces as a rigid framework, it is necessary to consider the bending moments in the columns. For preliminary column design it may be assumed that the points of inflection are located at the middle of the columns, and the shears in the columns are distributed in proportion to the relative rigidity. More refined approximate methods of stress analysis for lateral forces are described in textbooks on structural analysis.

Structural behavior of multistory buildings subjected to lateral forces is complex and highly indeterminate. First, the dynamic nature of loads such as wind, earthquakes, or blast are quite unpredictable. Even if the dynamic characteristics of these loading conditions can be described, the responses of various structures to these dynamic loads are difficult to evaluate. Although analytical procedures for determining elastic and elastoplastic behavior of idealized structures have been developed, these procedures are subject to many limitations, such as deformation and damping characteristics of various types of construction. No general, satisfactory solution to the foregoing problems is now available, although rational analyses are possible in some cases. The best example of a quakeproof multistory building is probably the 43-story Latino-Americana Tower in Mexico City (Fig. 12-21), which survived a severe earthquake without any damage. Provision of tolerances in the walls and windows, as well as rational design of the steel framing, accounted for the excellent behavior under seismic forces.

The common practice in design is to replace the dynamic loading conditions by equivalent static loads. The use of static loads does not yield an accurate representation of the actual structural behavior. But the practice can be justified by its simplicity and the fact that structures designed for static lateral loads usually perform satisfactorily under dynamic conditions. Of course, this cannot be valid generally, and considerable damage may result when the frequency of the loading may approach one of the critical frequencies of the structure.

Fig. 12-21   Latino-Americano Tower, Mexico City—43 stories high.   (Courtesy AISC.)

The conventional method of design assumes lateral forces due to wind and earthquake to be horizontal.   Forces acting on walls are transmitted to the floors and then to the frame.   Usually the floor system, consisting of beams and a slab, can be assumed to be rigid in the horizontal plane and the loads on the frame can be assumed to be concentrated at floor levels.   Then the vertical elements resisting these lateral forces may be of various types: truss frame, open frame, or shear wall.   Sometimes, a combination of all of these elements is used in a tall building.

Truss Frames.   Stresses and deformations in truss frames can be simply calculated assuming joints to be pin-connected; that is, neglecting secondary stresses. The rigidity of these frames depends on the cross-sectional areas of the members. Substantially rigid frames can be obtained with trusses. This high rigidity is very effective in resisting wind forces. For earthquake or blast loading the use of such rigid elements may result in high dynamic loads and thus overstress the framework.

Open Frames. Stresses and deformations in this type of frame can be calculated using such methods as slope deflection, energy, or moment distribution. Usually connections between the columns and girders are assumed to be rigid and have to be designed to transmit substantial moments. In some instances, flexible or semirigid connections may be used, in which case the connections will resist a moment consistent with the amount of rotation at the joint. The action of the frame under normal loading conditions, such as wind or minor earthquake, should be elastic. With exceptionally high lateral loads, such as severe earthquakes or extreme blast, plastic action may be considered. In preliminary design for wind loads, the moments at the beam-column connections may be estimated as follows. Consider a building frame of n tiers, k columns, with floor height h, width between adjacent frames b, and wind pressure p. Assume all columns to have equal rigidity, so that each carries the same shear, and in the bottom tier assume the point of inflection to be at midheight. The total lateral force producing shears and moments in the columns is $P = pbh(n - \frac{1}{2})$, the shear per column is $V = pbh(n - \frac{1}{2})/k$, and the average moment at the top and bottom of each column is

$$M = \frac{pbh(n - \frac{1}{2})}{k} \times \frac{h}{2} = \frac{pbh^2(n - \frac{1}{2})}{2k} \qquad (12\text{-}21)$$

This approximate formula may be used for preliminary design. It follows that, for medium tall buildings with a large ground area, with ratio $(n - \frac{1}{2})/k$ approximating unity, the moment M is approximately

$$M = \frac{pbh^2}{2} \qquad (12\text{-}22)$$

For tall buildings with a relatively small base and a large ratio of $(n - \frac{1}{2})/k$, the moment increases rapidly and open framing alone may require extremely heavy beam-column connections.

Shear Walls. Stresses and deformations in shear walls of masonry or solid concrete are more difficult to analyze. The wall rigidity is measured by both shear and bending deformations. In considering the rigidity of shear walls, effects of holes and cutouts must be considered, as well as the possibility of plastic behavior. When shear walls are used in combination with open frame, the interaction between them should be considered. Generally, within the elastic range the horizontal load is largely carried by the shear walls, but in the plastic range the open frames will supply high ultimate resistance.

The analysis of various types of frames subjected to lateral forces is complex and its treatment is beyond the scope of this book. Basically the shearing force at any one story is resisted by various elements in proportion to their rigidity. When the shear-resisting elements are symmetrical, the center of rigidity (or shear center), lies on the axis of symmetry. When the shear-resisting elements are not symmetrical, the shear center must be located analytically. If the shear force does not pass through the shear center, the building tends to twist, and this effect must be taken into account in calculating shear distribution.

12-13 FLEXIBLE, SEMIRIGID, AND RIGID CONNECTIONS

Frame connections may be classified according to their rotational characteristics as flexible, semirigid, and rigid connections. Flexible connections undergo large angles of rotation transmitting negligible moment (Fig. 12-22a). Deformations and moments in beams with flexible connections are taken as for ideally simply supported beams. Semirigid connections permit some end rotation but in so doing transmit appreciable end moment (Fig. 12-22b). A fully rigid connection ideally permits no rotation between the beam and the column, and transmits substantial end moment (Fig. 12-22c). The design of some of these connections has been discussed in Chapters 5 and 6.

In the analysis of framed structures connections are usually idealized as either type a or type c because this simplifies the calculation of moments in the members. This assumption does not result in the most economical beam size, however. As shown in Fig. 12-22, the proportioning of the beam in case a is governed by the simply supported beam moment $M = wL^2/8$; in case c, assuming columns and connections to be infinitely rigid, beam size is governed by "fixed-end" moment $M = wL^2/12$; in case b an appropriate choice of connection can make end moment M_E equal to center moment M_C, so that maximum moment $M = \frac{1}{2}wL^2/8 = wL^2/16$. Thus the section modulus required in case b is 50% of that in case a and 75% of that in case c. The rigidity of the connection is often defined as the percentage of the resulting end moment taken with reference to the "fixed-end" moment. Thus in the case just given, 75% rigidity of connection would give the desired moment distribution.

Using semirigid connections permits redistribution of moments in the frame, which often results in reduction of maximum moment values and use

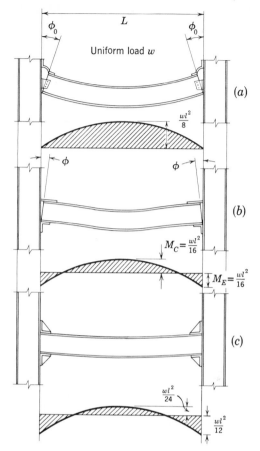

Fig. 12-22 Flexible, semirigid, and rigid connections. (*a*) flexible, (*b*) semirigid, and (*c*) rigid.

of lighter sections for the beams. Two problems arise in the analysis and design of a frame with semirigid connections. One deals with determination of moments for a given structure with known loads; the other deals with determination of moment–rotation characteristics of a particular connection.

The problem of analysis will not be treated here in detail. When the moment–rotation relationship of the connection is linear, the problem may be solved by modified slope-deflection or moment-distribution procedures.[16,17] When the moment–rotation relationship is nonlinear, or when it varies with the sign of the moment as well as with its magnitude, the general analytical problem becomes extremely complex and, when lateral forces must be considered, solution of the problem analytically may become impractical.

The moment–rotation characteristics of semirigid connections have been investigated experimentally.[18–22] In some simple cases,[16] the M–ϕ curve may be approximated by a straight line representing elastic behavior and another line representing post-yielding plastic behavior. Consider a typical top plate connection (Fig. 12-23). End rotation ϕ of the connection due to moment $M = Td$ is

$$\phi = \frac{e}{kd} \tag{12-23}$$

where $e = \epsilon L' = TL'/A'E$, $kd = $ the distance from the top plate to the center of rotation, and $A' = $ sectional area of top plate. The location of the center of rotation depends on deformation of the connection between the bottom flange of the beam and the support, and k varies between $\frac{1}{2}$ and 1. Neglecting the deformation of the support and assuming elastic behavior of all elements of the connection,

$$\phi = \frac{e}{kd} = \frac{TL'}{A'Ekd} \tag{12-24}$$

The slope of the M–ϕ curve is γ (Fig. 12-26):

$$\gamma = \frac{\phi}{M} = \frac{TL'}{A'Ekd\,Td} = \frac{L'}{A'Ekd^2} \tag{12-25}$$

When the stress in the top plate reaches yield value F_y, then rotation ϕ will increase without any further increase in moment. Thus the M–ϕ curve may be approximated as shown in Fig. 12-24. This approximation for the type of connection shown is in good agreement with test results.[22] For other types of connections, reported experimental results may be used.[19,22]

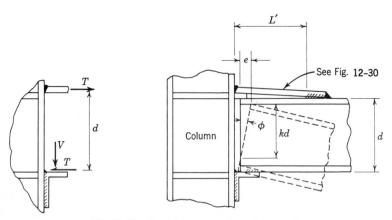

Fig. 12-23   Semirigid connection with top plate.

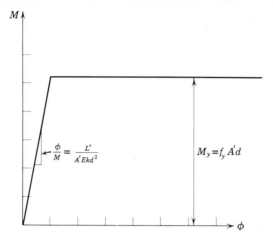

Fig. 12-24   Moment-rotation relationship of a semirigid connection with top plate.

The end moment M and rotation ϕ of the beam also depend on the load, span, and section properties of the beam. For the beam shown in Fig. 12-24, the well-known slope-deflection equation defines the end moment $M = M_F - 2EI\phi/L$, where M_F is the "fixed-end" moment corresponding to zero rotation. Since $M = Td$,

$$M = \frac{wL^2}{12} - 2EI\frac{\phi}{L} \qquad (12\text{-}26)$$

and

$$\phi = \frac{wL^3}{24EI} - \frac{ML}{2EI} = \frac{wL^3}{24EI} - \frac{TdL}{2EI} \qquad (12\text{-}27)$$

This relationship is shown graphically in Figs. 12-25 and 12-26. For $\phi = 0$, $M = M_F = wL^2/12$; for $M = 0$, $\phi = \phi_0 = wL^3/24EI$. If the beam is

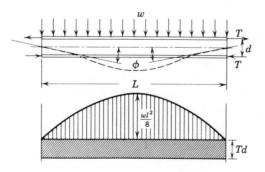

Fig. 12-25   End moments with semirigid connections.

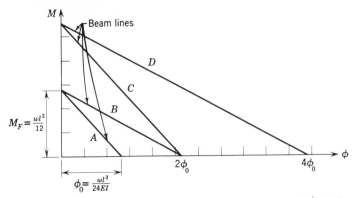

Fig. 12-26   Influence of span/depth ratio and load w on moment-rotation relationship of semirigid connections.   *Note: A—w, L/d.   B—$\frac{1}{4}w$, 2L/d.   C—2w, L/d.   D—$\frac{1}{2}w$, 2L/d.*

designed for a simply supported condition, then $M = wL^2/8 = f(I/C) = f(2I/d)$ and $\phi_0 = \frac{2}{3}(f/E)(L/d)$. Thus, for a given beam and given stress f, the amount of simple-span end rotation ϕ_0 is directly proportional to L/d, but the allowable fixed-end moment $M = f(I/C)$ is independent of L/d. Also, the fixed-end moment $M = wL^2/12$ and the end rotation of simple span $\phi_0 = wL^3/24EI$ are directly proportional to the load intensity w. The influence of the span/depth ratio and load w on the M–ϕ relationship for beams supported by semirigid connections is shown graphically in Fig. 12-26. The straight lines representing Eq. 12-26 are usually called "beam lines" to distinguish them from the connection M–ϕ curves shown in Fig. 12-24.

If the M–ϕ characteristics of the connection are known and the "beam line" can be defined for a particular structure, the corresponding values of end moment and rotation can be determined graphically, or analytically by solving Eqs. 12-25 and 12-26 simultaneously. Thus the point of intersection of the "beam line" and the M–ϕ curve for the connection define the values of M and ϕ, which must be the same for the beam and the connection. This solution assumes no discontinuity between end rotation of the beam and the connection.

Frame connections must be designed for strength—ability to carry anticipated reactions; rigidity—ability to develop desired restraint or freedom of deformation; and reserve capacity for strength and deformation. For semirigid connections these factors can best be evaluated by the use of M–ϕ diagram (Fig. 12-27). Note that connection A shown is suitable and adequate for both beams A and C, but that connection B is suitable and adequate only for beam A but does not have adequate reserve capacity for overload in case of beam C.

When the M–ϕ relationship for a connection may be defined analytically, as for a top plate connection, the expression for the required top plate area may be established by solving simultaneously Eqs. 12-24 and 12-27 as follows:

$$\phi = \frac{TL'}{A'Ekd} = \phi = \frac{wL^3}{24EI} - \frac{TdL}{2EI} \qquad (12\text{-}28)$$

Solving for T,

$$T = \frac{wL^3}{24I(L'/kAd + dL/2I)} \qquad (12\text{-}29)$$

The required top plate area $A' = T/f$ is

$$A' = \frac{wL^3}{24If(L'/kAd + dL/2I)} \qquad (12\text{-}30)$$

and, solving for A',

$$A' = \frac{wL^2}{12df} - \frac{2IL'}{d^2Lk} \qquad (12\text{-}31)$$

where k varies between $\frac{1}{2}$ and 1, depending on the position of the center of rotation. Conservatively k may be taken as 1.

In addition to the cross-sectional area A', the following factors should be considered in the design of the top plate, Fig. 12-28.

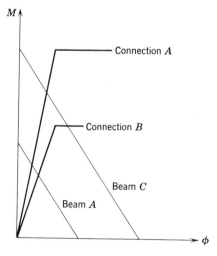

Fig. 12-27 Evaluation of end moments by moment-rotation relationships. Beam A—w, L/d. Beam C—$2w$, L/d.

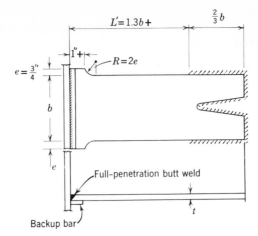

Fig. 12-28   Top plate details for a semirigid welded connection.

(*a*) All connection welds must be adequately designed for the forces calculated previously. No overstress should be allowed and stress concentrations are to be avoided.

(*b*) The length L' of unwelded portion of the plate must be sufficient to permit adequate deformation. This length must be at least 1.3 times the width of plate b (preferably greater).

(*c*) To reduce stresses on the butt weld between the top plate and the support, a flare may be used as shown in Fig. 12-28.

(*d*) To reduce stress concentration on the fillet weld between the top plate and the beam, a slot detail may be used, as shown in Fig. 12-28.

(*e*) The flanges and the web of the column (or girder) to which the connection is attached may be stiffened to prevent distortion under local concentrated forces.

(*f*) Under certain conditions—large lateral forces, or subsequent to yielding due to overloading—the top plate may be subjected to compression. To prevent buckling of the plate acting as a strut, the ratio of L'/t should not exceed 30.

12-14   DESIGN OF A MULTISTORY BUILDING*

Example 12-2.   The steel frame of a 25-story office building in San Francisco built for the Equitable Life Assurance Society (Fig. 1) has several unusual design features and is a good example of the imaginative solutions of

---

* This design was originally supplied through the courtesy of Paquette and Maurer, consulting structural engineers, San Francisco, Calif., but has been revised for A36 steel and 1963 AISC Specification.

problems encountered in the design of tall buildings which have to be framed for earthquake resistance.[23] The general arrangement of the framing and bracing systems is first described. Then sample designs for some of the columns and girders are shown.

(*a*) *General Description.* The building has a 15-story base with a rectangular plan of 168 × 122 ft, above which extends a 10-story tower of 168 × 80 ft (Figs. 2 and 3). To meet the requirements for earthquake-resistant design of the 1948 San Francisco Building Code and limitations on depth of floor construction, two different bracing systems are incorporated in the steel-frame design, one system in the 10-story tower and another in the 15-story base.

For the 10-story tower section, the horizontal force is carried by the entire tower framing, with all girders connected to the columns forming rigid

Fig. 1 Equitable Life Building under construction.

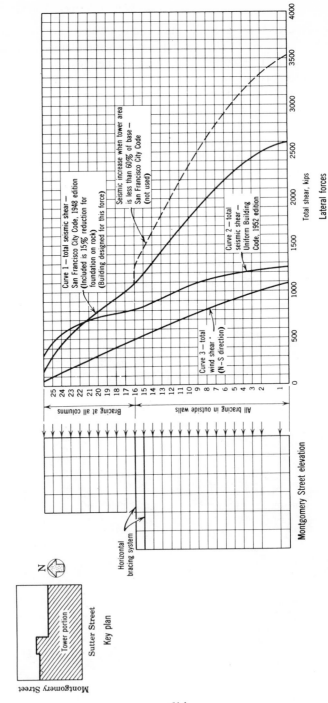

Fig. 2 Lateral forces for Equitable Life Building, San Francisco, Calif. *Note:* Figs. 2 through 5 are reproduced from the July 1, 1954 issue of *Eng. News-Record.*

614

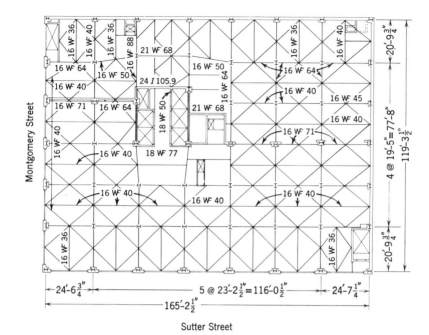

Fig. 3 Sixteenth floor framing incorporates horizontal bracing to transmit lateral forces from tower columns to exterior columns of the base section.

frames and transmitting bending moment. In order to provide adequate lateral resistance, the columns are turned to have some strong axes in each direction, as indicated in Fig. 3. The beams are limited to a maximum depth of 18 in. but are deepened at each end with a stub at the bottom. They are connected to the columns with a T-shaped section of I-beam at both top and bottom flanges (Fig. 4a).

The 15-story base of the building carries larger lateral forces, and it is not possible to brace the columns properly, using beams limited to 18 in. in depth. But greater structural depth can be obtained around the periphery of the building; hence all the lateral forces are carried in the exterior columns, framed with deep spandrel beams (Fig. 4b). From the 11th floor to the 16th, where moments are not too great, wide-flange sections connected to the columns by deep brackets serve as spandrel beams. Below the 11th floor, tapered spandrel beams resist the heavy bending moments. These beams are $5\frac{1}{2}$ ft deep at columns and 2 ft deep at midspan. The change in bracing systems from the complete tower framing to the exterior wall framing required a K system of horizontal bracing in the 16th floor. An additional secondary bracing is added in the 15th floor to obtain better continuity.

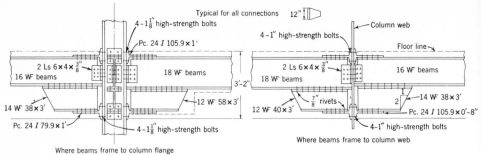

Typical for all connections   12″

4 – 1⅛″ high-strength bolts

Pc. 24 I 105.9 × 1′

4 – 1″ high-strength bolts

Column web

Floor line

2 Ls 6 × 4 × ⅜″
16 WF beams

18 WF beams

2 Ls 6 × 4 × ⅜″
18 WF beams

16 WF beams

3′-2″

14 WF 38 × 3′

12 WF 58 × 3′

12 WF 40 × 3′

⅞″ rivets

14 WF 38 × 3′

Pc. 24 I 79.9 × 1′

4 – 1⅛″ high-strength bolts

Pc. 24 I 105.9 × 0′-8″

4 – 1″ high-strength bolts

Where beams frame to column flange

Where beams frame to column web

Typical Tower Bracing Beams (16th to 25th Floors Inclusive)

(a)

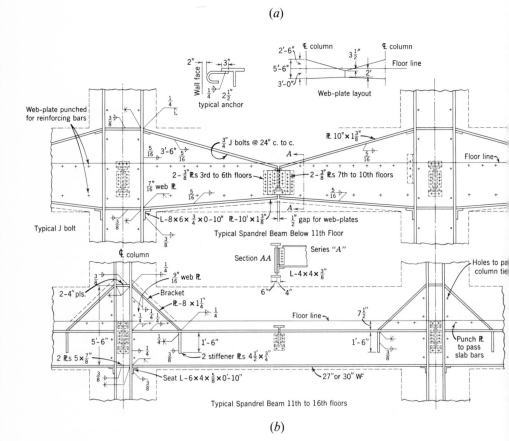

2′-6″   ℄ column   3½″   ℄ column

2″   3″

Wall face

1¼   2½″
typical anchor

5′-6″
3′-0″

Floor line

2′

Web-plate layout

Web-plate punched
for reinforcing bars

¼

⅜

¾″ J bolts @ 24″ c. to c.

℄ 10″ × 1⅜″

5/16   3′-6″   5/16

A

5/16

Floor line

2 – ⅜″ ℄s 3rd to 6th floors

⅞″ web ℄

5/16

2 – ⅜″ ℄s 7th to 10th floors

5/16

Typical J bolt

⅜

L-8 × 6 × ¾ × 0-10″   ℄-10′ × 1⅜″   ½″ gap for web-plates

⅜

Typical Spandrel Beam Below 11th Floor

℄ column

Section AA

Series "A"

¼   9″ web ℄
16

2 – 4″ pls.

Bracket

℄-8 × 1¼″

¼

L-4 × 4 × ⅜″

6″   4″

Holes to pa
column tie

Floor line

7½

5′-6″

1′-6″

1′-6″

2 ℄s 5 × ⅞″

2 stiffener ℄s 4½ × ¾″

Punch ℄
to pass
slab bars

⅜

Seat L-6 × 4 × ⅝ × 0′-10″

27″ or 30″ WF

Typical Spandrel Beam 11th to 16th floors

(b)

Fig. 4 Spandrel beams are of three types—bottom bracketed (top), butterfly (center), and top bracketed (bottom).

616

To avoid complex splicing of heavy column sections, when change of section occurs, the columns below the 14th floor have a continuous taper of $\frac{3}{32}$ in./ft. Exterior columns have a 42 in. web at the base, tapering up to a 12-in. web at the 14th story; flanges are 18 in. × 3 in. at the base and 16 in. wide at the top of the tapered section. Corner columns in the lower part of the frame are L-shaped members, with heavy flange plates at the extremities of the L, affording high resistance to lateral force from any direction (Fig. 5). All columns were erected in three-story lengths and spliced midway between spandrels.

Three methods were used for making connections: riveting, welding, and bolting. Tapered columns and beams were fabricated by welding plates together in the shop. All other shop connections were riveted. Spandrel beams in the base section of the building were field welded to the exterior columns to transmit heavy moments. Other field beam connections were made with high-strength bolts. Column splices were riveted.

Before being welded to the columns, the tapered spandrel beams were seated on angles bent to fit the slope of the bottom flange. Then the web and both flange plates were welded to the column flange. To relieve possible locked-up stresses due to these heavy field welds, the tapered beams were erected in two lengths and spliced at midspan with bolts.

(*b*) *Design Loads and Specifications*

Roof slabs: live load 20 psf.

Floor slabs: live load including partitions 100 psf.

Floor beams: live load including partitions 80 psf.

Columns: live load including partitions 47 psf.

Earthquake forces (according to the 1948 San Francisco Building Code and as shown in Fig. 2): The seismic coefficient varies from 0.068 for 1-story shear to 0.034 for 26-story shear. The Code allows $\frac{1}{3}$ increase in allowable stresses when considering earthquake forces; this is conveniently carried out by modifying the computed forces and moments with a ratio of $\frac{3}{4}$ for most of the calculations in this example.

Steel design stresses: AISC Specification, A36 steel.

Concrete: normal weight concrete with $f_c' = 3000$ psi up to and including the second floor; lightweight concrete with $f_c' = 2500$ psi and weighing 100 pcf for all concrete above second floor line.

A total of 5300 tons of structural steel was used for the 456,000 ft² of floor and roof, amounting to 23 psf. In addition, 3000 tons of steel piling were driven for the foundations. The total weight of the building is 92,000 kips. The maximum vertical load on any one column is 2500 kips.

(*c*) *A Typical Column for the Twentieth Story*

The vertical load on the column has been tabulated and computed to be 400 kips. The total horizontal shear in the story due to earthquake is 790 kips,

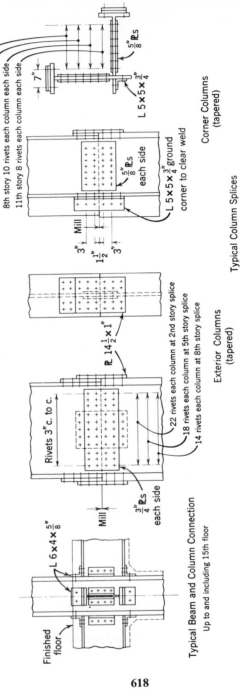

Fig. 5 Splices for heavily stressed tapered columns.

2nd story 14 rivets each column each side
5th story 12 rivets each column each side
8th story 10 rivets each column each side
11th story 8 rivets each column each side

$\frac{5}{8}$" ℞s

7"

L 5×5×$\frac{3}{4}$

Corner Columns
(tapered)

$\frac{5}{8}$" ℞s
each side

L 5×5×$\frac{3}{4}$ ground
corner to clear weld

3"

1$\frac{1}{2}$"

3"

Mill

Typical Column Splices

℞ 14$\frac{1}{2}$×1"

22 rivets each column at 2nd story splice
18 rivets each column at 5th story splice
14 rivets each column at 8th story splice

Rivets 3" c. to c.

$\frac{3}{4}$" ℞s
each side

Mill

Exterior Columns
(tapered)

Finished
floor

L 6×4×$\frac{5}{8}$"

Typical Beam and Column Connection
Up to and including 15th floor

618

69% of which will be carried by the 15 columns, acting along their major axes, at 36.5 kips per column. The remaining 31% of the shear is carried by columns acting along their minor axes. Since the net story height is 10 ft for the columns, the maximum moment in the columns can be computed, assuming the point of inflection at midheight:

$$36.5 \text{ kips} \times 5 \text{ ft} = 183 \text{ kip-ft} = 2200 \text{ kip-in.}$$

For 14 W^F column, the bending factor B_x as given by the AISC Manual, is about 0.185. Thus the equivalent direct load is

$$2200 \times 0.185 = 407 \text{ kips}$$

The total vertical load for design is therefore

$$407 + 400 = 807 \text{ kips}$$

Since one-third increase is permitted in the allowable stresses, this load may be modified by a factor of $\frac{3}{4}$ when using ordinary tables; thus

$$807 \times \tfrac{3}{4} = 605 \text{ kips}$$

This would require a 14 W^F 111 which carried 630 kips on an unbraced length of $Kl = 1.2 \times 10 = 12$ ft, (AISC Manual, p. 5-117, 3-18). Actually, a 14 W^F 150 was used to give additional rigidity to the building. This section will be used also for the 21st and 22nd stories, since the columns are spliced every three stories.

(*d*) *Exterior Column at Ground Floor*

Total seismic shear in first story = 2550 kips, Fig. 2.
Equivalent number of effective columns carrying equal shear = 14, Fig. 3.
Shear per column = 2550/14 = 182 kips.
Column height between spandrels = 20 ft.
Assuming the point of inflection at 13 ft, the maximum moment in column is

$$182 \times 13 = 2370 \text{ kip-ft}$$

Vertical load on column: Dead load = 1450 kips
                            Live load = 165 kips
                                        ─────────────
                            Total   = 1615 kips

Trial Section	Area, sq in.	I_x, in.^{4}	I_y, in.^{4}
Web 42 × 1 in.	42.0	6,160	3
2 flanges 18 × 3 in.	108.0	54,700	2920
Total	150.0	60,860	2923

$$\text{Section modulus} = \frac{60,860}{24} = 2540 \text{ in.}^3$$

$$r_{min} = \sqrt{\frac{2923}{150}} = 4.43 \text{ in.} \qquad r_{max} = \sqrt{\frac{60,860}{150}} = 20.10 \text{ in.}$$

Because of the presence of lateral loading the effective column length will be taken as the unbraced length, hence $K = 1.00$.

Effective slenderness ratios are

$$\frac{Kl}{r} = \frac{(1.00)(20)(12)}{4.43} = 54, \qquad \frac{Kl_b}{r_b} = \frac{(1.00)(20)(12)}{20.1} = 11.9$$

$$F_a = 17.99 \text{ ksi}$$

$$f_a = \frac{1615}{150} = 10.8 \text{ ksi}$$

$$\frac{f_a}{F_a} = \frac{10.8}{17.99} = 0.60 > 0.15$$

From Eq. 10-41:

$$\frac{f_a}{F_a} + \frac{C_m f_b}{\left(1 - \frac{f_a}{F_e'}\right)F_b} \leq 1.00$$

Assume $C_m = 0.85$. From Eq. 10-4:

$$F_e' = \frac{149,000,000}{(Kl_b/r_b)^2} = \frac{149,000,000}{(11.9)^2} = 1,050,000 \text{ ksi}$$

For seismic loading allowable stresses are increased by $\frac{1}{3}$. Then

$$\frac{10.8}{(1.33)(17.99)} + \frac{0.85(2370 \times 12/2540)}{\left[1 - \frac{10.8}{(1.33)(1,050,000)}\right](1.33)(22)}$$
$$= 0.45 + 0.33 = 0.78 < 1.0$$

At braced points, from Eq. 10-26

$$\frac{f_a}{0.6F_y} + \frac{f_b}{F_b} \leq 1.0$$

$$\frac{10.8}{(1.33)(0.6)(36)} + \frac{(2370 \times 12/2540)}{(1.33)(22.0)} = 0.37 + 0.38 = 0.75 < 1.0$$

Hence the column may be reduced in section and a final calculation made with loads and effective lengths determined from rational analyses.

The web shear can be approximately checked by $\frac{3}{4}(182,000)/42 = 3250$ psi.

(e) *Typical Spandrel Beam for Eleventh Floor.*

A detailed elevation of the beam is shown in Fig. 6. The vertical shear in the beam produced by seismic loading is computed as 51 kips. Assuming the

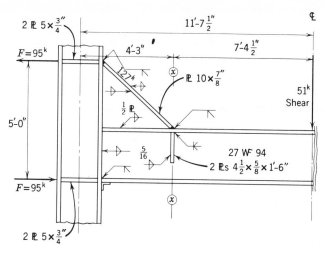

Fig. 6 Welding details for spandrel beam.

point of inflection at midspan, the moment at the critical section xx is

Seismic moment 51 kips $\times$ 7.37 ft = 375 kip-ft
Vertical load moment at xx $\qquad$ = $\underline{}$50$$ kip-ft (computation not shown)
$\qquad$ Total $\qquad\qquad\qquad\qquad$ = 425 kip-ft

A 27 W 94 has a section modulus of 243 in.[3]; the maximum bending stress is

$$425 \times \frac{12{,}000}{243} = 21{,}000 \text{ psi}$$

which is reduced to $21{,}000 \times \frac{3}{4} = 15{,}800$ psi, less than the allowable 24,000 psi. But the 27 W 94 will be used for rigidity and also to provide some extra area for resisting axial load in the beam.

The force F in the column stiffeners can be approximated as follows:

Seismic moments 51 kips $\times$ 10.62 ft = 542 kip-ft
Vertical load moments $\qquad\qquad$ = $\underline{}$90$$ kip-ft (computation not shown)
$\qquad$ Total $\qquad\qquad\qquad\qquad$ = 632 kip-ft

which is reduced to $632 \times \frac{3}{4} = 475$ kip-ft.

Neglecting moment resistance in the web, and for a moment arm of 5 ft,

$$F = \frac{475}{5} = 95 \text{ kips}$$

The top flange of the bracket has a stress of 127 kips and a vertical component of 85 kips, corresponding to a horizontal component of 95 kips. The required area of flange is approximately

$$\frac{127}{20} = 5.77 \text{ in.}^2$$

Use 10 in. × $\frac{7}{8}$-in. plate.

Web stiffener for 27 W carried 85 kips

$$A_s \text{ required} = \frac{85}{22} = 3.86 \text{ in.}^2$$

Use 2 plates $4\frac{1}{2}$ in. × $\frac{5}{8}$ in.

The length of the $\frac{5}{16}$ in. fillet weld required, at 3 kip/in., is

$$\frac{85}{3} = 28 \text{ in.}$$

Use 18-in-long plates, giving 36 in. of weld.

Check the flange of beam at the connection:

$$\text{Net } A_s = 9\frac{1}{4} \text{ in.} \times \frac{3}{4} \text{ in.} = 6.90 \text{ in.}^2$$

$$\frac{P}{A} = \frac{95}{6.9} = 13{,}700 \text{ psi}$$

Check the web shear; for maximum shear of 51 kips seismic load plus 15 kips vertical load for web thickness of 0.49 in. and depth of 27 in., the average shearing stress, reduced by factor $\frac{3}{4}$, is

$$\frac{\frac{3}{4}(51 + 15)}{27 \times 0.49} = 3740 \text{ psi}$$

(*f*) *Typical Tapered Spandrel Beam for the Ninth Floor.* A typical detailed design is shown in Fig. 7. The seismic shear acting on the beam is 67.5 kips. For the point of inflection at midspan,

Seismic moment 67.5 × 10.50 ft = 710 kip-ft
Moment due to vertical load   =   90 kip-ft
———————
Total   = 800 kip-ft

For combined loading, reduce to 800 × $\frac{3}{4}$ = 600 kip-ft. The force F is

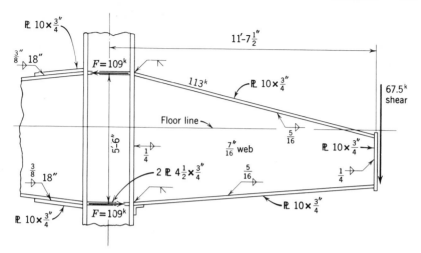

Fig. 7   Tapered spandrel beam.

computed with a lever arm of 5.5 ft.

$$\frac{600}{5.5} = 109 \text{ kips}$$

and the compression or tension in the inclined flange is 113 kips.

   To account for the lateral restraint afforded by the beam to the column, it is estimated that the lateral bending moment in the flange is 6 kip-ft. Try a 10 in × ¾-in. plate for the flange, with a lateral section modulus of $10^2 \times (0.75/6) = 12.5$ in.^{3} The maximum stress is

$$P/A = \frac{113}{7.5} \qquad = 15{,}000 \text{ psi}$$

$$M/S = 6 \times \frac{12}{12.5} = \ \ 5{,}700 \text{ psi}$$

$$\overline{\phantom{20{,}700 \text{ psi}}}$$

20,700 psi $<$ 22,000 psi

Use $\frac{7}{16}$ in. for the web plate, similar to other girders in the building.

PROBLEMS

   In the following problems, the statements are not intended to be complete. Any additional data required are to be assumed by the students. Refer to AISC Specification and local building codes. Use riveted, bolted, or welded connections as desired.

1. A steel mill building with bents spaced at 20-ft centers is to have a span of 80 ft, as shown in Fig. P-1. It is to be designed for a roof live load of 30 psf, a horizontal wind load of 20 psf, and a crane load having a concentration of 50 kips maximum (including impact) on each rail. Design the purlins, the truss, and the columns. Make proper assumptions for the type and weight of the roofing and the walls and the size of footings.

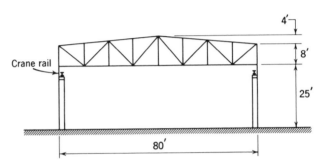

Fig. P-1

2. A three-hinged plate girder rigid frame shown in Fig. P-2 is to be designed for a shed. The frames are spaced 24-ft centers and carry purlins at 8-ft centers. Design the purlins and the frame, showing all details including the hinges. Make proper assumptions for the dead load and use wind load and live loads and earthquake forces consistent with the requirement of your locality. Estimate the amount of steel required per square foot of area covered.

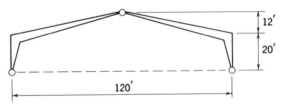

Fig. P-2

3. A section of a typical interior panel for a two story warehouse is shown in Fig. P-3. The ground floor has columns spaced at 30-ft centers in both directions; the upper floor columns are spaced 30 ft in one direction and 60 ft in the other. The second floor carries a live load of 300 psf on a 6-in.-thick reinforced-concrete slab. The slab is supported on steel beams at 10-ft centers, which in turn rest on the 30-ft girders. The roof carries 20 psf live load and 10 psf roofing. Design and detail the framing for a complete bay, including the beams, girders, and columns in both stories. The beams and girders may be simple, cantilever, or continuous spans. Assume the lateral forces not to be carried by the steel framing. Consider or neglect composite action of the slabs as desired.

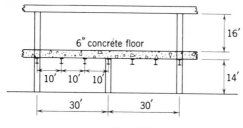

Fig. P-3

4. Solve Prob. 3 assuming earthquake forces to be carried by the steel framing. Refer to Sec. 12-5 or your local building code for the magnitude of the lateral forces.

5. A 10-story office building is 70 × 200 ft, rectangular in plan, and is to be designed with steel girders spanning the 70-ft direction, spaced at 20-ft centers. Each story is to have a clear height of 10 ft from the top of slab to the bottom of the girder, except the ground floor which will have a clear height of 18 ft instead. Design a typical interior girder above the ground floor. Also design the columns supporting the girder, assuming the columns to be continuous through the girder and spliced at the third floor. Detail the connections. The floors are a 6-in. reinforced concrete slab spanning 20 ft and carry a live load of 50 psf, in addition to a movable partition load of 20 psf. Assume that the horizontal forces on the building are not carried by the steel framing.

6. Solve Prob. 5 assuming the wind load on the building to be carried by the steel framing. Carefully design and detail the connection between the girders and the columns. Fix the base of the columns and use an approximate method for analyzing the wind-load moments.

REFERENCES

1. "Housing and Home Finance Agency Recommends More Realistic Roof Loads," *Eng. News-Record* **81** (July 3, 1952).
2. *Snow Load Studies*, U.S. Weather Bureau, Superintendent of Documents, U.S. Government Printing Office, Washington 25, D.C.
3. *American Standard Building Code Requirements of Minimum Design Loads in Buildings and Other Structures*, U.S. Department of Commerce, 1945.
4. *Live Loads on Floors in Buildings*, U.S. Department of Commerce, 1952.
5. "Wind Bracing in Steel Buildings," Final Report of Subcommittee 31, Committee on Steel of Structural Division, *Trans. ASCE* **105**, 1713 (1940).
6. "Vibrations in Buildings," *Building Research Digest*, N. 78, London, England (June 1955).
7. Davenport, A. G. "The Treatment of Wind Loadings on Tall Buildings," Symposium on Tall Buildings, University of Southampton, England, April 1966.
8. Housner, G. W., "Behavior of Structures During Earthquakes," *Proc. ASCE* **85**, No. EM4 (October 1959).

9. Clough, R. W., "Dynamic Effects of Earthquakes," *Proc. ASCE* **86,** No. ST4 (April 1960).

10. Penzien, J., "Dynamic Response of Elasto-Plastic Frames," *Proc. ASCE* **86,** No. ST7 (July 1960).

11. "Uniform Building Code," International Conference of Building Officials, Pasadena, Calif., 1967.

12. Murray, J. J., and T. C. Graham, "The Design of Mill Buildings," *AISC Proc.*, p. 76 (1959).

13. Lyse, I., and W. E. Black, "An Investigation of Rigid Frames," *Trans. ASCE* **107,** 127 (1942).

14. Bleich, F., "Design of Rigid Frame Knees," American Institute of Steel Construction, 1943.

15. Fisher, J. W., G. C. Driscoll, and L. S. Beedle, "Plastic Analysis and Design of Rigid Frame Knees," *Weld, Research Council Bull.* **39** (1958).

16. Johnston, B. G., and E. Mount, "Analysis of Building Frames with Semi-rigid Connections," *Trans. ASCE* **107,** 993 (1942).

17. Baker, J. F., *The Steel Skeleton*, Vols. I and II, Cambridge University Press, 1953, 1956.

18. Rathbun, J. C., "Elastic Properties of Riveted Connection," *Trans. ASCE* **101,** p. 524 (1936).

19. Lyse, I., et al., "Welded Beam-Column Connections," *Welding Journal* (October 1936, October 1937, October 1938).

20. Johnston, B., and G. Dietz, "Tests of Miscellaneous Welded Building Connection," *AWS Journal* (1942).

21. Hechtman, R. A., and B. G. Johnston, "Riveted Semi-rigid Beam-to-Column Building Connections," *AISC Progr. Rept.* **1** (1947).

22. Pray, R. F., and C. Jensen, "Welded Top-plate Beam Column Connections," *AWS Journal* (1956).

23. "Unique Framing for Earthquake Resistance," *Eng. News-Record* 32–34 (July 1, 1954).

13

Design of Bridges

ଟଟଟ

13-1 INTRODUCTION

The main function of a bridge is to carry vehicular or other traffic over a crossing, which may be a river, a canyon, or another line of traffic. Besides serving its specific purpose safely and economically, a bridge should be designed esthetically so that it will fit into and enhance the beauty of its surroundings.

A good bridge layout must take into account the geographical and geological conditions of the site. Clearance requirements, erection procedures, and the method of foundation construction will affect the type and span of the superstructure. The designer should consider all these factors during planning and design.

This chapter can not adequately acquaint the readers with the broad aspects of bridge planning and engineering. It is intended rather to provide the readers with an introduction to the structural principles and procedures involved in design of bridge framing, members, and connections.

13-2 DEAD LOADS

The dead load on a bridge can be divided into two parts: that of the floor and that of the main structure itself. The weight of a bridge floor can be determined by a preliminary design of the floor and then that weight can be used in computing the loads on the main structure. For a railroad bridge, the floor load consists of the tracks, the ties, and the ballast on a concrete or steel deck; in addition, there may be guard rails and miscellaneous items. For a highway bridge, the flooring may consist of a concrete deck with asphalt or a concrete wearing surface. In an orthotropic bridge the deck is a steel plate with a wearing surface of asphaltic materials and epoxies.

The weight of the main structure itself must be estimated before designing or analyzing the structure. The importance of a correct estimate increases with the span length of the bridge. For a short span, the weight of the main structure may represent only a minor portion of the total live and dead load. For a long span, that weight may constitute a major portion of the total load, and any error in the estimate would affect the computed stresses appreciably. Thus a reasonably close estimate is essential for a rapid completion of the design of long-span bridges.

For short- and medium-span bridges, previous engineering experience has enabled many designers to establish charts or formulas giving the over-all structural weights, which can be considered as fairly reliable. These formulas and charts, however, should be used with discretion because bridge specifications may differ with respect to their live load and impact, their methods for load distribution, and their allowable stresses. The weights of the floors may vary considerably, depending on the design. If the bases for the formulas are known, it is possible to make corrections for any particular case and arrive at reasonably close estimates.

A rather complete set of charts for the weight of bridge structures is given in Waddell's *Bridge Engineering*.[1] These charts were based on allowable stresses of 16 ksi and should result in weights about 10 to 15% greater than for bridges designed by the present AREA or AASHO Specifications. It should be noted further that the impact formulas and the type of live loads also differ from the present specifications.

For steel girder bridges, Waddell's charts indicate that the weight of metal per linear foot, for a given standard train, is approximately proportional to the span, for a span length between 40 and 100 ft; it is approximately proportional to the square root of the loading, that is, the class number. The class number used in Waddell's book is about equivalent to the number used for Cooper's E loading (see Sec. 13-3). For single-track railroad bridges, the weight of metal per linear foot of bridge is given by

$$w = kL\sqrt{C} \qquad (13\text{-}1)$$

where L = span length, feet
C = class number (between 40 and 80)
k = a constant, equal to 2.15 for deck bridges

For truss bridges, Waddell has developed a series of formulas for different types of truss design, giving the truss weight w in pounds per linear foot as a

Table 13-1   Formulas for Riveted Truss Weights

Type of Truss	Span, ft	Load Range p, kip/ft	Weight of Truss, kip/ft
Through Pratt with parallel chords	100–200	3–18	$0.180 + \dfrac{(L-50)p}{1480}$
Through Pratt with polygonal upper chords	200–300	3–18	$0.180 + \dfrac{(L-70)(p+0.3)}{1370}$
Deck Pratt with parallel chords	100–200	3–13	$0.180 + \dfrac{(L-30)p}{1590}$
	200–300	3–13	$0.180 + \dfrac{(L-80)p}{1130}$

function of the total load p in kips per linear foot and the span L. These formulas are listed in Table 13-1.

If properly adjusted, these formulas can be used for final design, or they can be used directly for a preliminary design, and a more accurate estimate of weight can be made from that preliminary analysis.

In addition to these weight formulas, the weight of a bridge structure can be estimated by determining the weight of a certain part of it. If the weight proportion of that part is known, then the weight of the entire structure can be approximated. For instance, it has been estimated that the average proportions between weights of the different components of a truss bridge are the following:

Bottom chords	20%
Top chords	25
Web members	25
Bracing	10
Connections	20
	100%

This method, although sometimes convenient and reasonably accurate, can lead to quite erroneous results in certain instances. For example, a truss with a high depth/span ratio would have proportionately more weight in the web, and a relatively shallow girder would have more weight in the flanges; hence the foregoing weight percentage may not apply. However, a knowledge of the proportional weight of different parts will help in the work of estimating.

13-3 LIVE LOADS AND IMPACT

Theoretically, it seems obvious that bridges should be designed for the actual live loads they are expected to carry. In practice, this is difficult to accomplish. First, there are all kinds of vehicles and combinations that may be carried by a bridge. It is often impossible to predict the weight of future vehicles. Hence it has become the practice to design bridges for "standard design loadings," which are representative of the nominally maximum vehicles. Allowable stresses with a proper factor of safety are used so that overloads may be carried without damaging the structure. In contrast to the design live loads for buildings, which are generally higher than the actual loads, design live loads for bridges in the United States are often lighter than the actual maximum vehicles.

The most common live loading for railway bridges is the Cooper's E loading, devised by Mr. Theodore Cooper as early as 1894 and still adopted as the standard loading in the United States as well as in many foreign countries. In the early days of railroading, E-35 and E-50 were the common standard. The present AREA Specification recommends the E-72, which is shown in Fig. 13-1, Fig. 13-1*a* is the usual loading, and Fig. 13-1*b* may control the design of short spans. Either *a* or *b* should be used, whichever gives the larger stress. An E loading of any other number can be directly proportioned from E-72 with the ratio of that number to 72.

As can be seen from Fig. 13-1*a*, this Cooper's loading assumes two heavy locomotives in tandem, with each locomotive followed by a tender, and together they haul a train of cars of a given uniform load intensity. Although actual weight and distribution of locomotives and trains may vary from the standard design loading, the proportioning of a bridge structure based on the standard loading is generally considered satisfactory. Modern diesel and electric locomotives, for example, differ appreciably from the E-loading, but their stress-producing effect can still be approximated by it.

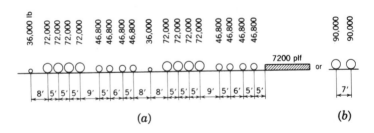

Fig. 13-1 Cooper's E-72 loading for railway bridges. Figures shown are for each track.

The impact of a moving train is generally expressed as a percentage of the static live load. The AREA Specification contains the following:

"Article 20. To the axle loads specified in Sec. A, Art. 18 there shall be added impact forces, applied at the top of rail and distributed thence to the supporting members, comprising:

1. The rolling effect:
Vertical forces due to the rolling of the train from side to side, acting downward on one rail and upward on the other, the forces on each rail being equal to 10 percent of the axle loads.

2. The direct vertical effect:
Downward forces, distributed equally to the two rails and acting normal to the top-or-rail plane, due, in the case of steam locomotives, to hammer blow, track irregularities, speed effect and car impact, and equalling the following percentage of the axle loads:

(*a*) For beam spans, stringers, girders, floorbeams, posts of deck truss spans carrying load from floorbeam only, and floorbeam hangers:

$$\text{For } L \text{ less than 100 ft} \quad 60 - \frac{L^2}{500}$$

$$\text{For } L \text{ 100 ft or more} \quad \frac{1800}{L - 40} + 10$$

$$(b) \text{ For truss spans} \quad \frac{4000}{L + 25} + 15$$

or due, in the case of rolling equipment without hammer blow (diesels, electric locomotives, tenders alone, etc.) to track irregularities, speed effect and car impact, and equalling the following percentage of axle loads:

$$\text{For } L \text{ less than 80 ft} \quad 40 - \frac{3L^2}{1600}$$

$$\text{For } L \text{ 80 ft or more} \quad \frac{600}{L - 30} + 16"$$

In these formulas, L, in general, denotes the loaded length of the member in question. For detailed definition of L and the reduction of live load and impact for bridges carrying more than one track, the reader is referred to the AREA Specification for Steel Railway Bridges.[2]

The standard live loading for highway bridges in the United States is the HS loading,[3] the letter H denoting highway trucks and S denoting semitrailer.

This loading started with the AASHO Specification in 1931 when H20 was the maximum standard, representing a truck weighing 20 tons. With the development of heavier truck–trailer combinations, the H20-S16 was introduced in 1941. It was slightly modified in 1944, and the standard at the present time is HS20-44. For simplicity in calculation, the lane loading is used for longer spans (Fig. 13-3) whenever it yields higher stresses than the single truck of Fig. 13-2.

Because of the fast growth of vehicular weight, the HS20-44 loading is

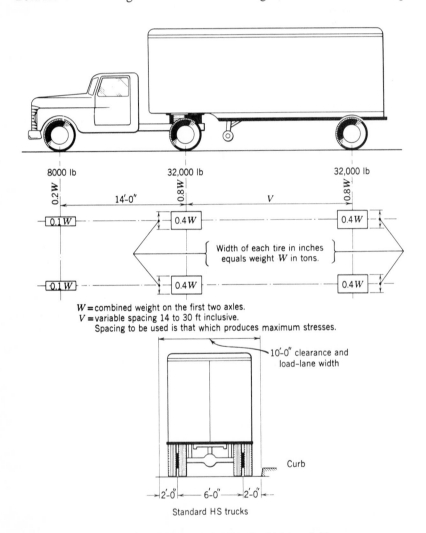

W = combined weight on the first two axles.
V = variable spacing 14 to 30 ft inclusive.
Spacing to be used is that which produces maximum stresses.

Fig. 13-2 HS20-44 truck loading for highway bridges.

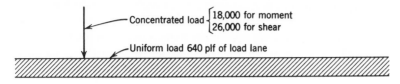

Fig. 13-3   HS20-44 lane loading for highway bridges.

often exceeded by actual vehicles. However, most bridges designed for HS20-44 will be able to carry these slightly heavier vehicles because of the factor of safety applied in the design. This factor of safety results from the less-than-yield stresses allowed in design, the conservative assumptions regarding the distribution of loads, the unlikelihood of simultaneous occurrence of heavy loads on all lanes and the predominence of dead load over live load. Hence it is possible to overload a bridge beyond its design live load without collapsing it, although repeated heavy loadings may result in fatigue failures, especially at details and connections.

The impact on highway bridges is given by the AASHO formula:

$$I = \frac{50}{L + 125}$$

with a maximum of 30%; $L =$ length in feet of the portion of the span that is loaded to produce the maximum stress in the member. Such simplified impact factors are necessarily highly approximate but have been found to be rather convenient for design. Live loads for long-span bridges must differ from those for short and medium spans. In fact, the AASHO Specification are intended for spans of not over 300 or 400 ft. Since the probability of loading a long span with a succession of the heaviest vehicles on all lanes is practically zero, the design live load can be of a smaller intensity. It is further noted that, because of the preponderance of dead load on long-span bridges, an increase in the actual live load will not affect the resulting stresses to any great extent.[4]

Both the weight of the design vehicles and the method of standard loadings vary in different countries. In Great Britain, for example, the common method is to apply an equivalent uniform loading for all designs. Depending on the span length and other conditions, the standard equivalent loading for British highways varies from 2420 psf for spans of 3 ft to only 40 psf for spans of 3000 ft.

The problem of live loading involves not only the weight and spacing of the vehicles and axles, but also the distribution of these loads over the supporting slabs and stringers which will obviously affect the design. Empirical formulas have been derived, based on theoretical and experimental studies, and presented in bridge specifications such as the AASHO[3] so that a definite and reasonably correct design can be developed. It must be remembered that such formulas are necessarily limited to the ranges for which they were derived. For unconventional designs and proportions, the engineer must use his own judgment and experience in interpreting and following these formulas.

13-4 WIND LOAD AND LOAD COMBINATIONS

Probably more uncertain than the magnitude of vertical live load is the action of wind load on bridges. For wide and short spans whose lateral stability is seldom a problem, the magnitude of wind load to be considered is immaterial since it will not affect the design of the bridge. The lateral and sway bracing of such a bridge should possess a reasonable amount of rigidity and will not likely be affected by the computed stresses resulting from wind.

For relatively long and narrow spans, the wind load used in design will not only control the design of the bracing but will often affect the main load-carrying members of the bridge. For flexible suspension bridges of long spans, it is not the static load of wind but rather its dynamic action that may cause trouble and even failure;[5] hence they must be designed to include dynamic strength and rigidity. Bridges of common types and spans, however, have been designed to withstand only static wind force and experience has shown that such an approach is usually satisfactory.

The magnitude of wind load to be used in design varies with different localities and depends on the specifications. The AASHO Specification for Highway Bridges calls for a transverse wind load of 50 to 75 psf and no less than 300 plf of bridge. Other clauses in the Specification spell out more details concerning the calculation and application of these loads under various conditions. A longitudinal wind force of 25 to 50% of the lateral is often assumed for design.

The AREA Specification for Steel Railway Bridges calls for a wind load of 300 plf on the train of one track, in addition to 30 psf on the exposed surface of the bridge (such as $1\frac{1}{2}$ times the vertical projection for girder bridges). For the unloaded bridge, a wind load of 50 psf is specified. In order to insure the stability of spans and towers, the live load to be considered simultaneously with wind is limited to 1200 plf on the leeward track, taken without impact.

Other forces considered in bridge design are the longitudinal forces produced by braking or traction. The AREA calls for 15% of the live load as the braking force and 25% of the weight on driving wheels as the traction force. The AASHO Specification assumes 5% of the live load in all lanes headed in the same direction. In all cases, impact is not to be included in the live load.

Centrifugal force of moving trains should be considered for railway bridges on curves, but it seldom affects highway bridge design. Thermal forces, resulting from variations in temperature, should be determined in accordance with the actual local conditions, although they are listed in some specifications. For bridge piers and abutments, forces of stream current, of floating ice and drift, and of earth pressure may often control the design.

Because of the rare occurrence of the assumed combination of lateral, longitudinal, and vertical forces acting on a bridge, it is generally agreed that higher allowable stresses would be permitted in the structure. Both the AREA and the AASHO Specifications allow a 25% increase in the basic allowable stresses when wind load is considered. For other details of design, the designer should refer to the corresponding specifications.

13-5 TYPES OF STEEL BRIDGES

Bridges can be classified either according to the service they perform or according to their structural arrangement. The majority of bridges are either highway or railway bridges. There are also bridges carrying a combination of traffic, such as a highway bridge with streetcars or pedestrian sidewalks, or a railway bridge carrying highway traffic at the same time. Occasionally there are bridges for pedestrians only, or bridges carrying canals and pipelines. Some bridges are movable (Fig. 13-4); they can be opened either vertically or horizontally so as to permit river traffic to pass beneath the structure.

Classified according to the cross section of the bridge, a deck bridge is one that has its floor resting on top of all the main carrying members, so that no bracing over the top of the traffic is required (Fig. 13-5). If the floor is connected to the lower portion of the load carrying members, thus placing the bracing over the traffic, it is called a through bridge (Fig. 13-6). If there is no overhead bracing and the main carrying members project above the floor level, it is called a semithrough or a pony truss bridge. A double-deck bridge is one in which there are decks on two different levels, both of which can be through decks, or one can be a through deck and the other an open deck. For economy in bracing, either the deck or the through type may have a triangular section where the roadway is supported by two inclined frames.

Classified according to the make-up of the main load-carrying members, an I-beam bridge has rolled I-beams as the main carrying members. When

(a)

(b)

Fig. 13-4 (a) Vertical lift bridge—raised. (b) Vertical lift bridge—lowered in position. Courtesy of Trygve Hoff & Associates, Cleveland.

636

Fig. 13-5 Continuous plate girder deck bridge. (Courtesy AISC.)

Fig. 13-6 Continuous through and deck truss bridge. (Courtesy AISC.)

spans exceed a certain limit, built-up plate girders are used (Fig. 13-5) and they are known as plate-girder bridges. Plate girders acting integrally with the steel deck are referred to as orthotropic bridges. For still longer spans, truss bridges are usually more economical (Fig. 13-6). Occasionally, the Vierendeel truss with quadrangular framing instead of the usual triangular framing is used. For very long spans, the suspension bridge is found to be economical, with high-tensile-strength steel cables carrying the main loads (Fig. 1-1). Suspension bridges are usually stiffened with trusses to obtain rigidity. In all these instances, the number of girders, trusses, or cables may vary from two to three or more, depending on the economy of layout or esthetics. Single-truss bridges have been occasionally advocated, although seldom built.

Classified according to the structural layout of the principal load-carrying members, most of the truss, girder, and beam bridges can have the following various arrangements. One common arrangement is the simple span type where the main carrying members span from one support to another but are discontinuous over the piers. It is sometimes economical to make the spans continuous (Figs. 13-5 and 13-6) to reduce the maximum positive moments. Such an arrangement may involve more calculations to analyze the internal stresses. It may be objectionable if the foundations are likely to settle unevenly, thus producing settlement stresses in the members. Some engineers prefer the cantilever arrangement (Fig. 13-45) to the continuous bridge because it has favorable moments along its length and is not subject to settlement stresses. The cantilever also is easier to analyze. However, a cantilever arrangement requires special hinge connections and is less rigid than a continuous one. The continuous layout has a higher ultimate loading capacity, as indicated by the theory of limit design.

The arch bridge is considered more esthetic than the simple truss spans. The arch itself may be made of girders or trusses, depending upon the span and surroundings. The most common design is to provide two hinges, one at each support, rendering the arch stresses statically indeterminate to the first degree. The three-hinge, one-hinge, and hingeless arches are sometimes employed. The horizontal arch thrust may be resisted by the abutments and piers or by ties along the roadway. This latter type is called a tied arch (Fig. 13-7).

Classified according to the type of connections, the great majority of steel bridges have been riveted. Welded and bolted bridges are currently being designed and constructed. Readers interested in more detail on welded bridges are referred to recent publications on the subject.[6] Some of the older bridges, especially in the United States, were pin-connected. It was thought that pin connections reduced the secondary stresses caused by the rigid joints of a riveted truss and could utilize effectively eye-bar tension members

Fig. 13-7 Tied-arch bridge—Fort Pitt Bridge, Pittsburgh. (Courtesy U.S. Steel Corp.)

of high-tensile-strength steel. However, pin connections are not being used at the present time. One main reason is that the pin joints are more difficult to maintain. Furthermore, they are not as free to rotate as they were previously believed to be.

13-6 BRIDGE FLOOR SYSTEMS

A steel bridge floor system consists of the roadway and the supporting members which transmit the loads to the main structure. Floor systems may be classified in accordance with the type of traffic carried, such as highway or railway floors; in accordance with the principal materials used, such as steel, timber, concrete, or masonry floors; or in accordance with the structural action of the floor, such as one-way or two-way slabs, composite concrete–steel floors, or orthotropic steel-plate floors. Selection of the appropriate floor system for a particular bridge is governed by the following considerations: quality of roadway surface, proper drainage, weight of floor system, required construction time, and over-all cost, including that of maintenance. Typical sections of highway and railway bridge floors are shown in Figs. 13-8 to 13-12.

Satisfactory roadway quality for highway bridges may be obtained with conventional concrete or asphaltic concrete or with open-grid steel flooring, which provides a durable, skid-resistant, smooth surface. For railroad

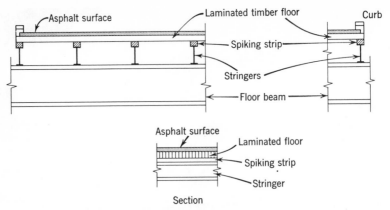

Section

Fig. 13-8   Timber floor for highway bridges.

bridges the roadway must provide a stable and easily maintained level base for the rails and sometimes must reduce vibration and noise during traffic. The latter condition usually requires use of ballast under the track ties (Fig. 13-9).

The floor should be properly drained in order to remove water from the

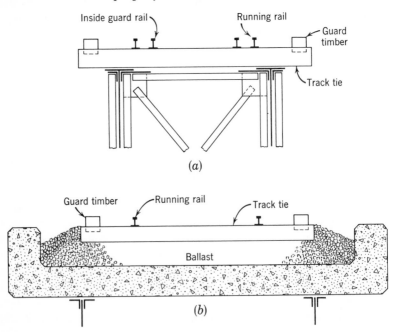

Fig. 13-9   Railway bridge floors: (*a*) open-timber deck floor, and (*b*) concrete ballasted floor.

structure as quickly as possible. With an open-grid steel floor, this is accomplished without any special drainage system (Fig. 13-10c). With other types of floors, the roadway must be "crowned" to collect the water in gutters, usually located at the curb, and frequent drains should be provided to discharge the water. Care must be taken to prevent dripping of drained water on the steel members of the bridge. Provisions also must be made for cleaning the drains in case they become clogged by debris.

Required construction time may be an important consideration in the choice of the floor system. For example, cast-in-place concrete slab requires adequate time for curing, during which time it cannot carry traffic. If delay in opening the bridge to traffic is costly, as for example when replacing some railroad bridges, a steel or prefabricated deck may be more economical.

Weight savings in bridge floor systems reduce the total dead load and result in lighter supporting structures, including the main structural members and the piers. For this reason substantial economies can be obtained, particularly in long span bridges. Weight reduction can be achieved by the selection of a proper structural system and using higher strength steels where appropriate.

A conventional steel bridge floor is supported on a series of beams, called stringers, which are generally placed in the direction of the span. The stringers are supported on transverse beams, called floor beams, which are connected to the main load-carrying members (Fig. 13-10a). In the design of a conventional floor slab supported on stringers, it is usually assumed that it acts as a one-way slab and that the wheel load is distributed over an effective width which varies with the span of the slab. The empirical definition of the effective width is based on theoretical studies and test results and is given in the appropriate specifications.

Bridge stringers are usually made of wide-flange or other standard sections. In the conventional design of stringers, it is assumed that there is no bond between the slab and the stringers, and the stringers are simply supported. The distribution of load on the stringers depends on the stiffness of the slab and of the supporting floor beams. Precise determination of the stringer load is difficult and therefore empirical rules for quick design are given in specifications.

Floor beams are made of either rolled shapes or built-up girders. Trussed floor beams are sometimes designed for wide bridges. For through bridges, stringers are connected to the web of floor beams in order to have the tops of both stringers and floor beams almost flush and ready to receive the concrete slab. Continuity is occasionally achieved for the stringers by plates welded or riveted to the top flange of the stringers over the floor beams. For deck bridges, floor beams may be made flush with the top flange of the main trusses or girders, and stringers often rest on top of the floor beams.

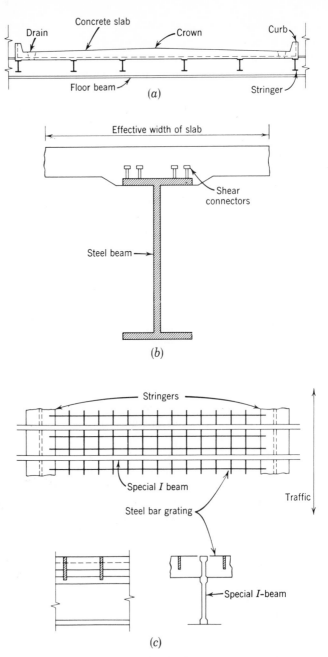

Fig. 13-10 Floors for highway bridges: (*a*) ordinary concrete floor, (*b*) composite beam construction, and (*c*) special steel grating.

Floor beams are almost always rigidly connected to the vertical members of trusses or to the main girders, thus giving lateral rigidity to the bridge as a whole. Hence floor beams are not simply supported but have some moment capacity developed at the ends. Yet, for the sake of simplicity in analysis, they are always designed as simple beams.

Live loads on the floor beams are computed from floor-beam reactions produced by some stringers. The transverse spacing of the vehicles is fixed for railway bridges but may be varied for the design of highway bridges. For highway bridges carrying more than two lanes, a reduction of live-load intensity is generally permitted. It must be realized that, in any case, the design is an engineering approximation, and good judgment must be exercised together with rational analysis and specifications in order to arrive at reasonably accurate results and practical designs.

The participation of floor slabs, stringers, and floor beams in carrying loads for the main truss or girder is usually neglected in design. Tests carried out by Professor C. Massonnet in Belgium indicated that, without special provision for integral composite action, such participation could reduce stresses in the main members as much as 10 or 20%. This means that the floor system will be subjected to additional stresses, and certain elements of the main members will be correspondingly relieved of some stresses.

It is possible, however, to design a floor system which will act both as a roadway resisting local traffic loads and as an integral part of the bridge girders or trusses. In such designs, the concrete slab may be made to act as an integral part of the compression flanges of the stringers, floor beams, and compression flanges of girders. Also, a welded steel-plate floor may be used with a lightweight asphalt or other wearing surface. The steel-plate floor is found particularly economical in long-span orthotropic bridges.

Effectiveness of the concrete slab as an integral part of a beam or girder depends on development of bond or shear connection between the steel and the concrete and typical shear connectors are shown in Fig. 13-11. Structures using such special shear connectors are commonly called "composite-beam construction."[7] In design of composite beams it is assumed that a portion of the concrete slab together with a steel beam forms a T-beam (Fig. 13-10b). The effective width of slab is usually taken as the smallest of the following: (a) distance center to center of beams, (b) one-fourth of the beam span, and (c) 12 times the least thickness of the slab. Assuming no slip between steel and concrete, the stresses in the composite beam may be determined by the method of transformed sections. Detail requirements for such a design are included in the AASHO Specification and other publications.

Typical design of a girder bridge using a steel plate floor is shown in Fig. 13-12 where the steel plate is used as a flange for the longitudinal ribs,

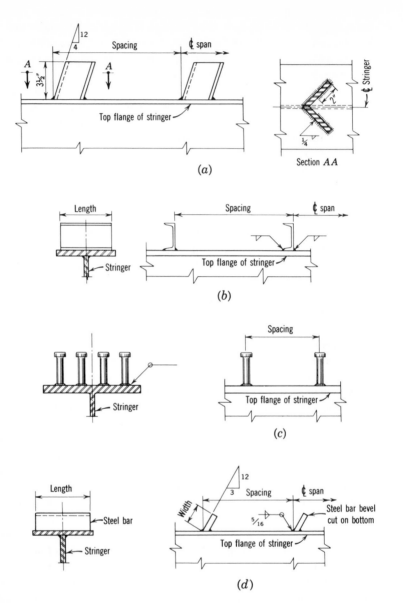

Fig. 13-11 Typical shear connectors: (*a*) angle shear connector, (*b*) channel shear connector, (*c*) stud shear connector, and (*d*) steel-bar shear connector. (Courtesy AISC.)

644

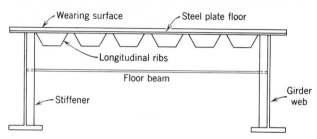

Fig. 13-12　Orthotropic plate floor for bridges.

the floor beams, and the main girders. The design of such a floor system is simplified by replacing it by a plate having some equivalent stiffness. However, since the stiffnesses in the longitudinal and lateral directions vary, the plate may be considered as an orthogonal anisotropic one, or as it is usually called, "orthotropic." Several steel bridges using orthotropic steel-plate floors have been built in Europe since 1947, with substantial saving in material and initial cost.[8]

Based on the theory of elasticity, several solutions for the orthotropic plates have been proposed.[9] Design of orthotropic plate bridges is discussed in greater detail in Chapter 15.

13-7　BRIDGE BRACING SYSTEMS

A bridge is actually a space structure which not only carries the vertical gravity loads to the supporting piers and abutments but also resists lateral and longitudinal forces such as those produced by wind, traction, etc. In order to obtain lateral and torsional rigidity of the bridge, horizontal and transverse bracings are provided. The analysis and design of a bridge, however, are simplified by assuming planar and linear components, such as the main trusses, floor beams, stringers, and bracing frames.

Consider first a deck bridge; the main lateral bracing is located in the loaded plane which is the plane of the upper flanges or the upper chord members. These flanges or chord members of the main vertical trusses or girders thus also serve as the flanges of the lateral truss and are connected together by the floor beams plus a system of diagonals or laterals, as they are called. These laterals may be single or double diagonals, or may be of the K type.

The lower chord of a deck bridge, the unloaded chord, carries little lateral load and may not have a complete lateral truss. The wind load is then transmitted to the top lateral truss by transverse bracing (or sway bracing) in the vertical planes containing the floor beams. Similarly, a half-through

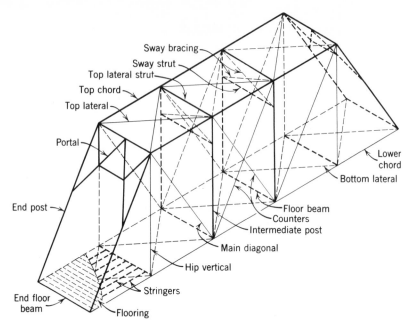

Sway bracing
Sway strut
Top lateral strut
Top chord
Top lateral
Portal
Lower chord
Bottom lateral
End post
Floor beam
Counters
Intermediate post
Main diagonal
Hip vertical
End floor beam
Stringers
Flooring

Fig. 13-13   Skeleton of a typical through truss highway bridge.

or pony bridge would have only a complete bottom lateral bracing. The floor beams are rigidly connected to the posts of trusses, or the stiffeners of girders, for the stabilization of the upper chords or flanges. On simple spans, it is usually more important to hold the top flanges or chords, since they are under compression and liable to buckle if not properly held.

For through bridges (Fig. 13-13) a system of top laterals is always employed, even though the live load is applied on the bottom plane. Such bracing will provide rigidity to the structure, stabilize the compression chord, and carry the main part of the wind load to the bridge portals. This top lateral system will consist of top lateral struts and diagonals. The portals of a through bridge cannot be cross-braced and are designed as rigid portal frames to transmit the load from the top lateral system to the bridge supports.

In order to ensure the torsional rigidity of the bridge, sway bracing is usually provided at all panels of a truss bridge. The proportion of wind load on the top chord that is carried through the sway bracing cannot be analyzed easily. The sway bracing is often designed more or less arbitrarily, with the real intention of adding rigidity, rather than strength, to the structure.

For curved chord trusses, the top lateral system is no longer a planar structure and should not be analyzed as such. However, when the stresses are low, the members are determined by requirements for minimum sizes and stress computation does not have to be more than an approximation.

Since there is a permissible increase in allowable stresses when considering wind, the design of the chords in vertical trusses may not be affected by lateral-force computations especially for wide and shallow bridges. For narrow and high bridges, and bridges of unusual proportions, wind load may become a problem and requirements of both lateral and torsional rigidity may influence the design.

13-8 TRUSS FORMS AND MEMBERS

The most common form of bridge truss is the Warren truss (Fig. 13-14*a*). As in most other trusses, the chords carry the bending moment and the diagonals carry the shear. The vertical members carry only the panel loads and can be economically designed. However, the Warren truss has relatively high secondary stresses, and a Pratt truss (Fig. 13-14*b*) is considered more desirable from that point of view.

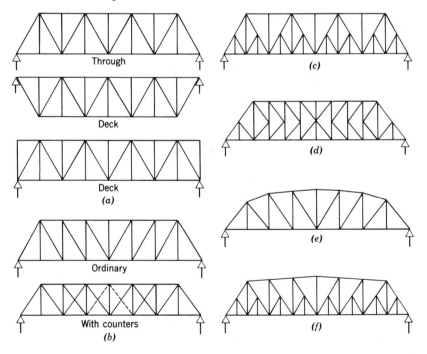

Fig. 13-14 Typical bridge trusses: (*a*) Warren trusses, (*b*) Pratt trusses, (*d*) subdivided Warren truss, (*d*) K-truss, (*e*) curved-chord Pratt truss, and (*f*) Pettit truss.

The economic height-to-span ratio of bridge trusses is about one-sixth to one-eighth, varying with the type of truss, loadings, span length, etc. It can be further shown that the optimum inclination of the diagonals is about 45°. Slight variation from these proportions will not noticeably affect the over-all weight; but excessive deviations may result in appreciable additional material for the truss. When truss spans are increased in length, their economical height will also increase. Thus both the Warren and the Pratt trusses will result in long panel lengths if the diagonal inclination remains about 45°. One way to shorten the panel length—in order to shorten the span of stringers—is to subdivide these trusses. A subdivided Warren truss is shown in Fig. 13-14c.

These subdivided trusses have the disadvantage of developing high secondary stresses, and K trusses may be preferred (Fig. 13-14d). K trusses will keep the desirable diagonal inclination, supply the required depth of truss, and at the same time limit the span of the stringers.

Truss chords may be curved to carry part of the shear and to reduce stresses in the diagonals (Figs. 13-14e and f). There is a slight increase in the cost of fabrication compared to a parallel chord truss, but for medium and long spans the additional cost may be more than balanced by the saving in material.

Trusses can be of either the single- or the double-plane type. A single-plane truss is one that has its gusset plates lying in one plane, that is, there is only one gusset plate at each joint. Such connections are suitable for light bracing and trusses with light loads and small members. For most bridge trusses, main members are composed of box sections and wide flange shapes; and gussets on two parallel planes will be needed. They are termed double-plane trusses.

Members of single-plane trusses are either bars or single and double angles (Fig. 13-15). Occasionally, four angles and single or double channels may be used. Double-plane trusses have wide-flange (W^F) sections, double channel, or built-up box sections (Fig. 13-16). To facilitate connections, the two planes of gusset plates must remain a constant distance apart; hence W^F sections are not as easily connected to a double-plane truss as built-up sections made of angles and plates.

The shape and size of the members are determined by their stresses as well as their connection requirements. Uniformity of shape is often desirable. For example, the entire top chord of a truss may be of inverted U shape (Fig. 13-16a); the bottom chord may be of the box shape (Fig. 13-16b); and the web members under tension may be of the I shape (Figs. 13-16c and d). The members can fit either inside or outside the gussets, but the distance a between the gusset plates should remain the same for a specific truss.

Bracing members are lighter than the main truss members, but they may

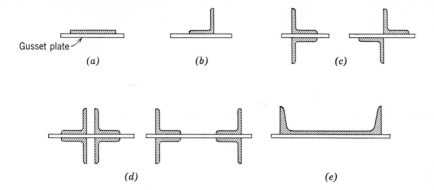

Fig. 13-15 Single-plane truss and bracing members: (*a*) bars, (*b*) single angle, (*c*) double angles, (*d*) four angles, and (*e*) channel.

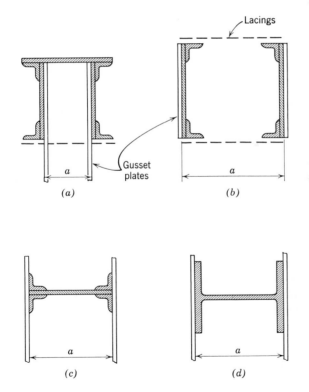

Fig. 13-16 Double-plane truss member: (*a*) inverted U section, (*b*) box section, (*c*) built-up section, and (*d*) Wᵀ section.

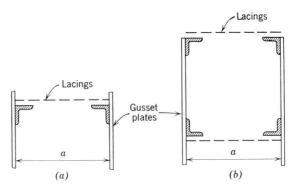

Fig. 13-17   Double-plane bracing members: (*a*) two angles, and (*b*) four angles.

need to possess a certain amount of rigidity. Hence it is often desirable to use latticed sections, such as shown in Fig. 13-17.

13-9   TRUSS CONNECTIONS

A common method of joining together the members of a truss is by gusset plates at the joints where the members meet. The members are connected to the gusset plates by riveting, bolting, or welding. Some shapes for gusset plates are illustrated in Fig. 13-52*b*.

The thickness of gusset plates is governed by several factors. First, a minimum thickness is necessary to develop the full strength of the rivets or bolts. If the rivets or bolts on the gussets are in single shear, a smaller bearing thickness is needed to develop their strength; whereas, if the rivets or bolts are in double shear, a greater thickness is required. For convenience in design and construction, it is preferable to use only one thickness of gusset plate for a truss. For light trusses, $\frac{3}{8}$ to $\frac{1}{2}$-in. gusset plates are used. For heavier trusses $\frac{5}{8}$ to $\frac{7}{8}$-in. plates are more commonly employed.

It is desirable to check the stresses in gusset plates to determine whether they are within allowable limits. The usual practice in design has been to consider critical sections in the plate using approximate methods. For each section the nominal stresses are determined by the usual beam theory, using direct load P and bending moment M:

$$f = \frac{P}{A} + \frac{Mc}{I}$$

If at the critical section, there is shear in addition to direct load and moment, it is evident that the shearing stresses should also be considered. This would necessitate the determination of the maximum principal stresses.

This method yields stress values which do not accurately represent the actual stresses. Determination of actual stresses in the gusset plates is not possible because of load concentrations, warping of plate sections, and local yielding. Compressive stresses along free edges of gusset plates may cause local buckling and stiffeners may be required to prevent this.

There are distinctly two different types of stress transmission through the gusset at a truss joint. The first occurs where the chord member is continuous through the gusset. Here the main portion of the stress in the chord is transmitted directly within the chord itself; only the difference of the chord stresses is carried through the gusset. This arrangement is often used in a truss in order to relieve the gusset plate of any excessive load. If chord splices are required, they are made outside of the joint in the lesser stressed member.

The second type of stress transmission in gussets occurs where the chord members are spliced right at the joints. The gussets at these joints are subjected to heavy stresses because they transmit the entire amount of the chord stresses. For compression chords, which bear against each other at the joints, the bearing surfaces are always milled so that a greater part of the load is transmitted directly through the member and not through the gusset. The AREA Specification provides that the splice, or the gusset in this instance, should be designed to transmit at least one-half the stress. Since these compression members are often spliced on all four sides in order to hold the abutting parts in alignment, these splice plates will help to carry some of the load and the gussets may be sufficient to carry the remainder. On the other hand, the gussets at tension splices cannot be easily designed without excessive additional material. Hence direct splice for tension chord is generally limited to small spans. For long cantilever spans, where the top chord is curved over the piers, pin connections can be advantageously used.

A valuable series of tests on stress distribution in gusset plates was conducted at the University of Tennessee.[10] The tests included only joints with continuous chords. For such gusset plates, it was found that the usual beam formulas are applicable to certain sections but not to others where high localized stresses exist.

For a curved chord truss, the members have to be fully spliced at the joint. This necessitates a heavy splice, and plates in addition to the gussets may be required. In this instance, more economical designs can be obtained by using shingle splices, similar to Fig. 11-38 whereby the joints are staggered so that one constituent can serve as splice plate for another. Field operations, however, are relatively complicated for such joints.

The design of welded trusses is basically similar to that of riveted trusses.

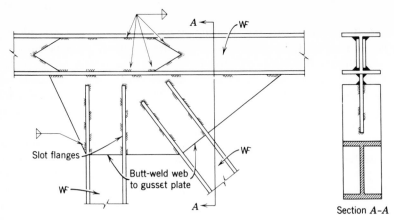

Fig. 13-18 A welded single-gusset connection for W^F members.

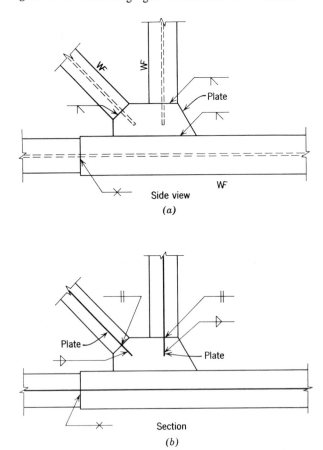

Fig. 13-19 A welded double-gusset connection for W^F members.

But, because of the difference in the connections, the choice of member sections is also affected. Rolled and built-up sections, which cannot be connected conveniently by fasteners and gussets, can sometimes be easily welded together. This is exemplified by the use of T sections, W^F sections, closed boxes, and pipes, in welded trusses.

For a light welded truss where a T section is used for the flange, angles can be welded directly to the stems, thus dispensing with gusset plates altogether. The main objection to such an arrangement lies in the stress concentration created in the stem of the tee at the welds.

A typical connection for a medium-sized truss is shown in Fig. 13-18 where the members are W^F sections connected by a single gusset. The gusset is welded to the bottom of the top chord, and the web members are slotted so the flanges straddle over the gusset and are welded to it. For W^F sections connected by two plane gussets, the flanges can be butt-welded to the gussets and the web is extended inside the gussets and fillet-welded to them (Fig. 13-19). Gusset plates are usually cut with straight edges to reduce costs, as shown in Fig. 13-20.

Although welded trusses are often used for light roof construction, their application to heavy bridgework has been rather limited. One problem is the existence of local moment and torque at welded connections which may produce high stress concentrations and, when repeated, could result in failure. With riveted or bolted connections, there is more "give" in the joint so that high localized stresses will be relieved before any serious failure results. However, safe and economical designs can be developed for heavy welded trusses.[6]

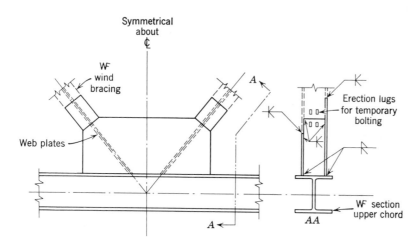

Fig. 13-20 Welded gusset-plate connection for wind bracings.

13-10   STRESS REPETITIONS AND REVERSALS

Several problems arise when a member of a truss is subject to stress reversal. First, the member and its connections must be designed so they can take stresses of both types. Second, the problem of fatigue may become important and must be considered. Stress fluctuations, even without reversals, may cause fatigue failures of members or connections at lower stresses than those at which they would fail under static loads. Such failures would be primarily due to stress concentrations introduced by the constructional details. It is important, therefore, that all details should be designed to avoid as much as possible stress concentrations produced by sharp corners and sudden changes in cross sections.

There are two common approaches to allow for the effect of fatigue. One is to reduce the allowable working stresses; another is to increase the computed forces. The allowance depends on the range of stress fluctuation, the number of repetitions, the quality of steel, and the type of connection.

British Standards for the design of highway bridges give values for the reduction of basic allowable stresses to take care of the fatigue effects. Values vary with the type of steel, the type of connection, the nature of stress, and the ratio of minimum to maximum stress.

American practice for fatigue allowance has been based on an increase in design forces except for welds, which are designed on reduced allowable stresses as described in Chapter 6. Both AASHO and AREA Specifications do not consider fatigue when there is no stress reversal, since basic allowable stresses are low enough to take care of such conditions. For members subject to reversal of stress during the passage of the live load, the following method of design is usually prescribed:

Determine the maximum stress of one type (tension) and the maximum stress of the opposite type (compression) and increase each by 50% of the smaller. Proportion the member so that it will be capable of resisting either stress so increased. The connections shall be proportioned for the sum of the maximum stresses.

To take care of stress reversals in the web members of a truss, counter members are used. The actual distribution of stress among the counters is a statically indeterminate problem. But, for simplicity, it can often be assumed that one member is acting at a time, taking tension only. This assumption is nearly correct when the members are slender so that they buckle under compression and hence take only a small proportion of the stress. The exact amount of compressive force in the counters can be computed if desired, taking into account the buckling and shortening of such columns.

There are three methods to arrange these counters at their intersections.

One is to let them pass by each other with no connection between them. This would result in a set of slender members, generally unable to carry compression. A second method is to connect them back to back on a gusset plate either by bolting or by welding, and the third is to splice one of them at the intersecting gusset.

13-11 SECONDARY STRESSES IN TRUSSES

Practically all triangular trusses are designed on the assumption that the members carry direct stresses only. This assumption is true when no transverse loads are applied along the length of the members, and no moments are applied or transmitted at the joints. Direct axial stresses so computed are termed primary stresses, in contrast to bending stresses which are termed secondary.

Secondary stresses may be produced in members of a truss by the following conditions:

(*a*) Eccentricities in the member connections—when the centroids of the sections at a joint do not intersect at one point, moments will be produced. Trusses are generally detailed so that such eccentricities will be avoided or minimized.

(*b*) Torsional moments—introduced by members not lying in the plane of the truss, such as floor beams in bridges. For usual designs, such stresses are neglected.

(*c*) Transverse loads on a member, such as the weight of the member itself. These are considered only when they are appreciable.

(*d*) Truss distortion and rigidity of joints which together induce the bending of the members.

The term secondary stresses in trusses, when used in its narrower sense, is often intended to denote only those produced by truss distortion and the rigidity of joints. The magnitude of these secondary stresses varies greatly. For common forms of trusses with members of high slenderness ratios, the secondary stresses generally range from 5 to 25% of the primary stresses. For subdivided trusses and Warren truss with verticals, certain members may have secondary stresses as high as 40 to 100% of the primary stresses.

Although the magnitude of secondary stresses can be high, their significance is not necessarily comparable to that of primary stresses. High secondary stresses exist only in some of the members and then only in the extreme fibers at the ends of the members. Even when these localized stresses reach the yield point, they may not cause collapse of the structure. With the low value of basic allowable stresses used in design, such high localized stresses do not become a problem unless repeated often enough, in which case fatigue failure may occur.

When it is desired to limit or reduce secondary stresses resulting from truss distortions, the width of the members in the plane of bending should be reduced relative to the length of the members. It is wise to choose truss types with low secondary stresses, avoiding the use of vertical hangers and sub-division. Attempts have been made to design trusses with pin-connected joints rather than bolted or welded ones. Unfortunately, friction in the pin holes may not permit free rotation of the connecting members. In one special instance, the trusses were fabricated to the distorted forms under a given loading and the members were forced together by jacks during erection so that for one loading condition the members would straighten and have no secondary stresses.[11]

Both the AREA and the AASHO Specifications for steel bridges contain the following clause regarding secondary stresses:

"The design and details shall be such that secondary stresses will be as small as practicable. Secondary stresses due to truss distortion or floor-beam deflection usually need not be considered in any member the width of which, measured parallel to the plane of distortion, is less than $\frac{1}{10}$ of its length. If the secondary stresses exceed 4000 psi for tension members and 3000 psi for compression members, the excess shall be treated as a primary stress."

This clause clearly indicates that there is a certain amount of reserve in the basic design stresses to allow for secondary stresses of the usual magnitude. Only when secondary stresses exceed approximately 20% of the primary ones, need they be considered in design.

The analysis of trusses for secondary stresses by classical methods has been considered a complex problem. Modern numerical methods and electronic computers simplify the work greatly. These methods are given in other textbooks[12] and will not be discussed here.

13-12 END BEARINGS AND HINGES

End bearings and articulations for bridges can be classified into four types:

(*a*) Fixed bearings
(*b*) Hinged bearings
(*c*) Sliding or expansion bearings
(*d*) Hinged, linked, and roller-jointed articulations

A fixed-end bearing is capable of supplying a vertical and a horizontal

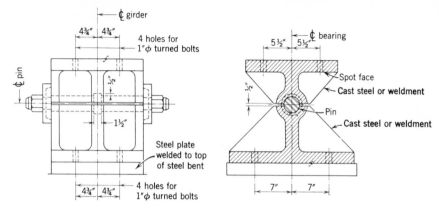

Fig. 13-21   Hinged bearing for medium spans. (Courtesy AISC.)

reaction plus a restraining moment. Because of the expense involved in fixing a heavy steel member at the ends, such a bearing is not usually designed for bridges.

A hinged bearing would permit rotation of the ends of the member. This is usually provided by a pin (Fig. 13-21). Hinges carrying heavy loads are provided with lubrication systems to reduce friction and ensure free rotation without excessive wearing.[13] A hinged bearing for spans under 100 ft is shown in Fig. 13-22.

Expansion bearings are of two types: the sliding joint and the roller joint. Sliding joints are used only for short spans and light loads. A typical

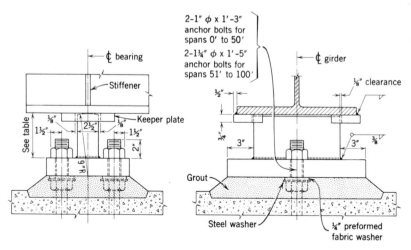

Fig. 13-22   Hinged bearing for short spans. (Courtesy AISC.)

example consists of a base plate bolted to the masonry foundation and a sole plate riveted with countersunk rivets or welded to the bottom flange of the girder or the lower chord of the truss. Although the two plates are generally planed and are supposed to slide freely on each other, the coefficient of friction may be high enough to develop a considerable amount of horizontal resistance, especially if dirt and rust accumulated between the plates. For greater stiffness and even distribution of bearing pressure, cast shoes or stiffened ribs may be employed.

To ensure free expansion of the bearings, rollers are installed. Two such bearings are shown in Figs. 13-23 and 13-24. Segmental rollers are preferred to round rollers, because they occupy less space than the round ones for a given diameter. Since the motion of the rollers is quite small relative to their radius, the arcs that are actually used are also small and a segmental roller is generally sufficient. Segmental rollers should be connected by two rods, which prevent the rollers from moving independently.

AASHO Specification, 1965, gives the following formula for allowable bearing p per linear inch on expansion rockers and rollers, for diameters d from 4 up to 25 inches:

$$p = \frac{F_y - 13,000}{20,000} 600d$$

where F_y = yield point in tension of steel in the roller or bearing plate whichever is the smaller.

When it is necessary to have a bearing that offers only normal reaction with no rotational or lateral restraint, a combination of hinge and roller bearing is used. The hinge permits rotation and the roller allows sliding.

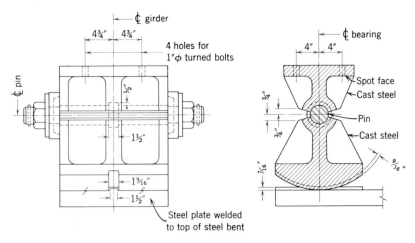

Fig. 13-23 Expansion bearing for medium spans. (Courtesy AISC.)

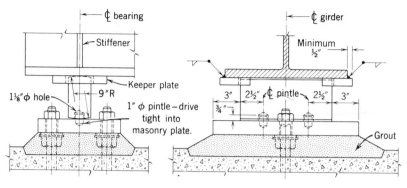

Note: See Fig. 11-22 for
additional dimensions

Fig. 13-24 Expansion bearing for short spans. (Courtesy AISC.)

For articulations within a truss or a girder, hinges are provided by either a pin or a link with two pins. A pin would permit rotational movement but transmitting both horizontal and vertical reactions. A link would transmit force only along the length of the link and gives a reliable means of controlling the direction of the reaction. Although it is possible to use an ordinary hinge or a roller joint in a big truss or girder, the link connection seems to be the more reliable, especially in view of the larger movement afforded by the link.

13-13 DESIGN OF A WELDED GIRDER BRIDGE

The use of all-welded steel girders and bents for the Division Street interchange structures of the San Francisco Freeway viaduct is described. Sample designs for an 82-ft span welded girder and a welded steel cap of a T-bent are given. This bridge was designed* by the Bridge Department, Division of Highways, State of California.

(*a*) *General Description.* Several miles of viaducts and distribution structures were built as part of the Bayshore Freeway system in San Francisco, by the State of California (Fig. 13-25). In order to conserve steel and to reduce the cost of construction, these were designed of all-welded simple-span girders, making use of their composite action with the concrete deck. Most of the spans were between 50 and 60 ft, using rolled beams. Cover plates were welded to the lower flanges of the rolled beams for spans from

---

* The actual bridge was designed for A7 steel but the design presented herein has been modified for A36 steel and 1965 AASHO Specification. (The original calculations were presented through the courtesy of Mr. Wendell Pond.)

Fig. 13-25   Steel viaduct—San Francisco Freeway. (Courtesy California State Division of Highways.)

70 to 85 ft. Welded plate girders were used for spans up to a maximum of 117 ft.

The bents supporting the girders were built of welded steel, having one, two, or three columns to a bent, depending on the length of the bent and its location. The single-column bents were preferred for their attractive appearance and were especially desirable when skew and curved crossings were encountered.

(*b*) *Loads and Specifications.* Live load: HS 20-44. In general, the following specifications are adhered to: Specifications for Highway Bridges, American Association for State Highway Officials; Specifications for Welded Highway and Railway Bridges, American Welding Society; additional specifications set up for welded plates over 1 in. thick.[14,15]

Typical design stresses are:

Tension in A36 steel 20,000 psi.
Reinforcing steel of intermediate grade designed for 20,000 psi.
Concrete 28-day cylinder strength $f_c' = 3000$ psi; allowable fiber stress in compression = 1200 psi: modular ratio $n = 10$.

(*c*) *Design of a Welded Girder.* A typical interior girder with a simple span of 82 ft will be designed, using composite action with its concrete slab of $6\frac{3}{4}$-in. thickness. The girders are spaced 8 ft-4 in. center to center (Fig. 13-26).

1. Design Moments
   Dead load:

Deck slab 0.563 × 8.33 × 0.150   = 0.705 kip/ft
Slab fillets   = 0.020
Steel girder, including diaphragms = 0.190   (estimated)

    Total dead load   = 0.915 kip/ft

$$\text{Maximum moment at midspan} = \frac{wL^2}{8} = \frac{0.915 \times 82^2}{8} = 770 \text{ kip-ft}$$

Live load plus impact: From Appendix A, AASHO Specification:

Maximum live load moment for one lane, 82-ft span = 1201 kip-ft

$$\text{Distribution of wheel loads to girder} = \frac{S}{5.5} = \frac{8.33}{5.5} = 1.52 \text{ wheel lines} =$$

$\dfrac{1.52}{2} = 0.76$ lane load

Maximum live-load moment per girder = 1201 × 0.76 = 915 kip-ft

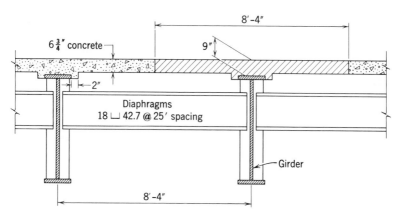

Fig. 13-26   Section of typical interior girders.

$$\text{Impact factor} = \frac{50}{L + 125} = \frac{50}{82 + 125} = 0.241$$

Maximum live load + impact moment = 915 × 1.241 = 1135 kip-ft

2. Section Properties of Girder, Midspan. For this design, it is assumed that temporary shoring or propping will not be placed under the steel girders before the deck slab concrete is placed. Hence the steel girder alone will carry the weight of the slab and the girder. The composite steel and concrete section will carry the live load and impact. Shrinkage and creep in concrete will affect the stresses in the structure but will be neglected in this design. Section for Dead load (Fig. 13-27): Try:

Top flange = 12 in. × $\frac{3}{4}$ in.
Web plate = 48 in. × $\frac{3}{8}$ in.
Bottom flange = 12 in. × $1\frac{3}{4}$ in.
Weight of girder section = 163 lb/ft
Weight of diaphragm and stiffener = 20

Total = 183 lb/ft   (190 assumed)

	Area, in.²	y	Ay	Ay^2	I_0
Top flange	9.00	50.12	451	22,600	—
Web plate	18.00	25.75	465	12,000	3460
Bottom flange	21.00	0.88	18	—	—
	$\sum A$ = 48.00 in.²		$\sum Ay$ = 934	34,600	3460

$$y_b = \frac{\sum Ay}{\sum A} = \frac{934}{48.00} = 19.5 \text{ in.}$$

$$\begin{array}{r} +3,460 \\ \hline 38,060 \end{array}$$

$$y_t = 50.5 - 19.5 = 31 \text{ in.} \qquad -y_b \times \sum Ay = 18,200$$

$$I_x = 19,860 \text{ in.}^4$$

Composite section for live loads (Fig. 13-28): To obtain properties of the composite girder, convert the effective area of the concrete deck to an equivalent area of steel. Use modular ratio $n = 10$.

Effective flange width = $\frac{1}{4}$ span length = $\dfrac{82 \times 12}{4}$ = 246 in.

or   = center-to-center girders   = 100 in.

or   = 12 × slab thickness = 12 × 6.75 = 81 in.   (use)

Equivalent slab area = 81 × $\frac{1}{10}$ × 6.75 = 54.68 in.²

$$I_0 = \frac{8.10(6.75)^3}{12} = 208 \text{ in.}^4$$

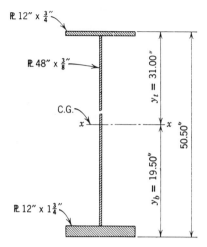

Fig. 13-27 Section for dead load at midspan.

	Area	y	Ay	Ay^2	I_0
Girder	48.00	19.50	934	18,200	19,860
Slab	54.68	55.37	3010	167,000	208

$$\sum A = 102.68 \qquad \sum Ay = 3944 \qquad 185,200 \qquad 20,068$$

$$y_b = \frac{\sum Ay}{\sum A} = \frac{3944}{102.68} = 38.30 \text{ in.} \qquad \begin{array}{r} +20,068 \\ \hline 205,268 \end{array}$$

$$y_t = 50.50 - 38.30 = 12.20 \text{ in.} \qquad -y_b \times \sum Ay = -151,000$$

$$I_x = \quad 54,268 \text{ in.}^4$$

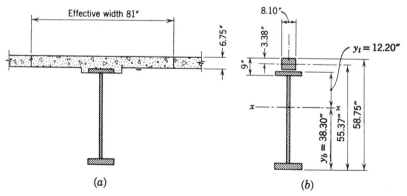

(a)   (b)

Fig. 13-28 Composite section at midspan: (a) actual section, and (b) transformed section.

3. Bending Stresses. Using the usual formula, $f = Mc/I$, we have the following:

	Top flange f_b	Bottom Flange f_b
Dead load	$\dfrac{770 \times 31.0 \times 12}{19,860} = 14.4$	$\dfrac{770 \times 19.5 \times 12}{19,860} = 9.1$
Live load	$\dfrac{1135 \times 12.2 \times 12}{54,268} = 3.1$	$\dfrac{1135 \times 38.3 \times 12}{54,268} = 9.6$
Total	$= 17.5$ ksi	18.7 ksi

$$\text{Allowable} = 20.0 \text{ ksi.}$$

Note that the bottom flange thickness may be reduced to $1\frac{5}{8}$ in.

4. Location of Flange Splices. For span lengths of this magnitude, it is economical to use a lighter steel section at the ends of the girder, since the plates would have to be spliced anyway. The usual practice is to use 40- to 50-ft maximum lengths of plate, and change flange plate thickness at the splices.

In this example the flange splices will be located approximately at the quarter point of the girder. The dead load and the envelope curve of live load plus impact moments are approximately parabolic in shape. The moments at the quarter point can therefore be assumed as three-fourths of the magnitude of the midspan moments. Using these moments a trial girder section can be obtained. The exact location of the splice can then be obtained graphically by plotting the resisting moment of the trial section on the dead load plus live load plus impact moment envelope curve.

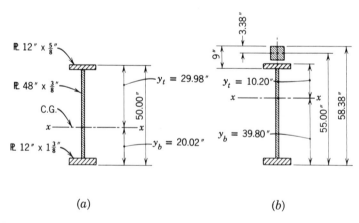

(a)   (b)

Fig. 13-29 Section at quarter span: (*a*) section for dead load, and (*b*) transformed composite section.

Approximate Moments and Trial Section

Dead-load moment $\quad = \frac{3}{4} \times 770 = 578$ kip-ft
Live load + impact moment $= \frac{3}{4} \times 1135 = 852$ kip-ft
Try: Top flange $\qquad\qquad = 12$ in. $\times \frac{5}{8}$ in. plate
$\quad$ Web plate $\qquad\qquad = 48$ in. $\times \frac{3}{8}$ in.
$\quad$ Bottom flange $\qquad\quad = 12$ in. $\times 1\frac{3}{8}$ in.   (Fig. 13-29)

Section for Dead Load

	Area	y	Ay	Ay^2	I_0
Top flange	7.50	49.69	373	18,500	—
Web plate	18.00	25.38	457	11,600	3460
Bottom flange	16.50	0.69	11	—	—

$$\sum A = 42.00 \text{ in.}^4 \qquad \sum Ay = 841 \qquad 30{,}100 \qquad 3460$$

$$y_b = \frac{\sum Ay}{\sum A} = \frac{841}{42.00} = 20.02 \text{ in.} \qquad \frac{3{,}460}{33{,}560}$$

$$y_t = 50.00 - 20.02 = 29.98 \text{ in.} \qquad -y_b \times \sum Ay = 16{,}800$$

$$I_x = 16{,}760 \text{ in.}^4$$

Composite Section for Live Load

	Area	y	Ay	Ay^2	I_0
Girder	42.00	20.02	841	16,800	16,760
Slab	54.68	55.00	3007	165,400	208

$$\sum A = 96.68 \qquad \sum Ay = 3848 \qquad 182{,}200 \qquad 16{,}968$$

$$y_b = \frac{\sum Ay}{\sum A} = \frac{3848}{96.68} = 39.80 \text{ in.} \qquad \frac{16{,}968}{199{,}168}$$

$$y_t = 50.00 - 39.80 = 10.20 \text{ in.} \qquad -y_t \times \sum Ay = 153{,}150$$

$$I_x = 46{,}018 \text{ in.}^4$$

Bending Stresses

	Top Flange f_b	Bottom Flange f_b
Dead load	$\dfrac{578 \times 29.98 \times 12}{16{,}760} = 12.4$	$\dfrac{578 \times 20.02 \times 12}{16{,}760} = 8.3$
Live load	$\dfrac{852 \times 10.20 \times 12}{46{,}018} = 2.3$	$\dfrac{852 \times 39.80 \times 12}{46{,}018} = 8.9$
Total	$= 14.7$ ksi	17.2 ksi

Theoretically it would be possible to reduce the size of the upper flange in

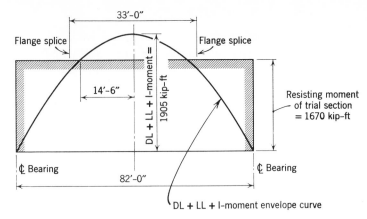

Fig. 13-30   Location of flange splices.

order to make the upper and lower flange stresses more nearly equal. However, experience indicates that girders having small flanges are subject to considerable distortion during fabrication and are too limber to handle during construction. Hence the trial section will be adopted.

Location of Flange Splices.   The resisting moment for 20.0 ksi = 20.0/ 17.1 × (578 + 852) = 1670 kip-ft. The location of the splice is determined graphically as shown in Fig. 13-30. Details of the splice are shown in Fig. 13-36.

5. *Shear Connectors.*   The shear connectors for connecting slab to girder will be $\frac{7}{8}$-in. stud bolts 4-in. long. The allowable load on each bolt will be 4.6 kips, based on Sec. 1.7.101, 1965 AASHO Specification using the formula $Z = 110d^2\sqrt{f_c'}$, with $d = \frac{7}{8}$ in. and $f_c' = 3000$ psi.

The spacing of connectors along the girder will be obtained graphically by plotting shear resistances of connectors on the live-load + impact shear diagram.

Shear Diagram.   At center line bearing, from Appendix A of AASHO Specification, shear for one lane = 63.8 kips.

V due to live load + impact = 1.241 × 0.76 × 63.8 = 60.3 kips per girder At midspan, from Fig. 13-31,

$$V \text{ due to live load for one lane} = \frac{32 \times 41 + 32 \times 27 + 8 \times 13}{82} = 27.8 \text{ kips}$$

V due to live load + impact = 1.241 × 0.76 × 27.8 = 26.3 kips per girder

Shear Resistance of Connectors.   The resistance value of one $\frac{7}{8}$-in. stud bolt = 4.6 kips. Use four studs in a group arranged as shown in Fig. 13-32. The resistance value of the group is 4 × 4.6 = 18.4 kips.

The resisting shear for a given spacing of connectors is found by rearranging the formula for longitudinal shear, $q = VQ/I$, where q equals resistance

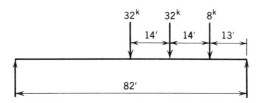

Fig. 13-31 Loading position for maximum shear at midspan.

value divided by spacing:

$$\text{Resisting shear } V = \frac{q \times I}{Q}$$

$$q = 18.4 \text{ kips}; \quad I = 46,018 \text{ in.}^4$$

$$Q = \text{area slab} \times \bar{y} = 54.68 \times 15.20 = 832 \text{ in.}^3$$

Resisting Shear V:

For 15 in. spacing: $\dfrac{18.4 \times 46,018}{15 \times 832} = 67.7 \text{ kips}$

For 18 in. spacing: $= 56.3 \text{ kips}$

For 24 in spacing: $= 42.4 \text{ kips}$

The layout of connectors is obtained from Fig. 13-33 and shown in Fig. 13-34.

In accordance with AASHO Specification, the maximum spacing allowed is 24 in.

The resisting shear of the connectors on the larger section used for the central portion of the span will be slightly greater than for the end sections. But the effect is small, and the foregoing calculation based on the end section will be considered satisfactory.

6. *Intermediate Stiffeners.* Intermediate stiffeners will be plates welded to only one side of the web. The design will be in accordance with Art. 1.7.72 of AASHO Specification.

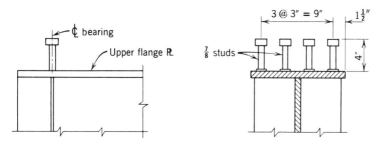

Fig. 13-32 Shear connector details.

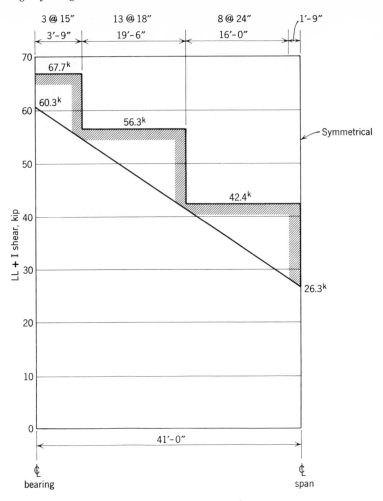

Fig. 13-33   Shear connector spacing.

Stiffener Spacing.   Shear at center line bearing:

$$\text{Dead load} = 0.915 \times 41 = 37.5 \text{ kips}$$
$$\text{Live load} + \text{impact} = 60.3$$
$$\text{Total} = 97.8 \text{ kips}$$

$$\text{Unit shearing stress in web } f_v = \frac{V}{th} = \frac{97.8}{0.375 \times 48} = 5.45 \text{ ksi}$$

$$\text{Allowable} = 11.0 \text{ ksi}$$

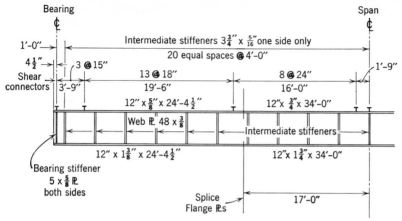

Fig. 13-34 Half-girder elevation.

Maximum stiffener spacing (use the smaller value)

= (1) 12 ft 0 ins.

(2) Unsupported depth of web = 4 ft 0 in.

$$(3) \quad \frac{11,000t}{\sqrt{f_v}} = \frac{11,000 \times 0.375}{\sqrt{5,450}} = 56 \text{ in.} = 4 \text{ ft 8 in.}$$

The value of 4 ft 0 in. controls and 20 equal spaces at 4 ft 0 in. will be used, with end spaces at 12 in. (Fig. 13-34).

Size of Stiffeners

Minimum width: $2'' + \dfrac{D}{30} = 2'' + \dfrac{48 + 2.6''}{30} = 3.7$ in., try $3\frac{3}{4}$ in.

Minimum thickness $= \frac{4}{16} = 0.25$ in.

From Art. 1.7.14, use $\frac{5}{16}$ in., minimum thickness. Check for required $I_{\min} = dt^3 J/10.9$

d = stiffener spacing = 48 in.

$t = 0.375$ in.

$$J = 25\left(\frac{D}{d}\right)^4 - 20 = 25\left(\frac{48}{48}\right)^4 - 20 = 5.0$$

$$I_{\min} = \frac{48(0.375)^3 \times 5.0}{10.9} = 1.16 \text{ in.}^4$$

$$I \text{ furnished} = \frac{bd^3}{3} = \frac{5}{16}\frac{(3.75)^3}{3} = 5.5 \text{ in.}^4$$

Use $3\frac{3}{4}$ in. $\times \frac{5}{16}$ in. plates.

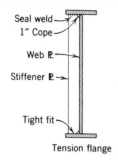

Seal weld
1" Cope
Web ℞
Stiffener ℞
Tight fit
Tension flange

Fig. 13-35 Typical girder section.

Size of Welds. Since there is no stress in the fillet weld connecting intermediate stiffeners to the web, the size of fillet will be nominal and in accordance with the AWS Specification. For $\frac{3}{8}$ in. plate thickness, the minimum weld is $\frac{3}{16}$ in. Use continuous $\frac{3}{16}$ in. fillet weld each side.

Transverse fillet welds should not be used on a tension flange, and so the bottom edge of stiffeners will be tightly fitted ($\frac{1}{16}$ in. maximum gap) to the upper surface of the bottom flange.

A seal weld will be used to fasten the upper edge of the stiffener to the upper flange (Fig. 13-35).

7. Bearing Stiffeners

Size of Stiffeners. Bearing stiffeners at the ends of girders are designed to carry the dead load plus live load plus impact girder reaction.

$$\text{Dead load} + \text{live load} + \text{impact reaction} = 97.8 \text{ kips}$$

$$\text{Area required for compressive stress} = \frac{97.8}{18} = 5.43 \text{ in.}^2$$

$$\text{Area required for bearing on flange} = \frac{97.8}{27} = 3.62 \text{ in.}^2$$

Use two plates 5 in. $\times$ $\frac{5}{8}$ in., one each side.

$$\text{Gross area for compression} = 6.25 \text{ in.}^2$$
$$\text{Net area for bearing (deduct 1 in. copes)} = 6.25 - 2 \times \tfrac{5}{8} = 5.00 \text{ in.}^2$$

Size of Welds. There will be a total of four fillet welds which must develop shear between web and stiffeners.

$$\text{Stress in weld} = \frac{97.8}{4 \times 48} = 0.51 \text{ kip/in. of weld}$$

$$\text{Minimum fillet for } \tfrac{5}{8}\text{-in. plate} = \tfrac{1}{4} \text{ in.}$$

From AWS Specification, $\frac{1}{4}$-in. fillet weld is good for 2.19 kip/in., which is more than required.

8. Flange-Web Weld. The flanges will be connected to the web plate by fillet welds on each side of the web (Fig. 13-36). The welds will be continuous as recommended by AWS Specification.

The welds will be designed to carry the longitudinal shear at the junction of flange and web.

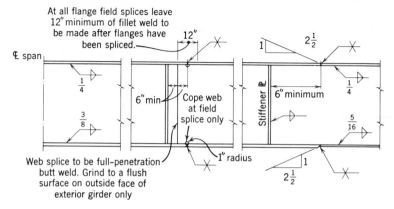

Fig. 13-36 Girder splice detail.

Top Flange
Dead load shear flow:

$$q = \frac{V \cdot Q}{I} \qquad \text{where} \quad Q = A\bar{y}$$

$$= \frac{37.5 \times 7.50 \times 29.67}{16{,}760} \qquad = 0.50 \text{ kip/in.}$$

Live load + impact:

$$q = \frac{60.3 \times (7.50 \times 9.89 + 54.68 \times 15.20)}{46{,}018} = 1.18$$

$$\text{Total } q = \overline{1.68} \text{ kip/in.}$$

Bottom Flange
Dead load shear flow:

$$q = \frac{37.5 \times 16.50 \times 19.33}{16{,}760} = 0.72 \text{ kip/in.}$$

Live load + impact: $q = \dfrac{60.3 \times 16.50 \times 39.11}{46{,}018} = 0.85$

$$\text{Total } q = \overline{1.57} \text{ kip/in.}$$

$$\text{Maximum stress on one weld} = \frac{1.68}{2} = 0.84 \text{ kip/in.}$$

From AWS Specification, the minimum fillet weld that can be used on $\frac{5}{8}$-in.-thick plate is $\frac{1}{4}$ in. The allowable shear on $\frac{1}{4}$-in. fillet is 2.19 kip/in. The minimum fillet weld on the $1\frac{3}{8}$-in.-thick plate is $\frac{5}{16}$-in. The minimum on the $1\frac{7}{8}$-in. plate is $\frac{3}{8}$-in.

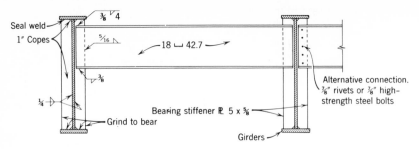

Fig. 13-37   End diaphragm.

9. End and Intermediate Diaphragms.   The design of these diaphragms will not be presented here.   Details are shown in Figs. 13-37 and 13-38.

(d) Design of a Steel Bent Cap.   The following calculations will illustrate the design of a steel cap of a T-bent used in the viaduct.   A general drawing of the bent is shown in Fig. 13-39.   Note that this bent supports one simple girder span of 59 ft 5 in. on each side.   It does not support the 82-ft span described in the foregoing Sec. (*c*).

1. Loads on Bent.   The dead load reactions from the girder on the bent are as follows: 81.5 kips for exterior girder, 59.5 kips for interior girder.

The live load + impact reaction on the bent will be computed as follows: Try truck loading (Fig. 13-40).

$$\text{Bent reaction} = 32 + \frac{45.4}{59.4}(8 + 32)$$

$$= 62.6 \text{ kips}$$

Try lane loading (Fig. 13-41):

$$\text{Bent reaction} = 26 + 0.64 \times 59.4$$

$$= 64.0 \text{ kips}$$

Lane loading is critical.

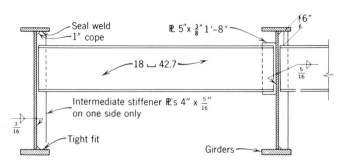

Fig. 13-38   Intermediate diaphragm.

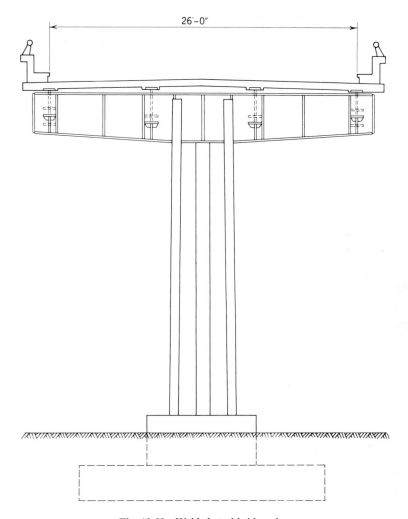

Fig. 13-39 Welded steel bridge pier.

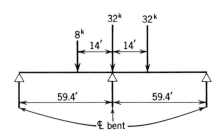

Fig. 13-40 Truck loading for maximum bent reaction.

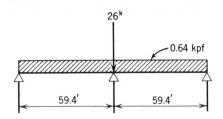

Fig. 13-41 Lane loading for maximum bent reaction.

Impact: (using $L = 59.4$ ft, conservatively)

$$I = \frac{50}{L + 125} = \frac{50}{59.4 + 125} = 0.271$$

live load + impact reaction = $1.271 \times 64.0 = 81.4$ kip/lane

The dead- and live-load plus impact reactions are shown in Fig. 13-42.

2. Moments and Shears

Dead-load shear: Girder reaction: $59.5 + 81.5 = 141.0$ kips

Weight of cap = $0.32 \times 11.75 = \underline{3.8}$

Total = $\overline{144.8}$ kips

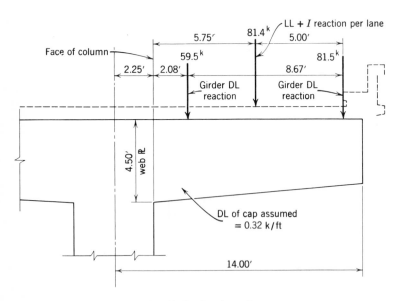

Fig. 13-42 Loads on bent.

Dead-load moment: Girder reaction: $59.5 \times 2.08 =$ 124 kip-ft

$$+81.5 \times 10.75 = \quad 877$$

$$\text{Weight of cap} = \frac{0.32(11.75)^2}{2} = \quad 22$$

$$\text{Total} = 1023 \text{ kip-ft}$$

Live load + impact shear = 81.4 kips

Live load + impact moment = $81.4 \times 5.75 = 468$ kip-ft

Summary

$$\text{Dead load} = 144.8 \text{ kips}$$

$$\text{Live load + impact shear} = \quad 81.4$$

$$\text{Dead load + live load + impact shear} = 226.2 \text{ kips}$$

$$\text{Dead-load moment} = 1023 \text{ kip-ft}$$

$$\text{Live load + impact moment} = \quad 468$$

$$\text{Dead load + live load + impact moment} = 1491 \text{ kip-ft}$$

3. Design of Maximum Section.   The cap will be designed for the stresses at the face of the column. Since both maximum moment and shear occur simultaneously at this section, principal tensile stresses must be investigated. The lower flange must also be checked for allowable compression.

In order to facilitate erection, it is desirable to use a narrow upper flange for the cap so that there will be ample clearance when lowering the girders into position on the beam seats. The width of the bottom flange, however, should be large as the allowable compressive stress in the lower flange is increased with greater width. Thus it is decided to use an upper flange narrower than the lower one.

A trial section is chosen, as shown in Fig. 13-43, with steel plates of equal areas but different dimensions for the upper and lower flanges. The neutral axis of this section is considered to

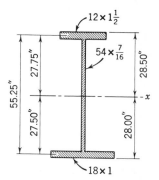

Fig. 13-43 Trial section for maximum moment.

be located at the middepth of the web plate, although actually slightly
above it.

$$\text{Web plate} = 54 \text{ in.} \times \tfrac{7}{16} \text{ in.}$$
$$\text{Upper flange} = 12 \text{ in.} \times 1\tfrac{1}{2} \text{ in.}$$
$$\text{Lower flange} = 18 \text{ in.} \times 1 \text{ in.}$$
$$\text{I of web} = 5,740 \text{ in.}^4$$
$$\text{I of flanges} = 2 \times 18 \times 27.6^2 = 27,470$$

$$\overline{}$$
$$I_x = 33,210 \text{ in.}^4$$
$$Q = 18 \times 27.75 = 500 \text{ in.}^3$$

Diagonal Tension in Web

$$f = \frac{Mc}{I} = \frac{1491 \times 27.00 \times 12}{33,210} = 14.5 \text{ ksi}$$

$$f_v = \frac{VQ}{It} = \frac{226.2 \times 500}{33,210 \times 0.438} = 7.8 \text{ ksi}$$

$$f_{\max} = \frac{f}{2} + \sqrt{\left(\frac{f}{2}\right)^2 + f_v^2} = \frac{14.5}{2} + \sqrt{\left(\frac{14.5}{2}\right)^2 + (7.8)^2} = 17.9 \text{ ksi}$$

allowable 20.0 ksi

Shear in Web

$$f_v = \frac{226.2}{54 \times 0.438} = 9.56 \text{ ksi} \qquad \text{allowable 12.0 ksi}$$

Compressive Stress in Bottom Flange

$$f = \frac{Mc}{I} = \frac{1491 \times 28.0 \times 12}{33,210} = 15.0 \text{ ksi}$$

$$\text{Allowable stress} = 20,000 - 7.5\left(\frac{L}{b}\right)^2$$

For computing the allowable compressive stress in a cantilever beam, it is
the usual practice to use L equal to twice the actual length of cantilever.

$$20,000 - 7.5\left(\frac{11.75 \times 12 \times 2}{18}\right)^2 = 20,000 - 2,050 = 17,950 \text{ psi}$$

The above stress computations indicate that the diagonal tension of 17.9 ksi controls the design. For an allowable tension of 20.0 ksi, the section may be reduced by about $\frac{1}{10}$ if desired.

Flange-Web Weld. The flanges will be connected to the web by continuous fillet welds on each face of the web. The welds must be designed to carry the longitudinal shear at the junction of flange and web.

$$q = \frac{VQ}{I} = \frac{226.2 \times 500}{33,210} = 3.42 \text{ kip/in.}$$

$$\text{Stress in one weld} = \frac{3.42}{2} = 1.71 \text{ kip/in.}$$

From AWS Specification, the minimum fillet weld that can be used on 1-in. and $1\frac{1}{2}$-in. thickness of plate is $\frac{5}{16}$-in. The allowable shear on a $\frac{5}{16}$-in. fillet is 2.74 kip/in. Use $\frac{5}{16}$-in. weld.

13-14 DESIGN OF A CANTILEVER TRUSS BRIDGE*

This example illustrates some main design features of the second Carquinez Bridge, California, which incorporated the use of high-tensile-strength steels and the combination of welding and bolting. Sample designs for a stringer, a floor beam, a tension and a compression member of the main truss are shown.

(*a*) *General Description.* The second Carquinez Bridge was built in 1957–1958 (Fig. 13-44) to parallel the first one. It is located about 200 ft from the first and hence has virtually the same layout. The main portion of the super-structure is 3350 ft long, consisting of two end anchor arms each 500 ft, two central spans each 1100 ft, and a central tower of 150 ft (Fig. 13-45). Each main span has two cantilevers at 333 ft each, supporting a suspended span of 433 ft.

The trusses are spaced 60 ft between centers and support a 52-ft four-lane concrete roadway with two 1-ft $10\frac{1}{2}$ in. steel curbs. The roadway slab for the two anchor arms is of normal-weight concrete; that for the remaining portion is of lightweight concrete at 100 pcf.

Three types of steel were used in the design of the actual superstructure: structural low-carbon steel A7, low-alloy steel A242, and high-tensile-strength steel T1, produced by the U.S. Steel Corporation. In the following

---

* This bridge was designed by the Bridge Department, Division of Highways, State of California. The calculations presented here are based on design data supplied by the courtesy of A. L. Elliott, Bridge Engineer.

Fig. 13-44   Carquinez Straits Bridge. (Courtesy U.S. Steel Corp.)

calculations, A36 steel has been substituted for the A7 steel and the 1965 AASHO Specification is followed where applicable. The use of T1 steel for the critical members of the truss reduced the size of the members considerably, and hence the secondary stresses were brought within limits. A substantial saving was thus effected.

When the first Carquinez Bridge was built in 1927, only riveted connections were used. Later experiences indicated the feasibility and economy in fabrication by welding. For welded truss members, the number of member shapes was reduced to three plates for H sections and to four or five plates for box sections, whereas many angles and plates would be required for riveted sections. Continuous automatic fillet welds were economically substituted for stitch rivets. Maintenance was made easier by the smooth surfaces free from rivet heads, lacing bars, and other details. Hence for this second bridge all major fabrication in the shop was done by welding.

All field connections at truss joints were made of high-tensile-strength bolts. It was believed that the high-tensile-strength bolts would provide reliable field joints at reduced cost.

For certain shop connections, where either riveting or bolting would be required, the contractor was allowed the option of using one or the other. The contractor, American Bridge Division, chose bolting, thus eliminating riveting from the entire job.

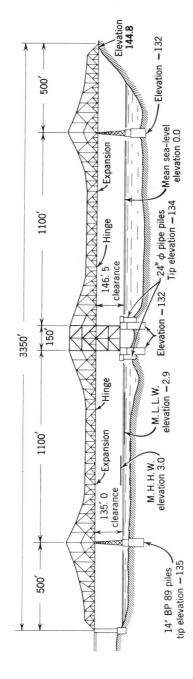

Fig. 13-45 Elevation of second Carquinez Bridge.

Elevation 144.8

Elevation −132

Mean sea-level
elevation 0.0

Expansion

Hinge

146.'5
clearance

24" φ pipe piles
Tip elevation −134

Elevation −132

M.L.L.W.
elevation −2.9

Hinge

Expansion

135.'0
clearance

M.H.H.W.
elevation 3.0

14" BP 89 piles
tip elevation −135

500'

1100'

150'

1100'

500'

3350'

(b) *Specifications and Allowable Stresses.* The bridge was designed for HS 20-44 loading. In general, the AASHO Specification for Highway Bridges was followed. Since the Specification was not meant for such long-span bridges, and did not provide for T1 steel, certain additional clauses had to be provided.

The allowable tensile stresses are as follows:

A36:   20,000 psi for all thicknesses.
A242: 27,000 psi for less than $\frac{3}{4}$ in. thickness.
        25,000 psi for $\frac{3}{4}$ in. to $1\frac{1}{2}$ in.
T1:   45,000 psi for all thicknesses.

The design of compression members, assuming riveted or bolted ends, was based on the formulas in the AASHO Specification for A36 and A242 steels. For high-strength alloy steel T1 the following formula was used:

$$\frac{P}{A} = 36{,}000 - 1.75\left(\frac{L}{r}\right)^2, \qquad \text{valid for } \frac{L}{r} \text{ values from 0 to 75}$$

In order to avoid local buckling of thin plates, formulas were derived for limiting the width-to-thickness ratios similar to values given in the AASHO Specification and other references.[16] For the purpose of this example criteria defined in Sec. 10-7, Table 10-5, will be used.

(c) *Design of an Interior Stringer.* A cross section of the roadway is shown in Fig. 13-46. An interior stringer with span of 27.78 ft will be designed for a panel of the cantilever arm.

Dead load:

7-in. slab of 100 pcf concrete	= 58.3 psf
$\frac{1}{2}$-in. future wearing surface	= 6 psf
	64.3 psf

For stringer spacing at 6 ft, 6 × 64.3 = 386 plf
Weight of stringer, assuming 24 W^F 76 = 76

Total dead load   = 462 plf

$$\text{Maximum moment} = \frac{wL^2}{8} = 0.462 \times \frac{27.78^2}{8} = 44.6 \text{ kip-ft}$$

Live load: Maximum moment per lane = 252 kip-ft (from AASHO table).

Distribution of wheel load for stringer at 6-ft spacing:

$$\frac{S}{5} = \frac{6}{5} = 1.2 \text{ wheel line} = 0.6 \text{ lane}$$

$$0.6 \times 252 = 151.2 \text{ kip-ft}$$

$$\text{Impact: } I = \frac{50}{27.78 + 125} = 32.8 \quad \text{use } 30\%$$

$$0.30 \times 151.2 = 45.4$$

Total dead load = 44.6 kip-ft
Live load   = 151.2
Impact   = 45.4

Total   = 241.2   kip-ft   use 241.

$$\text{Section modulus required} = \frac{241 \times 12,000}{20,000} = 145 \text{ in.}^3$$

Use 21 W⸏ 73 with section modulus of 150.7 in.[3]

(*d*) *Design of a Floor Beam.* A typical interior floor beam will now be designed for the cantilever span. The floor beam spans 60 ft and carries the 27.78-ft stringers (Fig. 13-46).

Dead load from stringers:

Interior stringers: 64.3 × 6 × 27.78 ft = 10.72 kip
Stringer weight   73 × 27.78 = 2.04

                                 12.76 kip

Exterior stringers: 64.3 × 5 ft × 27.78 = 8.93 kip
Stringer weight: 2.04

                                 10.97 kip

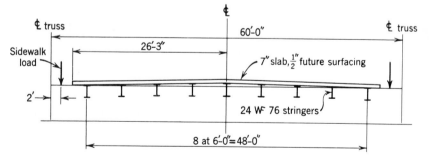

Fig. 13-46 Dead load on floor beam.

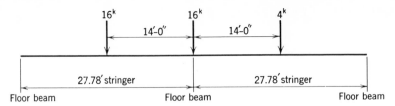

Fig. 13-47 Live load for one wheel line producing maximum reaction on floor beam.

Sidewalk load is computed as 4 kips. Assuming weight of floor beam at 320 plf, the maximum dead load moment and shear can be computed as

Maximum shear $V = 70.0$ kips,   maximum moment $= 1120$ kip-ft

Live load and Impact: The maximum floor beam reaction from each stringer is computed from Fig. 13-47. For each wheel line,

$$R = \frac{13.78}{27.78} \times 16 + \frac{13.78}{27.78} \times 4 + 16 = 25.94 \text{ kips}$$

The impact factor for length of 60 ft is 30%; hence,

R for live load + impact $= 1.3 \times 25.94 = 33.75$ kip/wheel line

If it is assumed that each interior stringer will have the maximum reaction, then the live load and impact on the floor beam will be as shown in Fig. 13-48. A reduction of 25% is allowed by the AASHO Specification for four lanes. The resulting maximum moment and shear are, therefore,

$V = 4 \times 33.75 \times 0.75 = 101$ kips
$M = [4 \times 33.75 \times 28 - 33.75(18 + 12 + 6)] \times 0.75 = 1920$ kip-ft

If only three lanes are considered to be loaded, omitting the extreme right lane in Fig. 13-50, a reduction of 10% is applied to the loading. We obtain

$V = 3.566 \times 33.75 \times 0.90 = 108$ kips
$M = [3.566 \times 33.75 \times 28 - 33.75(18 + 12 + 6)] \times 0.90 = 1940$ kip-ft

which turns out to be greater than when all four lanes are loaded.

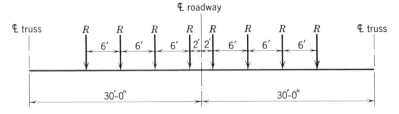

Fig. 13-48 Live load for maximum moment and shear in floor beam.

The maximum total moment for the girder is therefore $1120 + 1940 = 3060$ kip-ft and maximum end shear is $70 + 108 = 178$ kip. Note: the fact that the maximum dead- and live-load moments do not exist at the same point is ignored, thus yielding slightly conservative values.

The floor beam is assumed to be 66 in. deep, and the required minimum web thickness is

$$t = \frac{D}{140} = \frac{66}{140} = 0.47 \text{ in. or } \tfrac{1}{2} \text{ in.}$$

The required web area. using A242 steel. is $V = 178/15 = 11.9$ in.² Use 66 in. × ½-in. web.

For an approximate design of the flange, we can proceed as follows. Flange area required, including $\tfrac{1}{6}$ web, and assuming effective depth = 67 in., is

$$\frac{3070 \times 12,000}{67 \times 24,000} = 22.9 \text{ in.}^2$$

$$\text{Reduced by } \tfrac{1}{6} \times 66 \times \tfrac{1}{2} = 5.5$$

$$\text{Required area of each flange} = 17.4 \text{ in.}^2$$

In the final design, 16 in. × $1\tfrac{1}{4}$-in. plates are adopted for the central portion of 30 ft, and 16 in. × $\tfrac{7}{8}$-in. plates for the end quarter spans.

Since high-tensile-strength steel is used, it may be necessary to check for the deflections of the floor beam, to be sure that sufficient stiffness is obtained. These detailed calculations are not presented here.

Typical detail for the floor system is shown in Fig. 13-49.

(*e*) *Design of a Compression Member.* A lower chord member of the anchor span adjacent to the main tower will be designed. The computed stresses are as follows

Dead load	$= -2532$ kip
Live load	$= -963$
Impact	$= -54$
Dead load + live load + impact	$= -3549$ kip
30 psf wind load on bridge	$= \pm1484$ kip
Wind load on live load	$= \pm380$ kip
80% (dead load + live load + impact + wind load)	$= 0.80(5413) = -4330$ kip

The wind-load stresses are computed according to the AASHO Specification. Since an increase in the allowable stresses of 25% is permitted when considering wind load, the corresponding stresses are reduced, using a factor of $1/1.25 = 0.8$. Impact is computed to be 5.6% for a loaded length of 767 ft.

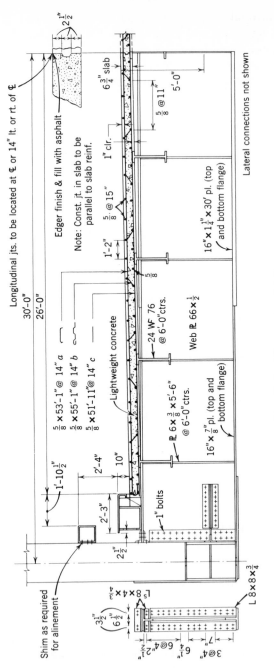

Fig. 13-49 Floor system details.

684

Secondary stresses resulting from the deflections of the truss are calculated and found to be insignificant and hence are neglected in the design.

The chosen section is shown in Fig. 13-50, using T1 steel,

2 plates 36 in. × 1¼-in.

      = 36 × 2.5 =   90.0 in.²

1 plate 24½-in. × 1 in.

      = 24.5 × 1 =   24.5

2 perforated plates 25 in.

   × ¾-in. with 14 in. × 28 in.

   perforations = 11 × 1.5   =   16.5

                       ‾‾‾‾‾

Total net area   = 131.0 in.²

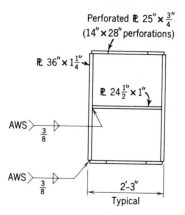

Perforated ℞ 25″ × ¾″
(14″ × 28″ perforations)

℞ 36″ × 1¼″

℞ 24½″ × 1″

AWS 3/8

AWS 3/8

2′-3″
Typical

Fig. 13-50 Section of a compression member.

Using the table from the AISC Manual, the moments of inertia are computed as follows:

$$I_x = 8.25 \times 678 \;=\; 5{,}570$$
$$2.50 \times 3888 = 9{,}720$$
$$\overline{}$$
$$15{,}290 \text{ in.}^4$$

$$I_y = 45.0 \times 331.5 = 14{,}920$$
$$1.0 \times 1227 \;=\; 1{,}230$$
$$1.5(1302.1 - 228.7) = 1{,}610$$
$$\overline{}$$
$$17{,}760 \text{ in.}^4$$

Minimum r is controlled by I_x

$$r = \sqrt{\frac{I_x}{A}} = 10.8 \text{ in.}$$

$$\frac{L}{r} = \frac{27.78 \times 12}{10.8} = 30.8$$

$$F_a = 36{,}000 - 1.75\left(\frac{L}{r}\right)^2 \text{ for T1 steel}$$

$$= 36{,}000 - 1.75 \times 30.8^2 = 34{,}340 \text{ psi}$$

$$P/A = \frac{4330}{131} = 33{,}100 \text{ psi.}$$

Check for plate thickness (see Table 10-5, Sec. 10-7).

Perforated Plates. One edge free, another hinged, permissible $w/t = 12$.

$$\frac{w}{t} = \frac{\frac{1}{2}(25 - 14)}{0.75} = 7.33 < 12$$

Side Plates and Web Plates. Both edges hinged, permissible $w/t = 36$.

side plates: $\dfrac{w}{t} = \dfrac{\frac{1}{2}(36)}{1.25} = 14.4 < 36$

web plates $\dfrac{w}{t} = \dfrac{24.5}{1.0} = 24.5 < 36$

(*f*) *Design of a Tension Member.* An upper chord of the anchor span adjacent to the main tower will be designed. The computed stresses are as follows:

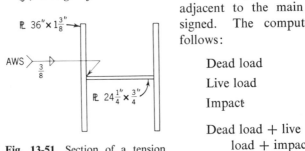

℞ 36″ × 1⅜″

AWS ⅜

℞ 24¼″ × ¾″

Fig. 13-51 Section of a tension member.

Dead load	$= +3862$ kip
Live load	$= + 965$
Impact	$= + 54$
Dead load + live load + impact	$= +4881$ kip

A 30 psf wind load produces a stress of only ±289 kips in this member and hence will not affect the design.

The chosen section is shown in Fig. 13-51, using T1 steel,

2 plates 36 in. × 1⅜-in. = 99.0 in.²

1 plate 24¼-in. × ¾-in. = 18.2 in.²

Gross area   = 117.2 in.²

The minimum moment of inertia is about the horizontal axis and is closely approximated by

$$I_x = \frac{2.75 \times 36^3}{12} = 10{,}700 \text{ in.}^4$$

$$\text{Minimum } r = \sqrt{\frac{10{,}700}{117.2}} = 9.56 \text{ in.}$$

$$\frac{L}{r} = \frac{120.75 \times 12}{2 \times 9.56} = 75.7 < 200$$

$$f_t = \frac{4881}{117.2} \times 1000 = 41{,}600 \text{ psi} < 45{,}000 \text{ psi}$$

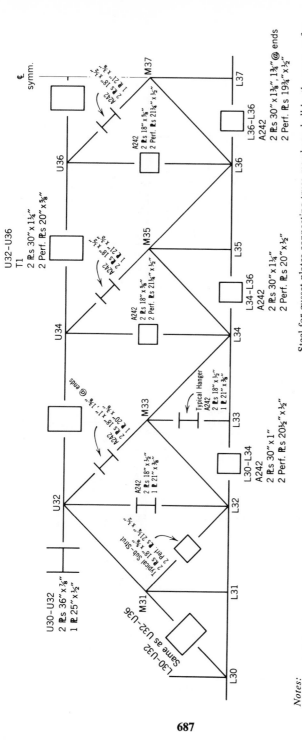

Notes:

All bolts shall be high-strength bolts.

All bolts in truss 1", $\frac{7}{8}$" except as otherwise shown.

All bolts in bracing $\frac{7}{8}$" except as otherwise shown.

The connections of diagonal U32-L34 & chords L34-L36 & L36-L37-L36 are to be accomplished for bolt holes. This is to be accomplished by butt welding a heavier flange plate ending 1'-0" from the first line of connecting bolts.

Steel for gusset plates connecting truss members shall be the same as for the member with the highest yield point steel at respective joints.

All other gusset plates shall be A7 steel.

All perforations in main truss members are to be 12 × 24 at 4'-0" min. ctrs.

Except as noted, the clear distance between the end perf. and the end of the cover shall be 2'-3" min. There shall be no perf. within 2'-3" of joints L31, L33, L35 and L37.

Fig. 13-52 Details of suspended span, Carquinez Bridge. (*a*) Member make-up.

687

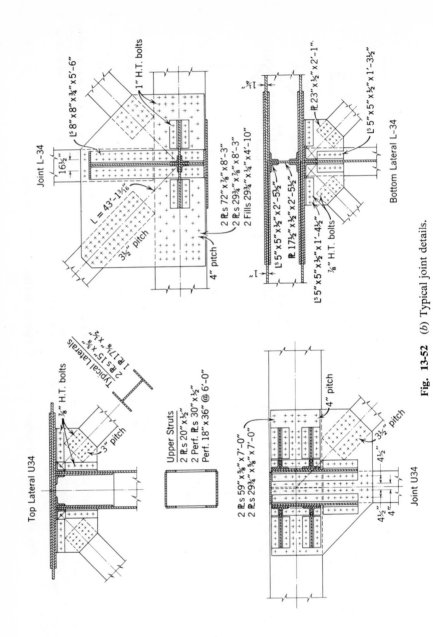

Fig. 13-52 (b) Typical joint details.

688

To compensate for loss of section at the connections, the end of the member is built up by welding on a heavier flange plate. This is accomplished by a 100% penetration butt weld at a distance of 1 ft 0 in. beyond the first line of connecting bolts. This weld is subjected to radiographic inspection. The make-up of this end section is as follows:

2 plates 36 × 1¾	= 126.00
1 plate 23½ × ¾	= 17.61
Gross area	= 143.61
Hole deduction 8 × 1⅛ × 1¾ × 2 =	31.50 in.²
Net area	= 112.11 in.²

$$f_t = \frac{4881}{112.11} = 43,500 \text{ psi} < 45,000 \text{ psi}$$

(*g*) *Joint Details of Suspended Span.* Typical member sections and connections for the 433-ft suspended span of this bridge are shown in Fig. 13-52. These are intended to illustrate the various connection details involved in a truss bridge.

PROBLEMS

In the following problems, the statements are not intended to be complete. Any additional data required are to be assumed by the students. Refer to AASHO Specification for Highway Bridges and to AREA Specification for Steel Railway Bridges. Unless specified, use riveting, bolting, or welding for the connections as desired.

1. A half cross section of a four-lane highway I-beam bridge is shown (Fig. P-1). The bridge has a simple span of 50 ft, carrying HS20-44 loading. Design and detail an interior beam, using a Wᶠ section of a desirable and economical depth, but not to exceed 30 in. Show diagrams connecting the beams and detail the beam supports.

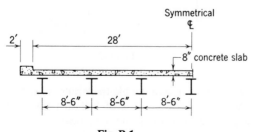

Fig. P-1

Use $\frac{7}{8}$-in. rivets. (*a*) Do not consider composite action between the beam and the slab. (*b*) Consider composite action and show the shear connectors.

2. A three-span steel I-beam bridge carries six lanes of highway traffic. It has cantilever spans with a profile shown in Fig. P-2. For the beams, it is decided to use rolled W sections spaced at 11-ft centers. Design the entire beam including the hinges. Use additional flange plates if required. Assume 9-in. concrete slab for the roadway. Neglect composite action. HS20-44 loading; lane loading will be considered sufficiently accurate for computation.

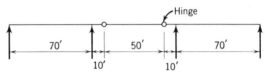

Fig. P-2

3. Figure P-3 shows a half cross section of a steel-deck plate-girder highway bridge, carrying HS20-44 loading on a simple span of 150 ft. Each girder is to be made up of plates of different sizes welded together and is limited to a maximum depth of 6 ft. Consider composite action with the slabs and suggest suitable shear connectors. The maximum length and weight of each piece of the girder is determined by transportation and erection requirements of the particular location. For example splices may be required either at midspan or at the third-points. Use high-tensile-strength bolts for field connections. Design an interior girder, complete with diaphragms, or cross-bracings, and support details.

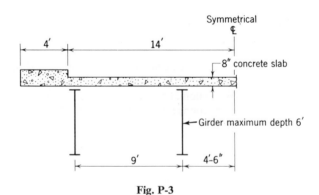

Fig. P-3

4. A steel truss bridge carries a double-track railroad and has a profile as shown in Fig. P-4. The floor system consists of ballasted tracks on a 10-in. concrete slab, which is supported on stringer and floor beams. Choose a suitable stringer spacing. Design an interior stringer and a floor beam. Detail the connections. Use

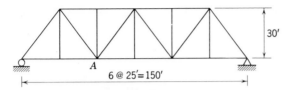

Fig. P-4

Cooper's E-72 loading. Compute dead, live, and impact stresses for the truss members meeting at joint A. Design and detail these members including the bottom lateral bracing. Use $\frac{7}{8}$-in. rivets.

5. Solve Prob. 4 for a highway bridge carrying two 14-ft lanes of HS20-44 traffic and two sidewalks each 4 ft wide. Reduce the depth of truss from 30 to 25 ft.

6. The elevation of a steel trestle for a bridge is shown in Fig. P-6. Each bent is to carry a vertical load P which may vary from a minimum of 1000 kips to a maximum of 1500 kips on each column, independently of the load on the other. In addition, maximum horizontal forces on the bent are shown and can act in either

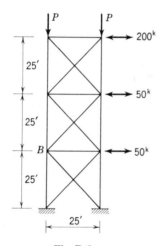

Fig. P-6

direction in the plane of the bent. Neglect the horizontal force acting perpendicular to the plane of the bent. Assume the diagonal members to carry only tension. Design the members of the bent meeting at joint B and detail the connections. Assume suitable sections for the members in the plane perpendicular to the bent.

7. The truss of a cantilever highway deck bridge (Fig. P-7) is to carry a uniform dead load of 4 kpf and a moving live load of 1.5 kpf including impact. The two trusses are spaced 40-ft centers carrying a four-lane roadway totaling 66 ft wide. Design the stringers and floor beam for a typical panel. Compute the

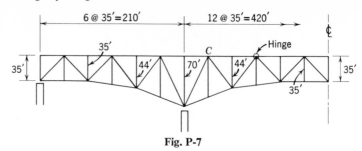

Fig. P-7

stresses in the members meeting at joint *C* and design them. Detail the joint, assuming suitable sizes for the lateral bracing members meeting at the joint.

REFERENCES

1. Waddell, J. A. L., *Bridge Engineering*, John Wiley and Sons, 1916.
2. Specifications for Steel Railway Bridges, American Railway Engineering Association.
3. Specifications for Highway Bridges, Association of American State Highway Officials, 1965.
4. Ivy, R. J., T. Y. Lin, S. Mitchell, N. C. Raab, V. J. Richey, and C. F. Scheffey, "Live Loading for Long-Span Highway Bridges," *Trans. ASCE* **119**, 981 (1954).
5. Farquharson, F. B., F. Smith, and G. S. Vincent, "Aerodynamic Stability of Suspension Bridges," *Univ. Wash. Eng. Expt. Sta. Bull.* **116**, in 5 parts (1949–1954).
6. *Welded Deck Highway Bridges*, 1950 and *Welded Highway Bridge Design*, 1952, James F. Lincoln Arc Welding Foundation.
7. Viest, I. M., R. S. Fountain, and R. C. Singleton, *Composite Construction in Steel and Concrete*, McGraw-Hill Book Co., 1958.
8. Wolchuk, R., "Orthotropic Plate Design for Steel Bridges," *Civil Eng.* **29**, No. 2 (February 1959).
9. Design Manual for Orthotropic Steel Plate Deck Bridges, AISC, 1963, New York.
10. Whitmore, R. E., "Experimental Investigation of Stresses in Gusset Plates," *Univ. Tenn. Eng. Expt. Sta. Bull.* **16** (1952).
11. Hool, G. A., and W. S. Kinne, *Movable and Long Span Steel Bridges*, McGraw-Hill Book Co., p. 207, 1923.
12. Maugh, L. C., *Statically Indeterminate Structures*, John Wiley and Sons, 1946.
13. *Highway Bridges of Steel*, American Institute of Steel Construction.
14. Hollister, L. C., "Design and Welding Practice for California's All-Welded Viaduct," *Civil Eng.* **22**, 181–185 (September 1952).
15. Binder, R. W., "Sequence and Continuity Mark Modern Welding Practice," *Civil Eng.* 186–191 (September 1952).
16. Sanders, A. L. R., "East St. Louis Veterans Memorial Bridge," *Trans. ASCE* **118**, 838 (1953).

14

Design of Light Gage Members

୨୨

14-1 INTRODUCTION

Light gage steel was first used in building construction in the United States about 1850, but it did not evolve as a building material until 1930. In its earliest uses light gage steel was limited to building components. Today it finds many applications in all types of industries, such as automotive, truck and trailer bodies, home appliances, railway cars, and various pieces of equipment.

Light gage steel members are used widely in structures subjected to light and moderate loads or short span lengths. For such structures the use of conventional hot-rolled shapes is often uneconomical because the stresses developed in the smallest available shape may be very low. The advantage of light gage members lie in the ease of forming a variety of shapes designed to use the material effectively and to simplify and speed up construction operations. Substantial economy can be achieved with mass production of standardized structural elements.

The shape of the light gage steel members varies with its application and engineers have learned to adapt this versatility to advantage in design of roof and floor panels, joists, and other individual structural members.

14-2 TYPES OF LIGHT GAGE STEEL MEMBERS

Light gage structural framing members are cold-formed from steel sheets or strips with thicknesses ranging from 18 gage (0.048) to about $\frac{1}{4}$ in. The usual shapes are channels, zees, angles, hat sections, tubular members, tees, and I sections (Fig. 14-1).

693

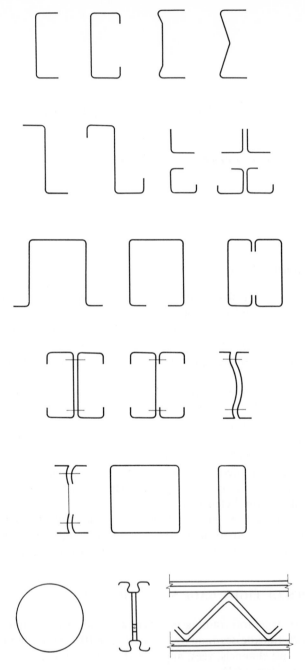

Fig. 14-1 Individual cold-formed structural sections.

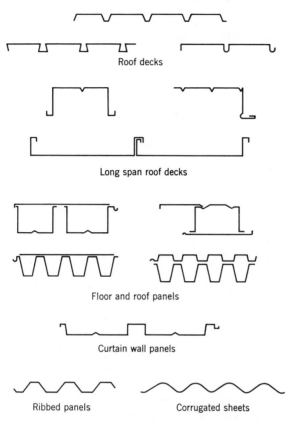

Roof decks

Long span roof decks

Floor and roof panels

Curtain wall panels

Ribbed panels   Corrugated sheets

Fig. 14-2   Decks, panels, and corrugated sheets.

These sections, 2 to 12 in., in depth can carry substantial loads and are used as primary framing members in buildings up to six stories in height.

Another category includes cold-formed sections generally manufactured in panel sizes to be used for roof and floor decks, siding, and walls. Several of these sections are illustrated in Fig. 14-2. Other shapes and configurations are possible and depend only on the use and cost of the roll forming equipment. The thicknesses used generally range from 26 gage (0.018 in.) to 14 gage (0.075 in.) for depths of panels of $1\frac{1}{2}$ in. to $7\frac{1}{2}$ in. The thickness of the uncoated sheets is designated by a nominal gage number, Table 14-1.

Table 14-1   Gage and Thickness of Uncoated Sheets

Gage no.	10	12	14	16	18	20	22	24	26	28	30
Thickness, in.	0.135	0.105	0.075	0.060	0.048	0.036	0.030	0.024	0.018	0.0149	0.0120

14-3 DESIGN CONSIDERATIONS

The use of thin light gage materials and the cold-forming process introduce a few additional factors to be considered in the design of structural components.[1,2] Unlike heavy steel construction which uses the hot-rolled structural shapes, the light gage members are so thin compared to their widths that buckling at low stress values will result under compression, shear, bending, and bearing. The critical buckling is generally of a local nature and precedes general buckling of the member.

Light gage design criteria are based on the post-buckling strength of the member after local buckling has occurred. Tests have verified that members will not necessarily fail when the elastic buckling stress is reached, but they will continue to carry additional load.

In compression members, a form factor Q is introduced which represents the effect of local buckling in reducing the column strength of the cross section. Because the cross sections are made of thin material with high width-to-thickness ratios and usually of an open section, they may be subject to torsional buckling or to torsional-flexural buckling, depending upon the relationship of the shear center to the centroid of the section.

The use of intermediate stiffeners or edge stiffeners can increase the load-carrying capacity of a cross section an appreciable amount. The effect of these stiffeners has been evaluated by tests. The function of a stiffener in a compression member is to increase the effective area of cross section by providing a reinforcement to a large width, thereby reducing its width-to-thickness ratio and increasing its critical stress.

The section properties of these cross sections are based on a reduced effective area for load determination.

As in conventional structural design, local effects such as type of connections, end bearing, and utilization of cold work must be considered.

The shapes which can be cold-formed are many and varied so that on occasion calculations for safe load capacity cannot be made. In these instances, the AISI Specification permits load tests to be substituted for calculations to determine the structural performance of the section.

Section properties of light gage steel shapes may be based on a linear or midline method. In this method the material is considered to be a line and area elements are considered as line elements. The thickness t is introduced after all linear computations are made.

Plastic design techniques do not apply to the light gage steel shapes because the width-thickness ratios are greatly in excess of those required for plastic hinges to develop, and local buckling would occur first. In addition, the stress-strain curve for the light gage steel grades does not exhibit the same characteristics as that of the structural grade steels.

14-4 ALLOWABLE STRESSES

The AISI Specification establishes a "basic design stress" F_b applicable to direct tension and compression in bending to be determined by dividing the yield point of the steel F_y by a factor of safety of 1.65:

$$F_b = \frac{F_y}{1.65}$$

This basic design stress, however, is applicable only to those conditions when the members fail by yielding. When the strength of a member is controlled by buckling, a reduced allowable stress or a reduced effective area must be used.

14-5 STRENGTH OF COLUMNS

The column strength of thin plates is defined as the load-carrying capacity of the member controlled by one or a combination of the following four types of failure: (a) crushing, (b) local buckling of thin-plate elements of the section over a short length of the column, (c) over-all or primary column buckling by lateral bending over the unsupported length of the member, and (d) torsional buckling or twisting of the section about a longitudinal axis.

Crushing Failure.   This type of failure can occur only in very short members, say L/r less than 20, with thickness of plate elements sufficient to prevent local buckling. Crushing begins when the stress in the column reaches the yield stress for the material. Light gage compression members are not likely to fail by crushing, and usually one or a combination of the other types of failure will govern the design.

Local Buckling Failure.   Pure local buckling failure occurs only in very short columns when L/r is less than 20. For columns of intermediate length, L/r varying approximately from 20 to 120, initial local buckling may occur at loads less than the ultimate, but the final failure is due to combined effects of local buckling and over-all primary column buckling.

Local buckling failure of short thin-wall columns occurs when all the plate elements of a column have developed their full load-carrying capacity. The latter is defined by the post-buckling strength for each "stiffened" plate element or by excessive distortion of an "unstiffened" element. Satisfactory prediction of the strength of thin compression elements is based on the results of an extensive experimental program.[3] Observation of the general behavior of thin-wall members tested in compression suggested the

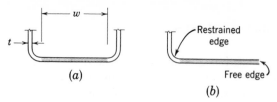

Fig. 14-3 Thin compression elements: (*a*) element stiffened along both edges and (*b*) unstiffened element.

classification of thin compression elements into two types: "stiffened" and "unstiffened" (Fig. 14-3). A "stiffened" element is a thin flat-plate element with both edges parallel to the direction of stress restrained by connection to web, flange, stiffening lip, etc. (Fig. 14-3*a*). An "unstiffened" element is free along one edge parallel to the direction of stress and restrained along the other (Fig. 14-3*b*).

Stiffened elements with small values of w/t fail by yielding without any evidence of buckling. Therefore, full area is effective for stresses up to yield value F_y. Stiffened elements with values of w/t greater than some limiting value $(w/t)_{\text{lim}}$ buckle slightly at loads below the yield. This buckling does not impair the load-carrying capacity of member, and failure occurs, at loads considerably greater than the initial buckling load, when one of the waves suddenly develops into a definite permanent "kink." Ultimate loads for stiffened elements can be predicted satisfactorily by replacing the actual flange width w by an effective width b_e subjected to uniform compressive stress f_m equal to maximum edge stress, as follows:

$$P_u = f_m A_e = f_m b_e t \qquad (14\text{-}1)$$

If the edge stiffener possesses adequate rigidity, the value of f_m may be taken equal to yield stress F_y.

The value of b_e may be determined from an empirical equation:

$$b_e = Ct\sqrt{\frac{E}{f_m}} = 1.9\left[1 - 0.475\sqrt{\frac{E}{f_m}}\,(t/w)\right]t\sqrt{\frac{E}{f_m}} \qquad (14\text{-}2)$$

Determination of b_e from Eq. 14-2 can be simplified by plotting a family of curves of b_e/t vs. w/t for various values of f_m (Fig. 14-4).

In order to provide adequate stiffening along a longitudinal edge, the stiffener—a web, flange, or lip—must have sufficient rigidity. This rigidity may be defined as a minimum moment of inertia I_m of the stiffener about its centroidal axis parallel to the stiffened element. The moment of inertia of a stiffening web should be not less than

$$I_m = 1.83t^4 \sqrt{\left(\frac{w}{t}\right)^2 - 144} \qquad (14\text{-}3)$$

but not less than $9.2t.^4$ Where the stiffener consists of a simple bent lip, the minimum depth d, (Fig. 14-5a), shall be not less than

$$d = 2.8t \sqrt[6]{\left(\frac{w}{t}\right)^2 - 144} \qquad (14\text{-}4)$$

but not less than $4.8t$.

In stiffened panels with large w/t ratios, much of the material is ineffective. In order to increase the effectiveness of such panels, intermediate stiffeners may be provided between webs (Fig. 14-5b). To be effective in stiffening both adjacent elements, these intermediate stiffeners must have a moment of inertia not less than twice that of an edge stiffener, defined by Eq. 14-3, where w/t is replaced by w_s/t, the width–thickness ratio of a subelement.

The effectiveness of a panel with intermediate stiffeners, Fig. 14-5b, where the w_s/t ratio does not exceed 60 is the same as that of an edge-stiffened element with the same w/t ratio. When the w_s/t ratio exceeds 60, the effective

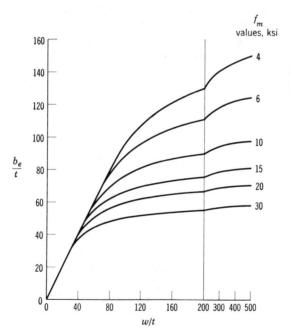

Fig. 14-4 Variation of effective width of stiffened elements. (From AISI Manual.)

width b_s' of subelement must be reduced as follows:

$$\frac{b_s'}{t} = \frac{b_s}{t} - 0.10\left(\frac{w_s}{t} - 60\right) \qquad (14\text{-}5)$$

where $b_s = b_e$ is the effective width of an edge-stiffened subelement defined by Eq. 14-2.

When more than one intermediate stiffener is used special limitations are indicated in the Specifications.[1]

Unstiffened elements with values of w/t ratios approximately less than 10 fail by yielding without any evidence of buckling. Therefore for such elements compressive strength f_m may be taken as F_y, provided that $(w/t)_1$ is less than 10 or $300,000/F_b$, whichever is smaller. This relationship between the compressive strength and w/t is shown in curve A, Fig. 14-6. In elements with w/t greater than 10 or $300,000/F_b$ but less than 25, local buckling may occur at stresses below the theoretical buckling stress. The compressive strength for such elements may be closely approximated by a straight line between yield stress F_y at $(w/t)_1$ and buckling stress f_{cr}, defined by Eq. 9-52 at w/t equal to 25, as shown by curve B in Fig. 14-6. In elements with w/t ratios between 25 and 60, gradual wavelike distortions may occur at stresses equal to or higher than the theoretical buckling stress f_{cr}, and the strength f_m of such elements can be closely approximated by f_{cr}, as shown by curve C in Fig. 14-6. In elements with w/t ratios greater than 60, the distortions even at low stresses are so pronounced that such elements should not be used in structural members.

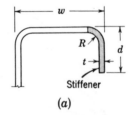

(a)

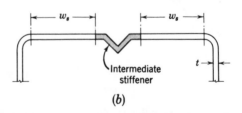

(b)

Fig. 14-5 Stiffened elements: (a) with simple bent lip and (b) with intermediate stiffener.

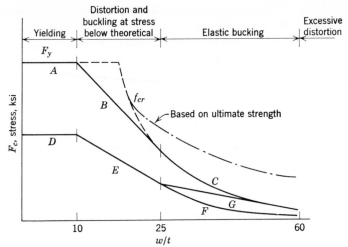

Fig. 14-6 Behavior of unstiffened elements.

The strength of unstiffened elements can therefore be taken as

$$P_u = f_m wt$$

where f_m values are defined by criteria shown in Fig. 14-6.

The **allowable axial load** on a short compression member (L/r less than 10) is governed by local buckling and may be obtained as follows:

A. For Stiffened Elements

$$P_a = F_b A_e = F_b \sum b_e t$$

where $F_b = F_y/1.65$ and b_e is defined as follows:

(*a*) Up to

$$(w/t)_{\lim} = 4020/\sqrt{f},$$

$b_e = w$.

(*b*) For w/t greater than the foregoing limit:

$$b_e = \frac{8040t}{\sqrt{f}} \left[1 - \frac{2010}{(w/t)\sqrt{f}} \right]$$

The value of f is the actual unit stress in psi in the compression element computed on the basis of the effective width. Since values of f and b_e are interdependent when w/t exceeds $(w/t)_{\lim}$, they are usually determined by successive approximations. Charts and tables are given in the Specification[1] to facilitate determination of effective width from the foregoing equations.

B. For Unstiffened Elements

$$P_a = F_c A = F_c \sum wt$$

where F_c is taken as $f_m/1.65$ for the weakest unstiffened element of the section, that is, the one with the greatest w/t ratio except for sections composed of combined stiffened and unstiffened elements having relatively large w/t ratios. The values of f_m are based on the criteria shown in Fig. 14-6 and the resulting values of F_c are defined as follows:

1. For w/t not greater than 10, provided that $F_y < 50,000$ psi

$$F_c = \frac{F}{1.65} \qquad \text{(see curve } D \text{, Fig. 14-6)}$$

2. For $25 > w/t > 10$, provided that $F_y < 50,000$ psi

$$F_c = A - B\left(\frac{w}{t}\right) \qquad \text{(see curve } E \text{, Fig. 14-6)}$$

where $A = 1.01F_y - 8640$, and $B = (\frac{1}{15})[(F_y/1.65) - 12,950]$

3. For $60 > w/t > 25$

(a) Section composed entirely of unstiffened elements:

$$F_c = 8.09 \times 10^6 \left(\frac{t}{w}\right)^2 \qquad \text{(see curve } F \text{, Fig. 14-6)}$$

(b) Section composed of combined stiffened and unstiffened elements:

$$F_c = 20,000 - 282\left(\frac{w}{t}\right)$$

(see curve G, which is based on the ultimate strength of the unstiffened elements and various factors of safety, Fig. 14-6).

C. *For Members Consisting of Both Stiffened and Unstiffened Elements*

$$P_a = F_c(A_{e,\text{stiff}} + A_{\text{unstiff}}) = F_c(\sum b_e t + \sum wt) = F_b Q A_{\text{gross}} \qquad (14\text{-}6a)$$

where F_c is taken as in B, $A_{e,\text{stiff}}$ is the effective area for stiffened elements taken from A–1, A–2 corresponding to w/t ratios and F_c values for the weakest element, A_{unstiff} is the area of all unstiffened elements, A_{gross} is the total cross-sectional area, and Q is a form factor equal to

$$Q = \left(\frac{F_c}{F_b}\right)\frac{A_{e,\text{stiff}} + A_{\text{unstiff}}}{A_{\text{gross}}} \qquad (14\text{-}6b)$$

Note that, for case A, $A_{\text{unstiff}} = 0$ and $Q = A_{e,\text{stiff}}/A_{\text{gross}}$, provided the effective area for stiffened elements is determined on the basis of F_b, and, for case B, $A_{e,\text{stiff}} = 0$ and $Q = F_c/F_b$.

Primary Column Buckling. The Euler buckling equation is valid for cross sections of light gage steel whenever the computed stress is below the proportional limit, and for stress above this limit the tangent modulus E_t is substituted for E,

$$F_{cr} = \frac{\pi^2 E_t}{(L/r)^2} \qquad (14\text{-}7)$$

Following the procedure recommended by the CRC this equation may be approximated by a parabola tangent to the Euler curve at $F_y/2$, so that:

$$F_{cr} = F_y - \left(\frac{F_y^2}{4\pi^2 E}\right)\left(\frac{L}{r}\right)^2 \qquad (14\text{-}8)$$

Allowable stresses may then be determined by dividing the critical stress by a factor of safety.

For

$$\frac{L}{r} \le \sqrt{\frac{2\pi^2 E}{F_y}} :$$

$$F_a = \frac{1}{\text{F.S.}} \times \left[F_y - \left(\frac{F_y^2}{4\pi^2 E}\right)\left(\frac{L}{r}\right)^2 \right]$$

For

$$\frac{L}{r} > \sqrt{\frac{2\pi^2 E}{F_y}} :$$

$$F_a = \frac{1}{\text{F.S.}} \times \left[\frac{\pi^2 E}{(L/r)^2} \right]$$

The value of $\sqrt{(2\pi^2 E/F_y)}$ is the limiting L/r ratio above which the column will fail in elastic buckling.

The effect of local buckling is included by introducing the form factor Q, which was previously explained. The foregoing column equations are adjusted to include this local effect factor and may be expressed as follows, using a constant factor of safety of 1.95 and $E = 29.5 \times 10^6$ psi.

For

$$\frac{L}{r} \le \frac{24,200}{\sqrt{QF_y}} :$$

$$F_a = 0.515 Q F_y - \left(\frac{Q F_y L/r}{47,500}\right)^2 \qquad (14\text{-}9)$$

For

$$\frac{L}{r} > \frac{24,200}{\sqrt{QF_y}} :$$

$$F_a = \frac{149,000,000}{(L/r)^2} \qquad (14\text{-}10)$$

where F_a is the allowable average axial stress in compression, other terms are as previously noted.

The form factor Q appears only in Eq. 14-9 when L/r is less than the limiting L/r and the critical stresses are above the proportional limit. For long columns primary buckling takes place before local buckling can occur.

A uniform factor of safety of 1.95 is used for all values of L/r because cold-formed sections of light gage steel are usually noncompact, eccentricities and initial crookedness have greater effects, and end connections are relatively flexible.

Typical curves for $F_y = 33$ ksi are shown in Fig. 14-7.

The maximum allowable ratio L/r of unsupported length L to minimum radius of gyration r should not exceed 200, except that during construction only, L/r may exceed this limit provided it is less than 300.

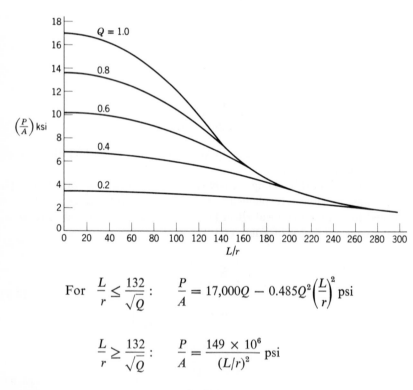

$$\text{For} \quad \frac{L}{r} \leq \frac{132}{\sqrt{Q}}: \quad \frac{P}{A} = 17{,}000Q - 0.485Q^2\left(\frac{L}{r}\right)^2 \text{ psi}$$

$$\frac{L}{r} \geq \frac{132}{\sqrt{Q}}: \quad \frac{P}{A} = \frac{149 \times 10^6}{(L/r)^2} \text{ psi}$$

Fig. 14-7 Permissible stress F_a for pin-ended columns, Eq. 14-9 for $F_y = 33$ ksi.

Torsional Buckling. As noted in Sec. 9-6, torsional buckling occurs when the torsional rigidity of the member is appreciably smaller than its bending rigidity, as in the case of thin-walled open sections. These sections may fail either by pure torsional buckling mode or combined torsional-flexural mode. The basic equation that determines the critical buckling load based on linear elastic behavior is given in Sec. 9-6 (Eq. 9-18), but the solution of this equation is somewhat cumbersome for usual design. Therefore, it is often helpful to know whether, for a given set of dimensions, the torsional-flexural buckling mode is critical. Solution of this problem for certain common shapes has been obtained and is illustrated in Figs. 14-8 to 14-10. In these figures, flexural buckling mode is critical in the region above or to the left of the curves shown. Below or to the right of the curve, the section shown will fail in the torsional-flexural buckling mode. For example, it can be seen from Fig. 14-9 that if b/a is less than 0.5, flexural buckling is critical when tL/a^z is greater than 0.5, and torsional-flexural buckling is critical only when tL/a^2 is less than 0.5.

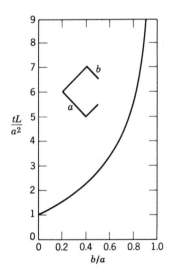

Fig. 14-8 Buckling modes for thin-wall angles (after Chajes, Fang, and Winter).

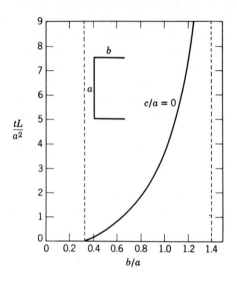

Fig. 14-9 Buckling modes for thin-wall channels (after Chajes, Fang, and Winter).

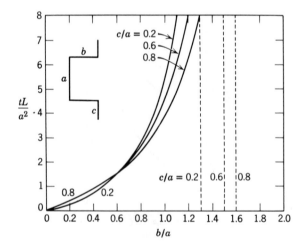

Fig. 14-10 Buckling modes for thin wall hat sections (after Chajes, Fang, and Winter).

Example 14-1. Select an appropriate section to carry an axial compression load $P = 30$ kips on a column with an unbraced length $L = 12.5$ ft made of sheet steel with specified yield point of 33 ksi. (Note: Typical light-gage sections are shown in Part IV of *Light Gage Cold-Formed Steel Design Manual*, 1962 Edition.)

Solution

As an initial trial, investigate a double channel section 8 × 6 × 10 ga. shown in Fig. 1.

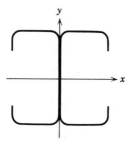

Fig. 1

(*a*) Properties of Full Section (from *Design Manual*):

$A = 4.00$ in.^{2}, $I_y = 7.90$ in.^{4}, $r_y = 1.41$ in., $I_x = 38.6$ in.^{4}, $r_x = 3.14$ in.

(*b*) Determination of Form Factor Q:

Since the member is composed of stiffened elements only:

$$Q = \frac{A_e}{A_g}$$

For stiffened elements:

$$b_e = w \quad \text{if} \quad (w/t) \leq 4020/\sqrt{f}$$

where f is P/A_e, and therefore Q must be found by trial.

Let $Q = 1$, then $f = P/A = 30/4.0 = 7.5$ ksi and

$$(w/t)_{lim} = 4020/\sqrt{7500} = 46.4$$

$$(w/t)_f = (2.5/0.1875) = 13.3 < 46.4$$

$$(w/t)_{wn} = (7.35/0.1875) = 39.9 < 46.4$$

Therefore $A_e = A_g$ and $Q = 1.0$.
Allowable value for the column:
  F_a is found from Eqs. 14-15 or from Fig. 14-7, using appropriate value of Q. In this case:

$$L/r = 150/1.41 = 106.5 \quad \text{and} \quad Q = 1.0:$$

$$F_a = 12 \text{ ksi and } P_a = F_a \times A_e = 12 \times 4.0 = 48 \text{ kips}$$

This value of allowable load exceeds the required capacity of 30 kips, and therefore a lighter section would be more economical.
Try double channel section $7 \times 5.5 \times 10$ gage.

(*a*) Properties of Full Section (from *Design Manual*)

$A = 3.54$ in.^{2}, $I_y = 5.90$ in.^{4}, $r_y = 1.29$ in., $I_x = 26.2$ in.^{4}, $r_x = 2.72$ in.

(*b*) Determination of Form Factor Q:

Try $Q = 1.0$, then $f = P/A = 30.0/3.54 = 8.5$ ksi

and
$$(w/t)_{lim} = 4020/\sqrt{8500} = 43.6$$

$$(w/t)_f = (4.5/0.135) = 16.7 < 43.6$$

$$(w/t)_w = (6.35/0.135) = 47 > 43.6$$

Then for the web:

$$b_e = (8040t/\sqrt{f})\left(1 - \frac{2010}{(\omega/t)\sqrt{f}}\right) = 6.3$$

Reduction in web area is negligible, and $Q = 1.0$. Then, allowable stress with $L/r = 150/1.29 = 116$, $Q = 1.0$ from Fig. 14-7 $F_a = 10.5$ ksi and $P_a = F_a \times A_e = 10.5 \times 3.54 = 37.2$ kips. This section still has about 20% excess capacity. A lighter section may be investigated.

Example 14-2. Determine the allowable compression load on the cross section with data indicated in Fig. 1.

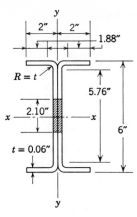

Fig. 1.

Given: $F_y = 33,000$ psi

$$(F_b = 20,000)$$

$L = 15$ ft.

The stud is adequately braced by wall sheathing about the y-y axis

Required: P_x (allowable load)

Solution

(a) *Properties of Full Section*

$$A = 1.159 \text{ in.}^2 \qquad I_y = 0.641 \text{ in.}^4 \qquad r_y = 0.74 \text{ in.}$$

$$I_x = 6.04 \text{ in.}^4 \qquad r_x = 2.28 \text{ in.}$$

(b) *Determination of Form Factor Q*

Since the member is composed of both stiffened and unstiffened elements

$$Q = \left(\frac{F_c}{F_b}\right)\left[\frac{(A_{e,\text{stiff}} + A_{\text{unstiff}})}{A_{\text{gross}}}\right]$$

(1) For unstiffened elements (flanges)

$$\frac{w}{t} = \frac{1.88}{0.06} = 31.3$$

From Sec. 14-5

$$F_c = 20,000 - 282(31.3) = 11,170 \text{ psi}$$

$$\frac{F_c}{F_b} = \frac{11,170}{20,000} = 0.559$$

(2) For stiffened elements (webs)

$$\frac{w}{t} = \frac{5.76}{0.06} = 96$$

and for

$$f = 11,170 \text{ psi}, \quad \frac{b}{t} = 61.0$$

The portion considered to be removed is $w - b = (96 - 61)t = 2.10$ in.

$$\left(\frac{A_{e,\text{stiff}} + A_{\text{unstiff}}}{A_{\text{gross}}}\right) = \frac{(1.159 - 2 \times 2.10 \times 0.06)}{1.159} = 0.783 \text{ in.}^2$$

(3) $Q = 0.559 \times 0.783 = 0.438$

(c) *Determination of Allowable Load,* P_x

$$\frac{L}{r_x} = \frac{15 \times 12}{2.28} = 78.9 < \frac{24,200}{\sqrt{0.438 \times 33,000}} = 201$$

$$F_a = 0.515 Q F_y - \left(\frac{Q F_y L/r}{47,500}\right)^2 = 6870 \text{ psi} \qquad \text{from Eq. (14-9)}$$

$$P_x = A F_a = 1.159 \times 6870 = 7960 \text{ lb.}$$

14-6 CYLINDRICAL TUBULAR COMPRESSION MEMBERS

Cylindrical tubes are economical for compression members because of their large ratio of radius of gyration to area which remains constant for all axes.

Thin-walled tubular members may fail by local and/or primary column buckling and both conditions must be investigated to determine the load capacity of the member.

The theoretical critical buckling stress for cylindrical tubes in the elastic range is discussed in Sec. 9.15.

Tests have indicated that actual values of buckling stress are much lower than the theoretical values, largely because of geometric imperfections in the section. The AISI Specification bases its provisions on the research of Plantema[5] and other tests conducted at the University of Illinois[6] and represented graphically in Fig. 14-11. The curve indicates that the ratio of ultimate stress at failure to yield point f_{ult}/F_y depends upon the parameter $Et/F_y D$. Based on this study the AISI Specification conservatively specifies that the ratio of the mean diameter to wall thickness D/t be

$$\frac{D}{t} \le \frac{3,300,000}{F_y} \tag{14-16}$$

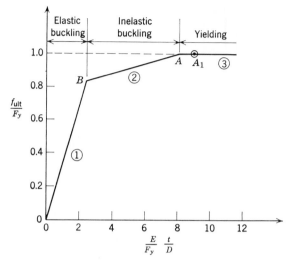

Fig. 14-11 Behavior of cylindrical tubular members.

On this basis, local buckling will not occur and the column formulas with Q equal to one apply.

14-7 DESIGN OF WALL-BRACED STUDS

Design of wall panels or partitions constructed with light steel studs faced on both sides with conventional wall boards, such as plywood and gypsum board, has been used widely in structures having floors framed of light steel. The steel studs, usually I-, Z-, or channel-shaped, are placed with the webs perpendicular to the wall (Fig. 14-12); thus usually the weak axis of the stud is also perpendicular to the wall. The buckling load about this weak axis depends on the lateral support provided by the wall board. Therefore for safe and economic design, it is necessary to evaluate the ability of the wall "skin" and its attachment to prevent failure of the steel studs in the plane of the wall. An extensive analytical and experimental study of this problem was conducted[7] involving tests to determine properties of various types of wall board, strength of studs with various types of lateral support, and strength of full-scale wall panels. This study indicated a rational procedure for determining the necessary characteristic of the wall material and attachment to prevent failure of the stud in the plane of the wall.

Three requirements must be satisfied in the design of such load-bearing panels: (*a*) wall material must be attached to both faces of the studs, (*b*) the spacing of the attachments and the rigidity of the wall-board material

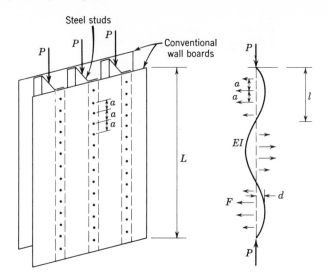

Fig. 14-12   Wall panel with steel studs.

must be sufficient to prevent lateral buckling or yielding of the studs in the plane of the wall, and (*c*) the strength of the attachment must be adequate to provide necessary restraint of the studs by the wall board.

Consider the wall panel shown in Fig. 14-12. A column of length *L* laterally supported by attachments to an elastic wall will buckle in several waves. The number of waves depends on the stiffness of the stud *EI*, the spacing of attachments *a*, and the elastic "spring" coefficient *k*. This coefficient is the force per unit length change of the wall strip having width *a*.

The coefficient *k* defines the load–deformation relationship of a particular stud, the type of attachment, and the wall-board material, and is best determined experimentally. Elastic restraint per wall face is one-half the total, that is, *k*/2.

If the spacing *a* of the attachments is small compared to the column length *L*, the restraints can be considered as a continuous elastic support. The buckling strength P_{cr} of such a column buckling in several half-waves may be approximated as follows:

$$P_{cr} = 2\sqrt{\frac{EIk}{a}} \quad \text{or} \quad P_{cr} = \frac{2\pi^2 EI}{l^2} \tag{14-17}$$

where *l* is the half-wave length of the buckled column (Fig. 14-12). Note that the critical load for a column with elastic supports is double that of a pin-ended column with length equal to half-wave length.

To develop a given required strength of stud of known proportions, that is,

P_{cr} and EI are known, the minimum required ratio k/a can be determined as follows:

$$\frac{k}{a} = \frac{P_{cr}^2}{4EI} \tag{14-18}$$

The spacing of attachments a may be selected as some convenient value, usually governed by practical considerations, or may be computed from the minimum ratio k/a if the elastic "spring" coefficient k of a given panel is known. This value of k is determined experimentally and a suitable test procedure has been outlined.[1] Some typical values of k are given in Table 14-2. It should be noted that Eq. 14-17 applies only if a is relatively small compared to L; otherwise Eqs. 14-17 and 14-18 must be modified.

Table 14-2

| Type of Wall Board | $\frac{3}{16}$-in. Bolts Tightened Firmly | | | |
| | $\frac{1}{2}$-in. Washers | | 1-in. Washers | |
	k, lb/in.	F, lb	k, lb/in.	F, lb
A	660	67.5	860	95
B	1550	125	1730	190

A $\frac{1}{2}$-in. standard-density wood fibre insulating board.
B $\frac{3}{8}$-in. gypsum board sheathing.

A convenient procedure for determining spacing a is to design the panel so that the critical slenderness ratio of the panel L/r_1 for buckling normal to the plane of the wall will be twice a/r_2, the slenderness ratio for buckling in the plane of the wall. Thus

$$2a = r_2 \frac{L}{r_1}$$

The value of k per wall face can be determined conservatively, using the condition that buckling in the plane of the wall does not occur before yielding; that is, P_{cr} shall not be less than $P_y = F_y A$.

From Eq. 14-18, using $P_{cr} = F_y A$,

$$k = a \frac{P_{cr}^2}{8EI} = a \frac{(F_y A)^2}{8EI} \tag{14-19}$$

With values of k and a selected for a particular design, it remains to provide an attachment sufficiently strong to prevent failure in the attachment at a load P below the buckling load P_{cr}. The force F which must be developed by the attachment depends on the deflection of the column d and the elastic "spring" coefficient $\frac{1}{2}k$ of each face of wall material:

$$F = \tfrac{1}{2}kd \qquad (14\text{-}20)$$

F shall not be less than

$$F_{min} = \frac{keP}{2\sqrt{EI_2k/a} - P}$$

where e is the stud length in inches/240, and P is the total design load on the stud. The other terms are as previously defined. If the minimum strength of the attachment between wall sheathing and stud is thus determined, it will be sufficient to resist the buckling of studs without failure of the attachment by tearing or loosening.

The minimum modulus of elastic support k shall be not less than

$$k = \frac{F_y^2 a A^2}{240,000,000 I_z} \qquad (14\text{-}21)$$

where I_z = moment of inertia of cross section of stud about axis perpendicular to wall, in.[4] For loads P below the buckling load, the deflection d of the column can be approximated very closely by Eq. 14-22:

$$d = e\left(\frac{P}{P_{cr} - P}\right) \qquad (14\text{-}22)$$

where e is the initial eccentricity or initial crookedness of the member. Without load eccentricity, the initial crookedness of thin sheet-steel studs may be approximated $L/240$. This is twice the crookedness allowed in usual practice, which is $\frac{1}{4}$ in. in 10 ft, or $L/480$.

Whether or not a given attachment satisfies the minimum strength requirement $F = kd$ must be determined experimentally. Some typical values of strength of bolt attachments based on one-fourth of ultimate are given in Table 14-2.

14-8   BENDING OF LIGHT GAGE BEAMS

The strength of a light gage cold-formed beam is limited by the smallest shear or bending moment that will produce yielding, buckling, or excessive distortion of any of its elements. Where the proper function of the beam depends on its deflection, the useful strength of the beam may be less than the value obtained by shear or moment limitation.

The stress distribution at failure depends on the symmetry of the cross section and the stress–strain characteristics of the material. For steels having a definite yield point, the effect of symmetry on stress distribution at failure can be illustrated as follows. The section shown in Fig. 14-13a

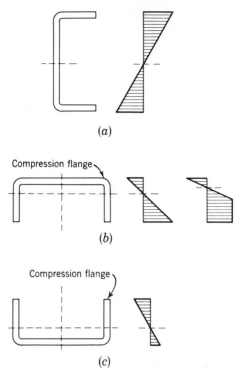

Fig. 14-13 Bending of light gage sections: (*a*) linear stress distribution, (*b*) plastic stress distribution, and (*c*) linear stress distribution.

bends about an axis of symmetry, and the tension and compression flanges reach yield stress approximately simultaneously. Thin compression flanges with w/t ratios of approximately 10 for unstiffened and 20 for stiffened elements are capable of developing yield stress without buckling but cannot deform very much beyond initial yield strain. Hence the stress distribution at failure is very nearly a linear one.

The section shown in Fig. 14-13*b* bends about a neutral axis not in the plane of symmetry and in this instance the tension flanges may yield, whereas the compression flanges are subjected to relatively low stress. Increase in load beyond initial yield of the tension flanges will result in shifting of the neutral axis upward toward the compression flange, and stress redistribution occurs, so that at failure both the tension and effective compression areas are stressed up to yield. It is important to note that, if the same section is bent in the opposite direction (Fig. 14-13*c*), the compression flanges may buckle or yield first. The failure occurs with very little further strain and the stress distribution is again very nearly a linear one.

Some steels do not have a definite yield point and if buckling does not occur first, these can be stressed considerably in excess of the nominal yield point, hence the ultimate strength of beams made with such steel may exceed the yield strength by as much as 25%.

The bending stress f_b is obtained using simple linear stress distribution so that $f_b = Mc/I$. The maximum permissible stress for stiffened elements of a laterally stable cross section is equal to $F_b = F_y/1.65$, and for unstiffened elements $F_b = F_c$ defined in Sec. 14-9 and shown in Fig. 14-5.

The effective width of stiffened elements must be known in order to determine the position of the neutral axis and the moment of inertia I of the effective cross section. Since the effective width depends on the magnitude of the bending stress f_b, and this stress depends on the section modulus I/c, one cannot be determined without the other. The solution of this problem is obtained by successive approximations. First stress f_b in the stiffened element is estimated, corresponding effective width b_e is either calculated from Eq. 14-2 using $f_m = 1.65f_b$, or scaled from Fig. 14-4, and corresponding section properties are calculated. The stress in the stiffened element is then recomputed. If it agrees closely with the value initially assumed, no further calculation is required. If it differs substantially, a revised value of f_b is used to recompute effective width, section properties, and corresponding new value of stress. This process is repeated until the value of f_b used to calculate effective width and the calculated stress based on section properties of the effective cross section agree closely.

In checking the permissible load or bending moment on the section, the effective width of the stiffened element is based on a stress 1.65 times the value at working load. In calculating deflections at working load, the effective width of the stiffened element should be taken as that corresponding to stress at working loads. Thus b_e for deflection calculation may be computed from Eq. 14-2 or taken from Fig. 14-4 using $f_m = f_b$.

Beam Webs.   Criteria for the structural safety of webs in light gage beams are somewhat different than for webs in rolled beams or plate girders. The problems in design of light gage beam webs differ from rolled beams because stiffeners cannot be used effectively with light gage beams, and partly due to the variation of yield-point stress for the various types of steel used. The maximum depth–thickness ratio h/t for unstiffened light gage webs is usually limited to 150.

Using Eqs. 9-55 and 9-56 to define initial buckling stresses and considering that the webs are long plates pin-supported on all four edges, the critical buckling stresses in pure shear and in pure bending are defined as follows:

$$\text{pure shear,} f_{v,\text{cr}} = 142 \times 10^6 \left(\frac{t}{h}\right)^2$$

and

$$F_{v,\text{allowable}} = 64 \times 10^6 \left(\frac{t}{h}\right)^2 \leq \tfrac{2}{3}F_b$$

and

$$\text{pure bending}, f_{b,\text{cr}} = 640 \times 10^6 \left(\frac{t}{h}\right)^2$$

and

$$F_{b,\text{allowable}} = 520 \times 10^6 \left(\frac{t}{h}\right)^2 \leq F_b \text{ (basic)}$$

Initial yielding of the webs subjected to pure shear occurs when the stress $f_v = F_y/\sqrt{3}$, and in pure bending when $f_b = F_y$. Post-buckling strength of thin webs in shear is not great; on the other hand, in bending considerable post-buckling strength is obtained.

Web Crippling. A theoretical determination of the web crippling or crushing strength of webs is an extremely complex problem. For thin webs an empirical solution of this problem was obtained based on results of an extensive experimental program.[8] The test program covered such variables as beam depth h, thickness t, flange width w, width of bearing B, single and double webs, web connectors, and type of loading. Two types of specimens have been investigated: one having single unreinforced webs and the other having double webs connected so as to provide a high degree of restraint against rotation of the webs (Fig. 14-14).

For specimen type a shown in Fig. 14-14 the principal parameters governing the crushing strength were B/t ratio and yield-point stress of the material. Thus the maximum allowable load, P_{cs} may be evaluated by the following:

End crushing:

$$P_{cs} = t^2 F_b(7.4 + 0.93\sqrt{B/t}) \tag{14-23}$$

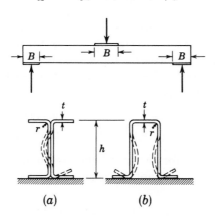

(a)   (b)

Fig. 14-14 Crushing strength of webs. (a) Single web and (b) double web.

Interior crushing:

$$P_{cs} = t^2 F_b(11.1 + 2.41\sqrt{B}/t) \tag{14-24}$$

For specimen type *b* (Fig. 14-14), the crushing strength of the web is a function of t/B, h/t, h/B, r/t, and F_y. The simplest expression representing the maximum allowable load for various yield points is:

End crushing:

$$P_{cs} = 100K_1 Bt[42 + 980(t/B) - 0.22(h/t) - 0.11(h/B)] \tag{14-25}$$

Interior crushing:

$$P_{cs} = 100K_2 Bt[23 + 3050(t/B) - 0.09(h/t) - 5(h/B)] \tag{14-26}$$

where

$$K_1 = (1.15 - 0.15r/t)(1.33 - 0.33F_y/33,000)F_y/33,000$$

and

$$K_2 = (1.06 - 0.06r/t)(1.22 - 0.22F_y/33,000)F_y/33,000$$

The expressions in Eqs. 14-25 and 14-26 represent the low extremes of web restraint likely to occur in practice and hence the values lie somewhat on the low side.

14-9 LATERAL BUCKLING OF COMPRESSION FLANGES

The possible failure of compression flanges due to lateral buckling was discussed in Chapter 9 with reference to standard rolled shapes. The basic equations are applicable to the thin-wall sections as well as to the rolled shapes, but the use of these equations is limited to instances where the material has not buckled locally. Assuming the effect of local buckling to be small, the gross section properties may be used in Eq. 9-42.

Because of the small torsional rigidity GK_t of the light gage section compared to the bending rigidity EK_b, Eq. 9-42 may be simplified by neglecting the term containing GK_t. For narrow I-beams this simplification leads to Eq. 9-45, which reduces to the following:

$$F_{cr} = \frac{\pi^2 E}{4(L/r_y)^2}\left(\frac{d}{r_x}\right)^2 = \frac{74,000,000}{(L/r_y)^2}\left(\frac{d}{r_x}\right)^2 \tag{14-27}$$

The allowable stress for I-shaped light gage beams is obtained using a factor of safety of 1.65 and noting that for usual shapes d/r_x is not greater than 2.5. Thus

$$F_c = \frac{F_{cr}}{1.65} = \frac{280,000,000}{(L/r_y)^2} \tag{14-28}$$

Assumption of narrow I-beams used in derivation of Eq. 14-27 is satisfactory when L/r_y is greater than $28{,}800/\sqrt{F_y}$. For an I or a channel shape, with L/r_y less than $12{,}900/\sqrt{F_y}$

$$F_c = \frac{F_y}{1.65}$$

For values of L/r_y between the above limits

$$F_c = 0.67F_y - \frac{(L/r_y)^2(F_y)^2}{2.46 \times 10^9}$$

In Z shapes with the same slenderness ratio L/r_y as in I shapes, lateral buckling occurs at a somewhat lower stress. Conservative estimates of this stress may be obtained as follows:

$$L/r_y > \frac{20{,}400}{\sqrt{F_y}}, \qquad F_c = \frac{140 \times 10^6}{(L/r_y)^2}$$

$$L/r_y < \frac{9100}{\sqrt{F_y}}, \qquad F_c = \frac{F_y}{1.65}$$

For values of L/r_y between the above limits:

$$F_c = 0.67F_y - \frac{(L/r_y)^2(F_y)^2}{1.23 \times 10^9}$$

Multiple-web shapes such as box, hat, and U shapes (Fig. 14-15) have greater lateral stability than single-web shapes of the same depth–width ratio. Box-shaped members are rarely critical in lateral buckling and beams with a span–width ratio of up to 100 do not buckle laterally. Flat hat sections with $I_y > I_x$ do not exhibit any tendency for lateral buckling; for deep hat sections with $I_y < I_x$, column formulas such as Eq. 14-7 or 14-8 may be used to estimate conservatively lateral buckling stress in the compression flange.

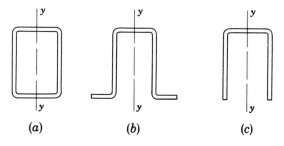

(a)   (b)   (c)

Fig. 14-15   Multiple-web shapes.

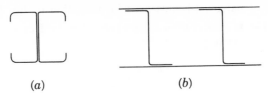

Fig. 14-16   Built-up light gage shapes.

14-10   BRACING OF BEAMS

Typical light gage shapes that are easily cold-formed are shown in Fig. 14-1. Other shapes (Fig. 14-16) can be fabricated using combinations of these common shapes. In using some of the single or built-up shapes, special consideration has to be given to bracing to prevent excessive twisting or lateral distortions, which may impair the proper function of the beams and result in additional secondary stress, which tend to reduce the load-carrying capacity of the member.

A transverse load acting on a channel in the plane of the web is eccentric with respect to the shear center and thus tends to twist the channel as shown in Fig. 14-17. Similarly a transverse load on a Z section does not lie in a principal plane and hence tends to produce a lateral displacement as well as a vertical one. When these sections are used as fully braced components of a rigid system, such as the built-up panels in Fig. 14-16, the shapes can be utilized effectively and special bracing may be required only during fabrication of the panel.

When these sections are not fully braced by other elements of the structure, special bracing must be used,[9] which should be spaced at sufficiently small

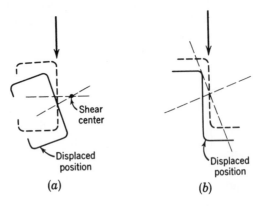

Fig. 14-17   Twisting and lateral displacements of beam sections.

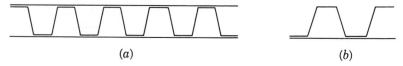

Fig. 14-18   Combinations of hat sections and sheet steel.

intervals and should have sufficient strength to limit distortion effectively. When two channels are joined together, the connections should serve the same purpose as the bracing and should be designed accordingly. Requirements for design of such connections and bracing are given in the AISI Specification. In many instances "hat" sections and sheet steel (Fig. 14-18) can be combined advantageously without any additional bracing.

14-11   EFFECT OF CURVATURE ON FLANGE STRESSES AND DISTORTIONS

Conventional analysis of I- and box-beams are based on the assumption that the flanges remain plane: that is, they do not distort in the direction normal to their plane. Figure 14-19 shows a part of an I-beam in pure bending. It can be seen that flange forces have components normal to the plane of the flange. The forces tend to "crush" the beam and actually produce distortions of the flange as shown in Fig. 14-19. For shapes where flange width–thickness ratios and beam curvature are both small, these "crushing" loads and flange distortions are negligible. For beams with wide, thin flanges designed for loads approaching yield, the distortions may no longer be negligible. The magnitude of the flange distortion may be determined approximately by considering a differential length ds of a beam subjected to a moment M, with flange loads P_f. The load P_f is equal to flange stress f times the effective flange area bt and f is equal to $(Mh/2)/I$. From similar triangles in Fig. 14-19, it follows that $P_f/R = P_c/ds$, and

$$P_c = \frac{P_f\,ds}{R} = \frac{(fbt)\,ds}{EI/M} = \frac{2f^2bt\,ds}{Eh} \tag{14-29}$$

and the intensity of the distributed "crushing" load w on the flange having width b,

$$w = \frac{P_c}{b\,ds} = \frac{2f^2}{E}\frac{t}{h}$$

The maximum flange distortion may be approximated as that of a cantilever plate of unit width and span $b/2$, loaded with uniformly distributed load w.[10] The deflection of such a plate can be obtained by conventional theory,

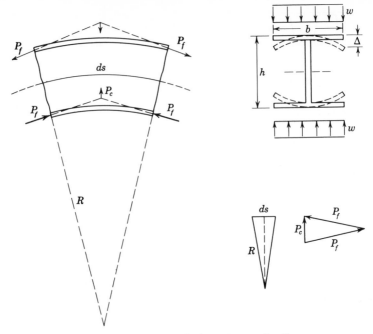

Fig. 14-19   Portion of I-beam in pure bending.

provided the stiffness of the plate is increased by a factor $1/(1 - \mu^2)$ to account for lateral restraint in a bent plate. The maximum deflection Δ of the flange is

$$\Delta = \frac{wL^4}{8EI}(1 - \mu^2) = \frac{wb^4}{128E}\left(\frac{1 - \mu^2}{t^3/12}\right) = \frac{6}{32}b\left(\frac{fb}{Et}\right)^2\frac{b}{h}(1 - \mu^2) \quad (14\text{-}30)$$

For design purposes, Δ may be limited to either an arbitrary limit, say 0.05 in., or a fraction of beam depth, $\Delta/h = 0.04$.

14-12   EFFECT OF SHEARING DEFORMATIONS ON FLANGE STRESS DISTRIBUTION

Conventional beam theory is based on the assumption that the effect of shearing deformations on stress distribution and deflection is negligible. For conventional structural members this assumption is satisfactory. Analytical and experimental studies[10] indicate that this assumption is not satisfactory for thin sheet-steel beams, particularly beams with wide flanges carrying concentrated loads on a relatively short span. The effect of shearing deformation on flange stress distribution is illustrated in Fig. 14-20.

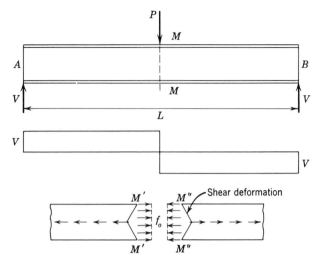

Fig. 14-20   Shearing deformation in flanges.

Consider the tension flange of beam AB (Fig. 14-20) showing the action of web shears. The segments of the flange on each side of the center section are shown in Fig. 14-20 in which the flange tension stresses are assumed to be uniform across the width. Because of flange shear stresses the section just left of MM tends to warp as indicated by $M'M'$, and the section just right of MM tends to warp as indicated by $M''M''$.
These distortions are incompatible because matching $M'M'$ and $M''M''$ results in a break in the flange at section MM. The incompatibility of deformation at section MM indicates that the assumed distribution of flange stresses must be incorrect. For the symmetrical case shown, it is apparent that section MM cannot warp and remain straight. To effect this condition of no warping, additional stresses are induced at the midflange and at the edges (Fig. 14-21). This system of stresses must be self-equilibrating, that is, the sum of compressive forces balances the sum of tensile forces. The final stress

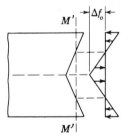

Fig. 14-21   Stresses in a beam section induced by shear deformations.

distribution at section MM is obtained by superposition of the uniform tensile stress and the additional induced stresses, as shown in Fig. 14-22. The magnitude of the secondary induced stresses varies from a maximum at a section where a sudden change in shear occurs to zero where the section is free to distort.

The mathematical analysis of this problem becomes rather complex and

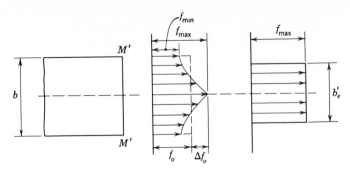

Fig. 14-22 Resultant stress distribution.

does not yield a simple relationship between stress and load. In general, the maximum stress depends on the type of loading conditions, the shape of beam cross section, and the end-support conditions. To simplify the determination of the maximum stress, the concept of effective width has been used. Consider stress distributions in the flanges of beams shown in Fig. 14-22. The total width b of the flange carrying a total force P_f with stress distribution varying from f_{max} to f_{min} can be replaced by effective width b_e' carrying the same total force P_f with uniform distribution of stress f_{max}. Analytical studies indicate that, for beams with thin wide flanges, within the practical range of flange-width–span proportions, the effect of shear deformation is negligible for distributed loads. For concentrated loads,

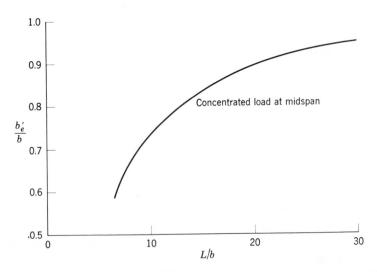

Fig. 14-23 Variation of effective width b_e for beam flange. (Courtesy NACA[11].)

or concentrated loads spaced at a distance greater than the flange width b, the effect of shear deformation is significant. An approximate relationship between b_e' and b, as a function of the span–width ratio[11] is given in Fig. 14-23.

The use of effective width to account for the shearing-deformation effect as well as for local buckling sometimes causes some confusion. The concept of effective width is merely a convenient method by which the stresses can be treated as uniform in situations where the actual stress distribution is not uniform. This concept has been most useful in two entirely distinct situations: (*a*) restrained buckling and (*b*) shear deformation as just described.

For *thin wide flanges* the shear-deformation effect has practical significance only in the reduction of the tension flange width b to b_e'. Effective width b_e of the compression flange in such instances is usually governed by the consideration of buckling.

14-13 COMBINED AXIAL AND BENDING STRESSES

Light gage cold-formed members subject to both axial compression and bending stresses shall be proportioned to resist both actions as an inter-related behavior. In this respect, the behavior of light gage members is similar to the hot-rolled structural shapes and the same general requirements are specified. The formulas for light gage members are modified to include the form factor Q for local buckling. They are expressed as follows:

(*a*) When $f_a/F_a \leq 0.15$:

$$\frac{f_a}{F_a} + \frac{f_b}{F_b} \leq 1.0$$

(*b*) When $f_a/F_a > 0.15$:

Between braced points

$$\frac{f_a}{F_a} + \frac{C_m f_b}{(1 - f_a/F_e')F_b} \leq 1.0$$

At braced points

$$\frac{f_a}{0.515QF_y} + \frac{f_b}{F_b} \leq 1.0$$

where f_a and f_b are axial unit stress and bending unit stress, respectively. F_a is the maximum allowable axial unit stress in compression where axial stress only exists. F_b may be equal to the basic design stress F_b, the allowable compressive stress for local buckling, or the allowable compressive stress for lateral buckling, whichever is smaller. However, at braced points lateral buckling need not be considered in the determination of F_b. The term C_m is a coefficient which depends on the type and distribution of the flexural loading. For most instances it can be taken as 0.85. F_e' is the Euler stress divided by a factor of safety and equal to $149,000,000/(L/r_b)^2$.

14-14   CONNECTIONS

The type of connections in light gage structures are essentially similar to those in conventional structures. Bolts, welds, and screws are used most commonly, and rivets and other special devices are used only occasionally. Usually ordinary unfinished bolts are used, although high-strength bolts also may be used.

For ordinary unfinished bolted connections the following criteria apply.[12] The ultimate *shearing strength* of the bolts is approximately equal to 0.6 tensile strength of bolt material based on the root-of thread section. Normally in design, an allowable stress of 10 ksi on gross section is used. In connection with sufficiently large edge distances, *bearing* failure may occur because of the bolt deforming the hole (Fig. 14-24). This failure may occur at a nominal stress $4.8F_y$ on area dt. Using a factor of safety of 2.3, the nominal allowable bearing stress is $2.1F_y$ or $3.5F_b$, where $F_b = F_y/1.65$.

For small edge distances in the line of stress (Fig. 14-25), failure occurs by *tear-out* at a nominal shear stress $\frac{2}{3}F_y$, on the shearing area $A = 2et$. Using a factor of safety of 2.3, the minimum permissible edge distance is

$$e = \frac{2.3P}{1.33F_y t} \cong \frac{P}{F_b t} \qquad (14\text{-}31)$$

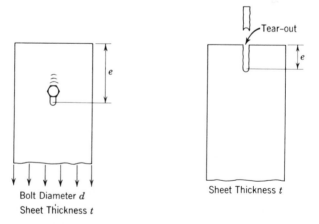

Fig. 14-24   Bearing failure of bolt hole.   **Fig. 14-25**   Tear-out failure of bolt.

In no case, however, should e be less than $1.5d$. For balanced bearing versus tear-out design $e = P/F_b t = 3.5F_b dt/F_b t = 3.5d$.

Failure across the net section in thin sheets is appreciably affected by stress concentration, particularly for small d/s ratios. The ultimate tensile strength f_m across net section was found to be

$$f_{tn} = \left(0.10 + 3.0\frac{d}{s}\right)f_{tu} \qquad (14\text{-}32)$$

where f_{tu} is the tensile strength of sheet steel. The reduction in ultimate strength is important only when $s/d = 3.33$, since f_{tn} can never be greater than f_{tu}. Tests indicate that for values of $s/d = 12.5$, f_{tn} values may be as low as $0.25f_{tu}$.

AISI Specification recommends reducing basic allowable tension stress $F_b = F_y/1.65$ by the factor $(0.10 + 3.0d/s)$ for calculating the allowable load on net section. Because the ratio f_{tu}/F_y for different sheet steels may vary from 1.2 to 1.6, the factor of safety against failure provided by this specification varies from $1.65 \times 1.2 = 1.98$ to $1.65 \times 1.6 = 2.64$.

The same criteria for bearing, tear-out, and net-section failures in connections using high-strength bolts may be employed.[13] Tests show that these criteria are slightly more conservative for the connections with high-tensile-strength bolts. Furthermore, the amount of slip in these connections is substantially reduced.

The ultimate shear strength of high-strength bolts can also be approximated by 0.6 of tensile strength of bolt material based on the root-section area. The strength of high-strength bolts may be taken as approximately twice that of ordinary unfinished bolts and thus a substantial reduction in number of bolts may be effected when the design is critical in shearing strength of bolts.

Fusion welding for connecting steel sheets of less than $\frac{1}{8}$-in. thickness is under investigation and standard procedures are not yet available.

For light gage materials, spot welds are usually used in accordance with *Recommended Practice for Resistance Welding* published by the American Welding Society. The safe tension or shear design load per spot, based on a factor of safety of approximately 2.5 recommended by the AWS, is given in Table 14-3. The following values are valid for low-carbon steels with F_y up to 70 ksi, provided the welding technique conforms to AWS Standards.

Table 14-3

Thinnest outer sheet, in.	0.010	0.020	0.030	0.040	0.050	0.060	0.080	0.109	0.125*	0.155*	0.185
Allowable load per spot, lb	50	125	225	350	525	725	1075	1650	2000	3000	4000

*Based on values for projection welds.

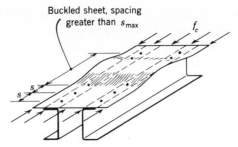

Fig. 14-26 Connector spacing for sheet steel under compression. $s_{max} = $ maximum permissible spacing required to prevent buckling, $s < s_{max}$.

In a light gage steel sheet subjected to compression and attached to the cross section as shown in Fig. 14-26, it is desirable to space the connectors in order to prevent buckling between them. The spacing s required in order to be on the verge of buckling at a critical stress f_{cr} can be obtained theoretically by assuming that the sheet acts as a restrained column between connectors, with end-fixity coefficient $C = 2.5$, and thus buckling stress f_{cr} is

$$f_{cr} = \frac{2.5\pi^2 E}{12(1 - \mu^2)}\left(\frac{t}{s}\right)^2 = 2.26E\left(\frac{t}{s}\right)^2 \qquad (14\text{-}33)$$

To provide a factor of safety of 1.65, f_{cr} must be not less than $1.65f_c$, or $1.65f_c = 2.26E(t/s)$. Solving for maximum permissible spacing s,

$$s_{max} = t\sqrt{\frac{2.26E}{1.65f_c}} = \frac{6300t}{\sqrt{f_c}} \qquad (14\text{-}34)$$

14-15 STRUCTURAL APPLICATIONS

The use of light gage steel members has grown from individual components to become an integral part of the structural framing system.

In panel form, the unit serves as siding and roofing attached to girts and purlins. Research has indicated that these panels have sufficient shear strength to serve as diaphragms,[14,15,16] thus eliminating additional bracing systems. When used as a roof, floor, or wall system, the panels prevent lateral buckling of beams and general buckling of columns.

Shell roof structures are also taking advantage of the inherent strength of light gage panels in the form of folded-plate and hyperbolic-paraboloid construction.[17,18] This effective use of panels serves the dual purpose of a structural component while at the same time satisfying architectural requirements.

The folded-plate roof (Fig. 15-47) is essentially plate girders acting together with common flanges. They can be used in many forms, for instance the sawtooth shape and the trapezoidal shape, in either longitudinal or circular structures. In general, plate width range from 7 to 12 ft, slopes of the plate vary from 20° to 45°, and spans are up to 100 ft. The light gage panels serve as the stiffened web of the folded-plate girder.

The hyperbolic-paraboloid roof structure (Fig. 15-52) uses the light gage panels as a membrane resisting in-plane shears·produced by the gravity loads. The paraboloid is a doubly curved surface produced by straight line generators. The panels are a natural material to form this surface.

Single-bay structures have been built using the light gage steel as the principal structural material to form a folded-plate rigid frame.

PROBLEMS

1. Determine the resisting moment of the two 8.0 × 4 unstiffened flanged channels in Fig. P-1 for a steel of $F_y = 36$ ksi. Channels are spot-welded together, laterally supported.

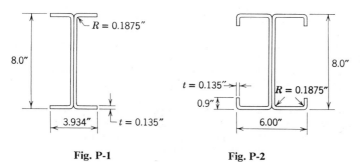

Fig. P-1Fig. P-2

2. Determine the uniform loading for a span of 12 feet for the section in Fig. P-2, $F_y = 40$ ksi, laterally supported. Calculate maximum deflection.
3. Determine the allowable uniform loading on the 16-foot beam in Fig. P-3,

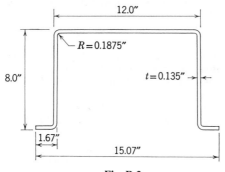

Fig. P-3

$F_y = 40$ ksi, calculate deflection at midspan. With a length of bearing of 1.25 in. check web buckling and crippling.

   4. Determine the loading if a stiffener fold is introduced in the center of the 12.0-in. flange of Prob. 3. Fold is 1-in. in depth, out to out, with radius of bend of 0.1875 in.

   5. For the section of Prob. 1, calculate the allowable column load for a length of 8 ft.

   6. Calculate column load for section in Prob. 2 for a length of 10 ft.

REFERENCES

   1. *Specification for the Design of Light Gage Cold Formed Steel Structural Members*, American Iron and Steel Institute, 1962.
   2. *Light Gage Cold Formed Steel Design Manual*, American Iron and Steel Institute, 1962. Also G. Winter, "Cold Formed Light Gage Steel Construction," *J. Structural Div. Proc. ASCE*, ST9 (1959).
   3. Winter, G., "Strength of Thin Steel Compression Flanges," *Trans. ASCE* **112**, 527 (1947).
   4. Chajes, A., Fang, P. J., and Winter, G., "Torsional Flexural Buckling, Elastic and Inelastic of Cold Formed Thin Walled Columns," *Cornell Eng. Res. Bull.* 66-1, 1966
   5. Plantema, F. J., "Collapsing Stresses of Circular Cylinders and Bound Tubes", Nat. Luchtvaart Lab. Report 5.280, Amsterdam, 1946.
   6. Wilson, W. M. et al., "The Strength of Cylindrical Shells as Columns," and "Tests of Cylindrical Shells," University of Illinois Eng. Expt. Sta. Bull. **255** (1933) and **331** (1941).
   7. Green, G. G., G. Winter, and T. R. Cuykendall, "Light Gage Steel Columns in Wall Braced Panels," *Cornell Univ. Eng. Expt. Sta. Bull.* **35,** Part II (1947).
   8. Winter, G., and R. H. J. Pian, "Crushing Strength of Thin Steel Webs," *Cornell Univ. Eng. Expt. Sta. Bull.* **35,** part I (1946).
   9. Winter, G., "Lateral Bracing of Columns and Beams," *Proc. ASCE* **84,** ST2 (March 1958).
   10. Winter, G., "Performance of Thin Steel Compression Flanges," International Association of Bridge and Structural Engineers, 3rd Congress, Preliminary publication, 1948.
   11. Winter, G., "Stress Distribution in an Equivalent Width of Flanges of Wide, Thin-Wall Steel Beams," *NACA TN* **784** (1940).
   12. Winter, G., "Tests on Bolted Connections in Light Gage Steel," *J. Structural Div. Proc. ASCE*, ST 2 (1956).
   13. Winter, G., "Light Gage Steel Connections with High-Strength High-Torqued Bolts," *Publ. Intern. Assoc. Bridge and Struct. Eng.* **16** (1956).
   14. Nilson, A. H., "Shear Diaphragms of Light Gage Steel," *J. Structural Div. Proc. ASCE* **86,** ST11 (November 1960).
   15. Bryan, E. R., "The Effect of Sheeting on Structural Design," *Building With Steel, J. of the British Constructional Steel Work Association* Vol. 3, No. 3 (August 1964).
   16. *Design of Light Gage Steel Diaphragms*, American Iron and Steel Institute, 1967.
   17. Nilson, A. H., "Folded Plate Structures of Light Gage Steel," *Trans. ASCE* **128,** Part II (1963).

18. Nilson, A. H., "Testing a Light Gage Steel Hyperbolic Paraboloid shell," *J. Structural Div. Proc. ASCE* **88,** ST5 (October 1962).

19. Yu, Wei-Wen., "Design of Light Gage Cold-Formed Steel Structures," Engineering Experiment Station, West Virginia University, Morgantown, West Virginia, 1965.

20. Luttrell, L. D., and Z. L. Moh, Unpublished Notes, Short Course, West Virginia University, Morgantown, West Virginia, June 1966.

15

Special Structures

J. W. GILLESPIE, J. F. McDERMOTT, AND W. PODOLNY Jr.*

15-1 INTRODUCTION

The previous chapters have been concerned with the analysis and design of components of steel structures, including some general features of the design of complete structural systems for conventional buildings and bridges. This chapter will discuss other types of buildings and bridges which represent new developments in structural design. These will be discussed in general terms with respect to the type of structure, its background, and theory associated with its development. Some unique current structures will be described and some applications for possible future developments will be proposed.

The rapid expansion of knowledge in the steel-making processes and the elaborate refinements of controlling the chemistry of the steels have brought forth many new grades for construction. Steels with greater strengths and corrosion resistance will continue to be developed. More efficient connecting materials and methods will undoubtedly be devised to keep pace with the development in the steels themselves. New fabrication and erection techniques will follow.

---

* J. W. Gillespie, Manager, Marketing Technical Services, United States Steel Corporation, Pittsburgh, Pennsylvania.

J. F. McDermott, Senior Research Engineer—Structural Mechanics, United States Steel Corporation, Applied Research Laboratory, Pittsburgh, Pennsylvania.

W. Podolny, Jr. Design Engineer, Marketing Technical Services, United States Steel Corporation, Pittsburgh, Pennsylvania.

The competition of materials in the construction industry will challenge engineers so that esthetic and economical structures will be designed to satisfy the requirements of the architectural and functional concepts.

The steel industry can look forward to greater achievements in the more efficient fabrication techniques by taking advantage of computer automated equipment. The development of prefabricated units for rapid shop fabrication and field erection will encourage structures that can be shipped and assembled easily.

Greater emphasis will be placed on space or three-dimensional structures. A principal objective will be the greater efficiency derived from a structure when all types of loadings may be supported by the same framework, since a three-dimensional structure can resist loads from all directions in addition to supporting gravity loads. For example, triangulated truss systems for roofs and floors, and unit systems for bridges on a curve, are proving to be economical.

Electronic computers are making it possible to analyze all types of structures easily and to solve the most complicated system accurately. More studies may be made to investigate the effect of each of the parameters affecting the design of the individual members. Thus the optimum design of structures with respect to weight, fabrication, erection, and cost may be possible.

Since 1946, the design of bridges has included the concept of a large plate girder with the complete width of the roadway as the compression flange. This type of bridge structure is called an "orthotropic" bridge. Long span girder bridges assisted by cables from pier masts has added another method to bridge design. Similar concepts have been successfully used in buildings required to cover a large area with column-free spaces.

The technique of prestressing steel beams, trusses, and structures has been introduced and studies are being made to determine their efficient utilization.

Steel shell structures include those which employ light gage diaphragms acting wholly or partially as membranes in conjunction with other steel components forming the outline of the configuration. Light gage steel which had been introduced as a roof and floor material has now found its place as a principal structural component of major structures. Shell structures may be hyperbolic paraboloids, folded plates, or other curved shapes of various sizes.

The development of square and rectangular tubing has added another hot-rolled shape to the family of wide flanges, I-beams, channels, and angles. Like the pipe, its efficiency as a column makes it a desirable architectural shape. Similar to the use of pipes to form trusses, space frames, and rigid frames, the square and rectangular tubes are just beginning to find their place in structural systems.

A more detailed discussion of the structures mentioned follows to acquaint the reader with the many possibilities of steel structures.

15-2 ORTHOTROPIC STEEL PLATE PANELS AND BRIDGES

Orthotropic steel plate panels are plates with stiffeners attached to one side, usually by welding (Fig. 15-1). These panels are frequently used for deck or wall components of ships, trucks, etc., and in decks of all-steel bridge spans.

The stiffeners of orthotropic panels between supports may extend in two directions at right angles, but usually the stiffeners extend in only one direction because details at stiffener connections incur high fabrication costs. The stiffeners are generally spaced close enough together so that, for examining the structural behavior, only the average panel stiffnesses per unit width

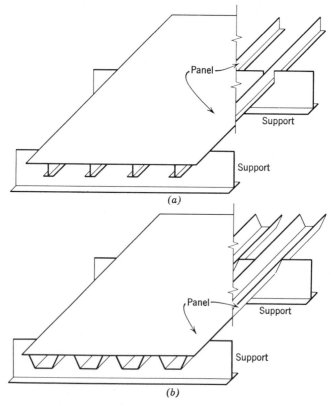

Fig. 15-1 Orthotropic plate panels: (*a*) open-rib stiffeners and (*b*) closed stiffeners.

need be considered. Thus, this panel can be analyzed and designed as an equivalent plate continuum without considering each stiffener as a separate structural element. "Orthotropic" implies that the stiffness per unit width of the panel is different in two perpendicular directions, being much greater in the direction of the stiffeners, as contrasted to "isotropic" plates, where the stiffness per unit width is the same in all directions.

As indicated in Fig. 15-1, there are two basically different types of orthotropic panels: (*a*) torsionally "soft" panels consisting of plates with open-rib stiffeners, such as bars, angles, structural tees, or I-shaped beams, and (*b*) torsionally stiff panels consisting of plates with closed stiffeners, such as U-, trapezoidal-, or Y-section stiffeners. Fabrication costs may be less with open-rib stiffeners. However, if the design is governed by concentrated loadings, the torsional rigidity provided by the closed stiffeners can result in significant material savings.

Orthotropic steel plate bridges consist of orthotropic panels acting monolithically with a supporting grillage of beams and girders (Fig. 15-2). In the past structural design of orthotropic bridges has consisted mainly of selecting steel shapes and plates on the basis of judgment and then performing a stress

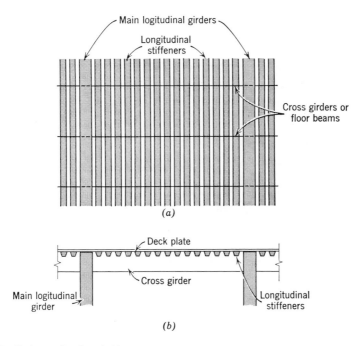

Fig. 15-2 Orthotropic plate bridge scheme: (*a*) plan section below deck plate and (*b* transverse cross section.

analysis. Now very useful design aids for conducting such analyses and an extensive list of references to theoretical and experimental work on ortho-tropic panels and bridges are included in the 1963 edition of the American Institute of Steel Construction, *Design Manual for Orthotropic Steel Plate Deck Bridges.*

Construction of orthotropic steel plate bridges started after World War II as a result of the extensive need for long span highway bridges over major European rivers. Some of these bridges utilized straight rib stiffeners (Fig. 15-3), or structural tee stiffeners (Fig. 15-4) mainly because of the judgment at that time that fabrication costs would be less and that corrosion protection would be easier with open-rib stiffeners. However, with improvements in welding techniques, such as the introduction of automatic and semi-automatic equipment, there is a growing trend to use closed stiffeners in orthotropic panels that are subject to concentrated loads, such as bridge decks. The superior torsional stiffness of closed stiffeners helps to spread the load over more stiffeners and results in less material required. Therefore, such stiffened plate designs as the U-stiffener (Fig. 15-5) and the Y-stiffener (Fig. 15-6) were developed.

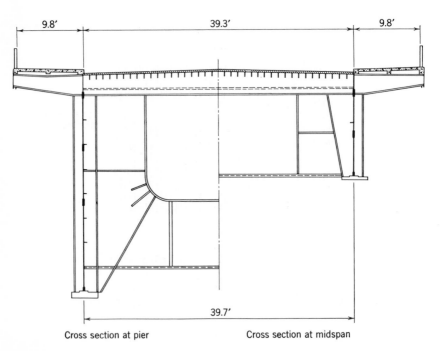

Cross section at pier   Cross section at midspan

Fig. 15-3 Save River Bridge in Belgrade, Yugoslavia, 1956, spans 246-856-246 ft. (Courtesy AISC.)

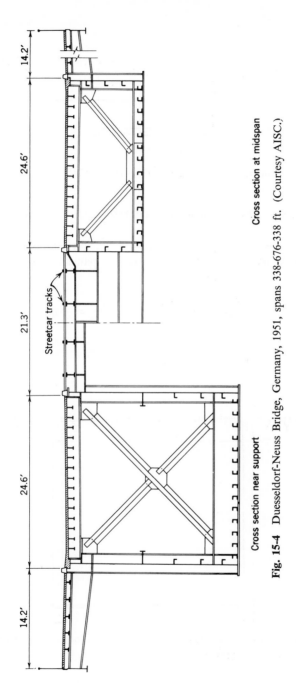

Cross section at midspan

Streetcar tracks

Cross section near support

Fig. 15-4 Duesseldorf-Neuss Bridge, Germany, 1951, spans 338-676-338 ft. (Courtesy AISC.)

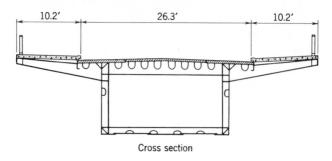

Cross section

Fig. 15-5   Weser Bridge, Porta, Germany, 1954, spans 209-255-348 ft.   (Courtesy AISC.)

15-3   ANALYSIS AND DESIGN OF ORTHOTROPIC PANELS

The basic differential equation relating deflection w and loading q of a panel with stiffeners so closely spaced that the panel behaves as an orthotropic plate is the Huber equation[1]

$$D_x \frac{\partial^4 w}{\partial x^4} + 2H \frac{\partial^4 w}{\partial x^2 \partial y^2} + D_y \frac{\partial^4 w}{\partial y^4} = -q \qquad (15\text{-}1)$$

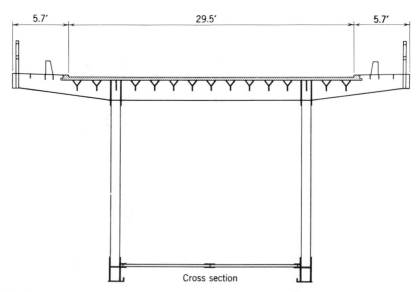

Cross section

Fig. 15-6   Fulda River Bridge, Bergshausen, Germany, 1961, spans 259-298-351-469-351-298-259 ft.   (Courtesy AISC.)

where D_x and D_y are the flexural rigidities in the x and y directions, respectively, and H is a measure of the torsional rigidy. This equation results from (a) expressing the in-plane strains, and then the stresses and subsequently the bending and twisting moments in terms of w and (b) substituting these expressions for moments into the equation for static equilibrium that relates the vertical loading, the vertical shears, and the bending and twisting moments. If y is the direction of the stiffeners, D_y is the product of the modulus of elasticity E and the average moment of inertia of the section I_y per unit width of the stiffened plate. In the x direction, transverse to the stiffeners, D_x is merely the plate flexural rigidity

$$D_x = \frac{Et^3}{12(1 - \mu^2)}$$

where t is the plate thickness and μ is Poisson's ratio. In many analyses of orthotropic panels D_x is so small in comparison to D_y that the first term can be neglected in Eq. 15-1. Then, Eq. 15-1 reduces to

$$2H \frac{\partial^4 w}{\partial x^2 \, \partial y^2} + D_y \frac{\partial^4 w}{\partial y^4} = -q \tag{15-2}$$

For closed stiffeners the AISC *Design Manual for Orthotropic Steel Plate Deck Bridges* states values of H calculated to reflect the effect of the elastic flexibility of the deck plate between stiffeners in reducing the resistance of the panel to twisting. The value of H can be generally considered zero for orthotropic panels having open-rib stiffeners, except as noted in the following.

Panels With Open Stiffeners. When D_x and H are both neglected, Eq. 15-1 for orthotropic panels with open-rib stiffeners is

$$D_y \frac{\partial^4 w}{\partial y^4} = -q \tag{15-3}$$

which is simply the differential equation for a beam. Equation 15-3 is satisfactory for uniform loading, but does not explain the ability of a panel with open-rib stiffeners to "spread" a loading to enable more than one stiffener to support a concentrated load when the load is placed directly over a stiffener. Equation 15-1 would have to be used to determine this spreading effect for panels with open-rib stiffeners. However, with the use of formulas and charts presented in the AISC *Design Manual* to evaluate this spreading effect, an open-rib panel may be analyzed as a series of parallel beams with each beam consisting of a stiffener and the portion of the plate acting with the stiffener.

Panels with Closed Stiffeners.   The concept of influence surfaces is useful in determining bending moments, shears, and reactions in orthotropic panels with closed stiffeners. This concept is illustrated for calculation of y-direction bending moments in a simply supported rectangular orthotropic panel with stiffeners extending in the y-direction as follows.

In Fig. 15-7,

(*a*) Assume that the panel is cut along the critical transverse section 2-2 and then rejoined with a continuous hinge.

(*b*) For any given value of n, a vertical line load

$$1 \cdot \sin \frac{n \pi x}{b}$$

along a given transverse section 1-1 would cause a net angle change at hinge 2-2 of

$$\theta_{21} \sin \frac{n \pi x}{b}$$

(*c*) If the line load were removed and horizontal jacks were installed at hinge 2-2 to cause a moment of

$$1 \cdot \sin \frac{n \pi x}{b}$$

in the panel on each side of the hinge, the net angle change at hinge 2-2 would be

$$\theta_{22} \sin \frac{n \pi x}{b}$$

and the net deflection at section 1-1 would be

$$w_{12} \sin \frac{n \pi x}{b}$$

(*d*) Because there cannot be an angle change at 2-2 in the actual panel, the actual moment at 2-2 due to the line load at section 1-1 must be

$$M_y = - \frac{\theta_{21}}{\theta_{22}} \sin \frac{n \pi x}{b} \tag{15-4}$$

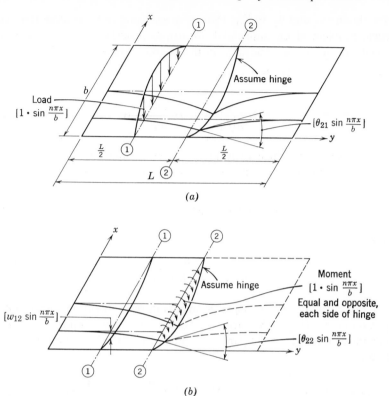

Fig. 15-7 Hypothetical hinge assumed at midspan: (*a*) distortions due to vertical line load and (*b*) distortions due to hypothetical jacking moment at midspan.

so that the hypothetical horizontal jacks exactly counteract the angle change at hinge 2-2 that would be caused by the vertical line load at 1-1 on the hypothetically hinged panel.

(*e*) From the law of reciprocal deflections,

$$\theta_{21} \sin \frac{n\pi x}{b} = w_{12} \sin \frac{n\pi x}{b} \tag{15-5}$$

Therefore, substituting Eq. 15-5 into Eq. 15-4,

$$M_y = -\frac{w_{12}}{\theta_{22}} \sin \frac{n\pi x}{b} \tag{15-6}$$

Equation 15-6 defines the ordinates along 1-1 to the influence surface for moments at 2-2 for the particular value of *n* considered.

To obtain w_{12} and θ_{22} in Eq. 15-6, it is necessary only to solve the hypothetical problem of the sinusoidal distortions resulting from the sinusoidal moments at the hinge. The steps in this solution are the following:

(a) Set $q = 0$ in Eq. 15-2, giving

$$2H \frac{\partial w}{\partial y^2 \, \partial x^2} + D_y \frac{\partial w}{\partial y^4} = 0 \qquad (15\text{-}7)$$

A Lévy solution, which conveniently gives a sinusoidal variation in the x direction of shears, moments, and distortions, is as follows:

$$w = \sum_{n=1}^{\infty} (C_1 \sinh \alpha_n y + C_2 \cosh \alpha_n y + C_3 \alpha_n y + C_4) \sin \frac{n\pi x}{b} \qquad (15\text{-}8)$$

where

$$\alpha_n = \frac{n\pi}{b} \sqrt{\frac{2H}{D_y}} \qquad (15\text{-}9)$$

(b) For calculation of midspan moments in the y direction, substitute the expression for w (Eq. 15-8) into a total of four different boundary conditions occurring at the ends of the panel span and also at the hypothetical hinge as follows:

$$
\left.
\begin{array}{l}
(1) \ w = 0 \\[2mm]
(2) \ \dfrac{\partial^2 w}{\partial y^2} = 0
\end{array}
\right\} \text{ for } y = 0
\qquad
\left.
\begin{array}{l}
(3) \ -D_y \dfrac{\partial^2 w}{\partial y^2} = 1 \cdot \sin \dfrac{n\pi x}{b} \\[2mm]
(4) \ \dfrac{\partial^3 w}{\partial y^3} = 0
\end{array}
\right\} \text{ for } y = \dfrac{L}{2}
$$

Simultaneous solution of these four equations yields numerical values to the four unknown constants.

For convenience of illustration, Fig. 15-7 applies specifically for moment at midspan, requiring four unknowns to be solved. For determining moment elsewhere, the Lévy equation must be evaluated separately both to the right and to the left of the hypothetical hinge, thus requiring eight equations to be solved. Boundary conditions similar to conditions 1, 2, and 3 above, can be written for each surface, thus giving six equations. The other two equations necessary for solution are the statements that at the hinge both surfaces have (1) the same value of w and (2) the same value of $(\partial^3 w/\partial y^3)$.

(c) With the general expression for w thus evaluated, w_{12} is known, and θ_{22} is the change in $\partial w/\partial y$ at the hypothetical hinge. Then M_y in Eq. 15-6 can be solved, giving the influence surface for moment at 2-2 due to a sinusoidal line loading

$$1 \cdot \sin \frac{n\pi x}{b}$$

at any station along the panel.

For any given sinusoidally distributed vertical load along a transverse line 1-1, the bending moment in the actual panel at some point $x = p$ along section 2-2 is equal to the product of (a) the maximum intensity of the line loading and (b) the ordinate z_{1pn} to the influence surface for moments (subscript 1 refers to line 1-1, subscript p refers to $x = p$, and subscript n refers to the sinusoidal variation of load, that is, the number of extreme values of load). However, most loadings are not sinusoidal, but any vertical load along a transverse line can be expressed by a Fourier series expansion

$$Q = \sum_{n=1}^{\infty} Q_n \sin \frac{n\pi x}{b} \tag{15-10}$$

where

$$Q_n = \frac{2}{b} \int_0^b Q \sin \frac{n\pi x}{b} \, dx \tag{15-11}$$

Thus every value of n results in an nth contribution to the value of Q; the maximum intensity of the nth line loading is Q_n as defined in Eq. 15-11. Consequently, because of a given line load Q along transverse line 1-1, the moment at $x = p$ along transverse line 2-2 is

$$M_y = \sum_{n=1}^{\infty} \left[\left(\frac{2}{b} \int_0^b Q \sin \frac{n\pi x}{b} \, dx \right) (z_{1pn}) \right] \tag{15-12}$$

For an area loading, such as a uniform loading q exerted by a truck tire between transverse lines 1-1 and 3-3, (line 3-3 is not shown in Fig. 15-7) the expression for moment at point $x = p$ on line 2-2 can be expressed as

$$M_y = \sum_{n=1}^{\infty} \left[\left(\frac{2}{b} \int_0^b q \sin \frac{n\pi x}{b} \, dx \right) (A_{13pn}) \right] \tag{15-13}$$

where on a vertical plane through $x = p$ and extending in the y direction A_{13pn} is an area bounded by (a) the influence surface for moments corresponding to that value of n (b) the horizontal reference plane for the influence surface ordinates, and (c) vertical planes in the x direction through lines 1-1 and 3-3.

The AISC *Design Manual* presents values of M_y and other pertinent information for the design of orthotropic plate panels subjected to highway truck loadings and outlines special procedures for analyzing panels that are continuous over several spans.

15-4 ANALYSIS OF ORTHOTROPIC BRIDGES

Orthotropic plate bridges generally consist of orthotropic panels acting monolithically with a supporting grillage that consists of cross girders supported by main longitudinal girders that rest on the bridge piers. For calculating stresses and distortions in orthotropic bridges, the Pelikan-Esslinger design method,[2] which consists of the following steps, can be used.

(*a*) By the procedure discussed above or with the design aids given in the AISC *Design Manual* analyze the orthotropic plate panels with the assumption that the supporting cross girders do not deflect. Calculate the reaction of the orthotropic panels on the cross girders.

(*b*) With the restriction against deflections removed, load the cross girders with the reactions from the orthotropic plate panels. The entire bridge deflects, and all structural components, both the girder grillage and the orthotropic plate panels, contribute to the support of the vertical loading on the cross girders.

The stresses of steps (*a*) and (*b*) are added to obtain the final stresses. Equations and charts are given in the AISC *Design Manual* for calculating the stresses of step (*b*).

Membrane Tension. Local stresses under a tire load due to transverse bending of the plate between stiffeners are usually not calculated in orthotropic plate analyses because field and test measurements have indicated that such stresses are generally not critical. They are considerably less than first-order theory (bending only) would indicate. This is explained by the fact that significant transverse membrane tension is developed in the plate as the portion of the plate under the load deflects relative to the rest of the plate when the loading is applied. Thus, for local stresses, second-order theory which involves both bending and membrane tension applies because there is generally a considerable extent of the panel beyond the vicinity of the loading that will offer sufficient in-plane resistance to the membrane tension.

15-5 PRESENT DESIGNS AND FUTURE DEVELOPMENTS OF ORTHOTROPIC BRIDGES

During the mid-1960s the first major orthotropic steel plate bridges were built in North America. Torsionally stiff U-shaped stiffeners were used in two Canadian bridges: the Port Mann Bridge at Vancouver, B.C. (Fig. 15-8), and the Concordia Bridge, Montreal (Fig. 15-9), which was built to handle traffic for the "EXPO 67" World's Fair. Similarly, the Poplar St. Bridge over the Mississippi River at St. Louis, Missouri, featured torsionally stiff trapezoidal-shaped stiffeners (Fig. 15-10). However, because only straight-rib geometry precludes any possibility of moisture entrapment, straight-rib stiffeners were used in the San Mateo Bridge over the southern arm of San Francisco Bay, California (Fig. 15-11) because unusually foggy conditions frequently occur in that location. These four bridges demonstrated the economy of orthotropic steel plate designs for long span bridges. During the construction of these bridges the only major unanswered technical question was determining the best type of wearing surface to place on top of the deck plates of orthotropic bridges.

The orthotropic steel panel concept has been used more widely in other structures of lesser magnitude, from ship bulkheads to floors of trucks. Figure 15-1 illustrates two alternate designs that were considered for the decking of a ship that would be subject to heavy fork lift truck loads. Because this was a roving load, the torsional stiffness provided by the trapezoidal stiffeners was beneficial, and the trapezoidal design was approximately 15 percent lighter than that using structural tees for the stiffeners.

The fact that the occurrence of membrane tension enables the designer to neglect localized bending stresses in the deck plate of an orthotropic panel suggests that, as the loading is applied, membrane tension transverse

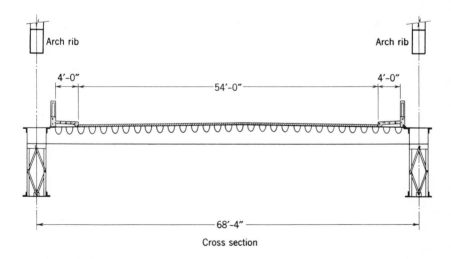

Cross section

Fig. 15-8 Port Mann Bridge in Vancouver, B.C., Canada, 1963, spans 360-1200-360 ft. (Courtesy AISC.)

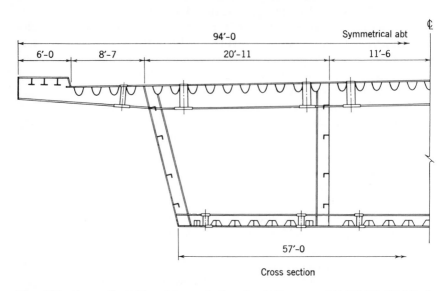

Fig. 15-9 Concordia Bridge, Montreal, Canada, 1965, spans 341-525-525-525-340 ft. (Courtesy of Dominion Bridge Company Limited, Canada.)

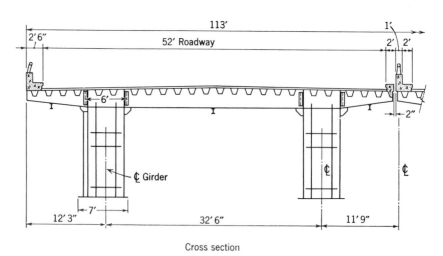

Cross section

Fig. 15-10 Popolar St. Bridge Over Mississippi River at St. Louis, Missouri, 1967, spans 300-500-600-500-265 ft. (Courtesy of *Engineering News-Record*, October 20, 1966.)

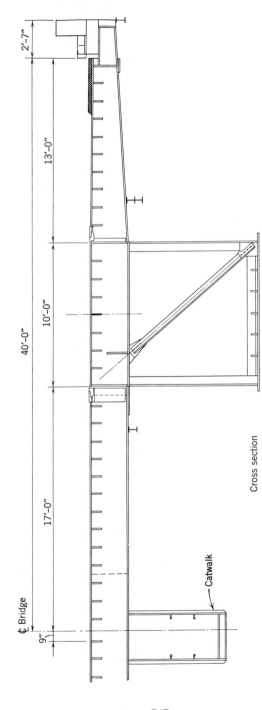

Fig. 15-11 San Mateo Bridge, San Francisco Bay, California, 1967, 6 spans at 292 ft, 375–750–375 ft, 6 spans at 292 ft. (Courtesy of the State of California, Department of Public Works, Division of Bay Toll Crossings.)

to the stiffeners will build up in the full width of deck plate between the longitudinal girders supporting the sides of the panel. Such extensive development of membrane tension would be quite beneficial in supporting the load, and its consideration could result in significant savings in material. However, unless there is suitable in-plane resistance offered to the membrane tension at the periphery of the panel the development of such membrane tension cannot be relied on. One possible way to insure such resistance would be to design the sidewalks on each side of a bridge to act as horizontal girders anchoring the membrane tension that occurs in the panels (Fig. 15-12). Such special transverse anchorage would be more economical in a wide bridge with several main longitudinal girders than in a narrow bridge with just two main girders. In orthotropic bridges full utilization of the superior tensile strength of steel will be realized only when provisions are made in design to utilize the beneficial effect of membrane tension in reducing material requirements.

At the present orthotropic steel plate spans are economical for long span bridges because of the significant savings in material requirements for the entire bridge that result from the lower dead weight of the all-steel spans. However, for orthotropic steel plate construction to be economical for short and medium spans, improved more economical fabrication techniques

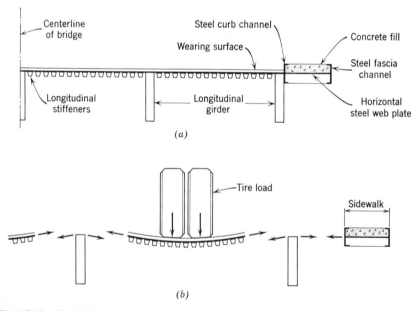

Fig. 15-12 Development of membrane tension beneficial in orthotropic bridge design: (*a*) transverse cross section and (*b*) exploded cross section showing membrane tension resisted by sidewalk.

particularly relative to welding must be developed. It is anticipated that such improvements will occur, with the result that a greatly increased percentage of new bridges will be orthotropic steel plate bridges.

15-6 CURVED BRIDGES

Most present-day highway bridges have their horizontal geometry determined by the desired roadway alignments rather than being determined by the most convenient crossing site. Consequently, many highway structures must be fitted to a curved roadway alignment. Depending on the curvature and span requirements, several possibilities exist for the structural configuration: simple spans on chords, continuous segmented girders, or continuous curved girders. The horizontally curved girder arrangement is generally considered to be the most pleasing esthetically.

Certain advantages accrue when curved girders are used for structures designed to fit curved highway alignments. The design and construction of the roadway slab are normally simplified because the girder spacing and the slab overhang width are constant over the length of the entire structure. In addition, if the structure is fitted to the alignment, certain structural efficiencies, such as longer spans and shallower depths, are usually obtained. Continuity of design, which is certainly desirable from a structural behavior standpoint, is easier to accomplish. This will result in the efficient use of longer spans and economical shallower construction.

A possible disadvantage of curved girder design is its potentially higher fabrication costs; however, the savings in material due to the increased structural efficiency and the other advantages mentioned above usually offset any increased fabrication costs.

Curved girder bridges can be classified as open framing, closed framing, or a combination of both. Open framing consists of curved girders rigidly connected to a system of diaphragms or floor beams, with no other bracing members. Closed framing is defined as girders tied together by diaphragms or floor beams and lateral bracing in the plane of the bottom flange. It is usually assumed that the top flange is restrained by the floor slab in both types of framing. A combination of open and closed framing consists of some of the girders in the system being tied together in accordance with open framing requirements, and the remaining girders being tied together in accordance with closed framing requirements.

An example of an open framing system under construction is shown in Fig. 15-13. Figures 15-14 and 15-15 illustrate a combination of open and closed framing. Note that the diaphragms are trusses in these cases. The framing between the outer girders and the adjacent girders is a closed system, whereas the construction of the interior girders is of the open framing type.

A rigorous analysis of a curved-girder bridge system can be accomplished either by the stiffness method or the flexibility method. However, if the structure satisfies the normal assumptions of a "grid" structure, the general analysis can be substantially simplified. These assumptions are: cross sections of members is symmetrical about a vertical plane, all members lie in one plane, and loads are applied normal to the plane of the structure.

A procedure for the approximate analysis for curved girder bridges of the open framing type has been proposed as follows:[3]

(*a*) Compute moments and shears on the assumed straight girders with spans equal to the developed lengths of the curved girders.

(*b*) Curve girders and compute the effect of curvature; determine end moments and shears on diaphragms required to restrain the girders from twisting; then compute secondary moments and shears on main girders due to these diaphragm end shears as indicated in step (*a*).

(*c*) Superimpose results and compute stresses due to all effects.

Good results, usually with an accuracy of less than 10 percent error, have

Fig. 15-13   Intercity Viaduct, Kansas City.

Fig. 15-14 Pasadena, Golden State Freeway, Los Angeles.

Fig. 15-15 North Freeway Overpass, Houston.

751

been obtained by this procedure. However, it is noted that for extremes of curvature, girder spacing, diaphragm spacing, etc., the approximate analysis may yield results more or less conservative. If the analyst lacks experience and has reservations about using an approximate method of analysis, he should resort to the more exact procedures.

A factor that warrants special consideration in the design of curved girders is torsion. Many curved girders are wide-flange beams or plate girders of I-section. These sections do not satisfy the assumptions normally imposed in a simple torsional analysis. Consequently, the simple torsional shear stress formula does not apply. The most significant torsional stiffness is derived from "warping" torsion. This type of torsional resistance is obtained through lateral bending of the top and bottom flanges.

The torsional moment T_b is resisted by shears F in the flanges equal to T/h. These flange shears cause lateral bending in the flanges which is dependent on the type of restraint offered to the flanges by the diaphragms and/or lateral bracing. For curved bridges it is normally assumed that the floor slab provides complete lateral restraint for the top flange for live loads which are applied after the floor slab is in place.

Improved techniques of analysis and design and increased emphasis on esthetics should result in broader acceptance of bridge structures on curved alignment. Several references are listed for those interested in pursuing further the subject of analysis and design of curved girders.

15-7 CABLE-SUPPORTED BRIDGES

Conventional steel bridges of short or intermediate span lengths have been built normally of wide-flange beams, plate girder, or trusses, whereas suspension bridges are built for exceptionally long spans. For longer spans of the medium range a cable-supported bridge, sometimes referred to as "stayed girder" bridge, may prove to be an economical and esthetical solution. A cable-supported bridge is a conventional plate girder or box-type bridge with intermediate supports provided by cables, hung from the top of one or more tower supports. This type of design permits shallower construction, which may be a significant factor.

The evolution of cable-supported bridges has taken place since 1950, although a similar form of bridge construction was proposed as early as 1784 by C. J. Löscher in Fribourg and about 1821 by Poyet in France. An approximately 110-ft long footbridge with sloping suspension members was built in England in 1817. A 256-ft span bridge constructed over the River Saale near Nienburg in 1824 collapsed the following year as a result of overloading by a crowd of people. At the time the knowledge necessary for designing and constructing such structures successfully was still inadequate. The suspension members consisted of forged tie bars or of chain links made of looped wires. A cable-supported bridge making use of multiple cables was built by Hatley around 1840. An even larger bridge of this general type is the Albert Bridge over the Thames, with a main span of 400 ft, constructed in 1873. In this structure the suspension system consisted of tie members converging at the tops of the towers. There were three sloping tie members on each side of the center span and four on each side of the end spans.[4]

A cable-supported bridge structure can be analyzed similar to a continuous beam on elastic supports. In such an analysis it is necessary to predetermine the sectional and geometrical properties of the cables, tower supports, and beam sections. After an initial analysis, it may be necessary to adjust certain of these properties to obtain a satisfactory state of stress. The effect of axial force in the beam sections due to the inclined cables must be considered in the final analysis. The extent to which the axial force may influence the design of the beam sections will depend on the cable arrangement. For the design of a cable-supported bridge, references 5 and 6 contain valuable information regarding the design of the cable-girder system and the design of the towers, respectively.

With this type of structure attention should be given to the possibility of generating any of the natural periods of vibration.[7] As with most cable-supported structures, because of their greater flexibility they are somewhat more susceptible to undesirable vibrations than conventional beam framing. Several methods, such as the use of mechanical dampers and/or the attachment of tie-down cables to the main cables, have been used successfully to obviate undesirable vibrations.

The 688-ft Pont des Iles Bridge in Montreal is an excellent example of a cable-supported bridge (Fig. 15-16). This structure consists of two simple spans bearing on the central pier and suspended at midspan by cables fixed at the top of a concrete tower. In this case, simple spans were selected so that they would contribute individually in causing bending in the tower and an unbalanced live load would not tend to release stress in the cables on one side of the tower. The spans consist of two longitudinal box girders supporting floor beams plus transverse box girders at the hanger points. The total width of the bridge deck is 94 ft.[8]

Another cable-supported bridge in the Montreal area is the Galipeault Bridge (Fig. 15-17). This structure is somewhat unusual because the cable supports are located at the end of the span on one side of the tower and at an intermediate point in the longer span on the other side.

The Severin Bridge at Cologne completed in 1959 has a main span of 991 ft, including a length of 400 ft without intermediate supports (Fig. 15-18). The A-shaped tower is about 204 ft high and all the cables are fixed directly

Fig. 15-16   Pont des Iles Bridge, Montreal. (Courtesy of CISC)

to the top. The tower is connected rigidly to the pier in the longitudinal direction of the bridge. The bridge is 97 ft wide and has two box-section main girders with a maximum depth of 15 ft 2 in. in the main span.

The Theordor Heuss Bridge (North Bridge) over the Rhine River at Düsseldorf was completed in 1958 (Fig. 15-19). It is the first cable-supported bridge built in Germany and has a center span of 853 ft. The structure is 87 ft 6 in. wide and has a cross section similar to the Severin Bridge of two box-section main girders. The towers are fixed to the girders and are 131 ft high.

An informative discussion of cable-supported bridge construction has been presented by Feige.[9]

15-8   CABLE-SUPPORTED ROOFS

Cable roof structures may be categorized into two basic types, suspension and cable supported.

In a suspension structure the cables are the main supporting elements

Fig. 15-17 Galipeault Bridge, Ste-Anne-de-Bellevue, P.Q., Canada. (Courtesy of CISC)

Fig. 15-18   Severin Bridge at Cologne. (Courtesy of AISC)

Fig. 15-19   Theordor Heuss Bridge over the Rhine River. (Courtesy of AISC)

for the roof deck and their curvature is a major consideration in the load-carrying capacity of the system.

Cable suspension structures may be subdivided into the following types:

(*a*) Single curvature

   (*a*) Parallel (Fig. 15-20)
   (*b*) Radial (bowl or dish shaped) (Fig. 15-21)

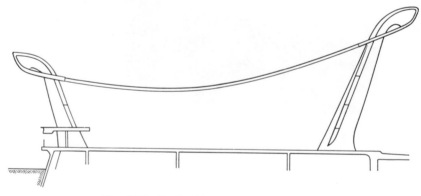

Fig. 15-20   Dulles Airport. Cable span 161 ft.

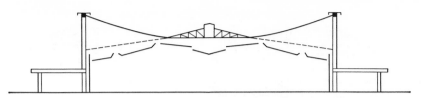

Fig. 15-21   Villita Assembly Building. Cable Span 132 ft.

(*b*) Double curvature

    (*a*) Hyperbolic-paraboloid (Fig. 15-22)

    (*b*) Conoid

(*c*) Double surface

    (*a*) Bicycle wheel (Fig. 15-23).

In a cable supported system (Figs. 15-24 and 15-25) the cable serves as the

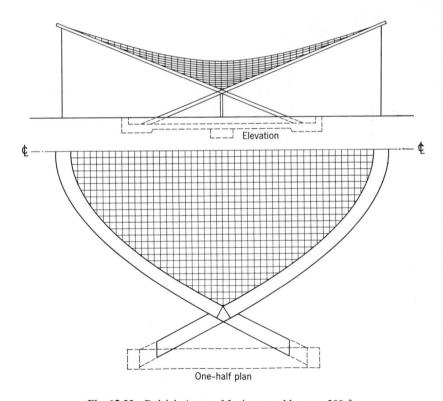

Fig. 15-22   Raleigh Arena. Maximum cable span 300 ft.

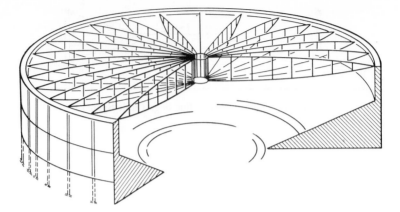

Fig. 15-23   Utica, New York, Auditorium.   Diameter 250 ft.

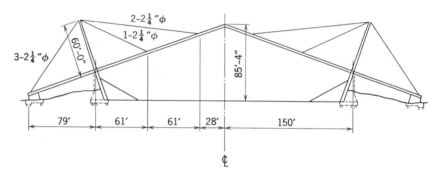

Fig. 15-24   Blyth Arena.

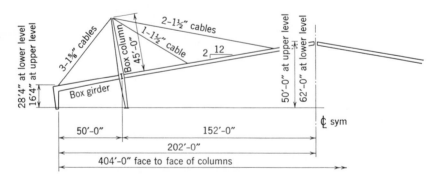

Fig. 15-25   Elevation of main frame.   Tulsa Exposition Center Building.

758

main tension element, the mast serves as a vertical compression member, and girders as flexural members. With this system the load-carrying capacity of the structure is not a function of the cable curvature which is practically straight, with the exception of sag due to its own weight.

The basic assumptions in cable supported roofs are as follows:

(*a*) Each half of the roof is designed as an independent cantilever roof.
(*b*) The column bases are assumed hinged.
(*c*) The roof girder is continuous at the mast and is hinged to the mast.
(*d*) The mast carries only axial stresses.

If the geometry of cable-supported structures, as defined, is specified as the dead load position, then the support provided to the roof girder at the connection of the cables is considered static under dead load. However, under live load these supports become elastic due to the stretching of the cables. The design of the roof girder becomes more complex for the condition of elastic supports.

With the advent of many new steels, new manufacturing methods, and new construction techniques, the technology of strand and cables has improved considerably.

The ultimate tensile strength of modern high strength cables in tension is four to six times that of structural carbon steel and at a cost per pound of only twice that of the steel. This naturally leads to an economy of materials.

During the past few years there have been constructed in various parts of the world as well as in the United States a considerable number of unusual roof structures using steel cables as the principal structural component.

The concept of bridging large spans with cables is not new. Primitive man constructed footbridges of hemp and bamboo cables. The early Chinese built suspension bridges of rope and iron chains. The Romans used wire strand and rope many centuries ago. A handmade specimen of their 1-in. by 15-in. long lay bronze rope is displayed at the Music Barbonico in Naples, Italy. The roof of the Roman Colosseum, built in 70 A.D., consisted of rope cables anchored at a center mast spaning radially to provide a sunshade cover for the arena.

Early attempts at the use of cables as tension members can be traced back to V. G. Shookhov, a Russian engineer, who in 1896 covered four pavilions at an exhibition at Nijny-Novgorod and used the same scheme at the Bary Boilerworks in Moscow. In 1933 a suspended structure for a locomotive roundhouse pavilion was constructed at the Chicago World's Fair. Suspended structures of greater antiquity can be found in tent structures from the days of Omar Khayám and earlier to the circus "Big Top" of not too many years ago.

15-9 ANALYSIS OF SUSPENSION ROOF SYSTEMS

The analysis and design of suspension systems involve the problem of determining the shape, maximum tension, and length of a cable under a given stationary load. If the loading is in the plane of the suspended cable, simple analytical solutions can be derived.

Once the cable geometry of length and curvature under dead load are known, the problem becomes one of calculating the tension, elongation, and change of shape of the cable under live load.

In Fig. 15-26a, let AB indicate a segment of cable supporting a vertical load of magnitude $p = p(x)$ which may or may not be constant, depending on whether the cables in a system are parallel, in which case p is uniform or radial, in which case a triangular distribution of p acts on the cable.

To maintain horizontal equilibrium the horizontal component H of the cable tension must be constant throughout the length of the cable.

For vertical equilibrium:

$$dV = p\, dx \tag{15-14}$$

However,

$$dV = V'\, dx \tag{15-15}$$

Then

$$V' = p(x) \tag{15-16}$$

By taking summation of moments about end A, the following equation results:

$$H\, dy + p\, dx\left(\frac{dx}{2}\right) - (V + V'dx)\, dx = 0 \tag{15-17}$$

which reduces to $Hy' - V = 0$ when dx^2 is ignored as being very small when compared to dx.

Then

$$Hy'' = p \tag{15-18}$$

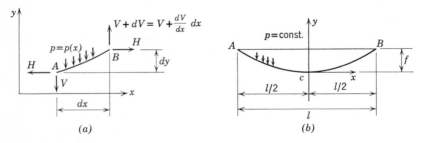

(a)   (b)

Fig. 15-26

from which

$$Hy = \int \int p \, dx \, dx + Ax + B \tag{15-19}$$

and with p equal to a constant this becomes

$$Hy \cdot \frac{px^2}{2} + Ax + B \tag{15-20}$$

The cable tension is given by

$$T^2 = H^2 + V^2 \tag{15-21}$$

or

$$T = (H^2 + V^2)^{1/2} = H\left[1 + \left(\frac{V}{H}\right)^2\right]^{1/2}. \tag{15-22}$$

But

$$y' = \frac{dy}{dx} = \frac{V}{H} \tag{15-23}$$

Then

$$T = H[1 + (y')^2]^{1/2} \tag{15-24}$$

Now consider a cable (Fig. 15-26b) on level supports A and B with its lowest point at C and with a constant load p per unit length of horizontal projection x.

At $x = 0$, $y = 0$ and $y' = 0$ and Eq. 15-17 reduces to

$$y = \frac{px^2}{2H} \tag{15-25}$$

and therefore

$$y' = \frac{px}{H} \tag{15-26}$$

For the special condition of level supports and with $x = l/2$ and $y = f$, these results yield

$$H = \frac{pl^2}{8f} \tag{15-27}$$

and

$$T = H\left[1 + \left(\frac{px}{H}\right)^2\right]^{1/2} = H[1 + 16n^2]^{1/2} \tag{15-28}$$

where n is defined as the sag ratio f/l.

The ordinate of the cable for this condition is determined as follows: with

$$H = \frac{pl^2}{8f} \tag{15-29}$$

then

$$y = \frac{px^2}{2H} = \frac{4f}{l^2} x^2 \qquad (15\text{-}30)$$

The length of the cable curve may be determined as follows:

$$S = \int ds = 2 \int_0^{l/2} \left[1 + \left(\frac{dy}{dx}\right)^2\right]^{1/2} dx = 2 \int_0^{l/2} \left[1 + \left(\frac{8fx}{l^2}\right)^2\right]^{1/2} dx$$

$$= 2 \int_0^{l/2} \left\{1 + \frac{64f^2x^2}{l^4}\right\}^{1/2} dx$$

$$= 2 \int_0^{l/2} \left(1 + 32\frac{f^2x^2}{l^4} - 8 \times 64\frac{f^4x^4}{l^8} + 4 \times 64^2\frac{f^6x^6}{l^{12}} \cdots\right) dx \qquad (15\text{-}31)$$

By binominal expansion and substitution of the sag ratio $n = f/l$

$$S = l[1 + \tfrac{8}{3}(n)^2 - \tfrac{32}{5}(n)^4 + \tfrac{256}{7}(n)^6 \cdots] \qquad (15\text{-}32)$$

For ratios of n of the order of $\tfrac{1}{5}$ sufficient accuracy may be achieved by truncating this expression to[10]

$$S = l[1 + \tfrac{8}{3}(n)^2 - \tfrac{32}{5}(n)^4] \qquad (15\text{-}33)$$

and for smaller values of n sufficient accuracy is given by[10]

$$S = l[1 + \tfrac{8}{3}(n)^2] \qquad (15\text{-}34)$$

By similar procedures equations may be derived for inclined cables subjected to uniform load and cables on level supports subjected to triangular loading.

A suspended roof structure presents a problem in aerodynamic instability to the engineer. The structure must be analyzed to resist not only static loads but also those dynamic effects due to wind, earthquake, and sonic waves. Because of their greater flexibility cable systems may be subjected to dynamic motions caused by any one of these effects. If the disturbing force is in the opposite direction to that of the static loading condition, the dynamic motion may be entirely or partially damped out; however, if the disturbing force is in the same direction as that of the static loads, the motion can become self-exciting unless the structural damping force is higher than the aerodynamic force.

If the structural damping is not of sufficient magnitude to eliminate the aerodynamic motion, the structure becomes dynamically unstable and the amplitude of vibration becomes larger and larger until the structure finally destroys itself. This phenomenon of vibratory motion is referred to as flutter. The wind velocity at which the structure becomes dynamically unstable is referred to as the critical wind velocity. This phenomenon is best illustrated by the spectacular disaster of the 2800 ft Tacoma Narrows Bridge, which was hit by a mild gale on November 7, 1940 and began to undulate and destroy itself in a relatively short period of time.

The mathematical analysis for describing the phenomenon of flutter and self-exciting vibrations and determining critical wind velocities in suspension roofs is rather complex. All structures have a natural frequency range such that if an externally applied force produces a vibration within this frequency range, the structure is caused to vibrate or flutter. Thus a vibrational condition may be reached when the effects of external and internal forces are in resonance and the structure collapses.

Aerodynamic forces which act on a suspension structure are dependent upon the shape and natural frequency of the structure, as well as the nature of the wind itself.

By designing such a structure that the critical wind velocity is sufficiently high, the danger of flutter may be eliminated for all practical considerations. Increasing the critical wind velocity may be achieved by (*a*) changing the shape of the structure, (*b*) increasing the structural damping characteristics, (*c*) increasing the natural frequency, and (*d*) combinations of these.

Because the effect of shape can be determined only by wind tunnel tests, the actual effect is usually not subject to control at the design stage, although previous tests can indicate generally what shape may be desirable to use.

Structural damping is dependent on the internal friction of the components, connections, anchorages, etc., and these are seldom known at the design stage.

The natural frequency is the only quantity which can be determined mathematically at the design stage, and is thus the only quantity over which the designer has control.

The frequency of a cable may be determined by[11]

$$W_n = n\left(\frac{\pi}{l}\right)\left[\frac{T}{(p/g)}\right]^{\frac{1}{2}} \tag{15-35}$$

where W_n = natural frequency
n = any integer (function of the mode of vibration)
l = span
T = maximum cable tension
p = load per increment of horizontal projection
g = acceleration due to gravity

In suspension structures, the light, high-strength cables acting in tension are extremely sensitive to unequal loadings so that vibration and flutter become major design considerations to be evaluated.

15-10   PRESENT DESIGNS AND FUTURE DEVELOPMENTS OF CABLE ROOF SYSTEMS

The Dulles International Airport at Chantilly, Virginia, built in 1962, is an example of a single curvature cable suspended structure with parallel cables (Fig. 15-20). The reinforced concrete pylons that act as anchorages for the cables are spaced 40 ft on centers. Two 1-in. steel cables spaced transversely on 10-ft centers support each transverse line of precast concrete roof panels. The cables were pretensioned and encased in a poured-in-place reinforced concrete rib, thus producing a continuous roof surface.

The idea of using single curvature radial tension members connected to compression and tension rings has been successfully used in the 110-ft diameter French Pavilion built at the Zagreb Fair in 1935 and the 308-ft diameter Municipal Stadium at Montevideo, Uruguay, built in 1959. The 132-ft diameter Villita Assembly Building (Fig. 15-21) is a steel-framed structure with the circular compression ring supported on 20 columns. From the compression ring two hundred $\frac{11}{16}$ in. diameter cables drape to a 40-ft diameter central tension ring which supports a trussed roof.

The first major modern application in cable suspended roof structures in the United States was in the early 1950s when the Livestock Judging Pavilion for the North Carolina State Fair at Raleigh (Fig. 15-22) was built. It was conceived by the late Matthew Nowicki, a young Polish architect. The basic principle in this structure is that of two arches on inclined planes supporting a transverse system of downward curving cables. A secondary system of upward curving cables placed at right angles to the main cables achieves a surface of double curvature referred to as a saddle-back hyperbolic paraboloid.

The Marie Thumas Pavilion built for the 1958 Brussels World's Fair expresses the principle of the conoid in its roof surface. A conoid surface is generated by a straight line that moves with one end along a straight line and the other end along a curved line, producing a surface of double curvature.

The United States Pavilion at the 1958 Brussels World's Fair, the New York State Pavilion at the 1964–1965 World's Fair in New York, and the 240-ft diameter roof for the Utica, New York, Auditorium (Fig. 15-23) are all examples of the double surface bicycle wheel concept.

The Blyth Arena at Squaw Valley, California (Fig. 15-24) and the Tulsa Exposition Center at Tulsa, Oklahoma (Fig. 15-25) are examples of cable-supported roof structures.

Man has long considered the possibility of enclosing under one roof vast areas, such as cities, where his environment could be controlled for year-round comfort in any geographical region on the earth. This dream may not be too far from reality when one considers that cable structures with relatively large spans have already been constructed or are in the planning stages. Roofs of spans in miles rather than feet are being considered.

15-11 SPACE STRUCTURES

Space structures refers to those classes of structures that resist applied loads by structural action in several directions simultaneously and are stable under any general system of loads. These structures may be pin-connected, rigidly framed, or a combination of both.

Many, and perhaps most, space roof structures are of a type that may be analyzed assuming pin-connected joints, bending being a truly secondary effect. A definition for this type of structure has been proposed as follows:

"The space truss is a three-dimensional framework of straight bars connected together by frictionless hinges in such a manner that it is stable and capable of resisting forces applied in any direction."[12]

Although it is recognized that actual structures do not have joints of frictionless spherical hinges, there are nevertheless many structures for which this assumption is valid in an analysis for the primary forces.

The analysis of space trusses has been based largely on experience and simplifying assumptions derived from the knowledge of coplanar trusses. For some types of space trusses this approach might give good results, but for the general case it cannot be relied upon for final analysis. With the increasing complexity of designs and with the development of new and higher-strength materials leading to more flexible structures a more exact solution is warranted. With the general availability of electronic computing equipment and the development of standard computer programs for general structural analysis accurate solutions to complex problems are readily available. However, these solutions are only as exact as the assumptions made by the analyst in the preparation of the computer program.

Because most space structures are highly statically indeterminate, it is necessary that the analyst makes initial "guesses" regarding member sizes. After the initial analysis is completed, the original member sizes can be adjusted as indicated by the results of the analysis. Depending on the closeness of the first "guess" and the discreteness of later adjustments, one or more analyses may be required.

Möbius[13] initiated the general theory of three-dimensional systems in 1837. He determined the required number of bars for a three-dimensional framework to be stable and he also discussed "critical" forms and developed the zero-load test for stability.

Möbius' work in three-dimensional systems remained virtually unknown to engineers. Thus, the theory of space trusses was developed independently. Probably one of the most outstanding individuals in developing the theory of space trusses was August Föppl.[14] Föppl's book written in 1892 was a collection of his work in space frameworks and in it he considered for the first time many important topics concerning space trusses. This book has served as the basis for much of the later work in this field.

Schwyzer[15] originated the method of plate analogy for analyzing space trusses in his dissertation in 1920. The work of Schwyzer was extended and published by Stüssi.[16,17,18] He also investigated curved space trusses and skewed space truss beams. The plate analogy of Schwyzer was summarized and published in the United States by Andersen and Nordby.[19] A brief description of the plate analogy method was presented by Niles and Newell.[20] Holloway[21] and Gillespie[22] further developed the plate analogy method and applied it to practical structures.

Tension coefficients were introduced into the analysis of space frameworks by Southwell.[23] A solution to the problem of analyzing space structures by substituting for the forces of the three-dimensional structure a corresponding system of coplanar forces was presented by Mayor.[24] A similar solution but with a slightly different approach had been published earlier by Mises.[25] The work of Mayor was simplified and published in the United States by Constant.[26]

The application of Castigliano's theorem and Saint-Venant's principle to space frameworks was illustrated by Southwell.[27] General discussions of various types of space structures and numerous illustrations of actual structures have been compiled and presented by Makowski.[28]

Two basic philosophies for general structural analysis are in existence: the stiffness (or displacement) method and the flexibility (or force) method. All other specialized approaches to structural analysis for certain types of structures are based on the fundamental concept of either of these methods. Both lead to the same result; however, the basic formulations of the mathematical models are reciprocal to each other. Depending on the familiarity of the analyst and the quantities, displacements or forces, desired, either the stiffness or the flexibility method could prove to be the most convenient approach to the analysis of any specific problem.

The stiffness method is formulated by expressing member end forces in terms of end displacements and applied loads. Then by establishing force equilibrium and displacement compatibility at each joint, a system of equations in terms of unknown joint displacements is obtained. This system of equations can be solved to obtain the joint displacements directly, which can be substituted into the force-displacement relationships to yield member forces.

The flexibility method is formulated by expressing member end displacements in terms of end forces and applied loads. Then by establishing displacement compatibility and force equilibrium at each joint, a system of equations in terms of unknown forces is obtained. This system of equations can be solved to obtain the member forces directly. These calculated forces can be substituted into the displacement-force relationships to obtain displacements if needed.

One of the largest flat space frame structures built in the United States is the Pauley Pavilion at the University of California at Los Angeles (Fig. 15-27). This structure provides a 300- by 400-ft clear span. A special detail (Fig. 15-28) was developed to reduce overall structural costs. The structural members in this design are wide-flange shapes. Although the final analysis was accomplished with the aid of an electronic computer, the first approximation of member sizes was accomplished by considering the space frame as a flat plate simply supported on four edges. Columns are spaced around the exterior at 33 ft 4 in. on centers.[29]

An exhibition hall in Denver covers an area 240 by 685 ft (Fig. 15-29). It consists of four equal bays, each supported on four space frame columns. Much of the framing is composed of 4-in. square tubular sections having a yield point of 50,000 psi. Some heavily loaded members have yield strengths

Fig. 15-27 Pauley Pavilion, University of California at Los Angeles. Courtesy of Welton Becket and Associates, Architects and Engineers

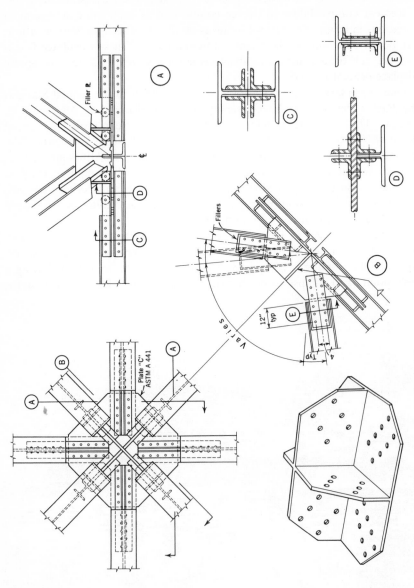

Fig. 15-28 Connection details, Pauley Pavilion. (Courtesy of Architectural Record.)

Fig. 15-29 Denver Exhibition Hall. (Courtesy of Engineering News-Record).

Fig. 15-30 Special connector, Denver Exhibition Hall. (Courtesy of Engineering News-Record.)

as high as 100,000 psi. A specially designed connector was utilized to reduce overall fabrication and erection costs (Fig. 15-30), and to accommodate up to twelve intersecting members.[30]

The trend toward providing controlled environments for sporting events, shopping complexes, etc., is demanding requirements for covering large areas. In many cases, column-free space is either highly desirable or required. This trend combined with improved techniques of analysis, design, fabrication and construction is encouraging greater use of space structures.

15-12 PRESTRESSED STEEL STRUCTURES

Prestressed steel structures can be generally classified as follows:

(a) Steel flexural members or frameworks that are bent after fabrication by the reaction of attached external tensioned bars or cables in a direction opposite to the anticipated service load direction of bending.

(b) Steel flexural members in which residual stresses opposite in sense to the anticipated service load stresses have been induced during the fabrication process.

(c) Steel struts intended for use as tension resisting members in which axial compressive stresses have been induced from the end-anchorage reaction of colinear tensioned strands or bars.

(d) Steel members which have been axially pretensioned by jacking for various special reasons.

The first three types of prestressing produce initial stresses in the member having a sense opposite to the sense of the stresses caused by the service loading so that the magnitude of service loading that will cause yielding will be increased. However, in all three types it must be realized that the ultimate strength of the structure is generally not affected by prestressing. The ultimate strength is merely the sum of the ultimate strengths provided by the various components of the member including the prestressing components.

The art of prestressing steel bridge spans and similar types of structures by external bars and cables has been practiced for many years. When roadway bridges first became subjected to loads comparable in general magnitude to modern truck loadings, such as the steam tractors with 10,000 lb wheel loads[31] which were in service during the early part of the twentieth century, the strength of new highway bridges had to be significantly greater than current designs, and the carrying capacity of some existing bridges has to be increased. Particularly in existing bridges, increased capacity was frequently achieved through "hog-tying"—extending cables or bars with turnbuckles from end to end of the bottom chords or flanges of the bridges and strutting them away from the chords or flanges (Fig. 15-31) at one, or more positions.

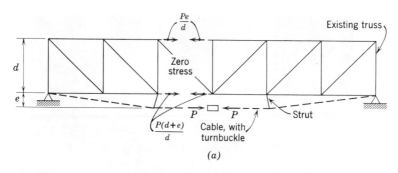

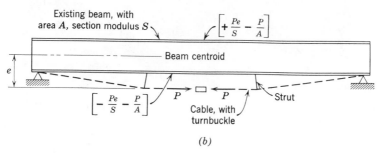

Fig. 15-31 Prestressing by bars or cables strutted from span. (a) Cable—prestress forces in truss. (b) Cable—prestress stresses in beam.

After the turnbuckles were tightened, the stresses thus induced enabled a heavier vehicle to cross the bridge without causing yielding of any members. External prestressing has been used similarly to increase capacity of roof arches (Fig. 15-32) and similar structures, which could not be provided economically otherwise.

During the early 1960's the Iowa State Highway Commission constructed an experimental 240-ft long three-span continuous bridge in which constructional alloy steel-cover plates were welded to structural carbon steel beams while the beams were held by jacks in a flexed position. This process which is described later, resulted in beneficially induced residual stresses in the fabricated beams.

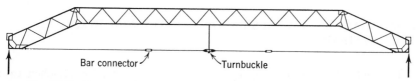

Fig. 15-32 Prestressed truss at a Harlow, England, Plant. (Courtesy AISC.)

Fig. 15-33   Detail of anchor tie being jacked against Wale. (Courtesy of E. D'Appolonia, Consulting Engineers Inc., Penn Hills, Pa.)

In foundation engineering it is frequently necessary to anchor the upper portions of sheet piling bulkheads to a back portion of the mass of earth and/ or rock that the bulkhead serves to retain. The conventional installation of "dead-man" anchorages to accomplish the tie-back is expensive and if the retained mass is a hillside may be virtually impossible. Instead, it is frequently convenient to drive steel H piles at a sloping inclination into the earth mass, stopping after sufficient friction has been developed over the pile surface. Then, the free end of each H pile may be jacked outward, reacting against a horizontal steel wale that is connected to the outside face of the sheet piling (Fig. 15-33). The jacks thus produce a pretension in the H piles. This serves to proof load the piles and results in a precompression of the earth immediately behind the bulkhead. This is beneficial because it increases the shearing strength of the soil, thereby tending to stabilize the earth mass behind the bulkhead.

15-13   ANALYSIS AND DESIGN OF PRESTRESSED STEEL SYSTEMS

The method of prestressing by tensioning a bar or cable attached to a member or structure has many variations (Figs. 15-31, 15-32, and 15-34), but the stresses or forces in the center portion where the bar or cable is parallel to the structure being prestressed depends only on the prestress force P and eccentricity e. The expressions for thrust and stress indicated in Fig. 15-31 are generally valid for evaluation of primary effects. Secondary effects, caused by beam-column action, must be considered separately. When the turnbuckles are tightened, the net effect of the forces of the prestressing cables or bars at any station on the structure is the same as applying an eccentric compression force on the beam or truss cross section. This action produces forces and stresses of a sense opposite to those caused by the superimposed live load.

In all cases such prestressing can result in an increase in the live load that initiates yielding if the prestressing elements are outside the "kern" of the section. The kern is a geometrically defined zone in the vicinity of the centroid of the member being prestressed. Where functional details permit, it is generally desirable to maximize the distance between the prestressing element and the centroid of the member being prestressed. The stress analysis consists of establishing the fact that the prestressing elements are not overstressed and that neither the tops nor bottoms of the beams or trusses are overstressed (*a*) with the prestress but without live load or (*b*) with the prestress and the maximum live load. If the structure is subject to repeated loading, fatigue analysis may impose additional limitations.

Two methods will be mentioned for inducing residual stresses by various

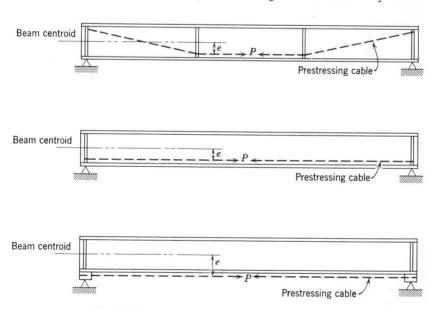

Fig. 15-34   Compact external prestressing of beams.

fabricating techniques in components of beams. One method is to weld the stem edge of an unstressed lower strength steel T-section to a higher-strength steel plate that is subjected to a high elastic tensile strain during the welding (Fig. 15-35). Release of the external tension after fabrication has the same affect as applying a precompression at the ends of the higher-strength plate of the composite beam. However, the heat of welding will result in some loss of prestress because it is not practical, and perhaps not possible to vary the plate stretching during the fabrication so that a constant tension is maintained in the plate until the composite beam has cooled.

A more easily controlled method of inducing residual stresses consists of jacking a lower strength steel beam in a direction which produces tension in one flange and compression in the other flange (Fig. 15-36). With the beam held in the jacked position, higher-strength plates are welded or otherwise suitably connected to the beam flanges. Finally, the release of the

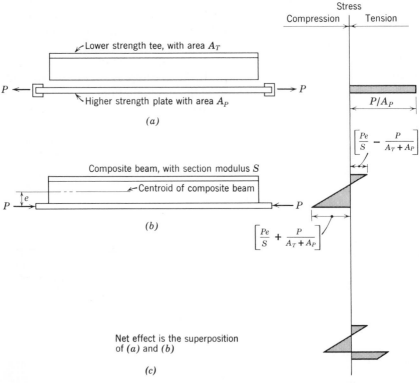

Fig. 15-35 Residual stresses from welding a prestressed plate to a tee: (*a*) pretension is maintained in plate while being joined to tee, (*b*) release of external restraint after fabrication is equivalent to applying end forces on composite member, and (*c*) resultant stresses are sum of these two effects.

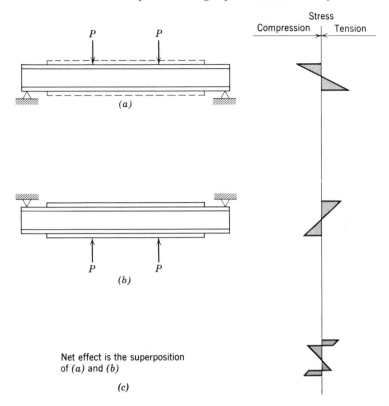

Fig. 15-36 Residual stresses from attaching cover plates on beam in jacked position: (*a*) beam jacked downward while higher strength plates are welded to flanges, (*b*) release of jacks of the fabrication is equivalent to jacking upward, and (*c*) resultant stresses from attaching cover plates on beam in jacked position.

jacks is equivalent to exerting forces equal and opposite to the jacking forces on the composite beam. This reduces the tension and compression in the original beam flanges, but produces precompression in the plate attached to the beam flange which was originally in tension, and pretension in the plate attached to the beam flange which was originally in compression. A variation of this jacking method is the patented "Preflex Technique" in which a concrete encasement instead of a steel plate is attached to the beam tension flange when the beam is in the jacked position.

In all of the prestressing methods which induce residual stresses it is necessary, as for external prestressing, to establish that no parts of the beam or attachments are overstressed either before or after the design load is super-imposed.

If in an existing structure additional high-strength strands or bars can be

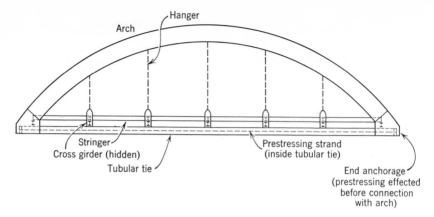

Fig. 15-37   Schematic elevation of tied arch bridge with prestressed ties.

used to reinforce lower-strength tension struts such as the ties of a tied arch (Fig. 15-37), it may be advantageous to cause a precompression in the strut such that under a service loading which produces tension the strain in the strut that can occur without yielding is more nearly equal to the strain in the reinforcing member that can occur without yielding. Thus, if a high-strength strand of area A_s and yield strength Y_s is used to reinforce a tubular tie member of area A_t and yield strength Y_t, the possible increase in load-carrying capacity of the reinforced tie, assuming that the tube does not yield in compression during the prestressing or that the strand does not yield in tension under service loading may be calculated as follows:

Condition	Ratio of Carrying Capacity without Yielding after Installation of Strand to Capacity before Installation of Strand
No prestress in strand	$$\left[1 + \left(\frac{A_s}{A_t}\right)\right]$$
Strand prestressed to fraction P of its yield strength	$$\left[1 + \frac{PY_sA_s}{Y_tA_t}\right] \times \left[1 + \left(\frac{A_s}{A_t}\right)\right]$$

It is desirable to use a high value for P in order to increase the ratio.

It can be shown that if the prestressing element is restrained from lateral movement with respect to the strut at sufficiently close intervals, there will be no tendency for the prestressing to cause column buckling of the strut. However, the possibility of local buckling of the strut has to be investigated as in conventional compression members.

15-14   CURRENT TRENDS OF PRESTRESSED STEEL SYSTEMS

With the "clean look" demanded of modern bridges the use of prestressing bars or cables exterior to the structure profile is generally not acceptable for new construction. However, serious studies are in process to investigate using external prestressing elements located generally within bridge beam profiles (Fig. 15-34). In such designs material savings can be readily demonstrated, but lowering the fabrication costs and providing adequate protection against corrosion of the prestressing elements appear to be the major problems to be solved.

As noted the procedure of jacking a steel beam and then welding a higher-strength steel cover plate appears to be the best way to lock in a beneficial prestress consisting of residual stresses. However, there is a problem of reducing fabrication costs so that the savings in material cost will be reflected in total cost savings.

Axial prestressing to induce uniform precompression in members that are components of structures is a proven method of prestressing steel as evidenced by its use in certain large structures, such as shown in Fig. 15-38. The advantages of prestressing individual members are that (*a*) it is possible to induce just the desirable prestress without undesirable secondary stresses in other members and (*b*) the prestressing elements need not be exposed.

Before it is decided to use prestressed steel in new construction, consideration should be given to the alternate solution of using higher strength steels in the main members, thus eliminating the need for prestressing. Prestressed steel is a possible solution for certain specialized design problems but does not, at present, appear to have general application.

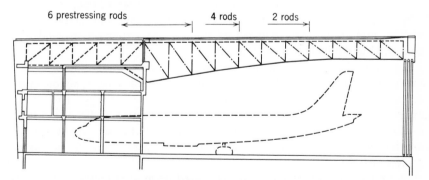

Fig. 15-38 Cantilevered hangar truss with prestressed steel rods along upper chord United Air Lines hanger at Chicago's O'Hare International Airport. (Courtesy AISC.)

15-15 TUBULAR STRUCTURES

Hot-rolled carbon steel hollow structural tubing is made in round, square, and rectangular shapes from a continuous welding or seamless process. Tubes are manufactured to applicable structural specifications ASTM A36, A501, and proprietary grades of high-strength low-alloy steels used for construction. Square tubes are produced in various sizes up to perimeters of 40 in., whereas rectangular tubes are limited to perimeters of 32 in. Pipe made to A53 Grade B is frequently used for structural design at values of A36 steel.

In the continuous weld process coils of steel called skelp are welded end-to-end to form a continuous band of steel and passed through furnaces. As the skelp exists from the furnace, it is formed into a round pipe and welded by pressure (Fig. 15-39). The pipe then goes to a stretch-reducing mill where it is brought to the desired diameter and wall thickness.

In seamless tubing a solid round bar of predetermined size is heated and then pierced by a mandrel while rotating at high speed (Fig. 15-40). The tube then proceeds to other rolling operations which bring it to the proper diameter and wall thickness. The round tube then passes to a sizing mill where it is formed into a square or rectangular shape.

The use of round and rectangular tubular members are becoming more generally adopted for structures. Tubes are of special interest to the architect from an esthetic viewpoint and to the engineer from a structural effectiveness viewpoint.

For a given weight there is no better section than that of the tube for torsional resistance. Under dynamic loading the tube has a higher frequency of vibration than any other section including a solid round one.

Resistance to wind of a round tubular section is lower than that of a plane section, and some codes will permit a one-third reduction in wind load compared to equivalent projected area.

A round tube may have as much as 30 to 40 percent less surface area than that of an equivalent rolled shape and thus reduces the cost of maintenance, cost of painting, fireproofing, and/or other protective coatings. The smooth external surface of the tube does not permit the collection of dirt or moisture, thus reducing the possibility of corrosion. If the tube ends are sealed, the interior surface is not subject to corrosion and therefore needs no further protection.

In the past the use of tubes was hampered because of connection details. A major contribution to the solution of this problem has undoubtedly been the development of fully automatic oxyacetylene tube-cutting machines which not only cut tubes to fit flat surfaces but also cut them to fit cylindrical

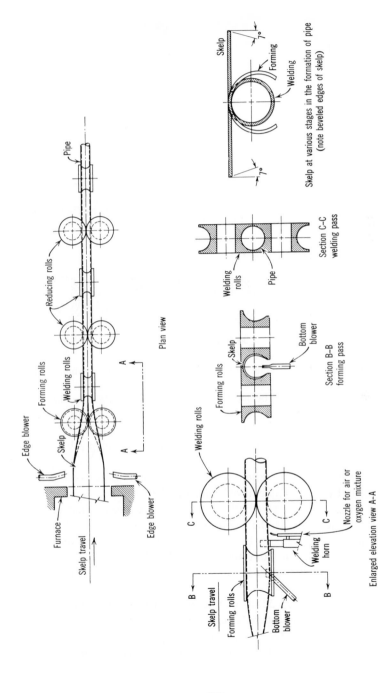

Fig. 15-39 Diagram depicting schematically the operations performed in a continuous forming and welding mill.

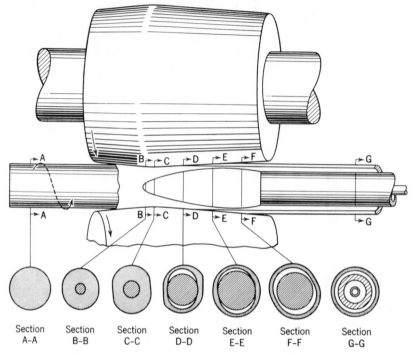

Fig. 15-40   Sketches illustrating action of rotary-piercing mill on the round billet.

surfaces such as tube to tube as well. The machine cuts the tubes to the correct profile with a beveled edge to simplify the welding process at the joint.

The use of tubes as structural components is not new—the Royal Albert Bridge in Saltash in England was built in 1859 using a riveted tubular section for the top chord.

Since World War II a large number of tubular structures have been constructed in Europe. There have been a number of applications of tubes in structures in England where mill buildings have been constructed with round tubular sections as the main frame. Trusses composed of tubular members have been erected with spans of 218 ft. An airplane hangar using tubular design has been built for the DeHavilland Engine Co. Square tubular columns were used in a building at the Langside Technical College in Glasgow, Scotland.

The French have built a 22 story building in Paris which employs concrete-filled tubular columns and tubular cross bracing. At Tjörn Island, Sweden, a 912-ft tubular arch bridge was constructed by the Germans in 1960. This bridge consisted of 12 ft 6 in. diameter tubes subassembled into 27 to 28 ft sections of $\frac{1}{2}$ to $\frac{7}{8}$ in. thick tubes.[33]

In the United States the giant Atlas and Saturn launching complexes at Cape Kennedy used round tubular sections as long bracing members.

The structural framing for the Air Force Academy Chapel at Colorado Springs designed by Architects-Engineers, Skidmore, Owings & Merrill, is composed of 100 welded tubular steel tetrahedrons. The main tetrahedron chords are 6 in. diameter tubes and the intermediate web members are 4 in. diameter tubes. All 100 tetrahedrons, connection members, and end-wall framing were erected in place and bolted, ready for field welding, in $89\frac{1}{2}$ working hours. After erection the tetrahedrons were given a final adjustment before final welding.

A cost comparison of tubular steel with rolled sections indicated that the increased unit cost for tubular members was offset by the increase in weight required for rolled shapes.[32]

15-16 DESIGN CONSIDERATIONS FOR TUBULAR STRUCTURES

The principal properties that govern the design of any structural member are its area, moment of inertia, section modulus, radius of gyration, and type of steel.

When a member is subjected to a simple tensile force, its effectiveness is dependent on its cross-sectional area, grade of steel, and its method of connection. Therefore a round, square, or rectangular tube, or any rolled shape such as a wide-flange beam with the same cross-sectional area and the same material, would have the same equivalent resistance to the force. In this type of application the tubular shape is of no advantage. Because of the higher cost of producing a tubular shape compared to a rolled section, the tubular section is at a disadvantage when tension is the only force under consideration.

For members in compression where buckling is critical the slenderness ratio l/r is the prime consideration for design. The radius of gyration r is dependent on the distribution of area about the center of gravity. Round tubes as well as square tubes have equal radii of gyration about the principal axes and are equally favorable with regard to column buckling. It may be well to point out that equal radii of gyration are only important when the unbraced or effective length is the same in all directions. For an unequal unbraced length a rectangular tube may be more economical.

The weight reduction of tubular columns compared with other rolled shapes varies and increases with increasing slenderness ratio. A comparison of tubular columns to columns of rolled shapes in the long column range where the slenderness ratio is greater than 120 is shown in Fig. 15-41.[33]

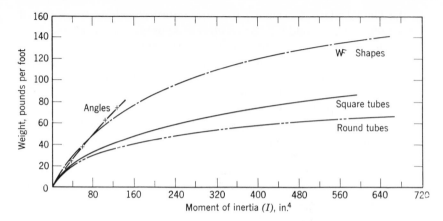

Fig. 15-41 Comparison of weights of structural members used as long columns.

The curves are plotted for varying moment of inertia because column strength varies as the moment of inertia of the cross section when column lengths and the modulus of elasticity are maintained. In Fig. 15-41 the weight of the section in pounds per foot is shown on the vertical axis and the moment of inertia on the horizontal axis corresponding to a slenderness ratio of 120 to 200. With columns of equal moments of inertia savings of about 30 percent are possible with square tubes when compared with angles and rolled shapes, whereas 50 percent savings can be achieved by using round tubes. Rectangular tubes are not considered in Fig. 15-41 because they would have less stiffness than a square tube for a given slenderness ratio.

Although the advent of welding and cutting equipment has done much to simplify the connection problems of tubular members, many design problems still remain. Considerable Research has been done by J. G. Bouwkamp[34,35,36] at the University of California, Berkeley, and by A. A. Toprac[37] at the University of Texas with respect to connections of round tubular members. The following summarize their findings.

Initially, the design of connections in round tube truss joints was patterned after practices used with standard rolled sections, for example, welding the tube member to a gusset plate (Fig. 15-42). This joint system has been found to be unsatisfactory for severe load conditions. The difficulty with this type of connection arises from the concentrated stress flow between the tube walls and the gusset plate even though the gusset plate is capable of transferring the member forces effectively.[35] The localized flow causes high stress concentrations in the tube at the start of the gusset plate.

A direct member-to-member joint connection is much more effective than the gusset plate connection because the stress distribution is more uniform in the tube walls. The parameters to be considered in a direct type joint

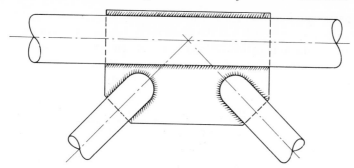

Fig. 15-42　Gusset-plate tube joint.

connection are as follows:

1. Ratio of the outer diameter d of the web member and the outer diameter D of the chord member d/D.

2. Ratio of wall thickness t and outer diameter D of the chord member t/D.

3. Length of weld between vertical and diagonal members at the joint related to joint eccentricity (Fig. 15-43).

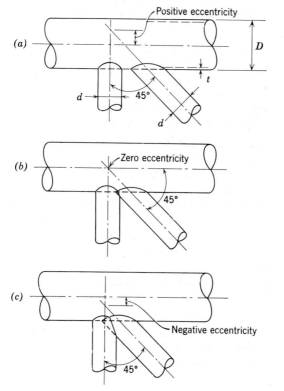

Fig. 15-43　Directly connected joints (Reference 35).

In a particular joint system the joint strength reduces as the d/D ratio decreases. Because of tension in the diagonal and compression in the vertical web members, the application of load to the chord tube wall becomes more severe with the decrease of d/D. This becomes particularly critical when the web members are not interwelded (Fig. 15-43a,b). In the situation where the web members are not interconnected (Fig. 15-3a) the total transfer of web member forces is through the wall of the chord member. Because the two web members produce opposite effects in the chord, large bending stresses result leading to early failure in the joint.

In a substantial interweld between the web members (Fig. 15-43c), a considerable portion of web member forces is transferred directly through the weld between the two web members. For an adequate length of intersection between the two web members, the incoming forces are transferred directly from one member to another without entering the chord tube, leaving the horizontal unbalanced load to be transfered into the chord wall by the weld connecting the web members to the chord member.[35]

Another problem has been the connection of rolled shapes to tubular columns. In an effort to solve this problem, research has been done at Cornell University[38] to determine the relative merits of five types of simple connections (Fig. 15-44).

The connection in Fig. 15-44a is the simplest and most economical in terms of fabrication; however, the rotation of the connection caused by bending in the beam due to an applied load causes severe distortion and buckling of the tube wall.

The type of connection (Fig. 15-44b) is of the same general one-sided type, with the exception that a rolled tee is used. By shop welding the edges of the tee as closely as possible to the corners of the tube, the stiffness of the unconnected faces can be employed and the flange of the tee distributes the load to the connected tube faces. Yielding developed rapidly in the lower part of the tee flange; however, the strains in the middle of the tube wall at the top and bottom of the connection were still in the elastic range and there was no distortion of the tube wall. Failure occurred in the weld.

The third type of connection (Fig. 15-44c), is the familiar unstiffened seat angle with a top angle. Top and bottom angles are of sufficient length to allow welding of the edges to the corner of the tube. The capacity of this type of connection is limited by the strength of the seat angle, particularly when the length of the angle is limited to the width of the tube column face.

A variation of the first type of connection is indicated in Fig. 15-44d. Yielding of the beam web in bearing in front of the top bolt was the first sign of distress. As loads were increased tube distortions manifested

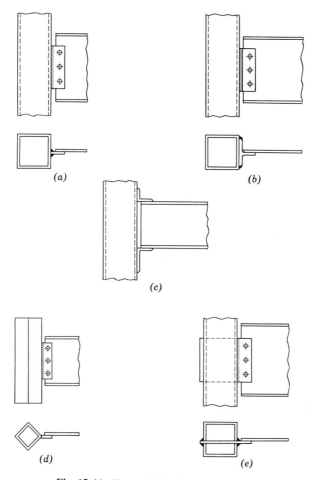

Fig. 15-44 Types of tubular connections.

themselves by distorting the original square cross section of the tube column to a trapezoidal shape.

The fifth type of connection investigated (Fig. 15-44e), was attached in the same manner as the first two types. This connection proved to be much more rigid than the others and provided very little deformation to the tube column walls.

Research on rigid-type connections of tube beams to tube columns has been conducted at McGill University, Canada.[39] Research is also being conducted at Drexel Institute of Technology on welded beam to column tubular connections. In this research five different welding procedures were evaluated. All faces of the beam tubes were chamferred at 45°. All vertical welds were ground flush with the face of the tube. The variables in the welding details are illustrated in Fig. 15-45. Joint W4 similar to W2 was annealed after welding to release any residual stresses caused by welding. The tests indicated that the different welding patterns have very little effect on the capacity of the connection except for the J joint. The capacity of the connection is limited by the local bending and buckling of the connected faces of the tube column (Fig. 15-46).

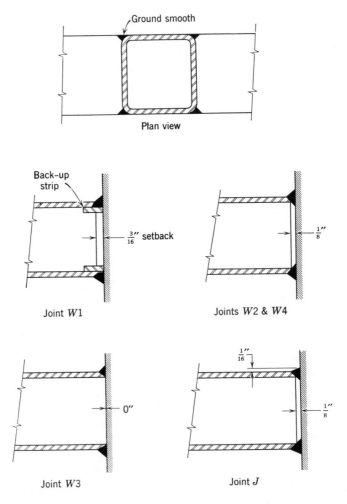

Fig. 15-45   Tubular weld joint details—beam to column.

Fig. 15-46 Tubular beam to column connection. (Courtesy of Drexel Institute of Technology.)

The analytical solution to the design of a rigid connection must correlate the following parameters:

(*a*) Width of column face to width of beam.
(*b*) Width of connected column face to the free column face.
(*c*) Width and thickness of connected column face.
(*d*) Thickness of column and beam tubes:
(*e*) Effect of column axial load.

Applications of Structural Tubing. Structural tubing may be used in a great variety of structures, such as tubular roof trusses for single-story industrial buildings, warehouses, and shopping centers where long span column free areas are required.

Tubes may be used to advantage in structures designed for material handling equipment such as bridge, derrick, and tower cranes where weight savings may be a very substantial economical consideration.

Tubular members may be used effectively in large space frame lattice structures for arenas, stadiums, and exhibition halls where appearance as well as weight becomes an important design consideration.

Such structures as offshore drilling installations, masts, and transmission towers are other examples where tubular sections may be utilized effectively.

Steel tubing is an efficient material that is adaptable to many different situations and in a great many instances is unsurpassed in its efficiency and therefore, holds a very dramatic promise for the future.

15-17 SHELL STRUCTURES

Shell structures are curved-shape constructions relatively thin in cross section that resist forces of low intensity applied over a large area. Roofs and canopies, subjected primarily to vertical loads, are often designed as shells, but frequently wall construction, components of ships, and other thin, curved barrier structures resisting lateral forces are also essentially shells.

The distinctive feature of shell structures is that lateral loads are resisted largely by in-plane, or "membrane" forces, rather than the usual bending that is typical of slabs and beams. Thus, because of its shape, the structural advantages of a shell are (a) the directions of certain membrane thrusts within the shell change along their paths enabling lateral loads to be supported by arch compression similar to the action of arch bridges and/or by suspension tension similar to the action of suspension bridge cables and (b) the in-plane shear resistance of a shell either supports the load directly or tends to redistribute stresses as the shell distorts so that the arching and/or suspension action can most effectively resist the loading with a minimal of bending in the shell.

Shells that are curved or bent in only one direction, such as folded plates, sag shells, and barrel vaults, can effectively utilize arch or suspension action in only one direction. Along any straight "free" edge perpendicular to the direction of the arch or sag the shell geometry does not provide the curvature desired for arch or suspension action. Consequently, free edges of such shapes generally tend to bend under loads placed near the free edges. Attached stiffeners, edge beams or bearing walls can minimize or prevent such localized bending along the outer longitudinal edges of these shells. Because one shell will give flexible support to a connected adjacent shell, localized bending must be considered. Generally, this local bending is not critical where adjacent folded plates or barrel vaults are joined together along fold lines.

Shells that are curved in two directions are much less sensitive to the presence of exterior edges, holes, and other discontinuities because there is generally either arch or suspension action along the free edge at the discontinuity to give local support to the loading. In such shells bending may be a design consideration, particularly near the edges, but the effect of edge bending does not usually affect the behavior of the shell as a whole. Shells with curvature in more than one direction are generally classified as (*a*) synclastic, positive Gaussian curvature (principal curvatures of the same sign at a given point) such as a dome, or (*b*) anticlastic, negative Gaussian curvature (principal curvatures of opposite sign at a given point) such as a hyperbolic paraboloid.

The most important considerations in the design of steel sheet shells are that they must be readily fabricated and not buckle under the design loading plus a suitable overload. Therefore corrugated steel sheets are frequently used in steel shell roofs with negative or zero Gaussian curvature because (*a*) the flexibility of the corrugations allows the sheets to conform to the warped

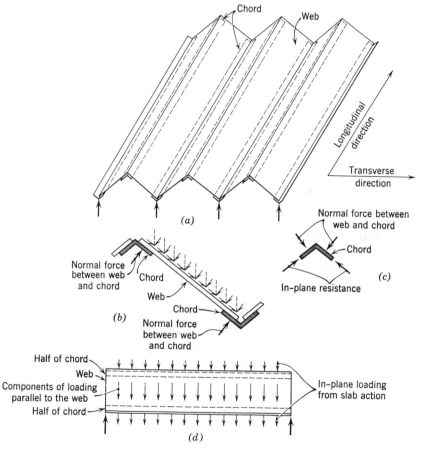

Fig. 15-47 Steel folded plate roof: (*a*) general view action of web, (*b*) beam slab, (*c*) forces from beam slab action cause in-plane loading, and (*d*) beam action of structure between fold lines.

shape of such shells, (*b*) the corrugations provide one-way bending strength and rigidity to the sheet to preclude over-all buckling, and (*c*) the corrugation curvature or the close spacing of corrugation bend lines can prevent premature local buckling.

One of the simplest and most attractive steel shell concepts is the folded plate (Fig. 15-47). A "pie-crust," "saw-tooth," or "parapet" shape of a roof cross section can be obtained by connecting inclined webs to top and bottom channel or bent-plate steel chords. Corrugated steel barrel vaults that are curved transversely but are straight longitudinally (Fig. 15-48) are structurally similar. In assemblies of each type of shell, co-adjacent shells behave essentially as parallel sets of longitudinal interconnected beams where the longitudinal bending and shearing stresses are the membrane stresses in the shell. However, stresses and deflections higher than those predicted by beam theory may occur in the vicinity of a free edge. Both types of shells require special provision for transverse bending stiffness. The webs of the folded plate must transmit the loading to the chords and therefore logically consist of corrugated steel sheets with the corrugations extending transversely. Barrel vault roofs may consist of corrugated steel sheets with the corrugations extending longitudinally. In this case, transverse stiffeners must extend across the barrel to maintain its arch shape so that all portions of the shell when viewed in a transverse cross section will deflect about the same amount.

Although not usually thought of as a shell, a sag shell (Fig. 15-49) is the most efficient method to use a steel sheet or plate because all the steel is essentially in tension. If a flat or sagged sheet is adequately restrained about its periphery, a lateral load against the sheet will be supported largely by membrane tension that develops as the load is applied. The effective use of steel sheets in containers, truck bodies, ship hulls, etc., often derives from this phenomenon.

One of the most popular types of shell roofs has been the hyperbolic paraboloid (Fig. 15-50). Because this shape, although parabolic in diagonal directions, is defined by families of straight lines in the two main directions, corrugated sheets are naturally adaptable. The corrugation bend lines must lie in one or both of the two main directions. Most long span steel hyperbolic paraboloids have been constructed of two mutually perpendicular corrugated steel sheets spot- or plug-welded together at their interfaces (Figs. 15-51 and 15-52). However, a simple layer of corrugated steel sheet can be satisfactory for shorter span hyperbolic paraboloid roofs.

Although steel sheets are generally used only for cladding domes, steel sheets have been used as the principal structure in other shells of revolution. Perhaps the most common are the conical roofs used for farm cylindrical storage containers. These roofs can be made from flat or corrugated steel sheets with the corrugations in a radial direction.

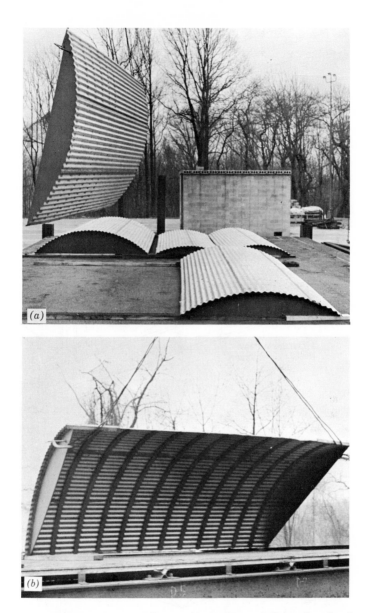

Fig. 15-48 Steel barrel vault test specimens: (*a*) exteriors and (*b*) interior, showing transverse stiffeners. (Courtesy of U.S. Steel Applied Research Laboratory, Monroeville, Pa.)

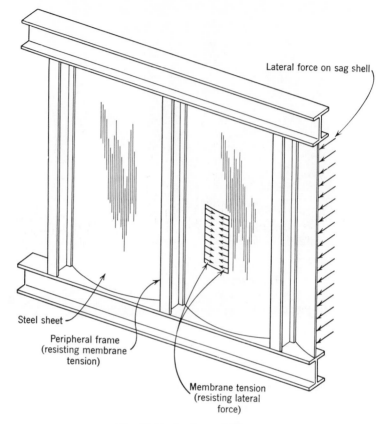

Fig. 15-49 Steel sag shell.

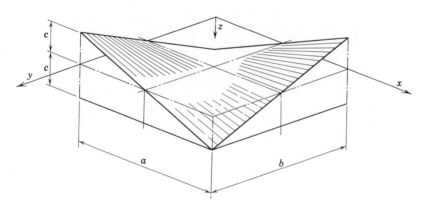

Fig. 15-50 Hyperbolic paraboloid geometry.

Fig. 15-51 Johnson and Hardin Printing Plant, Cincinnati, Ohio. (Courtesy of Truman P. Young and Associates, Engineers.)

Fig. 15-52 Frisch's Restaurant, Cincinnati, Ohio. (Courtesy of Truman P. Young and Associates, Engineers.)

15-18 ANALYSIS AND DESIGN OF STEEL SHELLS

Although "exact" shell theory is sometimes mathematically complex, practical solutions for the stresses in steel sheet shells often require only an understanding of the structural behavior of the shells and the use of simple, well-known equations of statics with certain subsequent refinements for forcing compatibility of strains and distortions.

Folded Plates. The structural behavior of steel folded plates is indicated in Fig. 15-47. As mentioned, the webs of folded plates must be designed to transmit loadings to the top and bottom chords. The net reactions of the webs on the chords must then be resisted by the entire folded plate acting as a series of beams with the chords alone resisting the bending moments by their longitudinal membrane thrusts and the webs alone resisting the in-plane shears. Usually, this relatively simple statics analysis is sufficient for the design of steel folded plate structures because where top and bottom chords do not deflect the same amount there is generally enough flexibility in the transverse direction at the web-to-chord connections to preclude the development of the significant transverse bending moments that would occur in a rigid structure. In the longitudinal direction the flexibility at the web-to-chord connections generally permits an isolation of the stress analysis within the chord. Thus hypothetical discontinuities evidenced by differences in chord stresses on either side of a fold line resulting from the statics analysis can be resolved by merely redistributing the stress within the chord alone. This may be accomplished by determining an average unit stress resulting in the same total thrust in the chord. This does not significantly effect stresses in the webs or remote chords.

Barrel Vault. As noted, barrel vaults can also be analyzed as longitudinal beams if the possible sagging of the edges is duly considered in the structural analysis. Exact solutions are quite complex, but approximate solutions can be attained by modifying the longitudinal beam solution stresses and deflections by empirically derived magnification factors.

Sag Shells. If a sagged or flat sheet is connected to a stiffened edge frame so that it is restrained in-plane at its boundaries, the sheet, under loads that are considerably in excess of those that would cause flexural yielding at the edge, will adequately support lateral loading by membrane tension as if it were piano-hinged at the edge. Thus the procedure outlined below can be used for analyzing the behavior under uniform load of both flat panels and sagged sheets with adequate edge restraint.

Summary of Simple-Supported Membrane Analysis[40]

Given: u = midspan lateral initial deflection (before loading).
       e = longitudinal deflection (movement of one edge toward the other).
       h = thickness of plate (or sheet).
       L = span length (short dimension of panel).
       q = uniform lateral loading per unit area.

1. $D = \dfrac{Eh^3}{12(1 - v^2)}$ where E = modulus of elasticity and v = Poisson's ratio

2. $w_0 = \left[\dfrac{5qL^4}{384D} + u + \dfrac{2}{\pi}\sqrt{Le}\right]$ = midspan deflection in absence of membrane tension

3. α is obtained from the following equation:

$$[1 + \alpha]^2\left[\alpha + \frac{12L}{\pi^2 h^2}\left(e + \frac{\pi^2 u^2}{4L}\right)\right] = \frac{3w_0^2}{h^2}$$

4. $p = \sqrt{\dfrac{\pi^2\alpha}{4}}$

5. Membrane tension $S = \dfrac{4Dp^2}{L^2}$

6. Midspan bending moment $M_c = \dfrac{qL^2}{8}\left[\dfrac{1\text{-sech } p}{p^2/2}\right]$

7. Axial unit stress $f_a = \dfrac{S}{h}$

8. Maximum bending unit stress $f_b = \pm\dfrac{6M_c}{h^2}$

9. Maximum combined unit stress $f_t = f_a + f_b$

10. Actual midspan deflection under load $w_0 = \left(\dfrac{w_0}{1 + \alpha}\right)$

Hyperbolic Paraboloids. Because the stiffnesses of two-ply (that is, two interconnected layers at right angles), corrugated steel sheet hyperbolic paraboloids are comparable to those of reinforced concrete shells, they have been designed satisfactorily with the equations generally used for designing reinforced concrete hyperbolic paraboloids. This assumes that a uniform vertical loading is resisted entirely by membrane shearing forces within the shell, which are at least approximately defined by

$$S = \frac{qab}{8c}$$

where S is the membrane shear per unit width of shell assumed constant throughout the shell, a and b are the plan dimensions between centerlines of support in the two main directions, and c is the vertical distance between the

Fig. 15-53 Canopy fabricated from a single thickness of 20-gage corrugated steel sheet. (Courtesy of American Bridge Division, U.S. Steel Corp.)

center of the shell and a straight line through either pair of diagonal corners (Fig. 15-50).

Because it appeared that smaller roofs such as canopies could be more economically designed in steel if only one layer rather than two of corrugated steel sheets were used, lateral loading tests[41] were conducted on model, subsize, and full-size shells the latter of which is shown in Fig. 15-53. The tests demonstrated that such roofs are structurally feasible, but they must be designed by a theory more complex than that of the preceding equation because significant bending stresses in addition to the in-plane shearing stresses occur under uniform loading as a result of the greater flexibility of the single ply.

15-19   FUTURE DEVELOPMENTS OF SHELLS

There have recently been significant technological innovations in steel sheet production and fabrication such as attaining the capacity to roll very wide steel sheets. As a consequence, there has been a significant improvement in the economics of steel sheet shell structures since these structures involve relatively large continuous areas of steel sheets joined to suitable steel edge members. Therefore increased use of steel shells is anticipated as architects and engineers come to realize the potentialities of such structures. Small structures are obvious applications: single-layer steel hyperbolic paraboloids for canopies of all types, park shelters or parking lot covers, steel barrel vaults for drive-in banks or school interbuilding passageways, and sag shells for side panels of railroad hopper cars or exterior panels of highrise buildings or blast-resistant structures. In addition, steel shells such as these can be used as component parts of steel-framed roofs in which the steel frame is a skeleton hyperbolic paraboloid, barrel vault, sag shell, or other curved configuration. Steel folded plates certainly offer excellent opportunities for economical design and artistic expression for roofs of buildings.

15-20 SEMI-MONOCOQUE HIGH-RISE BUILDINGS

In a monocoque structure the total structure participates in resisting load. An example is an aircraft fuselage or wing where the skin acts together with the frame to resist shear, bending, and torsional forces. In the automotive industry the same philosophy of total participation of a structure is termed unitized construction.

The monocoque structural theory approach is not yet evident in building construction. However, three-dimensional structural frameworks to resist lateral loads such as wind or earthquake are currently appearing. Until recently no concerted attempt has been made in building design to include a structural skin to act in conjunction with the frame to resist wind forces. For this discussion, a three-dimensional externally braced or rigid space frame to resist lateral and gravity loads acting on a high rise structure will be considered a semi-monocoque structure. It could also be termed "box-action."

Building structures may be classified into the following six categories:

1. Simply connected structures.
2. Semirigid structures.
3. Rigid frame structures.
4. Structures with shear trusses.
5. Structures with interacting frame and shear trusses.
6. Rigid box-type structure.
   (*a*) Closely spaced exterior columns (Fig. 15-54).
   (*b*) Closely spaced diagonals as exterior walls (Fig. 15-55).
   (*c*) Optimum column-diagonal-spandrel for exterior walls (Fig. 15-56).

Categories 4, 5, and 6 can be broadly interpreted as semi-monocoque structures as previously defined.

Diagonal Truss Exterior Wall.   In the IBM building in Pittsburgh, Pennsylvania (Fig. 12-33), all wind loads acting upon the structure are transferred through the floors acting as rigid diaphragms to the wall trusses. No wind forces are transmitted to the central core because of the greater stiffness of the exterior walls. A horizontal bracing truss connects the wall trusses to the floor system, thus providing lateral support to the wall trusses and transmitting wind load from the windward truss into the leeward and side trusses. Wind loads are carried by the wall trusses in direct stresses of tension and compression which efficiently carry the forces to the foundations.

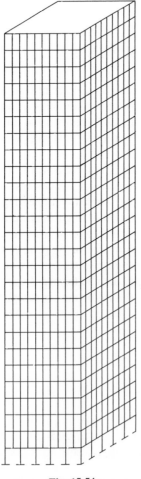

Fig. 15-54   Fig. 15-55

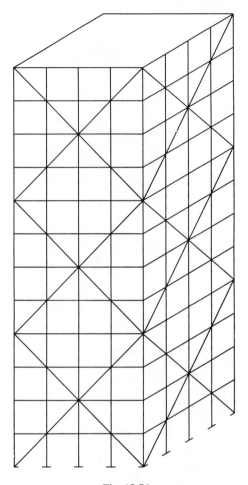

Fig. 15-56

Circular Rigid Frame.   One type of a semi-monocoque structure is a circular or polygonal framing system (Fig. 15-57).[43] In the circular framing system the major supporting columns are located on one or more concentric circles with the major girders in the tangential direction. All exterior columns do not need to be located on the circumference of the circle, but may be placed on the chords as well. Spandrel beams may be straight or curved. Several typical floor layouts of circular and polygonal arrangements are illustrated in Fig. 15-58.

The spandrel girders are rigidly connected to the columns to form a continuous moment resistant frame around the entire circumference of the

Fig. 15-57

building. Spandrel members require sufficient depth to resist economically the resulting drift from wind or other lateral loads that may be applied. Radial floor members act only to transmit gravity floor loads to the exterior and interior columns. It is advisable to have the columns so oriented that their strong axes resist the moments which act on the spandrel girders.

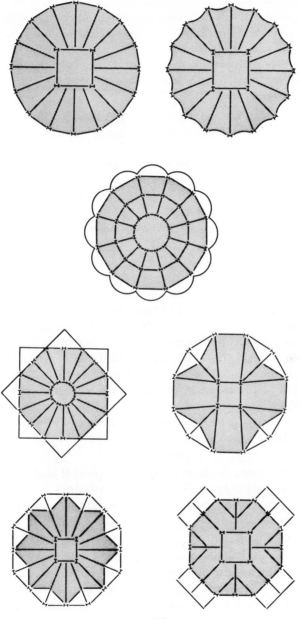

Fig. 15-58

If the structure has a very large diameter, two or more concentric rings may be required. In this case, one, several, or all rings may be designed to resist lateral loads. The ratio of building height to the diameter and the magnitude of the gravity and lateral loads are the governing criteria to select a single or multiple ring design and the extent of their interaction.

The basic assumptions in a lateral load analysis on a semi-monocoque frame are the following:[42,43]

(*a*) Lateral forces are concentrated at floor levels.

(*b*) Floors act as rigid diaphragms to transmit all horizontal forces in the plane of the floor.

(*c*) Axial stresses in the columns are proportional to their distance from the centroidal axis of the column areas.

(*d*) Column inflection points are taken at midheight between floors, except for lower transition stories.

(*e*) Beam inflection points are taken at midspan.

The assumption that column inflection points occur at midheight of the column length is not always valid for the lower stories. When the columns are fixed at the base, they are restrained from rotation and the joints at the story above would rotate a certain amount. The amount of this rotation is a function of the column-to-girder stiffness ratio. In the lower few stories the inflection points will be above the midheight of the columns. To correct this condition the spandrel girders in the selected transition stories are reproportioned to force the return of the inflection point to the midheight of the column. The number of stories selected to represent the transition range may be taken at 10 percent of the total number of stories in the building.

Lateral deflection or drift (Δ) is composed of the summation of three separate and distinct parts which may be calculated by the usual methods of structural theory:

$$\Delta = \delta_g + \delta_c + \delta_c'$$

where δ_g = deflection due to bending of the spandrel girders

δ_c = deflection due to bending in the columns

δ_c' = deflection due to axial deformation in the columns

A circular framing system offers an advantage of the least perimeter for a given floor area leading to economies of the exterior wall. Because of the circular shape most building codes will permit a reduction of wind pressures that consequently leads to an economy of materials.

A cylindrical framework may also be designed using diagonal bracing instead of rigid moment resistant connections. Diagonal bracing may be provided on the exterior framework of the structure so that it provides architectural esthetic expression as well as structural strength and rigidity (Fig. 15-59).

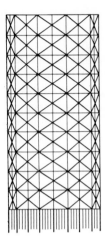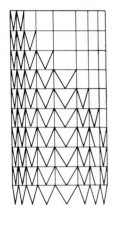

Fig. 15-59

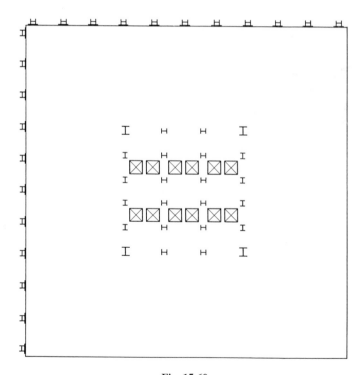

Fig. 15-60

Latticed Box Frame.   Another three-dimensional type of highrise structure is referred to as "The Latticed Box Frame" configuration. The exterior of the structure consists of a network of deep horizontal trusses with closely spaced exterior columns. The outer "tube" is then connected to an interior braced service core by means of long span composite beams and slab that act as diaphragms at each floor level. A necessary consideration in this type of construction is the symmetry of the floor and column arrangement. The central utility and elevator core must be placed to develop the symmetry of the structure and to provide a torsional balance to resist the lateral loads on the buildings (Fig. 15-60). As in conventional construction the composite floor system acting as a shear diaphragm transmits all gravity loads to the outer and inner core columns.

The basic philosophy of design for lateral loads in this type of structure is that of a cantilevered tube that resists the horizontal displacements by the usual theory of flexure. The alignment of the columns on the external faces

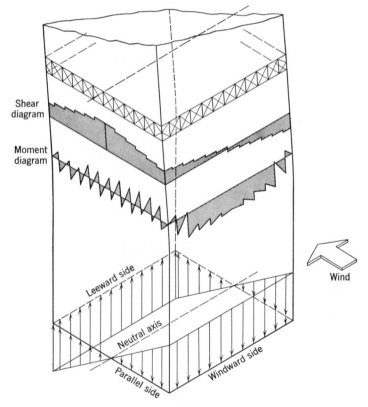

Fig. 15-61

of the building must be such that the strong axis of the column is effective on the sides that are parallel to the wind direction (Fig. 15-61). The exterior windward and leeward columns are thus arranged in such a position to have the weak axis in bending perpendicular to the direction of the lateral forces. These columns are assumed not to resist bending about the weak axis but instead to resist bending about the strong axis. These bending moments about the strong axis are induced by the horizontal shear flow of the windward and leeward spandrel trusses (Fig. 15-61).

The concept of semi-monocoque highrise buildings is in its infancy and more sophisticated developments will undoubtedly take place. These developments will be brought about by the extended use of electronic computers and the advantages to be gained from the economical use of the higher-strength steels. Plastic design theories will also be introduced into the stress analysis of three-dimensional building frameworks as herein described.

Until engineers develop a structural wall system which will resist gravity and lateral loads, the full monocoque action of a highrise building cannot be realized. At this writing there are a few structural wall systems which are known to stiffen the building for drift control but are not designed to support the loads. The engineering profession can look forward to more challenging structural designs as the results of research on the behavior of full-and semi-monocoque structures unfolds.

REFERENCES

1. Huber, M. T., *Probleme der Statik Technisch Wichtiger Orthotroper Platten*, Warsaw (1929).
2. Pelikan, W., and M. Esslinger, "Die Stahlfahrbahn, Berechnung and Konstruktion," *M.A.N. Forschungsheft* No. 7 (1957).
3. Chapter 12, "Horizontally Curved Girder," *Highway Structures Design Handbook*, Vol. I, U.S. Steel Corporation (1965).
4. Mehrtens, *Eisenbrückenbau*, Vol. I, Verlag Englemann, Leipzig (1908).
5. Homberg, "Einflusslinien von Schrägseilbrücken," *Stahlbau* **24** (1955).
6. Klöppel, Esslinger, and Kollmeier, "Die Berechnung eingespannter und fest mit dem Kabel verbundener Hängebrückenpylonen bei Beanspruchung in Brückenlängsrichtung," *Stahlbau* **34** (1965).
7. Klöppel and Weber, "Teilmodellversuche zur Beurteilung des aerodynamischen Verhaltens von Brücken," *Stahlbau* **32** (1963).
8. "Montreal Hosts a Double Bridge Spectacular in the St. Lawrence," *Eng. News-Record* (Aug. 5, 1965).
9. Feige, "The Evolution of German Cable-Stayed Bridges: An Overall Survey," *Acier Stahl Steel*, No. 12 (Dec. 1966).
10. Shaw, F. S., "Some Notes on Cable Suspension Roof Structures," *Journal Institution Engineers*, Australia (April-May, 1964).
11. Zetlin, Lev., "Steel Cable Creates Novel Structural Space System," *AISC Eng. Journal* **1,** No. 2 (Jan. 1964).

12. Gillespie, J. W., "Analysis of Curved Trusses in Space," *Ph.D. Dissertation*, Oklahoma State University, Stillwater (1961).
13. Möbius, A. F., *Lehrbuch der Statik*, Vols. 1, 2, Leipzig, 1837, Chapters 4, 5.
14. Föppl, A., *Das Fachwerk im Raume*, B. G. Teubner, Leipzig (1892).
15. Schwyzer, H., "Statische Untersuchung der aus ebenen Tragflächen zusammengesetzten Tragwerke," Dissertation E.T.H., Zurich (1920).
16. Stüssi, F., "Zur Berechnung von Stahlbrücken mit gekrümmten Hauptträgern," *Denkschrift der E.T.H. zum hundertjährigen Bestehen des S.I.A.*, Zurich (1937).
17. Stüssi, F., "Ausgewählte Kapitel aus der Theory des Brückenbaues," *Taschenbuch für Bauingenieure*, edited by F. Schleicher, Springer-Verlag, Berlin, 1955, pp. 905–963.
18. Stüssi, F., *Baustatik II*, Basel (1954).
19. Anderson, P. and G. M. Nordby, *Introduction to Structural Mechanics*, Ronald Press, New York 1960, pp. 235–282.
20. Niles, A. S. and J. S. Newell, *Airplane Structures*, John Wiley and Sons, New York (1938).
21. Holloway, C. J., Jr., "Analysis of Truss-Plate Structures by Plate Analogy," *M.S. Thesis*, Oklahoma State University, Stillwater (1963).
22. Gillespie, J. W., "Analysis of Truss-Plate Structures," *Journal Structural Division*, ASCE **91**, No. ST2, Proc. Paper 4300 (April 1965).
23. Southwell, R. V., "Primary Stress Distribution in Space Frames," *Engineering* **109** (Feb. 6, 1920).
24. Mayor, B., *Introduction à la Statique Graphique des Systémes de l'Espace*, Payot, Lausanne (1926).
25. Mises, R. von, "Graphische Statik räumlicher Kräftesysteme," *Zeitschrift für Mathematik und Physik* **64**, No. 3 (1916, p. 209).
26. Constant, F. H., "Stresses in Space Structures," *Transactions*, ASCE, **100**, Paper No. 1911 (1935). (Discussions by W. R. Osgood., L. E. Grinter, C. M. Spofford, A. H. Finlay, and F. H. Constant.)
27. Southwell, R. V., "On Castigliano's Theorem of Least Work and the Principle of Saint-Venant," *Phil. Mag.* **45**, S.6, No. 465 (Jan., 1923, pp. 193–212).
28. Makowski, Z. S., *Steel Space Structures*, Michael Joseph, London (1965).
29. "Space Frame Costs Less than $4 a Square Foot," *Architectural Record* (March 1966).
30. "Exhibition Hall for the Future Draws Inspiration from the Past," *Engineering News-Record* (September 22, 1966).
31. From a private communication with Mr. Neil Welden, former chief bridge engineer for the Iowa State Highway Commission.
32. "Tubular Tetrahedrons," *Progressive Architecture* (Sept. 1961, pp. 183–187).
33. Anderson, George C. and Roland R. Graham, "The Manufacture and Use of Hot-Rolled Carbon Steel Hollow Structural Tubing," paper presented at the 69th General Meeting of the American Iron and Steel Institute (May 24, 1961).
34. Bouwkamp, J. G., "Behavior of Tubular Truss Joints under Static Loads," Report No. SESM-65-4, Department of Civil Engineering, University of California, Berkeley.
35. Bouwkamp, J. G., "Concept of Tubular-Joint Design," *Journal Structural Division*, ASCE **90**, ST2 (April 1964).
36. Bouwkamp, J. G., "Applications of Tubular Structures in Buildings," presented at the Building Research Conference on New Metal Structures and Finishes in Washington, D.C., April 25, 1963.
37. Toprac, A. A., L. P. Johnston, and J. Noel, "Welded Tubular Connections: An Investigation of Stresses in T-Joints," *Welding J.* (Jan. 1966).
38. White, R. N. and Pen Jeng, Fang, "Framing Connections for Square Structural

Tubing," *Journal Structural Division ASCE*, **92,** No. ST2, Proc. paper 4782 (April 1966).

39. Redwood, R. G., "The Behavior of Joints Between Rectangular Hollow Structural Members," *Civil Engineering and Public Works Review* (Oct. 1965).
40. McDermott, J. F., "Theoretical and Experimental Study of Steel Panels in Which Membrane Tension Is Developed," *ASME Publication* 65-MET-15.
41. McDermott, J. F., "Load Tests on Corrugated Steel Sheet Hypar Canopies", International Congress of the International Association for Shell Structures, Mexico City (1967).
42. "Circular Steel Framing for Highrise Buildings," U.S. Steel Corporation, ADUSS 27-2565.
43. Scalzi, J. B., J. F. Fleming, and K. H. Chu, "Analysis of Circular Steel Multi-Story Frameworks," *Journal Structural Division, ASCE* **93,** No. ST. 1, Proc. paper 5095 (Feb. 1967).

Selected References for Curved Bridges

1. Fickel, H. H., "Analysis of Curved Girders," *Journal Structural Division, ASCE* **85,** No. ST7, Part 1 (Sept. 1959).
2. Hogan, M. B., "The Derivation of Two Five Moment Theorems for Continuous Plane Curved Beams," *Utah Engineering Experiment Station, Bull.* **31,** University of Utah (April 1947).
3. Lyse, I, and Johnston, B. G., "Structural Beams in Torsion," *Trans. ASCE* **101** (1936).
4. Michalos, J., "Numerical Analysis of Frames with Curved Girders," *Trans. ASCE*, **121** (1956).
5. Pippard, A. J. S., *Studies in Elastic Structures*, Arnold and Company, London, 1952.
6. Velutini, B., "Analysis of Continuous Circular Curved Beams," *Proceedings, ACI*, **22,** 3 (November 1950).
7. Volterra, E., "Deflections of a Circular Beam Out of Its Initial Plane," *Trans. ASCE* **120** (1955).
8. Yonezawa, H., "Moments and Free Vibrations in Curved Girder Bridges," *Journal Eng. Mechanics Division, ASCE* **88,** No. EM1 (February 1962).
9. Reddy, M. N., "Influence Lines for Continuous Curved Members with Lateral Loads," *M.S. Thesis*, Oklahoma State University, Stillwater (1963).

Appendix A

🦋🦋🦋

LIST OF STRUCTURAL DESIGN AND MATERIAL SPECIFICATIONS AND STANDARDS

1. AISC—Specification for the Design, Fabrication and Erection of Structural Steel for Buildings, New York, 1963.
2. AISC—Code of Standard Practice for Steel Buildings and Bridges, New York, 1963.
3. AASHO—Standard Specifications for Highway Bridges, Washington, D.C., 1965.
4. AREA—Specifications for Steel Railway Bridges, Chicago, 1965.
5. AISI—Specifications for the Design of Light Gage Cold Formed Steel Structural Members, New York, 1962.
6. AISC—Specification for Architecturally Exposed Structural Steel, New York, 1960.
7. Research Council on Riveted and Bolted Structural Joints of the Engineering Foundation—Specifications for Structural Joints Using ASTM A325 or A490 Bolts, New York, 1966.
8. AWS—Code for Welding in Building Construction, American Welding Society, AWS D1.0-63 and Addenda 1965, New York.
9. AWS—Specifications for Welded Highway and Railway Bridges, American Welding Society, AWS 02.0-63 and Addenda 1965, New York.
10. SJI—Standard Specifications and Load Tables, Open-Web Steel Joists, Steel Joist Institute, Washington, D.C., 1963.
11. AISI—Tentative Criteria for Structural Applications of Steel Cables for Buildings, New York, 1966.
12. NBS—Minimum Design Loads in Buildings and Other Structures, A.58.1, 1955.
13. AISI—Fire Protection Through Modern Building Codes, New York, 1961.
14. CSA—Steel Structures for Buildings, CSA, Standard S16-1965, Canadian Standards Association, Ottawa, Canada.
15. CSA—Deisgn of Highway Bridges, CSA Standard C6-1966, Canadian Standards Association, Ottawa, Canada.
16. AISE—Specifications for Electric Overhead Traveling Cranes for Steel Mill Service, American Iron and Steel Engineers, Pittsburgh, 1949.
17. AWS—Gas Metal, Arc Welding with Carbon Dioxide Shielding, Special Ruling by AWS Structural Welding Committee, American Welding Society, New York.
18. ASTM—Standards 1967, Parts 1, 3, and 4.

809

Appendix B

Table B-1 Torsional Properties*

Nominal size	Weight per ft, lb	K_t, in.^{4}	a, in.	Nominal size	Weight per ft, lb	K_t, in.^{4}	a, in.
Wide Flange Sections							
$36 \times 16^{1/2}$	300	68.80	119.1	30×15	210	30.69	112.5
	280	56.43	125.8		190	22.85	121.9
	260	44.78	133.8		172	17.01	132.0
	245	37.16	141.0	$30 \times 10^{1/2}$	132	10.35	99.8
	230	30.94	147.9		124	8.52	105.1
36×12	194	23.69	109.9		116	6.85	111.1
	182	19.61	115.8		108	5.35	117.7
	170	16.03	122.3		99	4.04	125.6
	160	13.23	128.6	27×14	177	21.57	103.2
	150	10.77	135.5		160	16.08	111.8
	135	7.49	147.3		145	12.15	120.8
$33 \times 15^{3/4}$	240	39.29	122.0	27×10	114	7.86	92.6
	220	30.33	130.8		102	5.64	101.4
	200	22.78	141.4		94	4.35	108.5
					84	3.00	118.6
$33 \times 11^{1/2}$	152	13.18	115.2				
	141	10.36	122.7	24×14	160	17.82	99.9
	130	7.87	130.4		145	13.24	108.3
	118	5.69	141.7		130	9.46	118.4

*From "Torsional Analysis of Rolled Steel Sections," Bethlehem Steel Corp., Bethlehem, Pa.

Table B-1   Torsional Properties—*continued*

Nominal size	Weight per ft, lb	K_t, in.^{4}	a, in.	Nominal size	Weight per ft, lb	K_t, in.^{4}	a, in.
			Wide Flange Sections				
24 × 12	120	8.84	100.9	16 × 7	50	1.62	58.31
	110	6.92	108.0		45	1.19	63.46
	100	5.24	116.5		40	0.85	69.62
24 × 9	94	5.58	80.7		36	0.59	76.10
	84	3.98	88.5	14 × 16	426	338.60	33.32
	76	2.89	96.3		398	278.71	34.79
	68	1.99	105.5		370	226.99	36.43
21 × 13	142	15.27	82.4		342	181.48	38.39
	127	10.97	90.6		314	142.60	40.65
	112	7.57	100.3		287	109.96	43.33
21 × 9	96	6.86	64.9		264	86.86	45.94
	82	4.32	73.6		246	70.93	48.40
					237	63.63	49.77
21 × 8¼	73	3.23	74.7		228	57.10	51.19
	68	2.62	79.1		219	50.72	52.83
	62	1.97	85.3		211	45.49	54.39
	55	1.33	93.9		202	40.22	56.21
18 × 11¾	114	9.93	71.42		193	35.21	58.27
	105	7.82	76.19		184	30.80	60.40
	96	6.02	81.80		176	26.87	62.74
					167	23.08	65.45
18 × 8¾	85	5.82	57.97		158	19.79	68.33
	77	4.42	62.53		150	16.96	71.37
	70	3.32	67.60		142	14.39	74.73
	64	2.56	72.60				
18 × 7½	60	2.40	62.71	14 × 16	320	136.87	41.00
	55	1.78	68.37	14 × 14½	136	13.64	71.17
	50	1.34	73.87		127	11.23	75.24
	45	0.96	80.63		119	9.32	79.41
16 × 11½	96	6.75	68.94		111	7.57	84.32
	88	5.19	73.89		103	6.09	89.88
					95	4.80	96.41
16 × 8½	78	5.08	51.64		87	3.72	104.0
	71	3.86	55.61				
	64	2.85	60.35	14 × 12	84	4.48	76.66
	58	2.13	65.25		78	3.57	81.86

Table B-1   Torsional Properties—*continued*

Nominal size	Weight per ft, lb	K_t, in.^{4}	a, in.	Nominal size	Weight per ft, lb	K_t, in.^{4}	a, in.
Wide Flange Sections							
14×10	74	3.92	63.04	$12 \times 6^{1/2}$	36	0.90	48.50
	68	3.06	67.73		31	0.58	54.58
	61	2.22	74.30		27	0.39	60.96
				10×10	112	15.31	32.02
14×8	53	1.97	57.90		100	11.05	34.85
	48	1.47	62.97		89	7.88	38.14
	43	1.06	69.09		77	5.19	42.76
					72	4.23	45.25
$14 \times 6^{3/4}$	38	0.86	58.72		66	3.32	48.45
	34	0.61	64.24		60	2.53	52.35
	30	0.41	70.93		54	1.87	57.14
					49	1.39	62.13
12×12	190	49.96	35.02				
	161	31.20	39.45	10×8	45	1.53	45.22
	133	17.95	45.61		39	0.98	51.47
	120	13.13	49.63		33	0.59	59.26
	106	9.23	54.71	$10 \times 5^{3/4}$	29	0.62	38.64
	99	7.54	57.92		25	0.40	43.52
	92	6.09	61.49		21	0.23	50.26
	85	4.87	65.61				
	79	3.90	69.91	8×8	67	5.14	26.99
	72	2.98	75.61		58	3.37	30.17
	65	2.21	82.42		48	1.99	34.86
					40	1.13	40.76
					35	0.78	45.46
12×10	58	2.14	65.96		31	0.54	50.47
	53	1.61	71.56				
				$8 \times 6^{1/2}$	28	0.54	40.52
					24	0.35	43.89
12×8	50	1.82	51.78				
	45	1.34	56.48	$8 \times 5^{1/4}$	20	0.23	38.28
	40	0.97	62.03		17	0.16	40.11
Light Beams							
$16 \times 5^{1/2}$	31	0.496	59.92	12×4	22	0.301	37.23
	26	0.280	68.80		19	0.189	41.95
					16.5	0.114	46.77
14×5	26	0.378	50.73				
	22	0.222	57.93				

Table B-1 Torsional Properties—*continued*

Nominal size	Weight per ft lb	K_t, in.^{4}	a, in.	Nominal size	Weight per ft lb	K_t, in.^{4}	a, in.
Light Beams							
10×4	19	0.239	33.22	8×4	15	0.140	30.60
	17	0.160	36.73		13	0.089	33.86
	15	0.106	40.16	6×4	16	0.230	20.40
					12	0.092	25.79
Joists							
12×4	14	0.072	52.70	8×4	10	0.044	41.8
10×4	11.5	0.050	49.18	6×4	8.5	0.034	33.71
Wᶠ Shapes and Light Columns							
6Wᶠ	25	0.478	28.52	5 *BS*	18.9	0.348	17.50
6×6	20	0.251	34.21	4 *BS*	13	0.156	15.10
	15.5	0.118	41.88				
5Wᶠ	18.5	0.300	20.63				
5×5	16	0.195	23.22				
American Standard Beams							
$24 \times 7^{7/8}$	120.0	13.15	46.76	12×5	35.0	1.10	27.75
	105.9	10.63	50.17		31.8	0.920	29.58
24×7	100.0	7.72	46.56	$10 \times 4^{5/8}$	35.0	1.31	19.45
	90.0	6.14	50.63		25.4	0.617	25.52
	79.9	5.01	54.41	8×4	23.0	0.555	17.04
20×7	95.0	8.55	37.25		18.4	0.342	20.23
	85.0	6.76	40.45	$7 \times 3^{5/8}$	20.0	0.455	14.08
$20 \times 6^{1/4}$	75.0	4.66	39.28		15.3	0.246	17.57
	65.4	3.58	43.10	$6 \times 3^{3/8}$	17.25	0.379	11.12
18×6	70.0	4.19	33.66		12.50	0.172	14.85
	54.7	2.41	41.26	5×3	14.75	0.334	8.40
$13 \times 5^{1/2}$	50.0	2.15	31.54		10.00	0.116	12.05
	42.9	1.57	35.28	$4 \times 2^{5/8}$	9.50	0.122	8.12
$12 \times 5^{1/4}$	50.0	2.85	21.55		7.70	0.074	9.59
	40.8	1.78	25.33	$3 \times 2^{3/8}$	7.50	0.093	5.50
					5.70	0.045	7.00

Table B-2   Approximate Values of K_b

	$$K_b = \frac{b^3 h^2 t_f}{24}$$
	$$K_b = \frac{b^3 h^2 t_f}{12}\left(\frac{2h t_w + 3b t_f}{h t_w + 6b t_f}\right)$$
	$$K_b = \frac{b^3 h^2 t_f}{6}(1 - 3p + 3p^2) + \frac{h^3 b^2 p^2 t_w}{4}$$ where $p = \dfrac{b t_f}{2b t_f + h t_w}$
	$K_b = 0$
	K_b of composite section in which all components have a common axis of symmetry y. Shear center O of section is defined as center of inertias about y axis, e.g.: $$C_{20} = \frac{I_3 C_{23} - I_1 C_{21}}{I}$$ where I_1, I_2, I_3 are moments of inertia about y axis, C is distance between respective shear centers, and $I = I_1 + I_2 + I_3$ $K_b = K_{b1} + K_{b2} + K_{b3} +$ $$\frac{I_1 I_2 C^2{}_{12} + I_1 I_3 C^2{}_{13} + I_2 I_3 C^2{}_{23}}{I}$$ where K_{b1}, K_{b2}, K_{b3} are torsion bending constants of respective components (Ref: Belyaev, N. M., *Resistance of Materials*, Moscow, 1951—in Russian)

Case 1. Concentrated end torque T on member with free ends

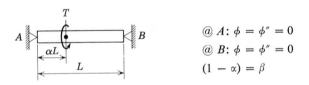

@ A: $\phi = \phi'' = 0$

@ B: $\phi'' = 0$

$$\phi = \frac{Tz}{GK_t}$$

Case 2. Concentrated end Torque T on member with fixed ends

@ A: $\phi = \phi' = 0$

@ B: $\phi' = 0$

$$\phi = \frac{Ta}{GK_t}\left[-\sinh\frac{z}{a} + \tanh\frac{L}{2a}\cdot\cosh\frac{z}{a} + \frac{z}{a} - \tanh\frac{L}{2a}\right]$$

Case 3. Concentrated torque T on member with pinned ends

@ A: $\phi = \phi'' = 0$

@ B: $\phi = \phi'' = 0$

$(1 - \alpha) = \beta$

For $0 < z < \alpha L$

$$\phi = \frac{TL}{GK_t}\left[\frac{a}{L}\left(-\frac{\sinh\dfrac{\beta L}{a}}{\sinh\dfrac{L}{a}}\right)\sinh\frac{z}{a} + \frac{\beta z}{L}\right]$$

For $\alpha L < z < L$

$$\phi = \frac{TL}{GK_t}\left[\frac{a}{L}\frac{\sinh\dfrac{\alpha L}{a}}{\tanh\dfrac{L}{a}}\cdot\sinh\frac{z}{a} - \frac{a}{L}\sinh\frac{\alpha L}{a}\cosh\frac{z}{a} - \frac{\alpha z}{L} + \alpha\right]$$

Table B-3 (*contd.*)

Case 4. Uniformly distributed torque m on member with pinned ends (m — torque per unit length)

@ A: $\phi = \phi'' = 0$

@ B: $\phi = \phi'' = 0$

$$\phi = \frac{ma^2}{GK_t}\left[-\tanh\frac{L}{2a}\sinh\frac{z}{a} + \cosh\frac{z}{a} - \frac{z^2}{a} + \frac{zL}{a} - 1\right]$$

Case 5. Concentrated torque T on member with fixed ends

@ A: $\phi = \phi' = 0$

@ B: $\phi = \phi' = 0$

$(1 - \alpha) = \beta$

For $0 < z < \alpha L$

$$\phi = \frac{Ta}{(H+1)GK_t}\left[-\sinh\frac{z}{a} - F_1\cosh\frac{z}{a} + \frac{z}{a} + F_1\right]$$

For $\alpha L < z < L$

$$\phi = \frac{TaH}{(H+1)GK_t}\left[F_2\sinh\frac{z}{a} + \frac{1 - F_2\cosh\frac{L}{a}}{\sinh\frac{L}{a}}\cosh\frac{z}{a} - \frac{z}{a} + \frac{F_2 - \cosh\frac{L}{a}}{\sinh\frac{L}{a}} + \frac{L}{a}\right]$$

where

$$H = \frac{\tanh\frac{L}{2a}\left(1 - \cosh\frac{\alpha L}{a}\right) + \sinh\frac{\alpha L}{a} - \frac{\alpha L}{a}}{\tanh\frac{L}{2a}\left(1 - \cosh\frac{\alpha L}{a}\right) - \sinh\frac{\alpha L}{a} + \frac{\alpha L}{a} - \frac{L}{a}}$$

$$F_1 = \left[(H+1)\frac{\cosh\frac{\beta L}{a} + 1}{\sinh\frac{L}{a}} - \tanh\frac{L}{2a}\right]$$

and

$$F_2 = \frac{(H+1)\cosh\frac{\alpha L}{a} - 1}{H}$$

816

Table B-3 *(contd.)*

Case 6. Uniformly distributed torque m on member with fixed ends (m — torque per unit length)

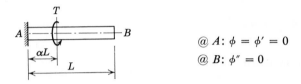

@ A: $\phi = \phi' = 0$

@ B: $\phi = \phi' = 0$

$$\phi = \frac{mLa}{GK_t}\left[-\sinh\frac{z}{a} + \tanh\frac{L}{2a}\cosh\frac{z}{a} + \frac{z}{a} - \frac{z^2}{aL} - \tanh\frac{L}{2a}\right]$$

Case 7. Concentrated torque T on member with one end fixed, one free

@ A: $\phi = \phi' = 0$

@ B: $\phi'' = 0$

For $0 < z < \alpha L$

$$\phi = \frac{Ta}{GK_t}\left\{ -\sinh\frac{z}{a} - \left[\tanh\frac{L}{a}\left(\cosh\frac{\alpha L}{a} - 1\right) - \sinh\frac{\alpha L}{a}\right]\cosh\frac{z}{a}\right.$$
$$\left. + \frac{z}{a} + \left[\tanh\frac{L}{a}\left(\cosh\frac{\alpha L}{a} - 1\right) - \sinh\frac{\alpha L}{a}\right]\right\}$$

For $\alpha L < z < L$

$$\phi = \frac{Ta}{GK_t}\left[\left(1 - \cosh\frac{\alpha L}{a}\right)\sinh\frac{z}{a} + \tanh\frac{L}{a}\left(1 - \cosh\frac{\alpha L}{a}\right)\cosh\frac{z}{a}\right.$$
$$\left. - \tanh\frac{L}{a}\left(1 - \cosh\frac{\alpha L}{a}\right) - \sinh\frac{\alpha L}{a} + \frac{\alpha L}{a}\right]$$

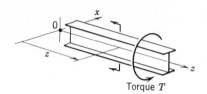

Table B-4

Angle of twist ϕ

$$\phi = \frac{T}{GK_t}\left(A \sinh \frac{z}{a} + B \cosh \frac{z}{a} + Cz^2 + Dz + E\right)$$

Flange lateral deflection u $u = \phi \frac{h}{2}$

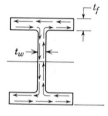

Rate of twist: $\dfrac{d\phi}{dz}$

Pure torsion: $T_t = K_t G \dfrac{d\phi}{dz}$

Torsional shearing stress: $f'_{tv} = Gt \dfrac{d\phi}{dz}$

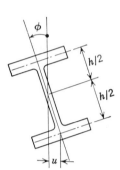

Flange torsion–bending (warping) moment:

$$M_f = EI_f \frac{du^2}{dz^2} = EI_f \frac{h}{2}\frac{d^2\phi}{dz^2} = EI_y \frac{h}{4}\frac{d^2\phi}{dz^2}$$

Flange torsion–bending (warping) stress:

$$f_{tb} = \frac{M_f}{I_f}x = \frac{Eh}{2}x\frac{d^2\phi}{dz^2}$$

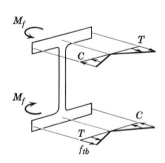

818

Table B-4 (*contd.*)

Flange torsion–bending (warping) shear:

$$V_f = \frac{dM_f}{dz} = EI_y \frac{h}{4} \frac{d^3\phi}{dz^3}$$

Flange torsion–bending (warping) shearing stress:

$$f_{tv}{''} = \frac{V_f Q_f}{I_f t_f} = E \frac{h}{2} \frac{d^3\phi}{dz^3} \frac{Q_f}{t_f}$$

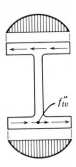

f_{tv}''

Table B-5 Hyperbolic Functions

Hyperbolic functions of ω				Hyperbolic functions of ω			
ω	$\sinh \omega$	$\cosh \omega$	$\tanh \omega$	ω	$\sinh \omega$	$\cosh \omega$	$\tanh \omega$
0.0	0.00000	1.00000	0.00000	1.6	2.37557	2.57746	0.92167
0.1	0.10017	1.00500	0.09967	1.7	2.64563	2.82832	0.93541
0.2	0.20134	1.02007	0.19738	1.8	2.94217	3.10747	0.94681
0.3	0.30452	1.04534	0.29131	1.9	3.26816	3.41773	0.95624
0.4	0.41075	1.08107	0.37995	2.0	3.62686	3.76220	0.96403
0.5	0.52110	1.12763	0.46212				
0.6	0.63665	1.18547	0.53705	2.1	4.02186	4.14431	0.97045
0.7	0.75858	1.25517	0.60437	2.2	4.45711	4.56791	0.97574
0.8	0.88811	1.33743	0.66404	2.3	4.93696	5.03722	0.98010
0.9	1.02652	1.43309	0.71630	2.4	5.46623	5.55695	0.98367
1.0	1.17520	1.54308	0.76159	2.5	6.05020	6.13229	0.98661
1.1	1.33565	1.66852	0.80050	2.6	6.69473	6.76901	0.98903
1.2	1.50946	1.81066	0.83365	2.7	7.40626	7.47347	0.99101
1.3	1.69838	1.97091	0.86172	2.8	8.19192	8.25273	0.99263
1.4	1.90430	2.15090	0.88535	2.9	9.05956	9.11458	0.99396
1.5	2.12928	2.35241	0.90515	3.0	10.01787	10.06766	0.99505
				6.0	201.7132	201.7156	0.99999

This table to be entered with a value of ω equal to L/a, $L/2a$, $L/4a$, etc., as required by the appropriate formulas from Table 8-2.

Values of ω from 0.1 to 3.0 will suffice for most construction involving rolled beams; if, however, ω should lie between 3.0 and 6.0, enter the table with $\omega/2$ and apply the following rules to obtain the function desired:

$$\sinh \omega = 2 \sinh \frac{\omega}{2} \cosh \frac{\omega}{2} \quad \text{Thus} \quad \sinh 4.0 = 2 \times 3.62686 \times 3.76220 = 27.28994$$

$$\cosh \omega = 2 \cosh^2 \frac{\omega}{2} - 1 \qquad \frac{d \sinh \omega}{d\omega} = \cosh \omega$$

$$\tanh \omega = \frac{2 \tanh \frac{\omega}{2}}{1 + \tanh^2 \frac{\omega}{2}} \qquad \frac{d \cosh \omega}{d\omega} = \sinh \omega$$

(Also $\cosh^2 \omega = 1 + \sinh^2 \omega$)

Appendix C

Table C-1 Section Moduli—Elastic
(See footnotes at end of table)

Elastic Modulus in³	Shape	b in	d/A_f	Elastic Modulus in³	Shape	b in	d/A_f
1105.1	36 W^F 300	16.7	1.37	541.0	36 W^F 160	12.0	2.94
				528.2	30 W^F 172	15.0	1.87
1031.2	36 W^F 280	16.6	1.40				
				502.9	36 W^F 150	12.0	3.19
951.1	36 W^F 260	16.6	1.52	492.8	27 W^F 177	14.1	1.63
				486.4	33 W^F 152	11.6	2.74
892.5	36 W^F 245	16.5	1.61				
				446.8	33 W^F 141	11.5	3.01
835.5	36 W^F 230	16.5	1.73	444.5	27 W^F 160	14.0	1.80
811.1	33 W^F 240	15.9	1.51				
				438.6	‡36 W^F 135	11.9	3.75
740.6	33 W^F 220	15.8	1.65	413.5	24 W^F 160	14.1	1.55
669.6	33 W^F 200	15.8	1.82	404.8	33 W^F 130	11.5	3.36
				402.9	27 W^F 145	14.0	1.97
663.6	36 W^F 194	12.1	2.39	379.7	30 W^F 132	10.6	2.87
649.9	30 W^F 210	15.1	1.53	372.5	24 W^F 145	14.0	1.71
621.2	36 W^F 182	12.1	2.55	358.3	‡33 W^F 118	11.5	3.88
586.1	30 W^F 190	15.0	1.69	354.6	30 W^F 124	10.5	3.08
579.1	36 W^F 170	12.0	2.73	330.7	‡24 W^F 130	14.0	1.93

Table C-1 (*contd.*)

Elastic Modulus in³	Shape	b in	d/A_f	Elastic Modulus in³	Shape	b in	d/A_f
327.9	30 WF 116	10.5	3.36	163.6	†14 WF 103	14.6	1.20
317.2	21 WF 142	14.1	1.49	163.4	12 WF 120	12.3	0.96
				160.0	20 I 95	7.2	3.03
299.2	30 WF 108	10.5	3.74	156.1	18 WF 85	8.8	2.28
299.2	27 WF 114	10.1	2.91				
299.1	24 WF 120	12.1	2.16	153.1	‡24 WF 68	9.0	4.55
284.1	21 WF 127	13.1	1.65	151.3	16 WF 88	11.5	1.77
274.4	24 WF 110	12.0	2.34	150.7	21 WF 73	9.3	3.46
				150.2	20 I 85	7.0	1.07
269.1	‡30 WF 99	10.5	4.23	144.5	12 WF 106	12.2	1.40
266.3	27 WF 102	10.0	2.37	141.7	18 WF 77	8.8	29.71
263.2	12 WF 190	12.7	0.66				
250.9	24 I 120	8.0	2.71	139.9	21 WF 68	8.3	3.73
249.6	‡21 WF 112	13.0	1.87	138.1	†14 WF 87	14.5	1.40
248.9	‡24 WF 100	12.0	2.58	134.7	12 WF 99	12.2	1.14
				130.9	‡14 WF 84	12.0	1.52
242.8	27 WF 94	10.0	3.61				
234.3	24 I 105.9	7.9	2.76	129.5	24 B 61	7.0	5.71
222.2	12 WF 161	12.5	0.75	128.2	18 WF 70	8.8	2.74
				127.8	16 WF 78	8.6	2.17
220.9	24 WF 94	9.1	3.07	126.4	21 WF 62	8.2	4.15
220.1	18 WF 114	11.8	1.58	126.3	20 I 75	6.4	3.97
216.0	14 WF 136	14.7	0.94	126.3	10 WF 112	10.4	0.88
				125.0	12 WF 92	12.2	1.21
211.7	‡27 WF 84	10.0	4.21	121.1	‡14 WF 78	12.0	1.63
202.2	18 WF 105	11.8	1.71	117.0	18 WF 64	8.7	2.99
202.0	14 WF 127	14.7	1.00	116.9	20 I 65.4	6.3	4.05
197.6	21 WF 96	9.0	2.50	115.9	16 WF 71	8.5	2.38
197.6	24 I 100	7.2	3.81	115.7	‡12 WF 85	12.1	1.29
196.3	24 WF 84	9.0	3.47	113.7	24 B 55	7.0	6.69
189.4	‡14 WF 119	14.7	1.06	112.4	10 WF 100	10.3	0.96
185.8	24 I 90	7.1	3.87	112.3	14 WF 74	10.1	1.80
184.4	18 WF 96	11.8	1.86				
182.5	12 WF 133	12.4	0.88	109.7	‡21 WF 55	8.2	4.85
176.3	‡14 WF 111	14.6	1.13	107.8	18 WF 60	7.6	3.48
				107.1	‡12 WF 79	12.1	1.39
175.4	24 WF 76	9.0	3.90	104.2	16 WF 64	8.5	2.63
173.9	24 I 79.9	7.0	3.94	103.0	14 WF 68	10.0	1.95
168.0	21 WF 82	9.0	2.93	101.9	18 I 70	6.3	4.17
166.1	16 WF 96	11.5	1.62	99.7	10 WF 89	10.3	1.06

Table C-1 (*contd.*)

Elastic Modulus in³	Shape	b in	d/A_f	Elastic Modulus in³	Shape	b in	d/A_f
98.2	**18 WF 55**	7.5	3.82	50.3	12 I 50	5.5	3.32
97.5	†12 WF 72	12.0	1.52	49.1	10 WF 45	8.0	2.04
94.1	16 WF 58	8.5	2.19	48.5	‡**14 WF 34**	6.8	4.58
93.2	**21 B 49**	6.5	6.00	47.0	**16 B 31**	5.5	6.49
92.2	‡14 WF 61	10.0	2.16	45.9	12 WF 36	6.6	3.45
89.0	18 WF 50	7.5	4.22	44.8	12 I 40.8	5.3	3.47
88.4	18 I 54.7	6.0	4.35	43.2	8 WF 48	8.1	1.53
88.0	†12 WF 65	12.0	1.67	42.2	‡10 WF 39	8.0	2.36
86.1	10 WF 77	10.2	1.20	41.8	‡**14 WF 30**	6.7	5.37
81.5	**21 B 44**	6.5	7.05	39.4	12 WF 31	6.5	3.98
80.7	16 WF 50	7.1	3.66	38.1	‡**16 B 26**	5.0	6.61
80.1	10 WF 72	10.2	1.28	37.8	12 I 35	5.1	4.34
78.9	‡18 WF 45	7.5	4.79	36.0	12 I 31·8	5.0	4.41
78.1	‡12 WF 58	10.0	1.90	35.5	8 WF 40	8.1	1.83
77.8	14 WF 53	8.1	2.63	35.0	†10 WF 33	8.0	2.83
73.7	10 WF 66	10.1	1.37	34.9	**14 B 26**	6.5	4.60
72.4	16 WF 45	7.0	4.07	34.1	‡12 WF 27		
70.7	‡12 WF 53	10.0	2.09	31.1	‡ 8 WF 35	8.0	2.05
70.2	14 WF 48	8.0	2.90	30.8	10 WF 29	5.8	3.52
68.3	**18 B 40**	6.0	5.68	29.2	10 I 35	4.9	4.12
67.1	‡10 WF 60	10.1	1.49	28.8	‡**14 B 22**	5.0	8.19
64.7	12 WF 50	8.1	2.35	28.9	‡ 8 M 34.3	8.0	2.18
64.4	**16 WF 40**	7.0	4.54	28.2	‡ 8 M 32.6	7.9	2.20
64.2	15 I 50	5.6	4.28	27.4	† 8 WF 31	8.0	2.31
62.7	‡14 WF 43	8.0	3.24	26.6	‡10 M 29.1	5.9	4.28
60.4	‡10 WF 54	10.0	1.63	26.4	**10 WF 25**	5.8	4.08
60.4	8 WF 67	8.3	1.16	25.3	**12 B 22**	4.0	7.20
58.9	15 I 42.9	5.5	4.39	24.4	10 I 25.4	4.7	4.37
58.2	12 WF 45	8.0	2.60	24.3	8 WF 28	6.5	2.66
57.9	**18 B 35**	6.0	6.88	23.6	‡10 M 22.9	5.8	4.42
56.3	‡16 WF 36	7.0	5.30	22.5	‡ 8 M 28	6.7	3.02
54.6	14 WF 38	6.8	4.06	21.5	‡**10 WF 21**	5.8	5.07
54.6	†10 WF 49	10.0	1.79				
52.0	8 WF 58	8.2	1.32				
51.9	‡12 WF 40	8.0	2.89				

Table C-1 (*contd.*)

Elastic Modulus in³	Shape	b in	d/A_f	Elastic Modulus in³	Shape	b in	d/A_f
21.4	**12 B 19**	4.0	8.69	10.1	† 6 W^F 15.5	6.0	3.72
21.1	‡10 M 21	5.8	5.09	10.1	6 B 16	4.0	3.84
21.0	‡ 8 M 24	6.5	3.09	9.9	‡ 8 B 13	4.0	7.87
				9.9	5 W^F 18.5	5.0	2.43
21.0	‡14 **B 17.2**	4.0	12.9	9.5	5 M 18.9	5.0	2.40
20.8	‡ 8 W^F 24	16.5	3.07				
18.8	10 B 19	4.0	6.47	9.3	***12 JR [10.6**		
				8.7	6 I 17.25	3.6	4.69
17.5	‡12 **B 16.5**	4.0	11.2	8.5	5 W^F 16	5.0	2.78
17.1	‡ 8 M 22.5	5.4	4.2				
17.0	8 W^F 20	5.3	4.08	7.8	‡**10 JR 9**	2.7	18.1
16.8	6 W^F 25	6.1	2.30	7.8	† 8 B 10	3.9	9.83
16.2	10 B 17	4.0	7.67	7.3	6 I 12.5	3.3	5.02
16.0	8 I 23	4.2	4.51	7.2	6 B 12	4.0	5.38
15.7	6 M 25	6.1	2.30				
15.5	‡ 6 M 18.5	5.3	4.33	6.5	***10 JR [8.4**		
15.2	‡ 8 M 20	5.4	4.89	6.0	5 I 14.75	3.3	4.67
				5.4	4 W^F 13	4.1	2.97
14.8	‡12 **B 14**	4.0	13.4	5.2	4 M 13	3.9	2.73
14.2	8 I 18.4	4.0	4.71	5.1	† 6 B 8.5	3.9	7.63
14.1	‡ 8 W^F 17..	5.3	4.95	4.8	5 I 10	3.0	5.11
14.0	‡ 8 M 17	5.3	5.00				
13.8	‡10 B 15	4.0	9.29	4.7	**8 JR 6.5**	2.3	18.6
13.7	‡ 6 M 22.5	6.1	2.60	4.4	***10 JR [6.5**		
13.4	‡ 6 W^F 20	6.0	2.81				
12.9	‡ 6 M 20	6.0	2.81	3.5	**7 JR 5.5**	2.1	18.7
				3.3	4 I 9.5	2.8	4.83
12.0	‡**12 JR 11.8**	3.1	17.4	3.0	4 I 7.7	2.7	5.13
12.0	7 I 20	3.9	4.62				
11.8	8 B 15	4.0	6.44	2.4	**6 JR 4.4**	1.8	19.0
				1.9	3 I 7.5	2.5	4.60
10.5	‡**10 B 11.5**	4.0	12.3	1.7	3 I 5.7	2.3	4.45
10.4	7 I 15.3	3.7	4.88				

† Identifies noncompact shapes for which bending stress F_b may not exceed 0.60 F_y in A36, A242, A440, and A441 steels (AISC Specification Section 1.5.1.4.1).

‡ Identifies non-compact shapes for which bending stress F_b may not exceed 0.60 F_y in A242, A440, and A441 steels (AISC Specification Section 1.5.1.4.1).

* Bending stress F_b may not exceed 0.60 F_y (AISC Specification Section 1.5.1.4.6).

Shapes subjected to combined axial force and bending moment may not be compact under Section 1.5.1.4.1, of the AISC Specification. Check all shapes for compliance with this Section.

Index